Lecture Notes in Computer Science 13356

More information about this series at https://link.springer.com/bookseries/558

Maria Mercedes Rodrigo ·
Noburu Matsuda · Alexandra I. Cristea ·
Vania Dimitrova (Eds.)

Artificial Intelligence in Education

Posters and Late Breaking Results, Workshops and Tutorials,
Industry and Innovation Tracks, Practitioners' and Doctoral Consortium

23rd International Conference, AIED 2022
Durham, UK, July 27–31, 2022
Proceedings, Part II

 Springer

Editors
Maria Mercedes Rodrigo
Ateneo De Manila University
Quezon, Philippines

Alexandra I. Cristea ⓘ
Durham University
Durham, UK

Noburu Matsuda
Department of Computer Science
North Carolina State University
Raleigh, NC, USA

Vania Dimitrova
University of Leeds
Leeds, UK

ISSN 0302-9743   ISSN 1611-3349 (electronic)
Lecture Notes in Computer Science
ISBN 978-3-031-11646-9   ISBN 978-3-031-11647-6 (eBook)
https://doi.org/10.1007/978-3-031-11647-6

This Springer imprint is published by the registered company Springer Nature Switzerland AG
The registered company address is: Gewerbestrasse 11, 6330 Cham, Switzerland

Preface

The 23rd International Conference on Artificial Intelligence in Education (AIED 2022) was hosted by Durham University, UK. It was organized in a hybrid face-to-face and online format. This allowed participants to meet in person after two years of running AIED online only, which was a welcome change. However, as the world was only just emerging from the COVID-19 pandemic and travel for some attendees was still a challenge, online participation was also supported. AIED 2022 was the next in a longstanding series of annual international conferences for the presentation of high-quality research on intelligent systems and the cognitive sciences for the improvement and advancement of education. It was hosted by the prestigious International Artificial Intelligence in Education Society, a global association of researchers and academics who specialize in the many fields that comprise AIED, including computer science, learning sciences, educational data mining, game design, psychology, sociology, linguistics, and many others.

The theme for the AIED 2022 conference was "AI in Education: Bridging the gap between academia, business, and non-profit in preparing future-proof generations towards ubiquitous AI." The conference hoped to stimulate discussion on how AI shapes and can shape education for all sectors, how to advance the science and engineering of intelligent interactive learning systems, and how to promote broad adoption. Engaging with the various stakeholders – researchers, educational practitioners, businesses, policy makers, as well as teachers and students – the conference set a wider agenda on how novel research ideas can meet practical needs to build effective intelligent human-technology ecosystems that support learning.

AIED 2022 attracted broad participation. We received 243 submissions for the main program, of which 197 were submitted as full papers, 37 were submitted as short papers, and nine were submitted as extended abstracts. Of the full paper submissions, 40 were accepted as full papers and another 40 were accepted as short papers. The acceptance rate for both full papers and short papers was thus 20%.

Beyond paper presentations and keynotes, the conference also included a Doctoral Consortium Track, an Industry and Innovation Track, Interactive Events, Posters/Late-Breaking Results, and a Practitioner Track. The submissions for all these tracks underwent a rigorous peer-review process. Each submission was reviewed by at least two members of the AIED community, assigned by the corresponding track organizers who then took the final decision about acceptance. The conference also included keynotes, panels, and workshops and tutorials.

For making AIED 2022 possible, we thank the AIED 2022 Organizing Committee, the hundreds of Program Committee members, the Senior Program Committee members, the AIED Proceedings Chair Irene-Angelica Chounta, and our Program

Chair assistant Jonathan DL. Casano. They all gave of their time and expertise generously and helped with shaping a stimulating AIED 2022 conference. We are extremely grateful to everyone!

July 2022

Maria Mercedes (Didith) T. Rodrigo
Noboru Matsuda
Alexandra I. Cristea
Vania Dimitrova

Organization

General Chair

Vania Dimitrova   University of Leeds, UK

Program Co-chairs

Noboru Matsuda   North Carolina State University, USA
Maria Mercedes (Didith)   Ateneo de Manila University, Philippines
   T. Rodrigo

Doctoral Consortium Co-chairs

Olga C. Santos   UNED, Spain
Neil Heffernan   Worcester Polytechnic Institute, USA

Workshop and Tutorials Co-chairs

Ning Wang   University of Southern California, USA
Srećko Joksimović   University of South Australia, Australia

Interactive Events Co-chairs

Dorothy Monekosso   Durham University, UK
Genaro Rebolledo-Mendez   Institute for the Future of Education, Mexico
Ifeoma Adaji   University of British Columbia, Canada

Industry and Innovation Track Co-chairs

Zitao Liu   TAL Education Group, China
Diego Zapata-Rivera   Educational Testing Service, USA

Posters and Late-Breaking Results Co-chairs

Carrie Demmans Epp   University of Alberta, Canada
Sergey Sosnovsky   Utrecht University, The Netherlands

Practitioner Track Co-chairs

Jeanine A. DeFalco   Medidata Solutions, Dassault Systemes, USA
Diego Dermeval Medeiros   Federal University of the Alagoas, Brazil
   da Cunha Matos

Berit Blanc   German Research Centre for Artificial Intelligence
                           (DFKI), Germany
Insa Reichow   German Research Centre for Artificial Intelligence
                           (DFKI), Germany

Panel Chair

Wayne Holmes   University College London, UK

Local Organizing Chair

Alexandra I. Cristea   Durham University, UK

Proceedings Chair

Irene-Angelica Chounta   University of Duisburg-Essen, Germany

Web Chair

Lei Shi   Durham University, UK

Online Activities Chair

Guanliang Chen   Monash University, Australia

Publicity Co-chairs

Elle Wang   Arizona State University, USA
Elaine Harada Teixeira de Federal University of Amazonas, Brazil
  Oliveira
Mizue Kayama   Sinshu University, Japan

Sponsorship Chairs

Craig Stewart   Durham University, UK
Ben du Boulay   University of Sussex, UK

Diversity and Inclusion Co-chairs

Eric Walker   Arizona State University, USA
Rod Roscoe   Arizona State University, USA
Seiji Isotani   University of São Paulo, Brazil

Senior Program Committee

Alireza Ahadi	University of Technology Sydney, Australia
Vincent Aleven	Carnegie Mellon University, USA
Laura Allen	University of New Hampshire, USA
Claudio Alvarez	Universidad de los Andes, Chile
Rafael D. Araújo	Universidade Federal de Uberlandia, Brazil
Tracy Arner	Arizona State University, USA
Luciana Assis	Universidade Federal dos Vales do Jequitinhonha e Mucuri, Brazil
Roger Azevedo	University of Central Florida, USA
Ryan Baker	University of Pennsylvania, USA
Tiffany Barnes	North Carolina State University, USA
Emmanuel Blanchard	IDÛ Interactive Inc., Canada
Nigel Bosch	University of Illinois Urbana-Champaign, USA
Steven Bradley	Durham University, UK
Christopher Brooks	University of Michigan, USA
Armelle Brun	Loria, Université de Lorraine, France
Maiga Chang	Athabasca University, Canada
Min Chi	BeiKaZhouLi, USA
Cesar A. Collazos	Universidad del Cauca, Colombia
Cristina Conati	University of British Columbia, Canada
Sidney D'Mello	University of Colorado Boulder, USA
Mihai Dascalu	Politehnica University of Bucharest, Romania
Jeanine A. DeFalco	Medidata Solutions, Dassault Systemes, USA
Michel Desmarais	Ecole Polytechnique de Montreal, Canada
Vania Dimitrova	University of Leeds, UK
Tenzin Doleck	Atlas Lab, USA
Benedict du Boulay	University of Sussex, UK
Márcia Fernandes	Federal University of Uberlandia, Brazil
Rafael Ferreira Mello	Federal Rural University of Pernambuco, Brazil
Kobi Gal	Ben Gurion University, Israel
Dragan Gasevic	Monash University, Australia
Sébastien George	LIUM, Le Mans Université, France
Ashok Goel	Georgia Institute of Technology, USA
Art Graesser	University of Memphis, USA
Peter Hastings	DePaul University, USA
Yusuke Hayashi	Hiroshima University, Japan
Bastiaan Heeren	Open University, The Netherlands
Neil Heffernan	Worcester Polytechnic Institute, USA
Laurent Heiser	Université Côte d'Azur, Inspé de Nice, France
Ulrich Hoppe	University Duisburg-Essen, Germany
Sharon Hsiao	Santa Clara University, USA
Lingyun Huang	McGill University, Canada
Seiji Isotani	University of São Paulo, Brazil
Yang Jiang	Columbia University, USA

David Joyner	Georgia Institute of Technology, USA
Akihiro Kashihara	University of Electro-Communications, Japan
Judy Kay	University of Sydney, Australia
Mizue Kayama	Shinshu University, Japan
Min Kyu Kim	Georgia State University, USA
Stefan Küchemann	TU Kaiserslautern, Germany
Jakub Kužílek	Charles Technical University in Prague, Czechia
Amruth Kumar	Ramapo College of New Jersey, USA
Susanne Lajoie	McGill University, Canada
H. Chad Lane	University of Illinois at Urbana-Champaign, USA
Nguyen-Thinh Le	Humboldt-Universität zu Berlin, Germany
Francois Lecellier	Xlim Laboratory, France
James Lester	North Carolina State University, USA
Shan Li	McGill University, Canada
Tong Li	Arizona State University, USA
Carla Limongelli	Università Roma Tre, Italy
Sonsoles López-Pernas	Universidad Politécnica de Madrid, Spain
Vanda Luengo	LIP6, Sorbonne Université, France
Collin Lynch	North Carolina State University, USA
Wannisa Matcha	Prince of Sonkla University, Pattani, Thailand
Noboru Matsuda	North Carolina State University, USA
Gordon McCalla	University of Saskatchewan, Canada
Kathryn McCarthy	Georgia State University, USA
Bruce Mclaren	Carnegie Mellon University, USA
Agathe Merceron	Beuth University of Applied Sciences Berlin, Germany
Eva Millan	Universidad de Málaga, Spain
Caitlin Mills	University of New Hampshire, USA
Tanja Mitrovic	University of Canterbury, New Zealand
Kazuhisa Miwa	Nagoya University, Japan
Riichiro Mizoguchi	Japan Advanced Institute of Science and Technology, Japan
Negar Mohammadhassan	University of Canterbury, UK
Phaedra Mohammed	University of the West Indies, Jamaica
Bradford Mott	North Carolina State University, USA
Roger Nkambou	Université du Québec à Montréal, Canada
Jaclyn Ocumpaugh	University of Pennsylvania, USA
Amy Ogan	Carnegie Mellon University, USA
Elaine H. T. Oliveira	Universidade Federal do Amazonas, Brazil
Andrew Olney	University of Memphis, USA
Luc Paquette	University of Illinois at Urbana-Champaign, USA
Bernardo Pereira Nunes	Australian National University, Australia
Niels Pinkwart	Humboldt-Universität zu Berlin, Germany
Kaska Porayska-Pomsta	University College London, UK
Thomas Price	North Carolina State University, USA
Mladen Rakovic	Monash University, Australia
Ilana Ram	Technion – Israel Institute of Technology, Israel

Martina Rau	University of Wisconsin-Madison, USA
Traian Rebedea	Politehnica University of Bucharest, Romania
Genaro Rebolledo-Mendez	Tecnologico de Monterrey, Mexico
Steven Ritter	Carnegie Learning, Inc., USA
Maria Mercedes (Didith) T. Rodrigo	Ateneo de Manila University, Philippines
Ido Roll	Technion - Israel Institute of Technology, Israel
Rod Roscoe	Arizona State University, USA
Jonathan Rowe	North Carolina State University, USA
Olga C. Santos	UNED, Spain
Kazuhisa Seta	Osaka Prefecture University, Japan
Sergey Sosnovsky	Utrecht University, The Netherlands
Namrata Srivastava	University of Melbourne, Australia
Richard Tong	Yixue Education Inc., China
Stefan Trausan-Matu	Politehnica University of Bucharest, Romania
Maomi Ueno	University of Electro-Communications, Japan
Masaki Uto	University of Electro-Communications, Japan
Kurt Vanlehn	Arizona State University, USA
Alessandro Vivas	UFVJM, Brazil
Erin Walker	Arizona State University, USA
Diego Zapata-Rivera	Educational Testing Service, USA

Program Committee

Mark Abdelshiheed	North Carolina State University, USA
Ifeoma Adaji	University of British Columbia, Canada
Bunmi Adewoyin	University of Saskatchewan, Canada
Seth Adjei	Northern Kentucky University, USA
Jenilyn Agapito	Ateneo de Manila University, Philippines
Kamil Akhuseyinoglu	University of Pittsburgh, USA
Bita Akram	North Carolina State University, USA
Samah Alkhuzaey	University of Liverpool, UK
Isaac Alpizar Chacon	Utrecht University, The Netherlands
Nese Alyuz	Intel, USA
Sungeun An	Georgia Institute of Technology, USA
Antonio R. Anaya	Universidad Nacional de Educacion a Distancia, Spain
Roberto Araya	Universidad de Chile, Chile
Esma Aïmeur	University of Montreal, Canada
Michelle Banawan	Arizona State University, USA
Ayan Banerjee	Arizona State University, USA
Jordan Barria-Pineda	University of Pittsburgh, USA
Shay Ben-Elazar	Microsoft, USA
Ig Ibert Bittencourt	Federal University of Alagoas, Brazil
Geoffray Bonnin	Loria, Université de Lorraine, France
Anthony F. Botelho	University of Florida, USA
Jesus G. Boticario	UNED, Spain

François Bouchet	LIP6, Sorbonne Université, France
Bert Bredeweg	University of Amsterdam, The Netherlands
Julien Broisin	IRIT, Université Toulouse 3 Paul Sabatier, France
Okan Bulut	University of Alberta, Canada
James Bywater	James Madison University, USA
Daniela Caballero	McMaster University, Canada
Alberto Casas-Ortiz	UNED, Spain
Francis Castro	New York University, USA
Geiser Chalco Challco	ICMC/USP, Brazil
Pankaj Chavan	IIT Bombay, India
Guanliang Chen	Monash University, Australia
Jiahao Chen	TAL Education Group, China
Penghe Chen	Beijing Normal University, China
Heeryung Choi	University of Michigan, USA
Andrew Clayphan	University of Sydney, Australia
Keith Cochran	DePaul University, USA
Ricardo Conejo	Universidad de Malaga, Spain
Mark G. Core	University of Southern California, USA
Alexandra Cristea	Durham University, UK
Mutlu Cukurova	University College London, UK
Maria Cutumisu	University of Alberta, Canada
Anurag Deep	IIT Bombay, India
Carrie Demmans Epp	University of Alberta, Canada
Diego Dermeval	Federal University of the Alagoas, Brazil
M. Ali Akber Dewan	Athabasca University, Canada
Tejas Dhamecha	IBM, India
Barbara Di Eugenio	University of Illinois at Chicago, USA
Daniele Di Mitri	DIPF—Leibniz Institute for Research and Information in Education, Germany
Darina Dicheva	Winston-Salem State University, USA
Mohsen Dorodchi	University of North Carolina at Charlotte, USA
Fabiano Dorça	Universidade Federal de Uberlandia, Brazil
Alpana Dubey	Accenture, India
Yo Ehara	Tokyo Gakugei University, Japan
Ralph Ewerth	L3S Research Center, Leibniz Universität Hannover, Germany
Fahmid Morshed Fahid	North Carolina State University, USA
Stephen Fancsali	Carnegie Learning, Inc., USA
Arta Farahmand	Athabasca University, Canada
Effat Farhana	Vanderbilt University, USA
Mingyu Feng	WestEd, USA
Reza Feyzi Behnagh	University at Albany - SUNY, USA
Carol Forsyth	Educational Testing Service, USA
Reva Freedman	Northern Illinois University, USA
Maurizio Gabbrielli	University of Bologna, Italy
Cristiano Galafassi	Universidade Federal do Rio Grande do Sul, Brazil

Lucas Galhardi	State University of Londrina, Brazil
Yanjun Gao	University of Wisconsin-Madison, USA
Isabela Gasparini	UDESC, Brazil
Elena Gaudioso	UNED, Spain
Michael Glass	Valparaiso University, Chile
Benjamin Goldberg	United States Army DEVCOM Soldier Center, USA
Alex Sandro Gomes	Universidade Federal de Pernambuco, Brazil
Aldo Gordillo	Universidad Politécnica de Madrid, Spain
Monique Grandbastien	Loria, Université de Lorraine, France
Floriana Grasso	University of Liverpool, UK
André Greiner-Petter	University of Wuppertal, Germany
Nathalie Guin	LIRIS, Université de Lyon, France
Sandeep Gupta	Arizona State University, USA
Binod Gyawali	Educational Testing Service, USA
Hicham Hage	Notre Dame University-Louaize, Lebanon
Rawad Hammad	University of East London, UK
Yugo Hayashi	Ritsumeikan University, Japan
Martin Hlosta	The Open University, UK
Anett Hoppe	TIB – Leibniz Information Centre for Science and Technology and L3S Research Centre, Leibniz Universität Hannover, Germany
Tomoya Horiguchi	Kobe University, Japan
Daniel Hromada	Einstein Center Digital Future and Berlin University of the Arts, Germany
Stephen Hutt	University of Pennsylvania, USA
Chanyou Hwang	Riiid, South Korea
Tomoo Inoue	University of Tsukuba, Japan
Paul Salvador Inventado	California State University, Fullerton, USA
Mirjana Ivanovic	University of Novi Sad, Serbia
Johan Jeuring	Utrecht University, The Netherlands
Srećko Joksimović	University of South Australia, Australia
Yvonne Kammerer	Stuttgart Media University, Germany
Shamya Karumbaiah	University of Pennsylvania, USA
Hieke Keuning	Utrecht University, The Netherlands
Rashmi Khazanchi	Open University of the Netherlands and Mitchell County School System, The Netherlands
Hassan Khosravi	University of Queensland, Australia
Jung Hoon Kim	Riiid, South Korea
Simon Knight	University of Technology Sydney, Australia
Kazuaki Kojima	Teikyo University, Japan
Emmanuel Awuni Kolog	University of Ghana Business School, Ghana
Tanja Käser	EPFL, Switzerland
Sébastien Lallé	Sorbonne University, France
Andrew Lan	University of Massachusetts Amherst, USA
Jim Larimore	Riiid, South Korea
Hady Lauw	Singapore Management University, Singapore

Elise Lavoué	LIRIS, Université Jean Moulin Lyon 3, France
Seiyon Lee	University of Pennsylvania, USA
Marie Lefevre	LIRIS, Université Lyon 1, France
Blair Lehman	Educational Testing Service, USA
Sharona T. Levy	University of Haifa, Israel
Fuhua Lin	Athabasca University, Canada
Qiongqiong Liu	TAL Education Group, China
Zitao Liu	TAL Education Group, China
Nikki Lobczowski	University of Pittsburgh, USA
Yu Lu	Beijing Normal University, China
Aditi Mallavarapu	University of Illinois at Chicago, USA
Leonardo Brandão Marques	University of São Paulo, Brazil
Mirko Marras	University of Cagliari, Italy
Daniel Moritz Marutschke	Ritsumeikan University, Japan
Jeffrey Matayoshi	McGraw Hill ALEKS, USA
Wookhee Min	North Carolina State University, USA
Sein Minn	Inria, France
Tsegaye Misikir Tashu	Eötvös Loránd University, Hungary
Merav Mofaz	Microsoft, Israel
Abrar Mohammed	University of Leeds, UK
Dorothy Monekosso	Durham University, UK
Kasia Muldner	Carleton University, Canada
Anabil Munshi	Vanderbilt University, USA
Tricia Ngoon	Carnegie Mellon University, USA
Nasheen Nur	Florida Institute of Technology, USA
Negar Mohammadhassan	University of Canterbury, UK
Phaedra Mohammed	University of the West Indies, Jamaica
Bradford Mott	North Carolina State University, USA
Marek Ogiela	AGH University of Science and Technology, Poland
Urszula Ogiela	AGH University of Science and Technology, Poland
Christian Otto	TIB – Leibniz Information Centre for Science and Technology University Library, Germany
Ranilson Paiva	Universidade Federal de Alagoas, Brazil
Rebecca Passonneau	Pennsylvania State University, USA
Rumana Pathan	Indian Institute of Technology Bombay, India
Prajwal Paudyal	Arizona State University, USA
Terry Payne	University of Liverpool, UK
Radek Pelánek	Masaryk University, Czech Republic
Francesco Piccialli	University of Naples Federico II, Italy
Elvira Popescu	University of Craiova, Romania
Miguel Angel Portaz	UNED, Spain
Shi Pu	Education Testing Service, USA
Ramkumar Rajendran	IIT Bombay, India
Insa Reichow	Deutsches Forschungszentrum für Künstliche Intelligenz, Germany
Luiz Antonio Rodrigues	UniFil, Brazil

Rinat Rosenberg-Kima	Technion, Israel
José A. Ruipérez Valiente	University of Murcia, Spain
Stefan Ruseti	Politehnica University of Bucharest, Romania
Shaghayegh Sahebi	University at Albany - SUNY, USA
Demetrios Sampson	Curtin University, Australia
Petra Sauer	Beuth University of Applied Sciences Berlin, Germany
Moritz Schubotz	Universität Konstanz, Germany
Flippo Sciarrone	Roma Tre University, Italy
Richard Scruggs	University of Pennsylvania, USA
Tasmia Shahriar	North Carolina State University, USA
Lei Shi	Durham University, UK
Jinnie Shin	University of Florida, USA
Daevesh Singh	Indian Institute of Technology, India
Sean Siqueira	Federal University of the State of Rio de Janeiro, Brazil
Caitlin Snyder	Vanderbilt University, USA
Srinath Srinivasa	International Institute of Information Technology, Bangalore, India
Merlin Teodosia Suarez	De La Salle University, Philippines
Thepchai Supnithi	NECTEC, Thailand
May Marie P. Talandron-Felipe	Ateneo de Manila University and University of Science and Technology of Southern Philippines, Philippines
Michelle Taub	University of Central Florida, USA
Pierre Tchounikine	Université Grenoble Alpes, France
Marco Temperini	Sapienza University of Rome, Italy
Craig Thompson	University of British Columbia, Canada
Armando Toda	University of São Paulo, Brazil
Hedderik van Rijn	University of Groningen, The Netherlands
Rosa Vicari	Universidade Federal do Rio Grande do Sul, Brazil
Maureen Villamor	University of Southeastern Philippines, Philippines
Thierry Viéville	Inria, Mnemosyne, France
Candy Walter	University of Hildesheim, Germany
Elaine Wang	RAND Corporation, USA
Zichao Wang	Rice University, USA
Chris Wong	University of Technology Sydney, Australia
Simon Woodhead	Eedi, UK
Sho Yamamoto	Kindai University, Japan
Amel Yessad	LIP6, Sorbonne Université, France
Bernard Yett	Vanderbilt University, USA
Ran Yu	GESIS - Leibniz Institute for the Social Sciences, Germany
Luyao Zhang	Duke Kunshan University, China
Ningyu Zhang	Vanderbilt University, USA

Qian Zhang	University of Technology Sydney, Australia
Guojing Zhou	University of Colorado Boulder, USA
Jianlong Zhou	University of Technology Sydney, Australia
Xiaofei Zhou	University of Rochester, USA
Stefano Pio Zingaro	Università di Bologna, Italy
Gustavo Zurita	Universidad de Chile, Chile

Additional Reviewers

Abdelshiheed, Mark
Afzal, Shazia
Anaya, Antonio R.
Andres-Bray, Juan Miguel
Arslan, Burcu
Barthakur, Abhinava
Bayer, Vaclav
Chung, Cheng-Yu
Cucuiat, Veronica
Demmans Epp, Carrie
Diaz, Claudio
DiCerbo, Kristen
Erickson, John
Finocchiaro, Jessica
Fossati, Davide
Frost, Stephanie
Gao, Ge
Garg, Anchal
Gauthier, Andrea
Gaweda, Adam
Green, Nick
Gupta, Itika
Gurung, Ashish
Gutiérrez Y. Restrepo, Emmanuelle
Haim, Aaron
Hao, Yang
Hastings, Peter
Heldman, Ori
Jensen, Emily
Jiang, Weijie
John, David
Johnson, Jillian
Jose, Jario
Karademir, Onur
Landes, Paul

Lefevre, Marie
Li, Zhaoxing
Liu, Tianqiao
Lytle, Nick
Marwan, Samiha
Mat Sanusi, Khaleel Asyraaf
Matsubayashi, Shota
McBroom, Jessica
Mohammadhassan, Negar
Monaikul, Natawut
Munshi, Anabil
Paredes, Yancy Vance
Pathan, Rumana
Prihar, Ethan
Rodriguez, Fernando
Segal, Avi
Serrano Mamolar, Ana
Shahriar, Tasmia
Shi, Yang
Shimmei, Machi
Singh, Daevesh
Stahl, Christopher
Swamy, Vinitra
Tenison, Caitlin
Tobarra, Llanos
Woodhead, Simon
Xhakaj, Franceska
Xu, Yiqiao
Yamakawa, Mayu
Yang, Xi
Yarbro, Jeffrey
Zamecnick, Andrew
Zhai, Xiao
Zhou, Guojing
Zhou, Yunzhan

International Artificial Intelligence in Education Society

Management Board

President

Vania Dimitrova   University of Leeds, UK

Secretary/Treasurer

Rose Luckin   University College London, UK

Journal Editors

Vincent Aleven   Carnegie Mellon University, USA
Judy Kay   University of Sydney, Australia

Finance Chair

Ben du Boulay   University of Sussex, UK

Membership Chair

Benjamin D. Nye   University of Southern California, USA

Publicity Chair

Manolis Mavrikis   University College London, UK

Tech and Outreach Officer

Yancy Vance Paredes   Arizona State University, USA

Executive Committee

Ryan Shaun Baker   University of Pennsylvania, USA
Min Chi   North Carolina State University, USA
Cristina Conati   University of British Columbia, Canada
Jeanine A. DeFalco   Medidata Solutions, Dassault Systemes, USA
Rawad Hammad   University of East London, UK
Neil Heffernan   Worcester Polytechnic Institute, USA
Christothea Herodotou   Open University, UK
Seiji Isotani   University of São Paulo, Brazil
Akihiro Kashihara   University of Electro-Communications, Japan
Amruth Kumar   Ramapo College of New Jersey, USA
Diane Litman   University of Pittsburgh, USA

Contents – Part II

Industry and Innovation Track

Workshops and Tutorials

Practitioner Track

Posters and Late-Breaking Results

Contents – Part I

Short Papers

Doctoral Consortium

The Black-Box Syndrome: Embracing Randomness in Machine Learning Models

Z. Anthis$^{(\boxtimes)}$

University College London, Gower St, London WC16BT, UK
qtnvzan@ucl.ac.uk

Abstract. Acknowledging the 'curse' of dimensionality, the educational sector has reasonably turned to automated (in some cases autonomous) solutions, in the process of extracting and communicating patterns in data, to promote innovative teaching and learning experiences. As a result, Learning Analytics (LA) and Educational Data Mining (EDM) have both been relying on various Machine Learning (ML) techniques to project novel and meaningful predictions. This inclusion has led to the need for developing new professional skills in the teaching community, that go beyond digital competence and data literacy. This paper seeks to address the issue of ML adoption in educational settings by using an interactive Exploratory Learning Environment (ELE) to test the potential impact of randomness on explainability. The goal is to investigate how misconceptions about stochasticity can lead to distorted projections or expectations, and potentially expose any lack of transparency (to teachers), indirectly affecting the overall trust in artificially intelligent tools.

Keywords: Learning Analytics (LA) · Educational Data Mining (EDM) · Machine Learning (ML) · Exploratory Learning Environment (ELE)

1 Introduction

As with many other fields, education policymakers around the world are rediscovering the potential impact of ML techniques on the learning process and outcomes, whilst striving to remain alert to their limitations. Applications in both Educational Data Mining (EDM) and Learning Analytics (LA) have already been showing promising results with regards to various aspects of performance such as achievement, correctness, engagement, participation, and reflection. Empirical evidence has confirmed that such ecosystems can provide acute computer-based support, across all levels of educational delivery. More recently, however, the educational community has come to the realization that, if ML is to constitute potential groundwork for systemic changes in practice, incorporating teachers' knowledge and expertise is of vital importance [1, 2]. This imposes the challenge of methods being potentially useful and, at the same time, ultimately understandable. Consistent with Baker [3], "the field should move towards greater interpretability, generalizability, transferability, applicability, and with clearer evidence for effectiveness". After all, apart from offering more interactive educational environments and optimizing

© Springer Nature Switzerland AG 2022
M. M. Rodrigo et al. (Eds.): AIED 2022, LNCS 13356, pp. 3–9, 2022.
https://doi.org/10.1007/978-3-031-11647-6_1

institutional proficiency, EDM and LA are being used to map computable student performance measures to explicit instructional practices [4]. With stakes that high (whereby derived outputs affect humans' lives), there is an emerging need for deeply understanding how these intricate decisions are furnished by AI [5]. That includes accounting for the intrinsic cognitive load connected with individual search approaches and problem-solving comprehension, especially when randomness and/or uncertainty is involved. In this paper, educational practitioners are gently introduced to the mathematical methods pertaining to model selection/validation, placing emphasis on explainability (and, by extension, trust) in AIEd.

2 Trustworthy Machine Learning

In the context of ML, the trustworthiness of a model refers to its ability of handling different plausible real-world scenarios, without continuous control. In recent years, attempts to relate considerations about trust from the social sciences to trustworthiness technologies proposed for such models are receiving increased attention [6–8]. Such endeavors fall within the developing field of so-called eXplainable AI (XAI), which is eventually expected to create "a suite of ML techniques that enables human users to understand, appropriately trust, and effectively manage the emerging generation of artificially intelligent partners" [9].

Explanations are considered enablers of human trust in computerized decision support. The European Union General Data Protection Regulation (enacted 2016) extended the automated decision-making rights in the 1995 Data Protection Directive to provide a legally disputed form of a "right to explanation"[1]. Explanations in ML can come in many forms, but a clear consensus regarding their looked-for properties is yet to arise. For one, the recent surge in Interpretable ML (iML) research as a standalone topic, has created transdisciplinary confusion in numerous fronts [10, 11]. In fact, the interchangeable misuse of interpretability and explainability continues to abound in the literature. The AI in Education (AIEd) sector could not be left unaffected. It is, nevertheless, necessary to appreciate the explanatory benefits of distinctive models, without any prior bias for their possible (lack of) interpretability and/or transparency.

Interpretability is loosely defined as the science of comprehending what a model did (or might have done), i.e., it is mostly about the extent to which a cause and effect can be observed within a system. Whereas transparency (often seen as the opposite of black- box-ness) merely stands for a model's capacity of becoming self-interpretable; sometimes posed as intelligibility [12], decomposability [13], or simulatability [14]. In simple terms, each of these traits can be perceived as being able to discern the internal mechanics involved, without necessarily knowing why. Explainability, on the other hand, is a much broader term, that encompasses the ability of transferring these mechanics into human-understandable language. It implies rendering a model dependable, with justifiability and accountability at its core. It is important to remember that, from a technical perspective, these concepts are closely interrelated (see Fig. 1).

---

[1] Regulation (EU) 2016/679 of the European Parliament and of the Council, *on the protection of natural persons with regard to the processing of personal data and on the free movement of such data*, and repealing Directive 95/46/EC (General Data Protection Regulation).

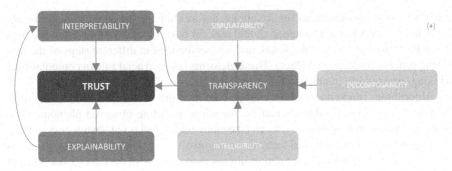

Fig. 1. A System Dynamics (SD) representation of contributing factors assumed to affect trust.

The success of EdTech often boils down to algorithmic simplicity and logical argumentation. Thus, the complexity of trust could be partly attributed to the simultaneity (mutual causation) among several contributing factors. For example, an explainable model is found to be (by default understandable and) almost always interpretable, although not necessarily vice versa. And, in some cases, upturn in one quality is achieved at the expense of another. Moreover, each of these properties could directly (or transitively) influence other aspects of model design, found to interfere with overall performance (e.g., accuracy or speed), thus enforcing additional programming trade-offs. Then again, there are cases where performance itself becomes more important for user trust, than explainability [15]. To add to the conundrum, sometimes the most accurate of explanations are not easily interpretable from people, to begin with. In fact, a substantial part of experimental research warns against judging systems by relying solely on human evaluations, as they tend to imply a strong bias towards simpler descriptions that can lead to "persuasive" (rather than transparent) systems [16, 17]. Although the integration of these principles as evaluative criteria during the design phase might appear inconsistent (adding complexity to deal with complexity), there is still evident value in exploring their conceivable linkage to post-hoc pedagogical use. As stated by Došilović et. al. [18], "Incompleteness in formalization of trust criteria is a barrier to straightforward optimization approaches. For that reason, interpretability and explainability are posited as intermediate goals for checking other criteria".

3 Randomness in ML Models

Models are basically useable representations constructed by the means of mathematical abstraction (signifying simple linear relationships and/or complex non-linear relationships). As such, they are naturally built in computer-tractable language, but also molded to work in partnership with people, toward human goals [19]. Psychological and didactical research suggest that paradoxes and controversies about the very meaning of randomness are indeed reproduced in human intuitions (built when faced with random situations), often contradicting traditional probabilistic reasoning [20, 21]. In this sense, misconceptions about the role of randomization in ML may act as perceptual barriers (especially for non-AI experts), clouding the educators' judgement via systematic biases and/or distorted projections and expectations. Before studying the applicability or repro-ducibility of any inferred result, one should seek for basic elements of randomness, such as random variables, functions, fields, and sets, embedded in different steps of the rea-soning process (inductive or other). These elements take a crucial role in computational thinking, particularly in understanding how the "learning" happens [22, 23].

Algorithmic randomness is an entropy-like measure of disorder which describes the possible effects of unknown causes, for when modeling observed phenomena or predicting future events. Thereby, it can take many forms, and is not always easy to track (even more so if non-analytical techniques are involved). However, the major sources of randomness in EDM tasks are almost always associated with the data (collection and preparation), validation scheme, and function approximation method. Table 1 provides an indicative example (task and method) for each described source.

Table 1. The four (4) main sources of Randomness occurring in EDM with examples.

	Task	Method
Data collection	Sampling	Stratified random
Data preparation	Dimensionality reduction	PCA
Model validation	Train-Test split	Holdout
Function approximation	Clustering	k-Means

4   Methodology

In this section, a feedback provision approach (drawn from AIEd postgraduate students) is presented, that is consisted out of three consecutive phases: attending an introductory course, interacting with an ELE, and taking an online survey. The overarching goal of this intervention can be summarized into tackling three research questions:

1. Does a teacher-led introduction to randomness increase explainability of applied ML models?

2. Does time spent on KIWI show statistically significant improvement in non-analytical problem solving?

3. Is there a statistically significant variation in respondents' (dis)trust in AIEd, based on LX feedback?

Introductory Course. First, all student participants are introduced to the basics (prin-ciples, concepts and ideas) of problem-solving search methods, via a single one-hour course that is teacher-centered, whereby particular emphasis is given to the clarity and thorough analysis of the content in relation to ML. Learners are essentially presented with the inherent function approximation aspect arising in both supervised and unsu-pervised learning, and are then exposed to a series of targeted analogies, illustrations and examples. Towards the end, the course becomes more open-ended, and students are encouraged to actively engage in discussion and exchange views, experiences, and ideas. This end-to-end process comprises the following five learning objectives:

1. Recognize the value of (and/or need for) non-analytical approaches in problem solving.
2. Appreciate randomness as a serviceable tool (or feature) in non-deterministic environments.
3. Conceptualize applied ML as a function approximation (search) problem.
4. Leverage stochastic (optimization) processes to introduce controlled randomness in ML models.
5. Understand the use of ensembles (over individual algorithms) for stability and robustness.

Exploratory Learning Environment. Secondly, participants are instructed to navigate to the Knowledge Inferencing Web-based Interface (KIWI) website, sign up and engage in a series of knowledge-centered tasks, all revolving around the classic combinatorial optimization problem known as Traveling Salesman Problem (TSP). Topics addressed include complexity increase, blind (partial and exhaustive) search, and (meta-)heuristic approaches. The e-learning platform essentially aims at building upon the previously acquired knowledge on randomness and uses Visual Interactive Simulation (controlled parametrization to conduct scenario analyses on-demand) as a channel to improve learner control. Assuming true comprehension of central mathematical programming concepts (e.g., optimality or efficiency) to be mainly contextual, it takes a constructionist approach, drawing from gamification principles (design patterns, aesthetics, and mechanics) and step-based hints, to increase user interest and motivation.

Learning Experience Survey. Eventually, participants are prompted to take an online (retrospective) survey, which concentrates on a quest to complement and add to previous research on the learning benefits of randomness comprehension (apart from those of exploratory interaction and adaptive visualization), in ML tolerance. The administered questionnaire is carefully designed to promote response rates and includes both open-ended and closed-ended answers options. The focus is placed on reinterpreting all undertaken activities, while reflecting on obtained results, to gain insight into the (meta-)cognitive processes that this activates. Questions reveal the progressive socio-epistemic knowledge of theoretical and experimental probability (e.g., local variability or stabilization of relative frequencies) in ML contexts, but also allow for policy-oriented descriptions of the current perspectives on trustworthy AIEd (and its implications in society), with explainability-by-design in mind.

References

1. Castañeda, L., Selwyn, N.: More than tools? Making sense of the ongoing digitizations of higher education. Int. J. Educ. Technol. High. Educ. **15**(1), 1 (2018). https://doi.org/10.1186/s41239-018-0109-y
2. Pedro, F., et al.: Artificial intelligence in education: challenges and opportunities for sustainable development (2019)
3. Baker, R.S.: Challenges for the future of educational data mining: the Baker learning analytics prizes. JEDM J. Edu. Data Min. **11**(1), 1–17 (2019)
4. Papamitsiou, Z., Economides, A.: Learning analytics and educational data mining in practice: a systematic literature review of empirical evidence. Educ. Technol. Soc. **17**, 49–64 (2014)
5. Goodman, B., Flaxman, S.: European union regulations on algorithmic decision-making and a "right to explanation." AI Mag. **38**(3), 50–57 (2017)
6. Ashoori, M., Weisz, J.D.: In AI we trust? Factors that influence trustworthiness of AI-infused decision-making processes. arXiv preprint arXiv:1912.02675 (2019)
7. Toreini, E., et al.: The relationship between trust in AI and trustworthy machine learning technologies. In: Proceedings of the 2020 Conference on Fairness, Accountability, and Transparency, Barcelona, Spain, pp. 272–283. Association for Computing Machinery (2020)
8. Ribeiro, M.T., Singh, S., Guestrin, C.: "Why should I trust you?" Explaining the predictions of any classifier. In: Proceedings of the 22nd ACM SIGKDD International Conference on Knowledge Discovery and Data Mining (2016)
9. Gunning, D.: Explainable artificial intelligence (xai). Defense Advanced Research Projects Agency (DARPA), nd Web, **2**(2) (2017)
10. Carvalho, D.V., Pereira, E.M., Cardoso, J.S.: Machine learning interpretability: a survey on methods and metrics. Electronics **8**(8), 832 (2019)
11. Doshi-Velez, F., Kim, B.: Towards a rigorous science of interpretable machine learning. arXiv preprint arXiv:1702.08608 (2017)
12. Lou, Y., Caruana, R., Gehrke, J.: Intelligible models for classification and regression. In: Proceedings of the 18th ACM SIGKDD International Conference on Knowledge Discovery and Data Mining (2012)
13. Eban, E., et al.: Scalable learning of non-decomposable objectives. In: Proceedings of the 20th International Conference on Artificial Intelligence and Statistics. PMLR (2017)
14. Bogdanov, D., et al.: High-performance secure multi-party computation for data mining applications. Int. J. Inf. Secur. **11**(6), 403–418 (2012)
15. Papenmeier, A., Englebienne, G., Seifert, C.: How model accuracy and explanation fidelity influence user trust. arXiv preprint arXiv:1907.12652 (2019)
16. Vandekerckhove, J., Matzke, D., Wagenmakers, E.-J.: Model comparison and the principle of parsimony. In: Busemeyer, J.R., Wang, Z., Townsend, J.T., Eidels, A. (eds.) Oxford Handbook of Computational and Mathematical Psychology, pp. 300–319. Oxford University Press, Oxford (2015)
17. Herman, B.: The promise and peril of human evaluation for model interpretability, p. 8. arXiv preprint arXiv:1711.07414 (2017)
18. Došilović, F.K., Brčić, M., Hlupić, N.: Explainable artificial intelligence: a survey. In: 2018 41st International Convention on Information and Communication Technology, Electronics and Microelectronics (MIPRO). IEEE (2018)
19. Abdul, A., et al.: Trends and trajectories for explainable, accountable and intelligible systems: an HCI research agenda. In: Proceedings of the 2018 CHI Conference on Human Factors in Computing Systems, Montreal QC, Canada, p. 582. Association for Computing Machinery (2018)

20. Borovcnik, M., Kapadia, R.: A historical and philosophical perspective on probability. In: Chernoff, E.J., Sriraman, B. (eds.) Probabilistic Thinking. AME, pp. 7–34. Springer, Dordrecht (2014). https://doi.org/10.1007/978-94-007-7155-0_2

21. Batanero, C.: Understanding randomness: challenges for research and teaching. In: CERME 9-Ninth Congress of the European Society for Research in Mathematics Education (2015)

22. Balasubramanian, V., Ho, S.-S., Vovk, V.: Conformal Prediction for Reliable Machine Learning: Theory, Adaptations and Applications. Newnes, London (2014)

23. Mitzenmacher, M., Upfal, E.: Probability and Computing: Randomization and Probabilistic Techniques in Algorithms and Data Analysis. Cambridge University Press, Cambridge (2017)

Developing an Inclusive Q&A Chatbot in Massive Open Online Courses

Songhee Han[✉] [ID] and Min Liu

The University of Texas at Austin, Austin, TX 78712, USA
song9@utexas.edu

Abstract. This study aims to examine massive open online course (MOOC) students' experiences with a natural language processing-based Q&A chatbot. Following the definition of 'inclusive learning' in MOOCs from the Universal Design of Learning approach, this study firstly compares students' behavioral intentions before and after using the chatbot. Next, this study investigates students' levels of several other learning experience domains after using the chatbot—teaching presence, cognitive presence, social presence, enjoyment, perceived use of ease, etc. After examining students' possible disparate learning experiences in these domains, this study investigates how age, gender, region, and native language factors influence students' learning experiences with the chatbot. Lastly, but most importantly, this study explores how demographic factors influence students' perception of chatbot interactions. If any are found, this study will focus on possible negative demographic factors that affect only certain groups of students to further examine how to improve a Q&A chatbot for inclusive learning in MOOCs.

Keywords: Inclusiveness · Massive open online course · Natural language processing-based (NLP) chatbot

1 Introduction

1.1 Background

To solve common problems with Q&A webpages, including text heaviness and cognitive overload, massive open online course (MOOC) providers are becoming interested in using natural language processing-based chatbots to provide more prompt responses to individual queries. Chatbots are software programs that communicate with users through natural language interaction interfaces [11–13]. Although chatbots are limited in their ability to correctly respond to every possible question, especially in the initial adoption phase, they can provide human-like responses through dialogue-based interactions which feel more immediate and customized compared to webpages. However, considering the distinctively broad spectrum in a MOOC student population, providing an inclusive learning environment becomes extremely important. Therefore, MOOC providers should make sure any demographic factors do not create inequitable learning experiences for certain groups of students upon making any technological changes,

© Springer Nature Switzerland AG 2022
M. M. Rodrigo et al. (Eds.): AIED 2022, LNCS 13356, pp. 10–15, 2022.
https://doi.org/10.1007/978-3-031-11647-6_2

including the utilization of chatbots. A study with a small sample size examined whether the learning experience from using an FAQ chatbot was disparate from using an FAQ webpage [6]. The results indicated a significant difference between the two interfaces in the level of perceived barriers (i.e., a higher level for the chatbot group) and the level of intention to use the assigned interface (i.e., a lower level for the chatbot group). However, their Q&A quality and enjoyment levels were equivalent. Among the demographic factors investigated, region and native language factors were significantly influential in creating disparate experiences. Most importantly, implications of the necessity of a chatbot interface and how to improve the students' Q&A experience with an FAQ chatbot were found in the study.

This study will primarily examine students' learning experiences with an enhanced Q&A chatbot with a larger sample size to build upon the previous study results. In detail, this study compares the levels of students' behavioral intentions before and after using the chatbot, following the definition of 'inclusive learning' in MOOCs [6] from the Universal Design of Learning (UDL) approach. Next, this study investigates students' levels of several other learning experience domains after using the chatbot—teaching presence, cognitive presence, social presence, enjoyment, perceived use of ease, and so on from the Community of Inquiry framework [4] and Technology Adoption Model [3]. After examining students' possible disparate learning experiences in these domains, this study investigates how age, gender, region, and native language factors influence students' learning experiences with the chatbot. Lastly, but most importantly, this study explores how demographic factors influence students' perception of chatbot interactions. Next, if any are found, this study will focus on possible negative demographic factors that affect only certain groups of students to further examine how to improve a Q&A chatbot for inclusive learning in MOOCs.

Notably, the meaning of inclusiveness will be further investigated from the UDL approach first in this study. The result of this conceptualization will be utilized in multiple aspects: evaluating students' experiences with the chatbot and examining better chatbot response design strategies to support students' equitable learning in MOOCs.

1.2 Research Questions

What demographic factors are influential in creating different learning experiences with a Q&A chatbot in massive open online courses, and what measures can be taken to mitigate possible negative learning experiences by changing chatbot response designs to promote an equitable learning environment?

Detailed Research Questions. There are three detailed research questions as follows:

- What is the relationship between students' demographic factors (i.e., age, gender, region, and native language) and their learning experience domains (i.e., teaching presence, cognitive presence, social presence, enjoyment, perceived use of ease, etc.) with a Q&A chatbot in massive open online courses?
- What is the structural relationship between their learning experience domains? Are there any differences among the demographic groups investigated in the study?

- From a phenomenological perspective, what are their lived learning experiences like with the chatbot when they possess demographic factor(s) identified as negatively influencing their interaction with the chatbot? How do they want the chatbot to respond to their questions?

2 Methods

This mixed methods explanatory sequential study has two phases: a user testing-oriented survey and interviews. For the user testing-oriented survey phase, I will enhance the FAQ chatbot developed for the previous studies [6, 7] to evolve into a learning assistant Q&A chatbot by taking the following two measures. First, a topic analysis will be conducted utilizing an unsupervised machine learning modeling technique, Latent Dirichlet Allocation, with relevant online forum content of the research site's courses provided for the recent three years. The extracted topics will be compared with the existing training set on the Dialogflow platform, and new topics and their content will consist of a new training set. Second, training set refinement (for the old training set) and language adjustment (for both old and new training sets) will be conducted based on the findings from the previous studies. In summary, the chatbot's knowledge base will be enhanced by a new training set from the accumulated forum posting contents of the research site in the site's learning management system for recent years. The manner of presenting the chatbot's response will be changed based on the findings from the two previous studies, which suggested some implications of promoting students' equitable learning experiences.

Once the chatbot on the Dialogflow platform is sufficiently trained based on the new and refined training sets, it will be deployed on a website enabling the study participants to interact by participating in the user testing-oriented survey. The survey questions will consist of five categories in the following order: initial behavioral intentions, user-testing, post-usage learning experience domain levels, expectations/challenges, and demographics. The questions regarding demographics and some regarding post-usage learning experience domains are designed to collect quantitative data, and the others of learning experience domains and expectations/challenges will collect qualitative data. Appendix A shows the tentative students' learning experience domains with the chatbot in the user-testing-based survey.

In the interview phase, phenomenological interviews will be conducted with a small group of students. Purposeful sampling will be conducted according to the quantitative analysis of the survey data to see the possible inequitable learning experiences influenced by students' demographic factors while interacting with the chatbot. With interview data, the students' self-reported elaborations about the challenges they experienced will also be utilized to see if any certain positioning influenced their perceptions in using the chatbot. Appendix B includes some examples of interview questions and prompts.

The recruited research site is located in the Southwestern U.S. and provides MOOCs for journalists' professional development. This site launches 15 courses on average per year, attracting students from 160 unique countries, and student numbers per course have ranged from 1000 to 5000 in the recent three years. Course(s) in this study will be delivered in English and last for four weeks in 2022 and 2023. The active students who make progress in the courses during the designated time period will be asked to participate in this study after receiving approval from an institutional review board. This study will aim to recruit over 100 students to attain statistical power in the quantitative data analysis for the survey. For interviews, students ($n =$ around 10) will be recruited. The preliminary quantitative data analyses will determine the criteria for recruiting interview participants.

3 Significance of the Study

According to a generally-accepted value proposition of MOOCs, MOOCs should support everyone willing to learn with open access by utilizing the flexible features of affordable courses [2]. Considering MOOC students often report they are not sufficiently guided by instructors, course providers, or peers in how to become successful learners with their course activities [c.f., 1, 8, 10], providing a Q&A chatbot could improve students' learning experience by adding one more tool for student support. More importantly, investigating students' learning experiences with new technology in MOOCs, such as chatbots, requires multiple aspects: learning theory, technology-enhanced learning environment, and technology acceptance. Considering a theory-agnostic research approach has been arguably pervasive in the MOOC study community due to its multidisciplinary nature [5, 9], this study will contribute to the field by providing what aspects to consider for inclusive learning when adopting new technologies in the MOOC space from concrete theoretical backgrounds. Moreover, the findings will suggest critical chatbot response design points that better serve diverse student needs in MOOCs.

Appendix A

Tentative student' learning experience domains with the chatbot.

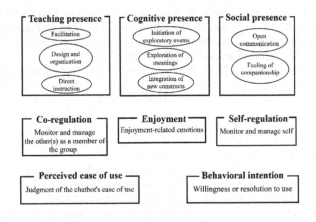

Appendix B

All interviews will be structured as open-ended conversations about their lived learning experiences with the chatbot. I will let participant responses shape the following questions asked during the interviews. Some examples of questions and prompts that I will use are as follows:

1. What was your experience like to use the chatbot?
2. Can you walk me through some of your thoughts about the chatbot's responses that you just received?
3. Can you tell me more about what you said when you described [...]?
4. What thoughts are standing out to you as you see the [erroneous] response here?
5. Earlier you mentioned that you were experiencing [...] Are you noticing the same thing here?
6. Some people describe [...] as they see this kind of response. Are you noticing something similar? Different? In what way?
7. Can you tell me more about this response that you received here? Do you find that it helps you or interferes with what you're seeking?
8. Can you describe any feelings that were generated as you used the chatbot?
9. Did you notice that the chatbot cannot respond to this kind of question? Do you think there should be a response to this kind of question?
10. I am going to observe you while you use the chatbot in this session. During the observation, I noticed [...]; did you realize that you were doing [...]? Can you help me to understand more about [...]?
11. Is there anything that you'd like to share about your experiences with the chatbot?

References

1. Bouchet, F., Labarthe, H., Bachelet, R., Yacef, K.: Who wants to chat on a MOOC? Lessons from a peer recommender system. In: Delgado Kloos, C., Jermann, P., Pérez-Sanagustín, M., Seaton, D.T., White, S. (eds.) Digital education: out to the world an back to the campus, pp. 150–159. Springer International Publishing, Cham (2017)
2. Coughlan, T., Lister, K., Seale, J., Scanlon, E., Weller, M.: Accessible inclusive learning: foundations. In: Ferguson, R., Jones, A., Scanlon, E. (eds.) Educational Visions: Lessons from 40 Years of Innovation, pp. 51–73. Ubiquity Press, London (2019)
3. Davis, F.D.: Perceived usefulness, perceived ease of use, and user acceptance of information technology. MIS Q. **13**, 319–340 (1989). https://doi.org/10.2307/249008
4. Garrison, D.R.: E-Learning in the 21st Century: A Community of Inquiry Framework for Research and Practice. Routledge, New York (2016)
5. Gasevic, D., Kovanovic, V., Joksimovic, S., Siemens, G.: Where is research on massive open online courses headed? A data analysis of the MOOC research initiative. Int. Rev. Res. Open Distrib. Learn. **15** (2014). https://doi.org/10.19173/irrodl.v15i5.1954
6. Han, S., Lee, M.K.: FAQ chatbot and inclusive learning in massive open online courses. Comput. Educ. **179** (2022). https://doi.org/10.1016/j.compedu.2021.104395.Article no. 104395
7. Han, S., Liu, M., Pan, Z., Cai, Y., Shao, P.: Making FAQ chatbots more Inclusive: an examination of non-native English users' interactions with new technology in massive open online courses. Revision Submitted (n.d.)

8. Julia, K., Peter, V.R., Marco, K.: Educational scalability in MOOCs: analysing instructional designs to find best practices. Comput. Educ. **161** (2021). https://doi.org/10.1016/j.compedu.2020.104054. Article no. 104054
9. Kovanović, V., et al.: Exploring communities of inquiry in massive open online courses. Comput. Educ. **119**, 44–58 (2018). https://doi.org/10.1016/j.compedu.2017.11.010
10. Liu, M., Zou, W., Shi, Y., Pan, Z., Li, C.: What do participants think of today's MOOCs: an updated look at the benefits and challenges of MOOCs designed for working professionals. J. Comput. High. Educ. **32**(2), 307–329 (2019). https://doi.org/10.1007/s12528-019-09234-x
11. Rubin, V.L., Chen, Y., Thorimbert, L.M.: Artificially intelligent conversational agents in libraries. Libr. Hi Tech **28**, 496–522 (2010). https://doi.org/10.1108/07378831011096196
12. Shawar, B.A., Atwell, E.S.: Chatbots: are they really useful? J. Lang. Technol. Comput. Linguist. **22**, 29–49 (2007)
13. Wambsganss, T., Winkler, R., Söllner, M., Leimeister, J.M.: A conversational agent to improve response quality in course evaluations. In: Extended Abstracts of the 2020 CHI Conference on Human Factors in Computing Systems, Honolulu HI USA, pp. 1–9. ACM (2020)

Graph Entropy-Based Learning Analytics

Ali Al-Zawqari[✉][iD] and Gerd Vandersteen[iD]

Department of ELEC, Vrije Universiteit Brussel, 1050 Brussels, Belgium
aalzawqa@vub.be

Abstract. The goal of this research is to strengthen the teaching strategy with quantitatively measured learning analytics. The entropy-based learning analytics aims to measure and understand students' progress by quantitatively measuring the difference between the content to be learned, the tutors' expectation of understanding, and the student's knowledge. This quantification will take similar steps than taken by Shannon for his information theory using a mathematical formalism to quantitatively measure knowledge (equivalent to Shannon's entropy) and knowledge transfer (equivalent to Shannon's mutual information). Knowledge graphs will be used to represent the content to be learned, the tutors' expectations, and the student's knowledge. Early results reveal that advanced analytical algorithms and graph entropy specified for educational applications is necessary for this research project to succeed.

Keywords: Probabilistic graphical models · Shannon theory · Graph entropy · Learning analytics

1 Introduction

The human learning process lies at the heart of many disciplines: pedagogy, psychology, neuroscience, linguistics, sociology, and anthropology [13]. Researchers of Artificial intelligence in education (AIEd) bring two interdisciplinary fields (AI and Education) together, intending to make explicit and precise computation forms of knowledge [14]. In addition, AIEd offers the tools to open up the "black box" of learning by providing a deeper understanding of its process [11]. This research project works towards using/developing AI techniques to accurately model and measure the students' learning process for better understanding and improvement of this process. The rationale is inspired by Shannon's work in information theory [15] which was the starting point of our modern communication society. He rigorously quantified the transferred information sent across a communication channel using entropy and mutual information concepts. This theory provided the tools to systematically improve and develop the communication techniques of our modern communication society. The long-term vision of this

This work was financially supported in part by the Vrije Universiteit Brussel (VUB-SRP19), in part by the Flemish Government (Methusalem Fund METH1) and in part by the Fund for Scientific Research (FWO).

M. M. Rodrigo et al. (Eds.): AIED 2022, LNCS 13356, pp. 16–21, 2022.
https://doi.org/10.1007/978-3-031-11647-6_3

project is to develop and demonstrate a new paradigm, an entropy-based learning analytic. Novel graph information-based learning analytics will be created which accurately measure and help to optimize the learning processes. Therefore, a strategy similar to the one Shannon followed will be considered using concepts such as entropy, mutual information [15] and interaction information [12].

2   Theoretical Framework and Methodology

2.1   Modelling the Learning Problems

When taken at the abstract level, the communication of knowledge from the teachers or learning platform to the students can be considered as a communication problem [15], especially when looking at the level of communicating and remembering things. In this abstraction, the knowledge about the studied subject acts as the transmitter, the teaching channel is the noisy communication channel, and the student knowledge level is the receiver, as shown in Fig. 1.

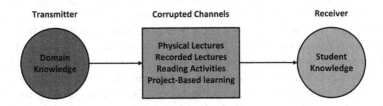

Fig. 1. Abstraction of the learning problem in communication model.

This approach suggests distributing the learning problem's modeling into three interactive models: the course subject model (domain knowledge), the teaching methods model (the channel), and the student model (student knowledge), as illustrated in Fig. 1. The learning rate can then be measured independently for each channel using mutual information between the transmitter (domain knowledge) and the receiver (student knowledge). Later, the "optimal" teaching channel can be personalized automatically to optimize the knowledge transfer to each student. However, in the educational context, the relation between different parts of information is essential, such as the connection between remembering the multiplication table and solving mathematical word problems. Shannon's model has no levels that provide information about these additional relationships. As a result, integrating Shannon's theory with pedagogy theory is a must for this work.

2.2   Knowledge Graph Representation for Education Sciences

As is established above, learning is more than transferring and storing information alone. It is also about understanding, applying, analyzing, etc. The different levels of learning were described by Bloom using his taxonomy [2] and later modified by splitting the taxonomy into a cognitive process dimension (remembering, understanding, applying, ...) and a knowledge dimension (factual, conceptual, procedural, ...) [10]. The knowledge to be learned can be expressed efficiently using graph-like representations in the scientific field of Knowledge Representation and Reasoning [3]. Here, the objective is to develop a probabilistic knowledge graph representation [8]. To this end, we need three graphs with the capabilities to capture different aspects of the problem: 1) the content graph: represents the knowledge to be learned (the content of the course); 2) the expectation graph: represents the learning levels/goals set by the teacher (using, e.g., Bloom's taxonomy); 3) the student's graph: represents the student's estimated knowledge. This contains a probabilistic measure of whether the learning goal is attained or not.

The objective is to define a common knowledge graph representation that captures these three aspects. This representation is necessary as it serves as an input to calculate the graph entropy that is needed to determine the (mutual and interaction) graph information. The research hypothesis is that it is indeed possible to represent all three aspects within a common graph representation. All processing can be performed with the inclusion of elementary operations only, which are the basis for more complex ones. The challenge of this knowledge graph for education sciences is that it demands the adaptation of known knowledge graphs [3,8] towards the requirements in education. This task requires an intensive collaboration with educational scientists to incorporate the various learning levels within the graph representation. Once it is possible to capture the different levels of learning (remembering, understanding, applying on examples), it becomes possible to measure these levels of learning using the theory described and developed in the following subsection.

2.3   Graph Entropy-Based Learning Analytics

The scientific objective is to develop a graph entropy-based approach to measure a student's progress in the learning process using mutual and interaction information. These are defined between two or more graphs using an entropy measure [12,15]. The entropy measures the information transport between the content graph, the expectation graph, and the student (learning) graph. Determining the entropy definition and its properties for the problem at hand will be pretty challenging. Various definitions already exist for information sources (with and without memory/correlation) and topologies of graphs. In addition, entropy at the level of remembering is identical to Shannon's entropy. However, there is currently no graph entropy that fulfills the following requirements: 1) the entropy of a single vertex equals Shannon's entropy for an information source; 2) the entropy of the union of independent graphs equals the sum of entropies of the individual graphs; 3) the expectation graph entropy must be able to measure the entropy of a subgraph for a particular learning level; 4) the student's graph entropy is proportional to the probability that the student obtains/constructs a knowledge vertex, similar to the entropy concept used for Markov processes [15].

These requirements deviate from the existing entropies of possibly correlated information sources [15], $H(v_i) = -\sum_i p_i \, log \, p_i$, and the entropies for graphs [5,6,9]. The latter entropies proposes a representative probabilistic function that contains the graph structure information [6] with $p_i = f(v_i)/\sum_i f(v_i)$ where $f(v_i)$ is a function of properties of the vertex v_i. These entropies do not satisfy the requirements mentioned above as this probabilistic representation does not measure the actual information entropy $H(v_i)$ of the vertex v_i. They only capture the graph's structure. Hence, an additional scientific objective is a derivation of supporting properties for this entropy: the mutual and interaction information. The research hypothesis is that a graph entropy that satisfies the above constraints can be found. The strategy used is similar to the work of Shannon, which starts by first defining the required properties/constraints for the entropy, followed by the determination of an entropy function that fulfills these constraints [15]. The challenge is that the needed entropy is somewhere in between the entropy definition of Shannon's information and the current graphs' entropies. Figure 2 shows the proposed integrated model of Bloom's taxonomy with the graph-based entropy.

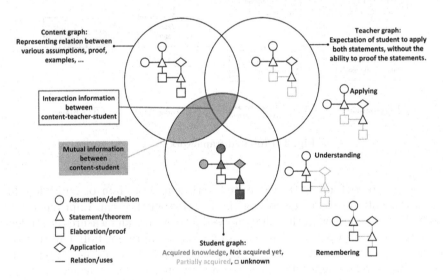

Fig. 2. Bloom's taxonomy and mutual information integrated model.

3   Early Results

The first step taken to examine the graph entropy-based learning analytics was to explore the possibility of measuring the student learning rate through probabilistic graphical models [1]. Figure 3 sums up the proposed framework. Each block represents a separate editable stage colored for different technology, namely: The **content graph** generates the graph of a specific course subject. It can be seen as part of a knowledge representation problem. The **teacher graph** adds probabilistic metrics to the knowledge graph. This block represents the rules and facts as the evaluation reference for what students will learn. The **teacher input of objectives and goals** offers power and control to the educators. The main output of this block will be probabilistic facts and vertex categorization based on Bloom's taxonomy. The **student input of answers to questions** results from specific tests. The **student learning parameters** are where the knowledge reasoning part takes the role. The student evaluation results can be used as an interpretation, which will help to generate the probabilistic parameters for a specific student. ProbLog [4] is offering the algorithm used for this task, namely the learning from interpretations (LFI) [7]. The **student graph** provides an estimate of the graph learned by the student. It allows the teacher to have a more in-depth look at what the student has learned and understood. The most probable explanations (MPE) [16] framework was applied to get more insight into the student's learning status.

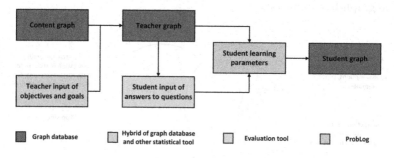

Fig. 3. Proposed framework in [1].

The proposed framework was tested with synthetic data of multiple educators/students in teaching/learning the De Morgan theorem. Implementation results have demonstrated that the proposed framework can measure the student learning rate even when the student evaluation data are not fully covering the course's objectives. However, it was shown that ProbLog could not provide the uncertainty level of a specific learned parameter in the case of incomplete data.

4   Summary

This project aims at introducing a new paradigm in the measurements of students learning in the academic setup. The rationale is inspired by Shannon's information theory work, which was the starting point of our modern communication society. However, the meaning of information in the educational context is much more complex than in the communication systems. This complexity motivated the proposal of using AI techniques, namely the probabilistic graphical models as a tool that captures/encoded the levels of information in the educational setup. The information theory, graph theory, and pedagogy theory are the main pillars of the proposed work. It is summarized by having three different probabilistic graphical representations: course content, teacher expectation, and student knowledge. Entropy-based analytics will be drawn from the interaction between these graphs to quantify the student's learning process. Early results showed that available probabilistic programming logic could serve as a tool for simple case scenarios with a complete evaluation of students' knowledge. In complex scenarios, these tools failed in providing stable results.

References

1. Al-Zawqari, A.: AI-Assisted learning: modelling and measuring learning rate. Master's thesis, Vrije Universiteit Brussel, Université Libre de Bruxelles (2020)
2. Bloom, B.S.: Taxonomy of educational objectives: the classification of educational goals. Cognitive domain (1956)
3. Chein, M., Mugnier, M.L.: Graph-based Knowledge Representation: Computational Foundations of Conceptual Graphs. Springer Science & Business Media, Heidelberg (2008). https://doi.org/10.1007/978-1-84800-286-9
4. De Raedt, L., Kimmig, A., Toivonen, H.: ProbLog: a probabilistic prolog and its application in link discovery. In: IJCAI, vol. 7, pp. 2462–2467. Hyderabad (2007)
5. Dehmer, M., Emmert-Streib, F., Chen, Z., Li, X., Shi, Y.: Mathematical Foundations and Applications of Graph Entropy. John Wiley & Sons, Hoboken (2017)
6. Dehmer, M., Mowshowitz, A.: A history of graph entropy measures. Inf. Sci. **181**(1), 57–78 (2011)
7. Fierens, D., et al.: Inference and learning in probabilistic logic programs using weighted Boolean formulas. Theory Pract. Logic Program. **15**(3), 358–401 (2015)
8. Koller, D., Friedman, N.: Probabilistic Graphical Models: Principles and Techniques. MIT press, Cambridge (2009)
9. Körner, J.: Coding of an information source having ambiguous alphabet and the entropy of graphs. In: 6th Prague Conference on Information Theory, pp. 411–425 (1973)
10. Krathwohl, D.R.: A revision of bloom's taxonomy: an overview. Theory Pract. **41**(4), 212–218 (2002)
11. Luckin, R., Holmes, W., Griffiths, M., Forcier, L.B.: Intelligence unleashed: an argument for AI in education (2016)
12. McGill, W.: Multivariate information transmission. Trans. IRE Prof. Gr. Inf. Theory **4**(4), 93–111 (1954)
13. Sawyer, R.K.: The Cambridge Handbook of the Learning Sciences. Cambridge University Press, Cambridge (2005)
14. Self, J., et al.: The defining characteristics of intelligent tutoring systems research: ITSs care, precisely. Int. J. Artif. Intell. Educ. **10**(3–4), 350–364 (1999)
15. Shannon, C.E.: A mathematical theory of communication. Bell Syst. Tech. J. **27**(3), 379–423 (1948)
16. Shterionov, D., Renkens, J., Vlasselaer, J., Kimmig, A., Meert, W., Janssens, G.: The most probable explanation for probabilistic logic programs with annotated disjunctions. In: Davis, J., Ramon, J. (eds.) ILP 2014. LNCS (LNAI), vol. 9046, pp. 139–153. Springer, Cham (2015). https://doi.org/10.1007/978-3-319-23708-4_10

Generating Narratives of Video Segments to Support Learning

Abrar Mohammed[(⊠)]

School of Computing, University of Leeds, Leeds, UK
`a.mohammed1@leeds.ac.uk`

Abstract. The predominance of using videos for learning has become a phenomenon for generations to come. This leads to a prevalence of videos generating and using open learning platforms (Youtube, MOOC, Khan Academy, etc.). However, learners may not be able to detect the main points in the video and relate them to the domain for study. This can hinder the effectiveness of using videos for learning. To address these challenges, we are aiming to develop automatic ways to generate video narratives to support learning. We presume that the domain for which we are processing the videos has been computationally presented (via ontology). We are proposing a generic framework for segmenting, characterising and aggregating video segments VISC-L which offers the foundation to generate the narratives. The narrative framework designing is in progress which is underpinned with Ausubel's Subsumption theory. All the work is being implemented in two different domains and evaluated with people to test their awareness of the domains-aspects.

Keywords: Learning videos · Domain ontology · Video segmentation · Video characterisation · Video aggregation · Video narratives

1 Problem Addressed

Videos have been widely used in various learning settings to facilitate independent learning and are becoming a key platform for digital learning [9,10]. However, there are major challenges that affect user engagement with videos. Learners' concentration span is reduced over time, which makes it hard to follow long videos [13,16]. Also, video content complexity could affect the engagement with videos and may cause confusion or boredom [15]. Consequently, learners may have to watch videos many times and may not be able to identify the most relevant key points in a video. This calls for finding new ways to identify the main points in a video and to direct learners to the corresponding parts in the video, and crucially, create narratives from these video parts to elaborate specific key points. These challenges are experienced at a scale with the increase of both the amount of video footage available and the number of learners who

---

A. Mohammed—Supervised by Vania Dimitrova.

use videos for learning. Previous researches had different attempts to address this issue by manually annotating important parts in the videos by teachers or learners [5,12]. To maintain the quality of the annotation some researchers use ontologies [8,17]. However the manual attempts did not scale the process; hence automated approaches are required to facilitate how to characterise video segments, especially if the domain of the videos is represented with an ontology (or a knowledge graph) [4,6]. Existing automatic ways for characterising videos do not offer domain related annotation, nor a link to the domain hierarchy of the concepts mentioned in the videos. Moreover, existing studies have not evaluated the impact of segmenting and characterising videos on learning.

To address these challenges, this PhD project poses the following research questions: **RQ1:** How to characterise video transcripts for learning by using domain ontology and past users' comments? **RQ2:** How to automatically segment videos to identify segments which are suitable for learning? **RQ3:** How to generate narratives from the characterised videos segments to support learning?

For **RQ1**, we were able to characterise predefined video segments by using the video's transcript, past users' comments and the domain ontology to apply semantic tagging . The output was characterised video segments with the focus topic/concept mentioned in both the transcript and the users' comments. This work was published in [14]. When there are no predefined video segments, we are proposing our generic framework for video segmentation, characterisation and aggregation to support learning (VISC-L), which addresses **RQ2** (see Fig. 1). The outcome of VISC-L, together with the domain ontology, will be used to generate video narratives following the subsumption theory for learning (**RQ3**).

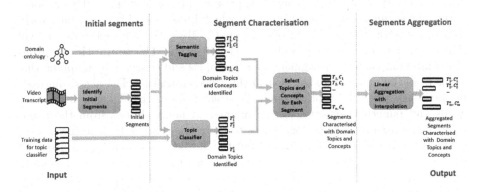

Fig. 1. Framework of videos segmentation and characterisation for learning VISC-L

2   Framework Outline and Methodology

2.1   Framework Outline and Theoretical Underpinning

Input. VISC-L is based on two assumptions. Firstly, it is assumed that the video transcripts relating to the domain to be learned is providing a description of what aspects of the domain are covered. The second assumption is that there is a domain ontology $\Omega = \{C, H\}$ which includes the relevant domain concepts C linked in a concept hierarchy H. We use $c_i \subset c_j$ to denote that c_i is a subclass of c_j. The top level concepts in the concept hierarchy define the main domain topics $\{T_1, ..., T_m\}$. In order to identify the main topics in the video, as part of the characterisation step, training data with domain topics as labels are needed. This can either be created with expert annotators or collected from past user interactions. When we applied our framework, we have used past user interactions in one domain, and we will explore expert annotations in the other domain.

Output. The output of VISC-L is a set of aggregated video segments with a start and end time in the corresponding video. Each video segment i is characterised with a set of domain focus topics (top concepts in Ω) and a set of concepts from the focus topics mentioned in the transcript of the video segment.

Initial Segments. Our video segmentation approach is inspired by text-tilling in text segmentation - starting with smaller units (e.g. sentences) and aggregating them to get larger coherent units (e.g. paragraphs). Hence, we include an initial segmentation step where the video transcripts are cut into small segments that are used as a starting point for aggregation. Initial segments can be done by using a certain number of text lines (e.g. we are using 6 lines) or by using pre-defined segments (e.g. as when we applied our semantic tagging algorithm into one of the domain where we have used high attention intervals from past interactions).

Segment Characterisation. In order to aggregate the initial segments, we need to identify what domain content is presented in each segment. This is done during the segment characterisation step which links each video segment i with a set of focus topics T_i and a set of concepts C_i. To do so, we propose to use two algorithms: semantic tagging and topic classification. The **semantic tagging** algorithm links each video segment to focus topics and concepts by mapping the terms from the ontology to the text in the video transcript. The algorithm first pre-processes the transcript through: (a) tokenisation; (b) cleaning from stop words and punctuation; (c) selecting nouns and noun phrases from the transcript; (d) matching the ontology terms to the noun phrases. If there is a match between the transcript noun phrases and the ontology, the ontology concept c_i will be identified (tagged to the text), noting also the path to reach a top-level concept. As a result, each segment i is linked to a set of focus topics and their corresponding concepts; we denote this as $<T_i^1, C_i^1>$ (where 1 indicates that this is an output from the first segment characterisation algorithm). A key challenge for this algorithm is word sense disambiguation - we need to disambiguate the topics based on the context, which is done with the second algorithm.

The second algorithm is a **topic classifier** which identifies a domain topic based on the context of that topic. Following the latest development in natural language processing, we use the Bidirectional Encoder Representations from Transformers (BERT) [7] as a topic classifier. BERT embeds pre-trained deep bidirectional representations from unlabelled text by jointly conditioning on both left and right context in all layers. Accordingly, it can be fine tuned with just one additional output layer to create state-of-the-art models for different language tasks, topic classification in this case. First, the BERT model is fine-tuned using training data with domain topic labels. The fine-tuned model is used as classifier to link each segment i to domain topics T_i^2 (2 indicates an output from the second segment characterisation algorithm). The last step in segment characterisation is to **combine the outputs from both algorithms**. For each segment i, the outcomes from both algorithms $<T_i^1, C_i^1>$ and T_i^2 are combined by intersecting the focus topics $Ti = T_i^1 \cap T_i^2$ and selecting the concepts C_i from C_i^1 that belong to T_i. Each segment is characterised by $<T_i, C_i>$ - a set of focus topics and their concepts.

Segments Aggregation. Following the text-tilling approach, small segments will be aggregated into larger segments. To maintain the flow of information within adjacent segments, we have developed an aggregation algorithm based on the **Thematic Progression Theory** [3]. This theory has been widely used for creating coherent text, and states that a good written text should have a relation between theme (which is the main clause) and rheme (which is the remainder of the text used to develop the theme). Three patterns for coherent text are suggested: Constant theme (when the first theme in one sentence is carried on and used at the beginning of the second sentence); Linear theme (the important message in a rheme of one sentence is carried in a theme in the second sentence), and Split theme (a development of a rheme with important information is used as themes in the subsequent sentences).

We adapt the Thematic Progression Theory when we aggregate adjacent segments to indicate coherent parts within videos. We associate the focus topic with the segment's theme and the focus concepts with the segment's rheme. We propose a **linear aggregation with interpolation** algorithm. The linear theme pattern was selected as the most appropriate, as it allows to keep a continuous focus topic and at the same time to take into account the specific concepts within that topic. Some segments can be without characterisation (i.e. it is not possible to link the video transcript to domain concepts), which can be because the speaker is silent or is digressing from the domain. If we look strictly for adjacent segments, these gap segments which break the topic flow will lead to starting a new aggregate. To smoothen the aggregation, we use interpolation. If the segments before and after a gap segment have common focus concepts, it is assumed that the common concepts spread across the three segments. Hence, the gap segment will be interpolated in the aggregated segment.

To generate video narratives, the video segments are combined following the **Ausubel's Subsumption Theory** for meaningful learning [2]. According to this theory, a primary process in learning is subsumption in which new material is related to relevant ideas in the existing cognitive structures derived from learning experiences. According to the subsumption theory, there are four types of subsumption: Derivative, Correlative, Super-ordinate and Combinational. We are aiming to automate the linking of video segments to generate narratives by following the focus topics and concepts in them using the hierarchy of the concepts in the domain ontology. Our narratives work using Ausubels' theory is motivated by its successful adoption for meaningful learning by using concept maps [1, 11] that allow learners to group information in related modules, making the connections between modules more apparent.

2.2 Methodology

We have adopted a data-driven approach to generate narratives from videos by first: segmenting videos, characterising video segments, aggregating adjacent segments and creating narratives from these segments. Our data set is either available videos collected by other researchers or to be collected using the search schema we have designed by utilising ontology terms to search for videos available on social learning platforms (i.e. YouTube) as follows: *<Domain Name, Topic Name, Concept Name>*. The input to our work is the video transcript and the domain ontology. Additionally, we need training data labeled with domain topics. Based on this, we can apply our segmentation, characterisation and aggregation framework (VISC-L). The output segments will provide the foundation to generate narratives of video segments. The narratives will be generated by applying the narrative's framework in two different domains and evaluate them with people to test their awareness of the domain aspects and their possible effect on their life.

2.3 Progress to Date

The work on characterising video segments using the domain ontology and the videos-past users' comments and videos transcript has been published in [14]. Additionally, we have applied VISC-L framework in the domain Presentation Skill. The result have been evaluated with learners by comparing the usability, perceived usefulness, mental demand and the learning impact of the characterised video segments generated by our work and by using the Google outcome. This work is submitted to another conference and has been accepted as a full paper.

The next step in this PhD research is to design, apply and evaluate the narrative framework in one domain (Presentation Skill). After that, VISC-L framework and the narrative framework will be applied and evaluated in another domain (i.e. health domain-COPD).

3   Expected Contribution

We propose a novel way to create narratives from video segments to support learning which is underpinned by pedagogical theories and utilises natural language processing. Our main contribution is the designing of two generic frameworks - one for videos segmentation, characterisation and aggregation for learning (VISC-L) and the other one for generating narratives from video segments. The work is being applied in two soft skills domains - presentation skills (giving pitch presentations) and healthcare (patient's quality of life needs assessment).

Acknowledgements. The application in the healthcare domain is funded by the European Union's Horizon 2020 research and innovation programme under grant agreement No 825750 (InADVANCE project).

References

1. Al-Tawil, M., Dimitrova, V., Thakker, D.: Using knowledge anchors to facilitate user exploration of data graphs. Semant. Web **11**(2), 205–234 (2020)
2. Ausubel, D.P.: A subsumption theory of meaningful verbal learning and retention. J. Gen. Psychol. **66**(2), 213–224 (1962)
3. Bloor, T., Bloor, M.: The Functional Analysis of English. Routledge (2013)
4. Cagliero, L., Canale, L., Farinetti, L.: VISA: a supervised approach to indexing video lectures with semantic annotations. In: 2019 IEEE 43rd Annual Computer Software and Applications Conference (COMPSAC), vol. 1, pp. 226–235. IEEE (2019)
5. Castro, M.D.B., Tumibay, G.M.: A literature review: efficacy of online learning courses for higher education institution using meta-analysis. Educ. Inf. Technol. **26**(2), 1367–1385 (2019). https://doi.org/10.1007/s10639-019-10027-z
6. Das, A., Das, P.P.: Semantic segmentation of MOOC lecture videos by analyzing concept change in domain knowledge graph. In: Ishita, E., Pang, N.L.S., Zhou, L. (eds.) ICADL 2020. LNCS, vol. 12504, pp. 55–70. Springer, Cham (2020). https://doi.org/10.1007/978-3-030-64452-9_5
7. Devlin, J., Chang, M.W., Lee, K., Toutanova, K.: BERT: Pre-training of deep bidirectional transformers for language understanding. arXiv preprint arXiv:1810.04805 (2018)
8. Dias, L.L., Barrére, E., de Souza, J.F.: The impact of semantic annotation techniques on content-based video lecture recommendation. J. Inf. Sci. **47**(6), 740–752 (2021)
9. Hsin, W.J., Cigas, J.: Short videos improve student learning in online education. J. Comput. Sci. Coll. **28**(5), 253–259 (2013)
10. June, S., Yaacob, A., Kheng, Y.K.: Assessing the use of Youtube videos and interactive activities as a critical thinking stimulator for tertiary students: An action research. Int. Educ. Stud. **7**(8), 56–67 (2014)
11. Katagall, R., Dadde, R., Goudar, R., Rao, S.: Concept mapping in education and semantic knowledge representation: an illustrative survey. Procedia Comput. Sci. **48**, 638–643 (2015)
12. Lagrue, S., et al.: An ontology web application-based annotation tool for intangible culture heritage dance videos. In: Proceedings of the 1st Workshop on Structuring and Understanding of Multimedia Heritage Contents, pp. 75–81 (2019)
13. Meseguer-Martinez, A., Ros-Galvez, A., Rosa-Garcia, A.: Satisfaction with online teaching videos: a quantitative approach. Innov. Educ. Teach. Int. **54**(1), 62–67 (2017)
14. Mohammed, A., Dimitrova, V.: Characterising video segments to support learning. In: Proceedings of the 28th International Conference on Computers in Education (2020)
15. Mongkhonvanit, K., Kanopka, K., Lang, D.: Deep knowledge tracing and engagement with MOOCs. In: Proceedings of the 9th International Conference on Learning Analytics and Knowledge, pp. 340–342 (2019)

16. Risko, E.F., Anderson, N., Sarwal, A., Engelhardt, M., Kingstone, A.: Everyday attention: variation in mind wandering and memory in a lecture. Appl. Cogn. Psychol. **26**(2), 234–242 (2012)

17. Schulten, C., Manske, S., Langner-Thiele, A., Hoppe, H.U.: Bridging over from learning videos to learning resources through automatic keyword extraction. In: Bittencourt, I.I., Cukurova, M., Muldner, K., Luckin, R., Millán, E. (eds.) AIED 2020, Part II. LNCS (LNAI), vol. 12164, pp. 382–386. Springer, Cham (2020). https://doi.org/10.1007/978-3-030-52240-7_69

Toward Improving Effectiveness of Crowdsourced, On-Demand Assistance from Educators in Online Learning Platforms

Aaron Haim[✉][ID], Ethan Prihar[ID], and Neil T. Heffernan[ID]

Worcester Polytechnic Institute, Worcester, MA 01609, USA
ahaim@wpi.edu

Abstract. Studies have proven that providing **on-demand assistance**, additional instruction on a problem when a student requests it, improves student learning in online learning environments. Additionally, crowdsourced, on-demand assistance generated from educators in the field is also effective. However, when provided on-demand assistance in these studies, students received assistance using **problem-based randomization**, where each condition represents a different assistance, for every problem encountered. As such, claims about a given educator's effectiveness are provided on a per-assistance basis and not easily generalizable across all students and problems. This work aims to provide stronger claims on which educators are the most effective at generating on-demand assistance. Students will receive on-demand assistance using **educator-based randomization**, where each condition represents a different educator who has generated a piece of assistance, allowing students to be kept in the same condition over longer periods of time. Furthermore, this work also attempts to find additional benefits to providing students assistance generated by the same educator compared to a random assistance available for the given problem. All data and analysis being conducted can be found on the Open Science Foundation website (https://osf.io/zcbjx/).

Keywords: Online education · On-demand assistance · Crowdsourcing

1 Introduction

As online learning platforms expand their content base, the need to generate on-demand assistance grows alongside it [6]. Crowdsourcing provides an effective method to generate new assistance for students [5,6,10]. As on-demand assistance generally improves student learning [3,5,10,12], educators and their assistance must be evaluated to maintain or improve the current level of quality and effectiveness [9].

In 2017, ASSISTments, an online learning platform [3], deployed the Special Content System, formerly known as TeacherASSIST. The Special Content System allows educators to create on-demand assistance for problems they assigned

© Springer Nature Switzerland AG 2022
M. M. Rodrigo et al. (Eds.): AIED 2022, LNCS 13356, pp. 29–34, 2022.
https://doi.org/10.1007/978-3-031-11647-6_5

to their students. On-demand assistance was known as **student-supports**, most commonly provided in the form of hints and explanations. Additionally, educators marked as *star-educators* had their *student-supports* provided to students outside their class for any problem the class's educator did not generated a *student-support* for.

While studies analyzed the effectiveness of educators who generated *student-supports* [9] using problem-based randomization, students learn cumulatively across problems [4], making it difficult to provide substantial claims on overall effectiveness in the platform. The first part of this work will develop and use an educator-based randomization, where all *star-educators* are ordered randomly for each student with a *student-support* provided from the top-most educator in the ordering who has generated a *student-support* for a problem, in place of problem-based randomization [6], where a *student-support* was provided randomly from the available *student-supports* for a problem, within the Special Content System to determine an educator's effectiveness.

Since an educator-based randomization will prevent students from receiving certain educators over the first study, benefits from other educators for a student may be unknown. A student may be put in an educator-based randomization where a certain *student-support's* effectiveness is poor compared to other *student-supports* on the problem. The second part of this work will develop an use a **reverse educator-based randomization**: a student uses the reverse order of educators from the first part of this work with a *student-support* provided from the bottom-most educator in the ordering who has generated a *student-support* for a problem.

Other benefits of educator-based randomization compared to problem-based randomization may also be revealed through additional analysis. After this work has collected the necessary data and determines which educators are the most effective, a comparison between previous measures of effectiveness across *student-supports* and educators will be conducted.

In summary, this work aims to answer the following research questions:

1. Which educators are the most effective at generating *student-supports*?
2. How did the effectiveness of the given educator ordering compare to reversed ordering?
3. Was there any hidden benefits from receiving educator-based randomization compared to problem-based randomization?

2   Background

In this work, ASSISTments will be used to conduct the studies. ASSISTments[1] is a free, online learning platform providing feedback and insights on students to better inform educators for classroom instruction [3]. ASSISTments provides problems and assignments from open source curricula, the majority of which is K-12 mathematics, which teachers can select and assign to their students.

---

[1] https://assistments.org/.

Students complete assigned assignments within the ASSISTments Tutor. For most problem types, students receive immediate feedback when a response is submitted for a problem, which tells the student whether the answer is correct [2]. When a *student-support* has been written by the assigning educator or a *star-teacher* for a problem, a student can request to receive the *student-support* at any time while completing the problem. *Student-supports* may come in the form of hints which explain how to solve parts of the problems [3,10], similar problem examples [5], erroneous examples [1,10], and full solutions to the problems [11].

By using the Special Content System, it found that delivering *student-supports* to students compared to immediately giving students the answer caused more student learning [6]. In addition, an analysis was conducted which reported evidence about which educators were generally more effective at improving student learning compared to other educators [9]. Those studies used problem-based randomization. The Special Content System will be modified to provide *student-supports* using educator-based randomization to investigate their effectiveness.

3   Methodology

This work will collect data over the course of the three months. During this time period, two studies each lasting a month will run a different selection mechanism. In between the two studies and after the final study has ended, there will be a two week interval where the selection mechanism will use problem-based randomization. These weeks will be treated as the dependent measure to determine a student's performance within the educator-based randomization.

The Special Content System will use the selection mechanism outlined in Table 1 during the associated time period. After the work has completed collecting data, the Special Content System will be restored to its original state before this work.

Table 1. Breakdown of work conducted

Name	Time period	Selection mechanism
Initial data	Before time period	Problem-based randomization
Study 1	1 month	Educator-based randomization
Mid-Test	2 Weeks	Problem-based randomization
Study 2	1 month	Reversed educator-based randomization
Post-Test	2 Weeks	Problem-based randomization

3.1   Study 1: Educator Ordered Selection

Study 1 will run over the period of a month. During this study, every student will be given a randomly ordered list of all available *star-educators* within the

ASSISTments platform. If an educator has a *student-support* written for a problem (Table 2 gives an example on the left where Educator A has a *student-support* for Problem Y while Educator B does not), the the student will be provided that educator's *student-support*. Otherwise, the next educator will be chosen to provide a *student-support* and so on until either an educator has written a *student-support* for the given problem or no educators have written a *student-support* (in which case none is provided). Using the example in Table 2, if Student 1 requested a *student-support* for Problem Y, the selection mechanism would determine that student would receive a *student-support* from Educator A. In contrast, Student 2 would receive a *student-support* from Educator B for Problem Y, as Educator C did not write a *student-support* and the next educator in the list, Educator B, has.

Table 2. An example of Educator Ordering data. Left: shows what educators wrote a student-support for certain problems where "Yes" means a educator wrote a studentsupport for a problem and vice versa for "No". Right: shows an ordering of all available educators (in this example) for each student from top to bottom.

	Problem X	Problem Y		Student 1	Student 2	Student 3
Educator A	Yes	Yes	Top	Educator A	Educator C	Educator D
Educator B	Yes	Yes		Educator B	Educator B	Educator A
Educator C	Yes	No		Educator C	Educator D	Educator B
Educator D	Yes	No	Bottom	Educator D	Educator A	Educator C

Benefits of an Educator Ordering. Since the ASSISTments platform is used to produce this work, providing each student an ordering of all available *star-educators* is favored over a single educator to better create educator-based randomization. *Student-supports* have been shown to improve student learning [6,9]; if an single educator has not written a *student-support* for a problem which other educators have, the application should still provide an available *student-support*. This is a common occurrence as nineteen *star-educators* have collectively generated 38,737 *student-supports*; however, the top five generated up over 50% with the top two generated approximately 37.6% of the available *student-supports*.

To validate the effectiveness of an educator ordering over a single educator, the ASSISTments Dataset [7,8] was used to simulate Study 1. There are 4,094,728 logged interactions where a *student-support* was selected for a given student on a problem. After pre-processing the data such that only interactions where a student has completed another problem after the current one and more than one *student-support* was available for selection, there are 2,226,779 logged interactions across 94,040 unique students.

As shown on the left of Fig. 1, about 90% of the students almost never received their top-most educator in the ordering, instead on average around 12.8% of the time. Those students would never receive a *student-support* if only single educator solution was used, which would stymie our ability to improve

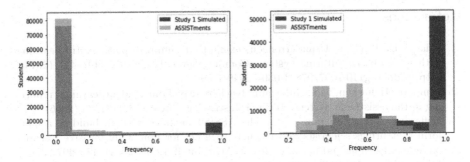

Fig. 1. A comparison of a simulated Study 1 compared to problem-based randomization method used by the ASSISTments platform. Shows the frequency students received their top ordered educator (left) and the frequency students received the educator which they were provided the most *student-supports* from (right).

student learning. On the right of Fig. 1, when using an educator ordering, more than 50% of the students nearly always received their most provided educator with the average around 82.4%. As such, an educator ordering is more effective at keeping students in an educator-based randomization while still maintaining improve learning standards within the ASSISTments platform.

3.2   Study 2: Reversed Educator Ordered Selection

Study 2 will run for a month following a two week interval after Study 1. Students will be provided a *student-support* from the lowest-most educator in the ordering determined from Study 1 who has written a *student-support* for the given problem. In the Table 2 example on the left, Student 1 will receive Educator D's *student-supports* first when available, then C's, then B's, then finally A's. As such, Student 1 will receive the *student-support* generated by Educator C for Problem Y while Student 2 will the *student-support* generated by Educator A.

3.3   Analysis Plan

As the data is currently under collection, no analysis has been formalized yet. Instead, a random 10% of the collected data will be used to attempt different modeling approaches and reduce noise. Afterwards the exact analysis method will be formalized and use the remaining 90% of the data.

Acknowledgements. We would like to thank the NSF (e.g., 2118725, 2118904, 1950683, 1917808, 1931523, 1940236, 1917713, 1903304, 1822830, 1759229, 1724889, 1636782, & 1535428), IES (e.g., R305N210049, R305D210031, R305A170137, R305A170243, R305A180401, & R305A120125), GAANN (e.g., P200A180088 & P200A150306), EIR (U411B190024 & S411B210024), ONR (N00014-18-1-2768), and Schmidt Futures. None of the opinions expressed here are that of the funders. We are funded under an NHI grant (R44GM146483) with Teachly as a SBIR.

References

1. Adams, D.M., et al.: Using erroneous examples to improve mathematics learning with a web-based tutoring system. Comput. Hum. Behav. **36**, 401–411 (2014). https://doi.org/10.1007/978-3-030-52240-7_69
2. Feng, M., Heffernan, N.T.: Informing teachers live about student learning: reporting in the assistment system. Technol. Instr. Cogn. Learn. **3**(1/2), 63 (2006)
3. Heffernan, N.T., Heffernan, C.L.: The ASSISTments ecosystem: building a platform that brings scientists and teachers together for minimally invasive research on human learning and teaching. Int. J. Artif. Intell. Educ. **24**(4), 470–497 (2014). https://doi.org/10.1007/s40593-014-0024-x
4. Lee, J.: Cumulative Learning and Schematization in Problem Solving. Universität Freiburg (2012)
5. McLaren, B.M., van Gog, T., Ganoe, C., Karabinos, M., Yaron, D.: The efficiency of worked examples compared to erroneous examples, tutored problem solving, and problem solving in computer-based learning environments. Comput. Hum. Behav. **55**, 87–99 (2016)
6. Patikorn, T., Heffernan, N.T.: Effectiveness of crowd-sourcing on-demand assistance from teachers in online learning platforms. In: Proceedings of the Seventh ACM Conference on Learning @ Scale, L@S 2020, pp. 115–124. Association for Computing Machinery, New York (2020). https://doi.org/10.1145/3386527.3405912
7. Prihar, E., Botelho, A.F., Jakhmola, R., Heffernan, Neil T.I.: ASSISTments 2019–2020 school year dataset, December 2021. https://doi.org/10.17605/OSF.IO/Q7ZC5
8. Prihar, E., Gonsalves, M.: ASSISTments 2020–2021 school year dataset, November 2021. https://osf.io/7cgav
9. Prihar, E., Patikorn, T., Botelho, A., Sales, A., Heffernan, N.: Toward personalizing students' education with crowdsourced tutoring. In: Proceedings of the Eighth ACM Conference on Learning @ Scale, L@S 2021, pp. 37–45. Association for Computing Machinery, New York (2021). https://doi.org/10.1145/3430895.3460130
10. Razzaq, L.M., Heffernan, N.T.: To tutor or not to tutor: that is the question. In: AIED, pp. 457–464 (2009)
11. Whitehill, J., Seltzer, M.: A crowdsourcing approach to collecting tutorial videos-toward personalized learning-at-scale. In: Proceedings of the Fourth (2017) ACM Conference on Learning@ Scale, pp. 157–160 (2017)
12. Wood, D., Bruner, J.S., Ross, G.: The role of tutoring in problem solving. Child Psychol. Psychiatry Allied Discip. **17**(2), 89–100 (1976)

Grading Programming Assignments with an Automated Grading and Feedback Assistant

Marcus Messer[✉][ID]

King's College London, London, UK
marcus.messer@kcl.ac.uk

Abstract. Over the last few years, Computer Science class sizes have increased, resulting in a higher grading workload. Universities often use multiple graders to quickly deliver the grades and associated feedback to manage this workload. While using multiple graders enables the required turnaround times to be achieved, it can come at the cost of consistency and feedback quality. Partially automating the process of grading and feedback could help solve these issues. This project will look into methods to assist in grading and feedback partially subjective elements of programming assignments, such as readability, maintainability, and documentation, to increase the marker's amount of time to write meaningful feedback. We will investigate machine learning and natural language processing methods to improve grade uniformity and feedback quality in these areas. Furthermore, we will investigate how using these tools may allow instructors to include open-ended requirements that challenge students to use their ideas for possible features in their assignments.

Keywords: Automated grading · Feedback · Assessment · Computer science education

1 Context and Motivation

Over the last few years, the number of students enrolled in computer science courses at universities has increased dramatically. As class sizes have grown, there has been an increase in coursework to mark and provide feedback on [9]. Assignments are often shared among numerous graders to cope with the increasing workload.

While having several graders allows assignments to be marked and distributed to students more quickly, it can do so at the expense of grade uniformity and feedback quality. The variation in grade uniformity is especially true in programming assignments when partially subjective criteria like readability, maintainability, and documentation are part of the rubric. Since there is no guarantee that the same graders will mark the same assignments, the variation in awarded grades is also relevant across several assignments and years.

© Springer Nature Switzerland AG 2022
M. M. Rodrigo et al. (Eds.): AIED 2022, LNCS 13356, pp. 35–40, 2022.
https://doi.org/10.1007/978-3-031-11647-6_6

Some courses use auto-graders to grade program correctness. These typically require students to follow a strict structure, such as fixed names for functions and the desired outputs. This approach introduces issues with resultant grades and pedagogical decisions when setting assignments. Additionally, it is pretty easy for students to get mismarked. A student may have committed a fundamental syntactic or logical error that, if manually marked, would result in a higher grade, as graders typically will give partial marks for correct logic [12].

Auto-grading of program correctness may impact instructors' pedagogical decisions when creating assignments. They may construct tasks specifically to work with auto-grading, limiting students' originality and possible enjoyment of the course. An example of this is Parsons Problems, which use a constrained problem space and pre-written code snippets to impose a lower cognitive load on students [14]. While specified requirements are necessary for all students, it is also critical to allow stronger students to investigate the assessed topic to expand their knowledge and creativity. Restricting the assignment to only grade correctness may also limit the ability of the instructor to grade other common learning objectives in novice programming courses, such as how students name variables and functions.

In addition to grade variation, the quality of feedback differs as well. Some graders give detailed and specific feedback, while others give little feedback. Even with numerous graders, the amount of time allocated to mark and provide feedback on an assignment is extremely short, making it difficult for graders to provide meaningful feedback.

The partially subjective elements of programming that we will be focusing on are maintainability, readability and documentation; their initial definitions are below and are subject to change, and will be referenced in the following sections as *the code criteria*.

Maintainability: This will focus on how well the student has designed their code. Some examples could be the maximum nesting level, how coupled the classes are and any repeated code blocks.

Readability: This will include how easy the student's code is to read. Some examples could be whether the student follows the code style conventions, white spaces between code blocks, and the functions and variables are named sensibly.

Documentation: The student's descriptions of their source code and any inline comments that the student has used. An example of student descriptions would be JavaDoc, and an example of inline comments would be a comment to explain a line of code that is hard to understand by just reading the code.

This research aims to develop a methodology and collection of automated tools using machine learning (ML) and natural language processing (NLP) to assist graders in marking and providing feedback on *the code criteria*.

2   Background and Related Work

There are various methods to evaluate source code, with metrics being one of the most common. Professional programmers utilise industry-standard technologies such as static code analysis and programming metrics [4,10] to verify if their code is readable and maintainable. Another use of metrics is to determine the readability of documentation [8]. Recent work by Nguyen et al. investigated how software engineering metrics could analyse code submissions [11]. They conclude that their results could lead to implementing data-driven and timely interventions, ultimately contributing to a scalable workflow for personalised student support.

In addition to code analysis, there has been research into how to grade programming assignments automatically. Previous research has focused on automatically determining whether a particular solution meets the given requirements and is error-free [6,12,13].

Parihar et al. created an automated grading and feedback tool called *GradeIT*. Before automatically grading, GradeIT uses program repair to automatically correct minor syntax errors, albeit with a suitable grade penalty. Any program that compiles successfully, with or without auto-repair uses weighted test cases to determine a grade. They automatically simplify the compiler error message to make it easier for beginner programmers to understand to provide feedback [12].

3 Problem Statement

Although the present method of having numerous graders allows for the marking of assignments for large courses in a short amount of time, concerns with grade uniformity and feedback remain. While program correctness auto-graders may aid in grade uniformity and the quality of feedback, there are still issues with misgrading and strict requirements. These problems contribute to student dissatisfaction [5], and the lack of formative feedback makes it difficult for students to develop their skills [15].

The strict structure and inability to grade *the code criteria* when using program correctness auto-graders reduce the scope of assignments. This reduction in scope limits the ability of instructors to include open-ended requirements that challenge students to use their creativity to complete and extend their assignments. These open-ended requirements allow grading *the code criteria*, especially readability and maintainability, as students have more freedom when writing their source code.

Our research will look into the following research questions to address these concerns.

RQ1. Can a set of metrics be defined to aid in grading readability, maintainability, and documentation?

RQ2. Can a tool be produced using the defined metrics to automatically grade readability, maintainability, and documentation, alleviating the need for human graders to mark these areas?

RQ3. Does assisting the grader in marking readability, maintainability, and documentation allow instructors to include open-ended requirements in their assignments?

RQ4. What impact does using an automated tool to aid in grading and producing feedback have on student satisfaction and performance?

RQ5. What effect does assisting the grader in marking readability, maintainability, and documentation have on grade uniformity and the quality of feedback for assessing program correctness?

4   Research Goals

The ability to automatically assist in grading and providing meaningful feedback has many benefits. Grading of *the code criteria* could be more consistent if done automatically.

Providing automated corrective feedback, in addition to grading, could help students master new skills [15]. According to Ferguson's research, students prefer written feedback that is timely and personalised to their work [5].

Another advantage of automating part of the grading and feedback processes is reducing the time between submission and feedback. Consistency of application of assessment criteria, the usefulness and promptness of feedback are all components in student satisfaction [7].

Assisting the grader by automatically grading *the code criteria* would allow more time to mark program correctness within the allocated time. Especially if the assignment contains open-ended requirements that are difficult to automatically grade, such as graphical user interfaces or features that students have proposed and implemented.

After developing the grading and feedback process, we will investigate the pedagogical impact of using this tool for both the instructors and the students. Specifically, we will investigate if using these tools encourages instructors to include open-ended requirements in their assignments, when and how to deliver feedback, and how much feedback to give.

5   Research Methods

The initial phase of this research will involve conducting a systematic literature review of current auto-grading and feedback systems. The review will examine how current auto-grading solutions in programming and other subjects produce grades and feedback using static code analysis and ML. The review's findings will serve as a foundation for concepts that can be combined or expanded to develop a set of tools.

The proposed solution will aid the grader by automatically marking the code criteria, allowing the grader to focus on marking and providing meaningful feedback on the program correctness. This section will discuss the planned implementation and evaluation in the rest of this section. All aspects of the project will investigate the integration of ML and NLP to extend the grading and feedback functionalities. As a base dataset for our development, we will use Blackbox, a large-scale dataset of novice Java code [3].

Maintainability: This will use a mix of existing, extended and new programming metrics and static code tools; existing metrics and tools include McCabe's Complexity Measure [10] as well as Chidamber and Kemerer's metrics suite for object-oriented design [4].

Readability: Similarly will use a mix of metrics, such as source lines of code and maximum nesting level, and static code tools, such as CheckStyle [1]. In addition, to metrics, we investigate how NLP can be used to detect and quantify how well students have named variables, functions and classes in relation to the code.

Documentation: We will focus on how readable comments (including JavaDoc) are using readability metrics [8]. Another factor in quality documentation is how comments (including JavaDoc) relate to the code. This research will investigate how NLP can detect and quantify the similarity between the comments and source code.

Feedback: Providing meaningful feedback is essential for student development. Utilising domain knowledge, auto-grading output, and NLP text generation could lead to a tool for producing feedback. A "human-in-the-loop" model will allow the predicted grade and feedback to be quality checked and included in any manual grading. This quality check can assist the model in learning what constitutes a good grade or feedback and what does not [2].

Evaluation: After development, we will evaluate the tool's output to examine grade consistency and feedback quality. To evaluate the grading of the code criteria and feedback quality, we will evaluate the output's validity, reliability, and objectivity. This evaluation could include comparing manually and automatically generated grades and feedback and confirming that the automatic elements are consistent across a wide range of submissions with similar issues. Finally, we will survey both the students and instructors to determine the impact of an automated tool to assist with grading. In addition to evaluating the tool's output, we will also investigate the pedagogical impact on both the students and instructors using automated tools to assist in grading. We may survey instructors to discover if utilising these tools to reduce grading workload encourages them to add open-ended requirements in their assignments.

6   Expected Contributions

The primary contributions of this work will include:

- An overview of state of the art in the form of a systematic literature review.
- A set of metrics and a methodology to assist in grading and providing feedback automatically using ML and NLP for grading readability, maintainability and documentation.
- An implementation of the methodology for Java programming assignments.
- An evaluation of the effect of auto-grading programming assignments has on course design.

References

1. checkstyle. http://checkstyle.sourceforge.io/. Accessed 14 May 2022
2. Bernius, J.P., Krusche, S., Bruegge, B.: A machine learning approach for suggesting feedback in textual exercises in large courses. In: Proceedings of the Eighth ACM Conference on Learning @ Scale (2021). https://doi.org/10.1145/3430895
3. Brown, N.C.C., Klling, M., Mccall, D., Utting, I.: Blackbox: a large scale repository of novice programmers' activity. In: Proceedings of the 45th ACM Technical Symposium on Computer Science Education (2014). https://doi.org/10.1145/2538862
4. Chidamber, S.R., Kemerer, C.F.: A metrics suite for object oriented design. IEEE Trans. Softw. Eng. 476–493 (1994). https://doi.org/10.1109/32.295895
5. Ferguson, P.: Assessment and evaluation in higher education student perceptions of quality feedback in teacher education. Assess. Eval. High. Educ. (2009). https://doi.org/10.1080/02602930903197883
6. Insa, D., Silva, J.: Semi-automatic assessment of unrestrained java code * a library, a DSL, and a workbench to assess exams and exercises. In: Proceedings of the 2015 ACM Conference on Innovation and Technology in Computer Science Education (2015). https://doi.org/10.1145/2729094
7. Kane, D., Williams, J., Cappuccini-Ansfield, G.: Student satisfaction surveys: the value in taking an historical perspective, 135–155 (2008). https://doi.org/10.1080/13538320802278347
8. Kincaid, J.P., Fishburn Jr., R.P., Rogers, R.L., Chissom, B.S.: Derivation of new readability formulas for navy enlisted personnel, February 1975. https://apps.dtic.mil/sti/citations/ADA006655
9. Krusche, S., Reimer, L.M., Bruegge, B., von Frankenberg, N.: An interactive learning method to engage students in modeling. In: Proceedings of the ACM/IEEE 42nd International Conference on Software Engineering: Software Engineering Education and Training (2020). https://doi.org/10.1145/3377814
10. Mccabe, T.J.: A complexity measure. IEEE Trans. Softw. Eng. 308–320 (1976). https://doi.org/10.1109/TSE.1976.233837
11. Nguyen, H., Lim, M., Moore, S., Nyberg, E., Sakr, M., Stamper, J.: Exploring metrics for the analysis of code submissions in an introductory data science course. In: ACM International Conference Proceeding Series, pp. 632–638, April 2021. https://doi.org/10.1145/3448139.3448209
12. Parihar, S., Das, R., Dadachanji, Z., Karkare, A., Singh, P.K., Bhattacharya, A.: Automatic grading and feedback using program repair for introductory programming courses. In: Annual Conference on Innovation and Technology in Computer Science Education, ITiCSE, pp. 92–97, June 2017. https://doi.org/10.1145/3059009.3059026
13. Rahman, M.M., Watanobe, Y., Nakamura, K.: Source code assessment and classification based on estimated error probability using attentive LSTM language model and its application in programming education. Appl. Sci. **10**, 2973 (2020). https://doi.org/10.3390/APP10082973
14. Shah, M.: Exploring the use of parsons problems for learning a new programming language (2020). www2.eecs.berkeley.edu/Pubs/TechRpts/2020/EECS-2020-88.html
15. Wisniewski, B., Zierer, K., Hattie, J.: The power of feedback revisited: a meta-analysis of educational feedback research. Front. Psychol. 3087 (2020). https://doi.org/10.3389/FPSYG.2019.03087

Scaffolding Self-regulated Learning in Game-Based Learning Environments Based on Complex Systems Theory

Daryn A. Dever[(✉)] and Roger Azevedo

University of Central Florida, Orlando, FL 32816, USA
{daryn.dever,roger.azevedo}@ucf.edu

Abstract. Several studies have attempted to capture and analyze the intersect of self-regulated learning (SRL) behaviors and agency (i.e., control over one's own actions) during game-based learning. However, limited studies have attempted to theoretically ground or analytically evaluate these constructs in appropriate theoretical assumptions that can discuss and aptly analyze SRL. As such, this paper argues that complex systems theory, which refers to SRL as a system that is self-organizing, interaction dependent, and emergent, should be integrated into theoretical models of SRL and be analyzed using nonlinear dynamical systems theory techniques to fully capture how learners' SRL behaviors can be captured and scaffolded during game-based learning. This paper guides future discussions and empirical research to understand how to better scaffold learners' SRL behaviors using restricted agency during game-based learning by: (1) understanding scaffolding SRL during game-based learning; (2) reviewing studies that review the intersection of SRL, agency, and game-based learning; (3) discussing the limitations within the field; (4) defining and defend SRL according to complex systems theory; and (5) discussing the open challenges in theoretically, methodologically, and analytically applying complex systems theory to SRL.

Keywords: Self-regulation · Game-based learning · Complex systems theory

1 Scaffolding SRL in GBLEs

At the intersect of playful learning and gamification, game-based learning environments (GBLEs) are used to simultaneously facilitate learning and promote interest and engagement as learners interact with instructional materials and concepts that are traditionally challenging (e.g., microbiology, scientific reasoning skills) [1, 2]. To increase engagement, GBLEs often afford learners full agency, i.e., control over their actions during interactions with features and elements within the GBLE [3–5]. However, several studies with GBLEs have shown that allowing most learners to facilitate and support their own learning within these environments are detrimental to learning outcomes [4,6–8]. Without the structure or support found in traditional instruction techniques, learners are unable to engage in self-regulated learning (SRL), or the monitoring and modulation of one's own cognitive, affective, metacognitive, and motivational processes [5,9, 10].

© Springer Nature Switzerland AG 2022

M. M. Rodrigo et al. (Eds.): AIED 2022, LNCS 13356, pp. 41–46, 2022.

https://doi.org/10.1007/978-3-031-11647-6_7

Effective and efficient deployment of SRL processes and strategies during learning is essential to self-initiate actions within a GBLE where SRL is required to identify, synthesize, integrate sources of information, and monitor dueling goals to efficiently complete the objectives of the GBLE [11]. Because of this as well as the limitation of learners' ability to engage in SRL during learning with an open-ended learning environments [12], it is critical for GBLEs to incorporate support in the form of scaffolds, such as restricted agency, which will maintain the primary goal of GBLEs in promoting learner engagement while simultaneously supporting learners' use of SRL strategies.

Current scaffolding of learners' SRL during learning guide learners in their interactions across several different types of advanced learning technologies including GBLEs [13–16]. Implicit scaffolds within GBLEs can covertly measure and support learners' knowledge of complex topics but commonly rely on learners' accurate use of SRL strategies. One such example includes learners' restricted agency which limits their interactions to encourage the use of specific strategies at time points throughout the learning process which best supports learning outcomes. However, there is a tradeoff effect between agency and engagement on learning where increased agency promotes learner engagement but may impede learning outcomes. The examination of how agency is related to both learning outcomes and learners' deployment of SRL strategies has been extensively studied with Crystal Island, a GBLE that simultaneously supports knowledge acquisition of microbiology content knowledge and scientific reasoning skills, incorporates implicit, fixed scaffolding in the form or restricted agency, or the limitation of one's own choices within the learning environment.

1.1 Crystal Island: An Example

Crystal Island takes place on a virtual island of researchers who have become ill from an unknown source where learners are tasked with identifying the illness [17]. To identify the illness, learners have two goals – to use scientific reasoning skills to complete the premise of the game and to learn as much information about microbiology as possible. This is accomplished using elements within the game including NPCs which act as sources of information (e.g., camp nurse who explains symptomology), a worksheet for the learner to synthesize information, a scanner to test food items for diseases, posters which hold diagrams and information about microbiology concepts, and books and research articles that are long-form texts about microbiology concepts.

Notably Crystal Island embeds scaffolding in the form of restricted agency. Specifically during their gameplay, learners are required to visit buildings throughout the environment in a pre-specified order such as visiting the camp infirmary followed by an NPC's residence. In addition to the structured movements, learners must also interact with all materials within the building before being allowed to leave to go to the next building. Both of these restrictions on learners' agency have been empirically tested to examine its relationship to learners' deployment of SRL strategies while gathering information throughout Crystal Island as well as their learning outcomes.

1.2 Prior Works on Restricted Agency in Crystal Island

Studies have been previously published that address how agency is related to how learners deploy SRL strategies. A study by Dever and Azevedo [6] collected eye-tracking data to examine how agency is related to learners' dwell times on dialogue with NPCs, posters, and books and research articles and their learning gains. Results from this study found that restricted agency as a scaffold significantly contributed to greater learning gains by facilitating greater dwell times on instructional materials. The study was further extended to identify how learners metacognitively monitored the information. A study by Dever et al. [4] used both log files and eye-tracking data to identify how learners differing in agency was related to their duration interacting with information that were relevant to the pre- and post-tests of the study. While results from this study found learners with restricted agency had greater learning gains there were no differences in the dwell times on relevant versus irrelevant text between agency affordances.

To further expound on the findings from Dever and colleagues [4, 6], Dever et al. [7] introduced both time and cognitive components to understand how learners' dwell times on instructional materials are related to learners' deployment of SRL strategies and moderated by the degree of agency afforded by then environment as well as their prior knowledge throughout their game play. Consistent with previous studies, Dever et al. [7] found that, regardless of prior knowledge, learners with restricted agency demonstrated greater learning gains on their microbiology content knowledge. Further a two-level growth model found that dwell times on relevant instructional materials decreased over time across agency conditions, learners in the full agency condition had greater dwell times on books and research articles and shorter dwell times on posters over their time in game.

Findings across all published studies show that restricted agency facilitates greater learning gains while also interacting with how learners deploy SRL strategies (i.e., content evaluations, information-gathering). However, several limitations and issues currently exist in current literature regarding how scaffolds have been measured, implemented, and assessed in supporting SRL, even including the aforementioned studies.

2 Defining SRL as a Complex System

Current issues in the measurement, implementation, and assessment of effective scaffolds in GBLEs arise in current literature due to the method in which learners' self-regulatory behaviors are modeled, traced, and assessed as they learn. More specifically, while current literature tends to use parametric, nonparametric, and time-series analyses to assess SRL, these techniques are limited in examining SRL as a complex, dynamic system which fluctuates over time, and could be influenced by prior actions that are not necessarily sequential. Additionally, current theoretical models of SRL (e.g., Winne [18]) do not support these assumptions of SRL behavioral dynamics demonstrated by learners, especially as they interact with an open-ended GBLE. Because of this, we propose and defend learners' SRL be modeled and analyzed as a complex system using complex systems theory and nonlinear dynamical systems theory respectively.

Complex systems theory refers to how systems' changing behaviors can be explained and predicted [19]. Complex systems are defined by three criteria: (1) *self-organization* which is defined as behavior where behavior is not controlled by a central programmer [20]; (2) *interaction dominance* where overall behavior arises from the interaction between components that make up the system [21]; and (3) *emergence* which refers to how the overall behavior cannot be broken down into individual components [19]. Consistent with these criteria, the components and strategies (e.g., task and cognitive conditions, operations) which structure SRL behaviors coordinate with each other to provide order in learners' demonstrated behaviors and are required to interact with each other to produce a behavior. Further, SRL cannot be broken down into individual strategies or compo-nents such as cognitive and metacognitive as the metacognitive control over one's own cognition is what elicits SRL behavior. In identifying SRL as a complex system, we can understand the overall health of learners' deployment of SRL strategies and understand how agency plays a role in promoting or inhibiting healthy behaviors to better integrate this scaffold into GBLEs.

2.1 Finding a Healthy Balance Between Scaffolding and Agency

Finding a healthy balance between too much agency and not enough has previously been attempted through past studies on degree of agency, but not through the lens of complex systems theory. The concept of far-from-equilibrium systems from complex systems theory states that the "health" of learners' SRL systems is demonstrated as a balance between the rigidity and adaptability of behaviors [22]. In other words, healthy SRL systems are found in the spectrum between SRL behaviors that are highly repetitive (e.g., interacting with a single instructional material) and too chaotic. As such, this paper assigns this concept to agency where more repetitive SRL behaviors are promoted through the complete restriction of agency, thereby diminishing engagement in instructional materials, and the allowance of full agency, promoting discovery-based learning which can lead to behaviors which are too chaotic for efficient and effective learning.

To the authors' knowledge, only one study has been published applying these concepts to SRL and agency during game-based learning. Dever et al. [8] used auto-Recurrent Quantification Analysis (aRQA), a nonlinear dynamical systems theory analytical method, to examine how learners assigned to either full or restricted agency interactions in Crystal Island differed in their information-gathering behavioral sequences throughout learning. Results from this paper demonstrates a need for GBLEs to both scaffold learners' interaction while simultaneously promoting diversity in their actions to improve learning outcomes. There still exists multiple open challenges in how to capture, interpret, and assess SRL behaviors during game-based learning using complex systems theory and its analytical methods.

3   Open Challenges and Future Directions

Several studies have attempted to examine the relationship between agency, SRL, and game-based learning, but limited studies have actually applied appropriate techniques to ground and analyze these complex processes. This paper draws attention to and defines

SRL as a complex system in its self-organization, interaction dominance, and emergent qualities that highlight humans and their learning processes as dynamic and complex. Several open challenges still exist in how to apply complexity science to SRL which leaves several questions for theoretical, methodological, analytical, and applied future directions: (1) How should contemporary models of SRL (e.g., Winne, [18]) integrate complexity science in their theoretical assumptions; (2) Can the dynamic nature of changing cognitive, affective, metacognitive, and motivational states be captured and related to other theoretical foundations of SRL including motivation and affect; (3) Can SRL be empirically defended as a complex system using nonlinear dynamical systems theory techniques; (4) How can restricted autonomy be further augmented as a scaffolding technique within GBLEs throughout domains while still promoting the use of diverse SRL strategies?

Acknowledgements. The research reported in the paper has been funded by the National Science Foundation (DRL #1661202 and DUE # 1761178). The authors would like to thank her dissertation advisor and committee members, and members of UCF's SMART Lab and NCSU's IntelliMedia Group for their numerous contributions. The authors would also like to thank Drs. Winne and Biswas for their mentoring of Ms. Dever through the AIEd Doctoral Consortium process.

References

1. Plass, J.L., Homer, B.D., Mayer, R.E., Kinzer, C.K.: Theoretical foundations of game-based and playful learning. In: Plass, J.L., Mayer, R.E., Homer, B.D. (eds.) The Handbook of Game-based Learning, pp. 3–24. MIT Press, Cambridge (2019)
2. Mayer, R.E.: Cognitive foundations of game-based learning. In: Plass, J.L., Mayer, R.E., Homer, B.D. (eds.) The Handbook of Game-based Learning, pp. 83–110. MIT Press, Cambridge (2019)
3. Bandura, A.: Social cognitive theory: an agentic perspective. Annu. Rev. Psychol. **52**, 1–26 (2001)
4. Dever, D.A., Azevedo, R., Cloude, E.B., Wiedbusch, M.: The impact of autonomy and types of informational text presentations in game-based environments on learning: converging multi-channel processes data and learning outcomes. Int. J. Artif. Intell. Educ. **30**, 581–615 (2020)
5. Taub, M., Sawyer, R., Smith, A., Rowe, J., Azevedo, R., Lester, J.C.: The agency effect: the impact of student agency on learning, emotions, and problem-solving behaviors in a game-based learning environment. Comput. Educ. **147**, 103781 (2020)
6. Dever, D.A., Azevedo, R.: Autonomy and types of informational text presentations in game-based learning environments. In: Isotani, S., Millán, E., Ogan, A., Hastings, P., McLaren, B., Luckin, R. (eds.) AIED 2019. LNCS (LNAI), vol. 11625, pp. 110–120. Springer, Cham (2019). https://doi.org/10.1007/978-3-030-23204-7_10
7. Dever, D. A., Banzon, A. M., Ballelos, N.A.M., Azevedo, R.: Capturing learners' interactions with multimedia science content over time during game-based learning. In: Proceedings of the 15th International Conference of the Learning Sciences – ICLS, pp. 195–202 (2021)
8. Dever, D., Amon, M. J., Wiedbusch, M., Cloude, E., Azevedo, R.: Analysing information-gathering behavioral sequences during game-based learning using auto-recurrence quantification analysis. In: Proceedings to be Presented at the 24th International Conference on Human-Computer Interaction (2022)

9. Sabourin, J., Mott, B., Lester, J.: Discovering behavior patterns of self-regulated learners in an inquiry-based learning environment. In: Lane, H.C., Yacef, K., Mostow, J., Pavlik, P. (eds.) Artificial Intelligence in Education. Lecture Notes in Computer Science (Lecture Notes in Artificial Intelligence), vol. 7926, pp. 209–218. Springer, Heidelberg (2013). https://doi.org/10.1007/978-3-642-39112-5_22

10. Taub, M., Azevedo, R.: Using sequence mining to analyze metacognitive monitoring and scientific inquiry based on levels of efficiency and emotions during game-based learning. J. Educ. Data Min. **10**, 1–26 (2018)

11. Taub, M., Mudrick, N., Bradbury, A.E., Azevedo, R.: Self-regulation, self-explanation, and reflection in game-based learning. In: Plass, J.L., Mayer, R.E., Homer, B.D. (eds.) The Handbook of Game-based Learning, pp. 239–262. MIT Press, Cambridge (2019)

12. Emara, M., Hutchins, N., Grover, S., Snyder, C., Biswas, G.: Examining student regulation of collaborative, computational, problem-solving processes in open-ended learning environments. Learn. Analytics **8**, 49–74 (2021)

13. Azevedo, R., Hadwin, A.F.: Scaffolding self-regulated learning and metacognition: implications for the design of computer-based scaffolds. Instr. Sci. **33**, 367–379 (2005)

14. Winne, P.H., Hadwin, A.F.: nStudy: tracing and supporting self-regulated learning in the internet. In: Azevedo, R., Aleven, V. (eds.) International Handbook of Metacognition and Learning Technologies. Springer International Handbooks of Education, vol. 28, pp. 293–308. Springer, New York (2013). https://doi.org/10.1007/978-1-4419-5546-3_20

15. D'Mello, S., Olney, A., Williams, C., Hays, P.: Gaze tutor: a gaze-reactive intelligent tutoring system. Int. J. Hum. Comput. Stud. **70**, 377–398 (2012)

16. Graesser, A.C., Hu, X., Nye, B.D., Sottilare, R.A.: Intelligent tutoring systems, serious games, and the generalized intelligent framework for tutoring (FIGT). In: O'Neil, H.F., Baker, E.L., Perez, R.S. (eds.) Using Games and Simulations for Teaching and Assessment, pp. 82–104. Routledge, New York (2016)

17. Rowe, J.P., Shores, L.R., Mott, B.W., Lester, J.C.: Integrating learning, problem solving, and engagement in narrative-centered learning environments. Int. J. Artif. Intell. Educ. **21**, 115–133 (2011)

18. Winne, P.H.: Cognition and metacognition within self-regulated learning. In: Schunk, D.H., Greene, J.A. (eds.) Educational Psychology Handbook Series, Handbook of Self-Regulation of Learning and Performance, pp. 36–48. Routledge/Taylor & Francis Group, New York (2018)

19. Favela, L.H.: Cognitive science as a complexity science. Wiley Interdisc. Rev. Cogn. Sci. **11**, e1525 (2020)

20. Heylighen, F.: Complexity and Self-Organization. In: Encyclopedia of Library and Information Sciences **3**, 1215–1224 (2008)

21. Holden, R.J.: People or systems? To blame is human. Prof. Saf. **54**(12), 34–41 (2009)

22. Veerman, F., Mercker, M., Marciniak-Czochra, A.: Beyond turing: far-from-equilibrium patterns an mechano-chemical feedback. Philios. Trans. Roy. Soc. A **379**(2213), 20200278 (2021)

Speech and Eye Tracking Features for L2 Acquisition: A Multimodal Experiment

Sofiya Kobylyanskaya[1,2(✉)]

[1] LISN-CNRS, Orsay, France
skobyl@limsi.fr
[2] Paris-Saclay University, Gif-sur-Yvette, France

Abstract. Spoken language variation analysis is increasingly considered in multimodal settings combining knowledge from computer, human and social sciences. This work focuses on second language (L2) acquisition via the study of linguistic variation combined with eye-tracking measures. Its goal is to model L2 pronunciation, to understand and to predict through AI techniques the related metacognitive information concerning reading strategies, text comprehension and L2 level. We present an experimental protocol involving a reading aloud setup, as well as first data collection to gather L2 speech with associated eye-tracking measures.

Keywords: L2 acquisition · Speech variation · Eye-tracking · Multimodality · Education

1 Introduction

"LeCycle" is a trilateral project (France, Japan, Germany) aiming to improve knowledge transfer in various domains of education. As part of this project, this PhD work focuses on second language (L2) acquisition and evaluation using a multimodal approach that combines eye tracking and speech. These metrics will be integrated into an AI-based system to provide a comprehensive analysis of speakers' reading strategies and to predict their challenges in L2 text processing and pronunciation during a reading aloud setup. Additionally, we aim to find reliable influential strategies (nudges) permitting to reinforce speakers' L2 skills by improving their learning behavior. This paper presents the state of the art on the combination of eye-tracking, speech and nudges applied to L2 learning, the experimental protocol and the platform implemented to obtain the first dataset from 40 participants. Finally, we summarize further research directions and challenges.

2 State of the Art

Multimodal Teaching Methods. Nowadays the application of CALL (computer assisted language learning) is widely spread as it can be beneficial at several

© Springer Nature Switzerland AG 2022
M. M. Rodrigo et al. (Eds.): AIED 2022, LNCS 13356, pp. 47–52, 2022.
https://doi.org/10.1007/978-3-031-11647-6_8

levels, *e.g.* it can stimulate the discussion among students [6], facilitate access to learning material, allow more flexibility in terms of study place and rhythm, provide instant feedback about the student's performance. CALL systems rely on a variety of automatic measures and AI techniques such as: facial recognition to identify the student's emotional state, attention and comprehension level [15], body temperature recognition for attention and emotion recognition[1], eye-tracking analysis for L2 level prediction [2], speech recognition to estimate pronunciation errors, etc. However, CALL systems have some disadvantages, such as for instance the lack of personalization and the poor error recognition accuracy [6]. By combining eye-tracking and speech measures, this work aims to contribute to the improvement of CALL systems.

L2 Pronunciation and Speech-Based Metrics. Previous work shows the interest of measuring speech features to assess the level of L2 mastering. For example, the verbal level can reflect the specificities of L2 pronunciation due to the speaker's L1 [8]. Hence, we can analyze realizations such as the voicelessness of consonants, the duration and the vowels' formants [19], etc. As for the paraverbal level, it can also reveal details about the level of comprehension, engagement, stress and other metacognitive states [26]. At this level, we can consider disfluencies such as pauses, hesitations and latencies that help the speaker to guide the interaction process [29]. They can also provide relevant information about L1-vs-L2 text processing strategies while reading aloud [14].

Speech and Eye-Tracking for L2 Teaching and Evaluation. Eye-tracking information can complement speech features. For example, [22] shows a correlation between eye movement and accented syllables in speech perception. Studies on object naming also highlight the correlation between speech planning and eye movement [13] and the correlation between word length and time spent on the acquisition of its phonological form [13]. According to [21], about one third of the words are skipped during silent reading, especially function words (usually shorter) that occur more frequently and are more predictable than content words [24]. Rare words require more time to be processed than frequent ones, therefore fixations on them are longer [23,24].

Eye-Tracking in Education and L2 Learning. Combining eye-tracking with machine learning can be used to understand students' mental state and motivation and aid in improving their learning achievements. For example, eye-tracking data can be used to classify emotional valence [16], predict co-occurring emotions [17], detect confusion [25] and predict educational goals while interacting with a pedagogical agent [16]. Eye-tracking can also be used in language learning to detect the language proficiency level [2,4] and to understand the mechanism of syntactic processing when reading in L2 [9]. To our knowledge, most studies on eye-tracking in L2 learning were conducted in a silent reading experimental setup. The present data represent a first attempt to combine spoken and

---

[1] https://www.techlearning.com/buying-guides/best-thermal-imaging-cameras-for-schools.

eye movement information, which can be a promising direction as it permits to capture both conscious and unconscious processes [10].

Nudges in Education. During the education process, it is crucial not only to understand learner's strategies and L2 acquisition challenges, but also to contribute to their facilitation. One possible solution may be the use of nudges. The term nudge, coming from economy theory, is defined as an influential tactic that modifies consumer's behavior in a discrete and indirect manner relying on their affective system [28]. It can also be used in the education sphere, but according to [27] only 4% of nudges are related to education. For example, social comparison nudges can contribute to grades' increase [3,11], those using extrinsic information such as rewards are efficient for younger children [12,18], and nudges relying on deadlines can improve self-discipline [30]. These strategies can also be found in L2 acquisition and pronunciation remediation, *e.g.* using facilitating contexts [5]. One of our goals is to highlight difficulties in L2 speaking and pronunciation and to apply appropriate nudging strategies to facilitate phonetics and phonology acquisition.

3   Experimental Protocol and First Results

We collected speech and eye-tracking data from 40 French native speakers. We used "Eye Got it" [7], a platform developed for the project that permits to record both eye-tracking and audio, while associating a forced aligner for speech.

Fig. 1. Experiment process: eye+speech recordings followed by eye-voice span calculation and forced alignment of speech+transcription

Data was recorded in natural indoor conditions in a silent room. We used Tobii Nano Pro for eye tracking and the microphone AKG Perception Wireless 45 Sports Set Band-A 500–865 MHz for speech (Fig. 1). Total duration per subject is around 30 min.

In the following sections we describe the experimental setup from preparing the volunteer to recording their post-experimental feedback.

Pre-experiment. All the participants are French native speakers, mostly students at different Parisian universities (>18 y.o.) and have at least a beginner level of English. They were asked to complete a survey concerning their linguistic background and vision issues (e.g. glasses). They were also invited to sign a consent form about the use of (anonymized) personal data.

Four texts are proposed, one in L1 French and 3 in L2 English: beginner, intermediate and advanced levels. The L1 text serves as a baseline for native pronunciation features. It contains declarative and interrogative sentences and is 456 words long. As for L2, texts were selected from the website "English For Everyone" devoted to English learning. Texts are written by professional English teachers, are adapted to different L2 levels and include multiple choice questions for text comprehension. The following criteria were taken into account for text selection: levels from mid-beginner to mid-advanced; number of words; format ("short stories"); types of sentences and readability measures.

As in [20], we computed lexical and syntactic complexity using [1] for L2 texts. A correlation between text level and lexical complexity is observed, as well as a relation between some texts and their syntactic complexity.

Prior to recording, participants are familiarized with the equipment and the eye-tracker is calibrated with "Eye Got it". The volunteer sits at 60 cm from the screen and is encouraged to maintain this position during the reading phase to avoid recalibration.

Experiment: Reading Aloud. All the participants read the same texts at their natural pace and volume and are free to use disfluencies. However, they are not allowed to look through the texts before the recording, in order to avoid pre-familiarization with potential unknown words, unexpected syntactic structures or any other lexical combinations in L2.

Post-experiment: Pronunciation/Comprehension Feedback. After reading each text, participants are asked to choose the words that were difficult to pronounce and/or to understand. The aim is to detect potential causes of non-canonical pronunciations and/or to correlate challenging words with disfluencies. Then, participants are invited to answer multiple choice questions about each text in L2. This task is aimed to combine the text comprehension level with the information provided in the survey in order to define the actual L2 level of the participants as labels for our future classification system. Note that the voice and the eye movement are not recorded during the post-experiment phase and the participants can take the time needed for the tasks.

Results of First-Step Data Collection. During the first stage of the experiments in February 2022, we collected data from 40 participants. Although we plan to extend the procedure to various socio-professional groups, the current volunteers are mainly academics and other staff members from different Parisian institutions.

Ages range from 18 to 35 and the participants have at least B1 level (according to their personal evaluation or to the score obtained at tests of English as foreign language). More than half of them wear glasses and have some vision problems. All the participants are native French speakers, around 14% of them are bilingual and >60% have exposure to other languages. Most of them (>60%) started learning English at the age of 6 y.o.–10 y.o. and around 30% of them have lived in an English-speaking country for at least several weeks.

4   Conclusion and Further Research

This paper focuses on an ongoing PhD work in the framework of the project "LeCycl". An experimental protocol has been built to gather eye-tracking and speech recordings for L2 acquisition have been described and here we describe the first results. Following work will focus on the contribution of the two modalities: machine learning algorithms will be applied to model L2 pronunciation, and nudging strategies will be added in order to facilitate L2 pronunciation acquisition. This innovative project involves many practical challenges, e.g. from eye-tracker calibration to L2 forced alignment and combination with eye-tracking measures. Ultimately, the most important challenge will concern appropriate machine learning techniques to efficiently combine speech and eye-tracking features.

References

1. Ai, H., Lu, X.: A web-based system for automatic measurement of lexical complexity. In: ALICO-2010, pp. 8–12, June 2010
2. Augereau, O., Fujiyoshi, H., Kise, K.: Towards an automated estimation of English skill via TOEIC score based on reading analysis. In: ICPR 2016, pp. 1285–1290 (2016)
3. Azmat, G., Iriberri, N.: The importance of relative performance feedback information: evidence from a natural experiment using high school students. J. Publ. Econ. **94**(7–8), 435–452 (2010)
4. Berzak, Y., Katz, B., Levy, R.: Assessing language proficiency from eye movements in reading. In: NAACL 2018: HLT, vol. 1, pp. 1986–1996. ACL, New Orleans, June 2018
5. Billières, M.: Méthode verbo tonale : diagnostic des erreurs surlaxe clair/sombre.www.verbotonale-phonetique.com/methode-verbo-tonale-diagnostic-erreurs-axe-clair-sombre/
6. Derakhshan, A., Salehi, D., Rahimzadeh, M.: Computer-assisted language learning (call): pedagogical pros and cons. Int. J. Engl. Lang. Lit. Stud. **4**, 111–120 (2015)
7. El Baha, M., Augereau, O., Kobylyanskaya, S., Vasilescu, I., Laurence, D.: Eye got it: a system for automatic calculation of the eye-voice span. In: 15th IAPR DAS (2022)
8. Flege, J.: Second language speech learning: theory, findings and problems, pp. 229–273 (1995)
9. Frenck-Mestre, C.: Eye-movement recording as a tool for studying syntactic processing in a second language: a review of methodologies and experimental findings. Second Lang. Res. **21** (2005)
10. Godfroid, A., Winke, P., Conklin, K.: Exploring the depths of second language processing with eye tracking: an introduction. Second Lang. Res. **36**(3), 243–255 (2020)
11. Goulas, S., Megalokonomou, R.: Knowing who you are: the effect of feedback information on short and long term outcomes. Economic rese, Department of Economics, University of Warwick (2015)
12. Guryan, J., Kim, J.S., Park, K.H.: Motivation and incentives in education: evidence from a summer reading experiment. Econ. Educ. Rev. **55**, 1–20 (2016)

13. Huettig, F., Rommers, J., Meyer, A.: Using the visual world paradigm to study language processing: a review and critical evaluation. Acta Psychologica **137**, 151–71 (2011)
14. Kang, S.: Exploring L2 English learners articulatory problems using a read-aloud task (2020)
15. Kumar, M.: Advanced Educational Technology. Sankalp Publication, Bilaspur (2020)
16. Lallé, S., Conati, C., Azevedo, R.: Prediction of student achievement goals and emotion valence during interaction with pedagogical agents. In: AAMAS 2018, pp. 1222-1231 (2018)
17. Lallé, S., Murali, R., Conati, C., Azevedo, R.: Predicting co-occurring emotions from eye-tracking and interaction data in MetaTutor. In: Roll, I., McNamara, D., Sosnovsky, S., Luckin, R., Dimitrova, V. (eds.) AIED 2021. LNCS (LNAI), vol. 12748, pp. 241–254. Springer, Cham (2021). https://doi.org/10.1007/978-3-030-78292-4_20
18. Levitt, S.D., List, J.A., Neckermann, S., Sadoff, S.: The behavioralist goes to school: Leveraging behavioral economics to improve educational performance. Am. Econ. J. Econ. Pol. **8**(4), 183–219 (2016)
19. Boula de Mareuil, P., Vieru-Dimulescu, B., Woehrling, C., Adda-Decker, M.: Accents étrangers et régionaux en francais: Caractérisation et identification. Traitement automatique des langues **49**(3), 135–163 (2008)
20. Novikova, J., Balagopalan, A., Shkaruta, K., Rudzicz, F.: Lexical features are more vulnerable, syntactic features have more predictive power. CoRR abs/1910.00065 (2019)
21. Rayner, K.: The 35th sir Frederick Bartlett lecture: eye movements and attention in reading, scene perception, and visual search. Q. J. Exp. Psychol. **62**(8), 1457–1506 (2009)
22. Reinisch, E., Jesse, A., M. McQueen, J.: Early use of phonetic information in spoken word recognition: lexical stress drives eye movements immediately. Q. J. Exp. Psychol. **63**(4), 772–783 (2010)
23. Roberts, L., Siyanova-Chanturia, A.: Using eye-tracking to investigate topics in L2 acquisition and L2 processing. Stud. Second Lang. Acquisit. **35** (2013)
24. Schotter, E., Fennell, A.: Readers can identify the meanings of words without looking at them: evidence from regressive eye movements. Psychon. Bull. Rev. **26** (2019)
25. Sims, S.D., Conati, C.: A Neural Architecture for Detecting User Confusion in Eye-Tracking Data, pp. 15–23. Association for Computing Machinery, New York (2020)
26. Stolcke, A., et al.: Automatic detection of sentence boundaries and disfluencies based on recognized words, January 1998
27. Szaszi, B., Palinkas, A., Palfi, B., Szollosi, A., Aczel, B.: A systematic scoping review of the choice architecture movement: toward understanding when and why nudges work. J. Behav. Decis. Mak. **31**(3), 355–366 (2018)
28. Thaler, R., Sunstein, C.: Nudge: Improving Decisions About Health, Wealth, and Happiness. Yale University Press, New Haven (2008)
29. Vasilescu, I., Adda-Decker, M.: Language, gender, speaking style and language proficiency as factors influencing the autonomous vocalic filler production in spontaneous speech, September 2006
30. Weijers, R.J., de Koning, B.B., Paas, F.: Nudging in education: from theory towards guidelines for successful implementation. Eur. J. Psychol. Educ. **36**(3), 883–902 (2021)

Using Open Source Technologies and Generalizable Procedures in Conversational and Affective Intelligent Tutoring Systems

Romina Soledad Albornoz-De Luise[1]([⊠]) [iD], Miguel Arevalillo-Herráez[1] [iD], and David Arnau[2] [iD]

[1] School of Engineering, University of Valencia, Valencia, Spain
{romina.albornoz,miguel.arevalillo}@uv.es
[2] Department of Didactics of Mathematics, Valencia, Spain
david.arnau@uv.es

Abstract. In the last years, the educational field has been influenced by technological advances. The digital transformation in educational environments allows the incorporation of virtual teaching-learning environments, which allow or facilitate learning opportunities for students, showing, for example, where they make mistakes and providing personalized help whenever they require it. In addition, these systems provide permanent access availability whenever it is possible to access the Internet. Traditionally, simultaneously many students learn word problem-solving skills in the classroom through instruction from only one educational professional. The Intelligent Tutoring System (ITS) Hypergraph Based Problem Solver (HBPS) is capable of tutoring the whole process of solving arithmetic-algebraic word problems, in a personalized way and without imposing any restrictions on the resolution path. Nevertheless, the student-system interaction is performed through a traditional interface by selecting items from a drop-down menu and clicking on buttons. Since dialogue is the fundamental communication mechanism for human-human, we propose use a framework to improve the interaction of the HBPS using a conversational user interface that allows performing the same actions more easily using natural language as the main means of interaction. My thesis research focuses on two main topics. The first one is related to the incorporation of a conversational agent using an open source machine learning framework that is fully configurable. The second one in concerned with testing and modifying different neural architectures to improve performance in intent classification and entity extraction, in such a way that it can be exported to other mathematical domains.

Keywords: Intelligent tutoring systems (ITS) · Interactive learning environments (ILE) · Conversational agents · Natural language processing (NLP) · Affective support · Algebraic word problems

© Springer Nature Switzerland AG 2022
M. M. Rodrigo et al. (Eds.): AIED 2022, LNCS 13356, pp. 53–58, 2022.
https://doi.org/10.1007/978-3-031-11647-6_9

1   Problem and Related Work

Historically, solving arithmetic-algebraic word problems has been a key component in elementary mathematics studies. A mathematical word problem is a coherent storytelling that presents information about the mathematical operations and equations involved. Depending on whether the process of translating text in natural language into an expression through which the problem can be solved, only involves data or involves an unknown, the solution of the problem is said to be arithmetic or algebraic nature [12].

The resolution of this type of problem exists in all aspects of daily life, related to education and professional planning, health, investments, as well as social challenges and is not limited to the academic activity of the mathematical area. Understanding these problems and dealing with them requires a level of mathematical literacy and thinking mathematically.

Traditionally, simultaneously many students learn word problem-solving skills in the classroom through instruction from only one educational professional. This implies that students do not necessarily receive immediate appropriate feedback, and they may not have the opportunity to know what they did wrong or how to correctly solve the problem.

Thanks to the increased presence of technology, the teaching and learning of word problem solving evolve in terms of the design of computational systems. Multiple computing systems have been developed, some of these are intended to replace the role of the teacher, others offer environments in which the solver can use different representation systems or can be freed from routine tasks such as the calculation of arithmetic operations.

For example, the interactive computer animation-based tutor *ANIMATE* [10] allows the animated and dynamic representation of word problems, through the construction of a network of equations that represents the formal mathematical structure of the problem situation. To evaluate the validity of the network of equations (selected from a palette of options), an animation is generated that, the student must evaluate whether or not the network of equations is correct, being able to reorganize the network and change the values of the variables until the expected animation is generated. Animate cannot give remedial feedback or determine the validity of the actions or offer assistance. *HERON* [13] is a computer-based tool that uses a graphical solution tree to represent the structure in order of the operations required to solve a problem, also is able to monitor the constructive activity of the student and provide feedback on different types of errors. The cognitive tutor *Practical Algebra Tutor* (PAT) [9] focuses on the process of symbolization, students can work on it representing the problem situations in tables, graphs, and symbols, and use these representations to answer the questions asked. If the program detects that the student has made a mistake, it provides feedback, also the system offers help requests. *Ms. Lindquist* [7] is an Intelligent Tutoring System (ITS) designed to carry out tutorial dialogs to supervise the construction of algebraic expressions, and is able to decide where to focus the conversation depending on the correct or incorrect content present in the answers of the students, adding multiple tutorial questions for each error.

Ms. Lindquits supports limited tutorial dialogs, as algebraic expressions are the only valid inputs for symbolizing word problems.

On the other hand, the ITS *MathCAL* [5] aims to tutor the resolution of a word problem, comprising the four stages of problem-solving mentioned by Pòlya [11], which is why the student is taken sequentially through four different interfaces. The program is capable of determining the validity of the solution and offering a correct solution on request. *AnimalWatch* [2] is a web-based ITS designed to solve arithmetic word problems with authentic scientific content, where each problem includes two hints that are available on request by the student, and an answer panel that allows enter responses and receive immediate feedback, but focusing on the outcome of the problem, and not on the entire resolution process. Finally, the ITS called Hypergraph Based Problem Solver (HBPS) [1] which is fully focused on the translation stage of the problem solving process, is able to supervise both the arithmetical and algebraic ways of solving word problems, checks the validity of inputs without imposing any restriction on the solution path adopted, and provide both solicited and unsolicited adapted feedback.

With the rise of smart devices and social networks, the text has now become the most common form of communication. The dialogue has been a popular topic in research on Natural Language Processing (NLP) that includes understanding a user's textual input, detection of sentiment and emotion in text, information extraction, and translation, among others. However, none of the systems cited above have incorporate dialogue system to aid improvement human-machine conversation.

Have been developed different types of conversational agents with different functions, using various technologies. There are some educational environments that are currently in use, and whose main objective is to generate knowledge as a human tutor would, on specific topics. Among them, we can mention Chatbot [3], an educational chatbot whose objective is to promote the learning of the study of computer science. Also, ScratchThAI is a conversational system [8] that teaches about the Scratch programming language and, Autotutor [6] a multidisciplinary intelligent conversational system. These educational environments use different methods to understand the goal in the user input, classify it according to predefined user intent, extract relevant attributes in the message and, generate an appropriate response. For example, Autotutor implements latent semantic analysis to relate terms to concepts, while Chatbot uses Freeling to provide natural language analysis capabilities and ScratchThAI uses NLP provided by DialogFlow, the Google NLP platform.

To our knowledge, there is still no ITS for solving arithmetic-algebraic word problems that incorporate conversational agents to improve student interaction in a more confortable and natural way, giving it dynamism and credibility, adding personality traits and emotions. Another important characteristic of systems development, is the usability, that is, that the system is easy to learn and use. Systems such as, ANIMATE, HERON, and PAT they lack usability, since the student spends more time learning to use the system than using it. Using a conversational system would allow us to offer a more natural interface, as well as to track new variables that could not be considered in the previous button-based interface, such as the inclusion of affective traits that can potentially be used to improve the capabilities of the current system.

To address these challenges, my research focuses on design a conversational system using an open source framework, and incorporate improvements on the selected neural architecture for the intent classifier and the entity extractor, in such a way that they can be exported to other mathematical domains.

2   Proposed Solution and Methodology

The purpose of this research is to incorporate a conversational dialogue system to HBPS, designing and building an education-oriented conversational agent that is capable of interacting with a human interlocutor through textual conversations in different languages, to improve usability, reinforce learning, arouse the motivation of students to learn, and incorporate emotional states in the learning process.

We propose use the open source Machine Learning (ML) framework called Rasa [4], taking advantage both of its free to use, and it is fully configurable architecture that allows the designer to customize its behavior through the use of a series of ready-to-use standard components to perform various tasks, including pre-processing, intent classification, entity extraction, and response selection, and allow to add other custom NLU components to perform other tasks not supported by default, such as spell checks and sentiment analysis. These components would allow existing ITS to be adapted to an interaction based on natural language and integrate textual emotion detection, taking into account three main requirements: implementation using open source software, ease of use by students, and real-time working.

At this point in the investigation, we have built a conversational agent. Since these types of toolkits work using a dataset of tagged sentences, it was necessary to carry out two data collection sessions, a total of 79 high school students (14–15 years of age) have participated, who have been asked to solve 3 algebraic word problems interacting through a system of chat. Chat logs from both sessions were processed to clean data, remove duplicate messages and typos, and close incomplete sentences. Then, we analyzed the messages to identify the possible labels that express what the user wants to do or achieve (intent) and identify key information to extract from the text (entities), to evaluate the user responses. To the left of Fig. 1 it can be see an example of a conversation between a teacher and a student, and to the right the intent and entities associated with each student message. Finally, we created a corpus in Spanish that has been translated into English language.

For intent classification and entity extraction, Rasa implements the DIET (Dual Intent and Entity Transformer) neural network architecture. For both the training and prediction stage, the text inputs must be represented in a numeric format to feed the DIET classifier. For this conversion Rasa offers by default

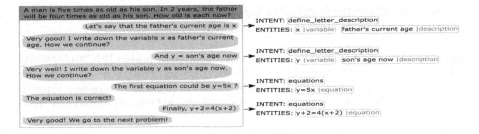

Fig. 1. Example of a conversation (left). Intent/entity labels associated with student messages (right).

different components. Using this default configuration provided by Rasa, we trained two models, one in Spanish and the other in English. At the time, we use Rasa only for natural language understanding since currently our system takes a user message as input and sends it to Rasa for the intent prediction and entity extraction, after Rasa returns the intent and/or the entities associated with the user message, the system analyzes the similarity between the extracted entities and those defined in the problem XML, and returns a predefined response accordingly.

Although we have obtained reasonable accuracy in intent classification and entity extraction, we plan to compare performances using other configurations as too others ML techniques. We also plan to modify neural architectures to provide improvements for these two tasks, so that they can be exported to other mathematical domains. Once the onboarding task of the conversational agent is finished, we plan to carry out further studies in the classroom, on the one hand, to use the collected writings to equip the agent with the ability to detect emotions and on the other hand another one we intend to improve the modeling of the student, to advance in the design of aids and optimal sequences of problems and adapted to each student in particular.

3 Expected Contributions and Impact

This doctoral project intends to contribute to AIed by providing complimentary research in interactive learning environments, in particular on conversational and affective intelligent tutoring systems for the resolution of word arithmetic-algebraic problems, problems that exist in all aspects of daily life, and not limited only to academic activity. We will strive in our research to make all proposed additions work successfully, developing exportable methods that can be reused in other mathematical domains, by combining the use of open source technologies and generalizable procedures.

We also aim to use natural language to be able to detect some relevant cognitive states, such as concentration and frustration. This would allow the system to react to such states by providing affective support. Finally, the proposed system could relieve the workload of human teachers, improve learning outcomes and the experience of students in virtual environments.

Acknowledgements. This research has been supported by project PGC2018-096463-B-I00, funded by MCIN/AEI/10.13039/501100011033 and "ERDF A way of making Europe"; project AICO/2021/019. funded by Valencian Regional Government (Spain); and grant PRE2019-090854, funded by MCIN/AEI/10.13039/501100011033 and "ESF Investing in your future".

References

1. Arevalillo-Herráez, M., Arnau, D., Marco-Giménez, L.: Domain-specific knowledge representation and inference engine for an intelligent tutoring system. Knowl.-Based Syst. **49**, 97–105 (2013). https://doi.org/10.1016/j.knosys.2013.04.017
2. Beal, C.R.: AnimalWatch: An Intelligent Tutoring System for Algebra Readiness. Springer, New York, New York, NY (2013). https://doi.org/10.1007/978-1-4419-5546-3_22
3. Benotti, L., Martnez, M.C., Schapachnik, F.: A tool for introducing computer science with automatic formative assessment. IEEE Trans. Learn. Technol. **11**(2), 179–192 (2018). https://doi.org/10.1109/TLT.2017.2682084
4. Bocklisch, T., Faulkner, J., Pawlowski, N., Nichol, A.: Rasa: Open source language understanding and dialogue management (2017)
5. Chang, K.E., Sung, Y.T., Lin, S.F.: Computer-assisted learning for mathematical problem solving. Comput. Educ. **46**(2), 140–151 (2006). https://doi.org/10.1016/j.compedu.2004.08.002
6. Graesser, A., Chipman, P., Haynes, B., Olney, A.: Autotutor: an intelligent tutoring system with mixed-initiative dialogue. IEEE Trans. Educ. **48**(4), 612–618 (2005). https://doi.org/10.1109/TE.2005.856149
7. Heffernan, N.T., Koedinger, K.R.: An intelligent tutoring system incorporating a model of an experienced human tutor. In: Cerri, S.A., Gouardères, G., Paraguaçu, F. (eds.) ITS 2002. LNCS, vol. 2363, pp. 596–608. Springer, Heidelberg (2002). https://doi.org/10.1007/3-540-47987-2_61
8. Katchapakirin, K., Anutariya, C.: An architectural design of scratchthai: a conversational agent for computational thinking development using scratch. In: IAIT 2018, Association for Computing Machinery, New York, NY, USA (2018). https://doi.org/10.1145/3291280.3291787
9. Koedinger, K., Anderson, J.: Illustrating principled design: the early evolution of a cognitive tutor for algebra symbolization. Interact. Learn. Environ. **5**, March 1998. https://doi.org/10.1080/1049482980050111
10. Nathan, M.J., Kintsch, W., Young, E.: A theory of algebra-word-problem comprehension and its implications for the design of learning environments. Cogn. Instr. **9**(4), 329–389 (1992)
11. Pólya, G.: How to Solve it: A New Aspect of Mathematical Method. Princeton University Press (1945)
12. Puig, L., Cerdán, F., Filloy, E.: Acerca del carácter aritmético o algebraico de los problemas verbales. Aprendizaje y Enseñanza del Álgebra 1, July 1990
13. Reusser, K.: Tutoring systems and pedagogical theory: representational tools for understanding, planning, and reflection in problem solving. Comput. Cogn. Tools **1**, 143–177 (1993)

A Context-Aware Approach to Personalized Feedback for Novice Programmers

Hemilis Joyse Barbosa Rocha[1]([✉]),
Patrícia Cabral de Azevedo Restelli Tedesco[1], and Evandro de Barros Costa[2]

[1] Federal University of Pernambuco, Recife, Brazil
{hjbr,pcart}@cin.ufpe.br
[2] Federal University of Alagoas, Maceió, Brazil
evandro@ic.ufal.br

Abstract. In this article, we propose the development of a context-sensitive tool for providing personalized 3I (informative, interactive and iterative) feedback to novice programmers during the programming problem solving process. To achieve this aim, we have carried out different research stages, where the first is to understand the provision of feedback for novice programmers, investing in carrying out a study in two perspectives, theoretical and experimental. Thus, this study was divided into three stages: systematic literature mapping, systematic literature review and an experiment. As one result of this study we organize the acquired knowledge and elaborate a Context-Aware Taxonomy for Feedback (TaFe). In addition, we designed a Conceptual Architecture Based on Multi-Agents and Computational Context, considering functional requirements identified in the knowledge represented in TaFe. For the next steps, we plan to validate TaFe and the architecture using appropriate methodological instruments. Finally, we intend to develop and validate a feedback solution based on a real problem identified in an experiment.

Keywords: Feedback · Context-aware · Novice programmers

1 Introduction

UNESCO has defined the problem-solving skill as one of the eight key competences for sustainability in the 21st century [17]. However, in the domain of computer programming, novice learners still have difficulties in problem solving activities. This is especially true in the stages of expressing, executing and evaluating the solution [1]. In these stages, situations may occur in which the student cannot finish their solution or even start one [7]. Therefore, it is essential to provide feedback to assist the student in the successful completion of the activity.

© Springer Nature Switzerland AG 2022
M. M. Rodrigo et al. (Eds.): AIED 2022, LNCS 13356, pp. 59–64, 2022.
https://doi.org/10.1007/978-3-031-11647-6_10

Considering that novice programmers are the most affected by the feedback provided [20] and aiming at a better understanding of the provision of feedback for the mentioned audience, we carried out a study with two perspectives, theoretical and experimental. Thus, this study was divided into three stages: i) **1 - Stage:** in order to identify the main approaches used to generate feedback during problem solving in the computer programming domain for beginners, we carried out a Systematic Mapping of Literature - SML [8]; ii) **2 - Stage:** we carried out a Systematic Literature Review - SRL to discover the main sources, forms of presentation, types of content and the level of adaptation of the feedback during learning problem solving in the computer programming domain for beginners; iii) **3 - Stage::** seeking to observe the behavior of novice programmers when solving programming problems - PPS, we carried out an experiment where students were submitted to four different moments of PPS. At first, students did not receive feedback; in the second, students could request feedback from the teacher at any time; in the third, they only received feedback at the end of the problem resolution; and in the fourth, the teacher observed the resolution of each student's problems and feedback was provided, if they deemed it necessary. We observed that students performed better in the second and, notably, in the fourth moment [9].

As a result of the mentioned study, we were able to identify the main solutions available in the literature and observe the still existing needs demonstrated by the students, especially in the stages of expressing, executing and evaluating the solution [1]. In addition, we were able to find and catalog the key components that should be present in providing feedback on problem solving activities for novice programmers. As a result, based on the knowledge acquired through the study and on the proposal of the framework developed by Narciss [2], we organized all knowledge into a first version of a taxonomy of feedback aimed at solving programming problems. For a second version of the taxonomy, we considered that the feedback must be adapted to the characteristics of the student and the task [2]. Thus we sought to enrich our taxonomy with contextual information [12]. In this way, by also representing contextual elements, we make our taxonomy viable to be used in the knowledge base by systems such as, for example, an online Judge, enabling the customization and adaptation of the feedback content [12]. And with that, we developed a Context Sensitive Taxonomy for Feedback (TaFe).

The use of TaFe could fill some gaps still present in the systems for teaching programming, since they still have difficulties in providing adequate and useful feedback to students [3]. Most systems provide poor quality feedback leading students to come up with an incorrect solution on the first try and not try a second time. Such a situation of demotivation can lead the student to abandon the course [10]. Thus, the idea is to provide feedback throughout the problem solving process [18]. In this process, during the execution of the aforementioned experiment, we observed that in 86% of cases of providing feedback, more than one round of interaction with the student was necessary. That is, when the teacher perceived the demand for help, several interactions could be generated with the student, and it was necessary to provide feedback for each one of them. Thus, we realized that when the student's solution presented an error, there was a cause that was not explicit or not understood and, therefore, the need for rounds of interactions in the search for the true cause of the error presented.

Given the above context, we propose the development of a Context Sensitive Tool for Providing 3I Feedback (informative, interactive and iterative) Personalized and Adaptive to Novice Programmers during the process of solving programming problems. The tool will be able to simulate the teacher's behavior, reported in the previous paragraph, generating iterated interactions and providing feedback adapted to the context of the student and the problem. This tool should be developed in the form of a plug-in in order to be coupled with online Judges [10]. To achieve this goal, we have designed an A context-sensitive multi-agent based architecture, considering functional requirements identified in the knowledge represented in TaFe. We chose the Multi-Agent technology aiming at a more flexible and open solution to be adapted to different forms of existing systems. In the next two sections we further detail all work completed, work in progress and our future plans.

2   Previous Research

In this section we present a description of the context and the current state of the research, highlighting what was obtained as a result on two fronts of contribution, namely: the proposal of a feedback taxonomy focused on the programming domain for novices and the outline of a feedback architecture aligned with the main elements of the taxonomy.

2.1   Context

Besides the investment in the literature, the main theoretical support of our feedback study was mainly focused on the work of Narciss [2]. In this work, the interactive, two-feedback-loop (ITFL) model is presented, which explains the main factors and effects of feedback in interactive instruction, essentially representing the articulation between internal and external feedback. External feedback is associated with the instructions provided by the teacher and internal feedback is based on the idea of self-regulation [11]. In the domain of teaching computer programming, there are researches in order to contribute to the development of the internal feedback of the student new to programming [4]. On the other hand, continuous external feedback and monitoring of student progress are still considered one of the main innovative approaches to teaching programming [6].

2.2   Context-Aware Taxonomy for Feedback - TaFe

TaFe is expressed in three levels of specification: categories, subcategories, and contextual elements, where i) Category: representations are the feedback elements and are at the highest level of the taxonomy; ii) Subcategories: are at the intermediate level and are category specifications; iii) Contextual elements: are the data, information or knowledge that characterize the TaFe categories and subcategories. TaFe is made up of six categories, each of which can have two and associated with the subcategories are the contextual elements of the feedback. The feedback categories represented in the taxonomy are: i) provider: entity that provides feedback; ii) receiver: entity that receives the feedback; iii) message: representation of the feedback content; iv) trigger: motivation to provide feedback; v) moment: moment when the feedback is provided; and vi) format: form of availability of the feedback.

To validate TaFe, we are setting up an experiment with introductory programming teachers through semi-structured interviews and application of a questionnaire using the Likert scale of self-report. Our next step is to run the experiment with teachers from at least three different Higher Education/High School.

2.3   Multi-agent Architectural Solution

In the literature there are several solutions for Intelligent Tutoring Systems - ITS aimed at teaching programming. In the work of Silva *et al.* [5] there is an ITS approach to provide adaptive feedback during the resolution of programming exercises. However, adaptation is based solely on the student's solution status. From there, the need arises to provide feedback considering the characteristics of the student, the characteristics of the [2] task and the characteristics that emerge when the student interacts with the activity. In this sense, we considered all the feedback knowledge represented in TaFe to develop a Multi-Agent and Context Sensitive Architecture [13] for a feedback tool.

To develop an architectural solution that meets the system requirements, identify its components, its functionalities, such as resource management, coordination and communication between agents, we follow the Agent Oriented Software Engineering - AOSE Tropos [15] methodology. Furthermore, it is worth noting that as agents behave according to different contexts, most agents have the function of serving a contextual focus [12]. The most abstract view of the architecture is composed of six modules: student model, problem model, instructor model, problem solving module, feedback model and the agent community module.

To validate the developed architectural solution, we are planning an experiment using the ATAM [16] architecture evaluation method where the main stakeholders are the teachers. So, the next step are to run the experiment.

3   Future Research Plans

In this section, we will discuss the next steps towards developing a solution for providing 3I feedback to novice programmers during the programming problem solving activity. Below we describe the four major steps to achieve this goal:

Step 1 - Validation of TaFe: The first step is to complete the validation of the context-based taxonomy for feedback (TaFe) through the experiment, mentioned in the previous section, with teachers of introductory programming courses. After this step, we will submit the results to the academic community in a conference appropriate to the area.

Step 2 - Architecture Validation: As described in the previous section, we will use the ATAM architecture evaluation method [16] to validate our Multi-Agent and Context Sensitive Architecture.

Step 3 - Plug-in development: One of the phenomena observed in the execution of the experiment during moments 2 and 4 of problem solving was, a priori, having a wrong perception of the reason for the feedback. In this way, we realized that when the student's solution presented an error, there was a cause that was not explicit and, therefore, it would take some iterations with feedback for the true cause to be revealed.

Given the mentioned scenario, the third stage of the development of this research is to develop a Context-Sensitive Tool for Providing Personalized and Adaptive 3I Feedback to Novice Programmers during the programming problem solving process. The main function of the tool is to simulate the teacher's behavior, being able to generate iterated interactions and provide feedback adapted to the context of the student and the problem. This tool should be developed in the form of a plug-in in order to be coupled with online judges.

Step 4- Experimenting with the plug-in solution: With the conclusion of the plug-in solution, we will experiment with groups of novice programmers in at least three different institutions.

References

1. Medeiros, R.P., Ramalho, G.L., Falcão, T.P.: A systematic literature review on teaching and learning introductory programming in higher education. IEEE Trans. Educ. **62**(2), 77–90 (2018)
2. Narciss, S.: Feedback strategies for interactive learning tasks. In: Spector, J.M., Merrill, M.D., Van Merrienboer, J.J.G., Driscoll, M.P. (eds.) Handbook of research on educational communications and technology, 3rd edn., pp. 125–143. Erlbaum, Mahwah, NJ (2008)
3. Santos, S.C., Borba, M., Brito, M., Tedesco, P.: Innovative approaches in teaching programming: a systematic literature review. In: CSEDU - 12th Conference on Computer Supported Education, 2020, Praga. Proceedings of the 12th Conference on Computer Supported Education, pp. 1–10 (2020)
4. Aureliano, Viviane C.O., Tedesco, Patricia C. de A.R., Caspersen, Michael E.: Learning programming through stepwise self-explanations. In: 2016 11th Iberian Conference on Information Systems and Technologies (CISTI), 2016, Gran Canaria. 2016 11th Iberian Conference on Information Systems and Technologies (CISTI), 2016. p. 1
5. Silva, P., Costa, E., de Araújo, J.R.: An adaptive approach to provide feedback for students in programming problem solving. In: Coy, A., Hayashi, Y., Chang, M. (eds.) ITS 2019. LNCS, vol. 11528, pp. 14–23. Springer, Cham (2019). https://doi.org/10.1007/978-3-030-22244-4_3
6. Santos, S. C., Tedesco, P. A., Borba, M., Brito, M.: Innovative approaches in teaching programming: a systematic literature review. In: Proceedings of the 12th International Conference on Computer Supported Education, vol. 1, pp. 205–214 (2020)

7. Kyrilov, A., Noelle, D.C.: Do students need detailed feedback on programming exercises and can automated assessment systems provide it? J. Comput. Sci. Coll. **31**(4), 115–121 (2016)

8. Rocha, Hemilis B.R., Tedesco, Patricia C. de A. R., Evandro, Costa B.: On the use of feedback in learning computer programming to novices: a systematic mapping study. Informatics in Education (2022)

9. Rocha, Hemilis B.R., Tedesco, Patricia C. de A.R., Evandro, Costa B.: (submitted). Exploring the effects of feedback on the problem-solving process of novice programmers. In: 18th International Conference on Intelligent Tutoring Systems (2022)

10. Pereira, F.D., et al.: Early dropout prediction for programming courses supported by online judges. In: Isotani, S., Millán, E., Ogan, A., Hastings, P., McLaren, B., Luckin, R. (eds.) AIED 2019. LNCS (LNAI), vol. 11626, pp. 67–72. Springer, Cham (2019). https://doi.org/10.1007/978-3-030-23207-8_13

11. Butler, D.L., Winne, P.H.: Feedback and self-regulated learning: A theoretical synthesis. Rev. Educ. Res. **65**(3), 245–281 (1995)

12. Vieira, V., Tedesco, P., Salgado, A.C.: Designing context-sensitive systems: An integrated approach. Expert Syst. Appl. **38**(2), 1119–1138 (2011)

13. Michel, F., Ferber, J., & Drogoul, A. (2018). Multi-agent systems and simulation: A survey from the agent commu-nity's perspective. In Multi-Agent Systems (pp. 17–66). CRC Press

14. Knapik, M., Johnson, J.: Developing Inteligent Agents for Distributed Systems. Computing McGraw-Hill, NY (1998)

15. Henderson-Sellers, B., Giorgini, P.: Agent Oriented Methodologies. Publisher: Idea Group Publishing.– B. Henderson-Sellers and P. Giorgini, editors. AgentOriented Methodologies. Idea Group Inc. (2005)

16. Kazman, R., Klein, M., Clements, P.: ATAM: Method for architecture evaluation. Carnegie-Mellon Univ Pittsburgh PA Software Engineering Inst. (2000)

17. UNESCO (2017). Education for sustainable development goals: Learning objectives

18. Nguyen, A., Piech, C., Huang, J., Guibas, L.: Codewebs: scalable homework search for massive open online programming courses. In: Proceedings of the 23rd International Conference on World Wide Web, pp. 491–502. WWW'14. ACM, New York, NY, USA (2014)

19. Perera, P., Tennakoon, G., Ahangama, S., Panditharathna, R., Chathuranga, B.: A Systematic Review of Introductory Programming Languages for Novice Learners. IEEE Access (2021)

20. Marwan, S., Gao, G., Fisk, S., Price, T.W., Barnes, T.: Adaptive immediate feedback can improve novice programming engagement and intention to persist in computer science. In: Proceedings of the 2020 ACM Conference on International Computing Education Research, pp. 194–203 (2020)

Towards Speech-Based Collaboration Detection in a Noisy Classroom

Bahar Shahrokhian[✉] and Kurt VanLehn

School of Computing and Augmented Intelligence, Arizona State University, Tempe, AZ 85287, USA

{Bahar.Shahrokhian,Kurt.Vanlehn}@asu.edu

Abstract. An Intelligent Orchestration System, such as our FACT [1], should act like an automated teaching assistant that helps teachers provide relevant, timely help. To do so, it needs to know what the students are doing and thus who needs help more than the others. This is especially important when students work in small groups and the teacher's ability to monitor every group frequently diminishes. This project is an attempt to investigate the feasibility and challenges of only using the students' speech to predict each group's collaboration status. We are using machine-learning techniques to build models that agree with our human annotator's collaboration status judgments.

Keywords: Collaboration detection · Educational data mining · Intelligent orchestration systems · Intelligent tutoring systems

1 Introduction

"Classroom Orchestration" refers to the teacher's real-time management of classroom activities, students, and information in classes that integrate small group work, individual work, and whole-class work [2–4]. To effectively manage the classroom flow, the teacher needs to be constantly aware of the activity performed by each student [5], which they often accomplish by reading visual cues or listening from across the room. However, the awareness is particularly difficult to maintain when students work in small groups since much of their interaction is spoken and inaccessible to the teacher.

One of the features of small group work teachers often want to monitor is whether students are collaborating properly [6]. Teachers often do not want one student to dominate and do all the work, nor do they want students to split up the work and work separately and silently. Given that collaboration is a 21st-century skill [7, 8], collaborative problem solving (CPS) is a critical and necessary skill not only across educational settings but also in the workforce [7]. While working in a group, students learn more if they collaborate closely, both working on the same part of the task and each turn of their conversation builds on the preceding turns, especially their partner's turn [9].

However, getting students to collaborate closely and effectively takes more than simply asking them to do so [10]. In a classroom with multiple groups working simultaneously, if teachers want to help groups engage in productive collaboration, they need to know which groups are already collaborating and which are not.

© Springer Nature Switzerland AG 2022
M. M. Rodrigo et al. (Eds.): AIED 2022, LNCS 13356, pp. 65–70, 2022.
https://doi.org/10.1007/978-3-031-11647-6_11

To meet this need, several systems have been developed to classify the small groups' dynamics based on the multi-modal data gathered from students, which could include actions, speech, video, screen recording, eye tracking, and more [11–15].

Most investigators have trained different machine learners using acoustic features [13, 16–20]. Lexical features (i.e., words) were not included because they would tend to make the detector specific to a particular task or even a particular problem-solving approach to the task. Because acoustic features carry little meaning, there is a greater chance that the classifier will generalize across tasks [21]. Acoustic features may also protect the students' privacy, provided that the raw audio is deleted after the acoustic features are extracted. Hence, the research question we are addressing is, using only acoustic features of students' speech, to what extent can collaboration be detected when small groups are working in authentic classrooms of 20 to 30 students.

There is a plethora of work focused on developing speech-based collaboration detection systems in an educational setting. One of the closest works to ours is [20]. They used speech input and trained collaboration classifiers for an orchestration system. Their input consists of primary non-lexical data such as temporal, acoustic, and voice activity features. But, unlike us, their study happened in a lab setting. Although their coding scheme, like ours, is based on ICAP coding [9], they choose a coarser coding scheme for collaboration. Another difference is their segmentation process. They placed segment boundaries between subtasks to avoid mixing up different patterns of interactions. Also, they placed a segment boundary after 4 min to avoid long segments. While our segmentation happened in two steps as described in pre-processing section. We believe that our approach, although more time-consuming, works better for keeping the related patterns of speech together.

2   Theoretical Framework and Methodology

2.1   Raw Data Collection and Pre-processing

This paper's raw data were gathered during a class trial FACT [1, 22, 23] in the spring of 2019. This trial consisted of six 50-min periods of 8th-grade students working on specific sets of mathematics lessons, called Classroom Challenges [24], using our FACT web-based platform [22].

Each of the 6 periods consisted of 9 to 16 groups of mostly two students. However, we only had permission to collect data from 31 groups total. We annotated 20 groups (64% of the data) due to low audio quality or class being too short for the others. 40.4% of all annotated students are male.

During this study, the teachers wore a lavalier microphone and carried a tablet to access our orchestration system to manage classroom lessons and students. While working in a group, students each used their Chromebooks to access a shared group workspace, typically with their heads down over the tablets while talking. Most students wore a noise-canceling headset microphone, Audio Technica PRO8Hex, connected to a four-track digital audio interface. In each session, four groups' interactions were captured using a video camera with its shotgun mic and a second channel for a boundary mic laid on the group's table.

ELAN software [25, 26] was used to synchronize all the media streams for a period. All the students' microphones in a group were connected to the same recording device, so their audio streams were synchronized at the hardware level. All other data sources were synchronized manually. We asked the teacher and students to clap very loudly at the same time at the start of the class and used that time as a reference point.

After all the data were synchronized, we extracted the time period of the class for each group where a group activity was assigned to them. First, we sectioned that time period by placing a section boundary when there was a long salience or the group's conversation changed topic. In the second step, these sections were divided into 30-s segments. Human annotators labelled each of these 30-s segments based on a lightly modified version of Chi's ICAP framework [9] which had 11 labels.

In the processing stage, we realized that the unbalanced label distribution of our data negatively impacted our classifier's performance. Hence, we decided to reduce the number of labels from 11 to 4 by focusing on frequently occurring categories that we deemed more important for the teacher, namely: 1ST (one person dominating the conversation), 2ST (the group is collaborating properly), and OFF (the group is off-task). We collapsed all other categories to Other. The Other label contains the following labels: 0ST (no one is sharing thinking), INDEP (working on different subtasks), VISIT (teacher is visiting and speaking with the group), STUCK (the students were waiting for help), DONE (The students were done with the task) and several others.

2.2 Processing

Before feature extraction, each segment's noise was removed using version 3.0.0 of Audacity(R) recording and editing software [27]. We then extracted 1582 acoustic and prosodic features from each 30-s frame using OpenSmile [28], and INTERSPEECH's emobase2010 feature set [29].

Since the number of extracted features was too big, we used PCA (Principal Component Analysis) approach among others to reduce our data dimension. PCA uses an orthogonal transformation to convert a set of correlated features into the smallest set of uncorrelated ones, referred to as principal components. We choose 26 top features whose variance would sum up to 80%. Then, we ran a pipeline with many classifier algorithms to choose the best possible model. Random Forest (RF), SVC, Decision Tree, and AdaBoost performed well with this task. RF models with a max depth of 2000 and two random states outperformed all other models.

RF algorithms consist of relatively uncorrelated models (trees) operating as an ensemble. Each tree makes a prediction and the one with the majority vote gets elected. Hence, it will usually outperform those individual models. RF algorithm tends to be less affected by outliers it is also known to work well with small noisy datasets such as ours. Since our labeled data was limited, deep learning did not render reliable results. We are planning to use pre-trained models for different subtasks in the future.

3 Result and Conclusion

Table 1 and Table 2 present the confusion matrix of our best classifier. The kappa score of 0.68 given the limited and noisy data with a sparse confusion matrix is expected. Our results are comparable with similar work in this field [19, 20].

Table 1. .

Predicted Class (Kappa score = 0.68, F1 = 0.75)						
True Class	Labels	1ST	2ST	OFF	Others	**All**
	1ST	7	9	0	1	**17**
	2ST	7	40	1	0	**48**
	OFF	0	1	0	1	**2**
	Others	8	27	1	35	**71**
	All	**22**	**77**	**2**	**37**	**138**

Table 2. .

Labels	Accuracy (%)	Precision (%)	Recall (%)	F1 (%)
1ST	82	32	41	58
2ST	67	52	83	91
OFF	97	0	0	N/A
Other	72	95	49	66
Weighted average	**60**	**71**	**60**	**60**

The performance of our classifier (Kappa score = 0.68, F1 = 0.75) was better than [20] (F1 = 0.70, k = 0.55). This is surprising given that they conducted their experiments in the noiseless lab. This can be due to differences in preprocessing, segmentation, and annotation procedure. Our result could show that a supervised classifier can agree with human annotators almost 70% of the time when it comes to predicting categories that the teacher cares about.

The F1 score for 2ST is very promising (91%). Also, the rate of false-negative segments for 2ST is low which is a good sign. This is because if we want some version of this classifier to automatically detect collaboration among groups, the last thing we want is to misclassify the group who is engaging properly and distract them when they are making progress by unnecessary system intervention. Classification for 1ST segments was less precise and misclassified the 1ST, VISIT, STUCK, DONE segment as 2ST. One explanation is that students are usually engaging in these segments however asymmetrical or non-educational. Hence, we are planning to add limited lexical features so we can add more context and help the classifier to make the distinction between different group engagements. We should investigate its effect on privacy concerns and lack of generality.

Given our limited amount of data and comparing it to similar works, our results were promising. However, we can see that acoustic features alone would not be able to reliably detect collaboration among students in a live classroom. To increase our speech classifier's performance, we are currently adding speech activity features to our set of features. We are planning to incorporate lexically and log data into our current feature set as well.

Acknowledgments. This research was supported by grant NSF FW-HTF 1840051.

References

1. VanLehn, K., et al.: Can an orchestration system increase collaborative, productive struggle in teaching-by-eliciting classrooms? Interact. Learn. Environ. **29**, 987–1005 (2021). https://doi.org/10.1080/10494820.2019.1616567
2. Dillenbourg, P.: Design for classroom orchestration. Comput. Educ. **69**, 485–492 (2013). https://doi.org/10.1016/j.compedu.2013.04.013
3. Dillenbourg, P., Prieto, L.P., Olsen, J.K.: Classroom orchestration. In: International Handbook of the Learning Sciences, pp. 180–190. Routledge, New York (2018). https://doi.org/10.4324/9781315617572-18
4. Ramakrishnan, A., Ottmar, E., LoCasale-Crouch, J., Whitehill, J.: Toward automated classroom observation: predicting positive and negative climate. In: 2019 14th IEEE International Conference on Automatic Face & Gesture Recognition (FG 2019), pp. 1–8. IEEE (2019). https://doi.org/10.1109/FG.2019.8756529
5. Dillenbourg, P., Jermann, P.: Technology for Classroom Orchestration. In: Khine, M., Saleh, I. (eds.) New Science of Learning, pp. 525–552. Springer, New York (2010). https://doi.org/10.1007/978-1-4419-5716-0_26
6. Martinez-Maldonado, R.: A handheld classroom dashboard: teachers' perspectives on the use of real-time collaborative learning analytics. Int. J. Comput.-Support. Collab. Learn. **14**(3), 383–411 (2019). https://doi.org/10.1007/s11412-019-09308-z
7. C. Graesser, A., Foltz, P.W., Rosen, Y., Shaffer, D.W., Forsyth, C., Germany, M.-L.: Challenges of assessing collaborative problem solving. In: Care, E., Griffin, P., Wilson, M. (eds.) Assessment and Teaching of 21st Century Skills. EAIA, pp. 75–91. Springer, Cham (2018). https://doi.org/10.1007/978-3-319-65368-6_5
8. Laal, M., Laal, M., Kermanshahi, Z.K.: 21st century learning; learning in collaboration. Procedia. Soc. Behav. Sci. **47**, 1696–1701 (2012). https://doi.org/10.1016/j.sbspro.2012.06.885
9. Chi, M.T.H., Wylie, R.: The ICAP framework: linking cognitive engagement to active learning outcomes. Educ. Psychol. **49**, 219–243 (2014). https://doi.org/10.1080/00461520.2014.965823
10. Dillenbourg, P., Järvelä, S., Fischer, F.: The evolution of research on computer-supported collaborative learning: from design to orchestration. In: Technology-Enhanced Learning: Principles and Products (2009). https://doi.org/10.1007/978-1-4020-9827-7_1
11. Segal, A., et al.: Keeping the teacher in the loop: technologies for monitoring group learning in real-time. In: André, E., Baker, R., Hu, X., Rodrigo, M.M.T., du Boulay, B. (eds.) AIED 2017. LNCS (LNAI), vol. 10331, pp. 64–76. Springer, Cham (2017). https://doi.org/10.1007/978-3-319-61425-0_6

12. Rodríguez-Triana, M.J., Martínez-Monés, A., Asensio-Pérez, J.I., Dimitriadis, Y.: Scripting and monitoring meet each other: aligning learning analytics and learning design to support teachers in orchestrating CSCL situations. Br. J. Edu. Technol. **46**, 330–343 (2015). https://doi.org/10.1111/BJET.12198

13. Martinez-Maldonado, R., Yacef, K., Kay, J.: TSCL: a conceptual model to inform understanding of collaborative learning processes at interactive tabletops. Int. J. Hum. Comput. Stud. **83**, 62–82 (2015). https://doi.org/10.1016/J.IJHCS.2015.05.001

14. Sümer, Ö., Goldberg, P., D'Mello, S., Gerjets, P., Trautwein, U., Kasneci, E.: Multimodal engagement analysis from facial videos in the classroom (2021)

15. Subburaj, S.K., Stewart, A.E.B., Ramesh Rao, A., D'Mello, S.K.: Multimodal, multiparty modeling of collaborative problem solving performance. In: Proceedings of the 2020 International Conference on Multimodal Interaction, pp. 423–432. ACM, New York (2020). https://doi.org/10.1145/3382507.3418877

16. Martinez, R., Wallace, J.R., Kay, J., Yacef, K.: Modelling and identifying collaborative situations in a collocated multi-display groupware setting. In: Biswas, G., Bull, S., Kay, J., Mitrovic, A. (eds.) AIED 2011. LNCS (LNAI), vol. 6738, pp. 196–204. Springer, Heidelberg (2011). https://doi.org/10.1007/978-3-642-21869-9_27

17. Gweon, G., Jain, M., McDonough, J., Raj, B., Rosé, C.P.: Measuring prevalence of other-oriented transactive contributions using an automated measure of speech style accommodation. Int. J. Comput.-Support. Collab. Learn. **8**, 245–265 (2013)

18. Richey, C., et al.: The SRI speech-based collaborative learning corpus. In: Proceedings of the Annual Conference of the International Speech Communication Association, INTERSPEECH, 08–12 September, pp. 1550–1554 (2016). https://doi.org/10.21437/Interspeech.2016-1541

19. 1Bassiou, N., et al.: Privacy-preserving speech analytics for automatic assessment of student collaboration. In: Proceedings of the Annual Conference of the International Speech Communication Association, INTERSPEECH, 08–12 September, pp. 888–892 (2016). https://doi.org/10.21437/Interspeech.2016-1569

20. Viswanathan, S.A., VanLehn, K.: Collaboration detection that preserves privacy of students' speech. In: Isotani, S., Millán, E., Ogan, A., Hastings, P., McLaren, B., Luckin, R. (eds.) AIED 2019. LNAI, vol. 11625, pp. 507--517. Springer, Cham (2019). https://doi.org/10.1007/978-3-030-23204-7_42

21. Zhang, B., Provost, E.M., Essl, G.: Cross-corpus acoustic emotion recognition with multi-task learning: seeking common ground while preserving differences. IEEE Trans. Affect. Comput. **10**, 85–99 (2019). https://doi.org/10.1109/TAFFC.2017.2684799

22. FACT. https://fact.asu.edu/. Accessed 26 Oct 2021

23. VanLehn, K.: How can fact encourage collaboration and self-correction. In: Deep Comprehension: Multi-Disciplinary Approaches to Understanding, Enhancing, and Measuring Comprehension, pp. 114–124 (2018). https://doi.org/10.4324/9781315109503

24. Mathematics Assessment Project. http://map.mathshell.org/index.php. Accessed 26 Oct 2021

25. Nijmegen: Max Planck Institute for Psycholinguistics, T.L.Archive.: ELAN (Version 6.2). https://archive.mpi.nl/tla/elan

26. Lausberg, H., Sloetjes, H.: Coding gestural behavior with the NEUROGES-ELAN system. Behav. Res. Methods **41**, 841–849 (2009). https://doi.org/10.3758/BRM.41.3.841

27. Audacity® software is copyright © 1999–2021 Audacity Team. The name Audacity® is a registered trademark.: Audacity. http://audacityteam.org

28. Eyben, F., Schuller, B.: openSMILE:). ACM SIGMultimedia Records. **6**, 4–13 (2015). https://doi.org/10.1145/2729095.2729097

29. Schuller, B., Steidl, S., Batliner, A., Burkhardt, F., Devillers, L., Müller, C., Narayanan, S.S.: The INTERSPEECH 2010 paralinguistic challenge. In: Eleventh Annual Conference of the International Speech Communication Association (2010)

Exploring Fairness in Automated Grading and Feedback Generation of Open-Response Math Problems

Ashish Gurung[(✉)][iD] and Neil T. Heffernan[iD]

Worcester Polytechnic Institute, Worcester, MA 01609, USA
{agurung,nth}@wpi.edu

Abstract. The rapid growth and development of NLP techniques have resulted in Computer-Based Learning Platforms (CBLPs) leveraging innovative approaches toward automated grading and feedback generation of open-ended problems. Researchers have explored these techniques in driving a varying range of interventions that range from assessing the quality of the work and recommending changes to the answers that can enhance the quality of the responses for students to automated grading and feedback generation of responses for teachers. A crucial aspect of the automated assessment of student response is identifying and addressing fairness and equity issues in an educational context, as academic performance can impact the types of opportunities available to the students. While prior works have conducted posthoc analysis exploring aspects of algorithmic fairness of various models, the assessment of open-ended answers is often subjective. Teachers leverage contextual knowledge such as the perception of the student effort or students' prior knowledge. While such factors exist, it is not obvious how data from the teacher can introduce biases or introduce measurable risks to the fairness and equity of the NLP models. In this paper, we build on our prior analysis of the grading behavior of teachers on open-ended math problems for middle school students and explore possible next steps we can take to expand on our work. First, we propose a simulation study to explore the various risks associated with Human-AI interaction in the automated grading of open-ended problems. Second, we propose an extensive study expanding on our work to generate grades for open responses when a student is anonymized vs. not anonymized.

Keywords: Open-ended problems · Fairness · Bias · Grading

1 Introduction

The integration of CBLPs into classrooms and the willingness of teachers to utilize them in various capacities in their classrooms has enabled researchers to explore the effectiveness of CBLPs through data-driven methods. Consequently, researchers have focused on developing their CBLPs in alleviating difficult or tedious tasks faced by teachers in their everyday classroom activities.

ⓒ Springer Nature Switzerland AG 2022
M. M. Rodrigo et al. (Eds.): AIED 2022, LNCS 13356, pp. 71–76, 2022.
https://doi.org/10.1007/978-3-031-11647-6_12

Researchers, however, have faced challenges in supporting open-ended problems due to the variance in the answers. While the recent advancement in NLP and machine learning have made progress towards automating the assessment of open-ended questions in various domains, the evaluation of open-ended responses remains a predominantly manual task for teachers. Writing is a critically important skill, and it facilitates students with an avenue to exhibit their thought processes and ability in formulating arguments and providing justifications for their work [11,26]. In mathematics, it enables teachers to gauge whether students have a strong comprehension of mathematical concepts. Furthermore, teachers can leverage open-ended problems to identify situations where students may be able to answer close-ended problems correctly by shallowly learning and applying procedural rules [17,23].

The assessment of open response problems is a largely subjective task. Giving concise responses to open-ended maths problems further underscores the subjective nature of grading open response problems. While grading of open-ended responses often relies on rubrics or other standardized procedures to help optimize the evaluation procedure, teachers often account for contextual factors. The students' past academic performance, persistence exhibited during lessons, or other qualities may affect the teachers' grades. It is important to emphasize that this does not necessarily mean that the grading is unfair. The subjective nature accounting for student ability can positively impact students through personalized feedback [13,14]. However, teachers usage of contextual information in the assessment of students' performance presents a unique challenge in automating the grading of open-ended responses and raises concerns about ensuring the fairness.

Our goal in this work is to build on our prior work and explore teacher grading behavior of open-ended problems and the role of student identity on the grades. Explore the effects of anonymized vs. non anonymized data in the automated grading and feedback generation of open response problems. As such, this paper aims to address the following research questions:

1. Does using anonymized grades in NLP models mitigate possible biases introduced by student identity?
2. What factors affect the teacher's perception of AI agents in automated grading of open-ended math problems?
3. How does teacher perception of AI agents influence their behavior?

2   Background

Growth and innovation in Education Technology (Ed-Tech) have influenced the adaption and regular usage of CBLPs in classrooms. Through ease of logging data, the adaption of CBLPs has motivated researchers to explore the effectiveness of various design paradigms, from traditional teacher-driven designs to self-paced learning, peer learning, discussion-oriented learning, demonstration-focused learning, and flipped classrooms. Several platforms often provide a selection of these features for teachers and students to leverage instead of simply focusing on a single one. Similar to the different design paradigms, researchers have also taken a varied approach in prioritizing the focus of their platform. Some provide a generic platform to host content and leverage crowdsourcing to address learner needs, such as generating problems and solutions [2,7], collecting hints and explanations [4,27]. Other platforms focus on specific domains such as writing skills [3,22], mathematics [5,12], programming [21] to facilitate learning by providing content that addresses the specific needs of learners. It is important to note that these two approaches of prioritizing focus are not mutually exclusive. Platforms often leverage a combination of designs that focus on a specific domain while also facilitating crowdsourcing features that address learner needs.

The automated grading and feedback generation of open-ended problems has been particularly challenging. Researchers have explored various approaches to provide real-time feedback and assessment of open responses to support students. Similar efforts have also been made to support teachers by automating the assessment of open-ended responses. Various approaches such as hand-crafted boutique pattern matching [24], and deconstructing grading rubrics into knowledge components [25]. The rapid growth and innovation of NLP have provided a significant advantage to automating the assessment of open-ended student responses. Researchers have explored NLP in evaluating a diverse range of responses from short-answer responses in mathematics [1,9] to long-form responses such as essays [3,16]. Neural network models such as Word2Vec [19], Glove [20], and BERT [8] have enables the ability to capture semantic and contextual information from responses. While using deep learning models has improved the NLP models' performance, they require a large corpus of data that often are not readily available or easy to compile.

Researchers have explored the effectiveness of NLP models in the automatic grading and feedback generation of responses; there is a requirement for examining the effectiveness of the automation while accounting for fairness. Most examinations of fairness revolve around the algorithm's performance [10,15] and model generalizabilityacross target groups to identify possible biases [6,18]. However, post hoc analysis of models can be rather challenging. These biases can only be mitigated if we are conscious of their existence beforehand or by detecting the existence of biases across certain aspects, such as genders or biases across ethnicity. We propose exploring the utility and effectiveness of NLP models when trained on anonymized data vs. when trained on non-anonymized data.

3 Teacher Grading Behavior

In prior work, we reported on a pilot study where we asked 14 teachers to grade anonymized open-ended responses of students who worked on three open response problems in the month prior to the study. Of the 14 teachers, only 9 completed the study. The data corpus only included the students of the 14 teachers in the pilot study. A random sample of 25 responses was generated per teacher, where we checked to ensure that at least 10 of the 25 responses were responses from their students. If the random sample had less than 10 open responses from their students, then additional open responses were selected for the teacher to grade by randomly selecting additional responses from their students. If a teacher did not have any of their students in the random sample, they were assigned an additional 10 responses, making the total number of problems they graded 35. Table 1 reports on the 9 teachers who completed the study along with the total number of problems they graded(N) non-anonymized beforehand and anonymized during the pilot study. Some teachers had less than 10 problems to grade because we had to remove duplicate answers (e.g., empty responses or answers of "I do not know") to ensure that the teacher graded a unique set of responses.

As shown in Table 1, we explored the teacher's grading behavior by applying Cohen's Kappa to measure the variation in their grading of student responses when anonymized vs. non-anonymized. We found the agreement coefficient to be as low as $k = 0.163$ and as high as $k = 0.67$, which was concerning as it indicated that the teacher disagreed with themselves when it came to scoring their students when their students across conditions. The grading behavior was lower than anticipated, indicating significant differences in teacher grading behavior when students were anonymized. Given that the grades are given on a 5 point scale, and the teacher's assessment may reasonably vary by a small degree, we also explored a relaxed calculation of Kappa. We computed the intra-rater reliability of each teacher with an off-by-one adjustment; if the absolute difference in score across conditions was one or less, then we treated it as equivalent. The adjustment resulted in notably higher kappas indicating that teachers have consistent general grading behavior. We also computed the average difference in the grades across conditions. While most teachers were more lenient graders when they knew the student's identity, some of the teachers were more lenient when the student was anonymized.

Table 1. Exploring the grading behavior of teachers when they had access to students' identity vs. when students were anonymized.

Teacher	N	Intra-rater reliability (Cohen's kappa)	Intra-rater reliability (Relaxed Cohen's Kappa)	Avg grade diff (initial - anonymized)
Teacher1	10	0.2857	0.8550	−0.2
Teacher2	10	0.6774	1.0000	0.2
Teacher3	10	0.2307	0.6666	−0.2
Teacher4	10	0.5161	0.8387	0.3
Teacher5	11	0.1630	0.5268	0.27
Teacher6	19	0.4264	0.7816	0.57
Teacher7	9	0.3793	0.3793	0.44
Teacher8	10	0.4366	0.5522	0.3
Teacher9	9	0.5344	0.8301	−0.66

3.1 Analysis Plan

Currently, we are designing a larger study expanding our pilot study to explore teacher grading behavior and investigate if the proportion of the behavior where some teachers are more lenient grader than others repeats itself across teachers. The more extensive study also provides the data to train the NLP models to compare the model performance when trained on anonymized grades versus non-anonymized grades.

Acknowledgements. We would like to thank the NSF (e.g., 2118725, 2118904, 1950683, 1917808, 1931523, 1940236, 1917713, 1903304, 1822830, 1759229, 1724889, 1636782, & 1535428), IES (e.g., R305N210049, R305D210031, R305A170137, R305A170243, R305A180401, & R305A120125), GAANN (e.g., P200A180088 & P200A150306), EIR (U411B190024 & S411B210024), ONR (N00014-18-1-2768), and Schmidt Futures.

References

1. Baral, S., Botelho, A.F., Erickson, J.A., Benachamardi, P., Heffernan, N.T.: Improving automated scoring of student open responses in mathematics. International Educational Data Mining Society (2021)
2. Bhatnagar, S., Lasry, N., Desmarais, M., Charles, E.: DALITE: asynchronous peer instruction for MOOCs. In: Verbert, K., Sharples, M., Klobučar, T. (eds.) EC-TEL 2016. LNCS, vol. 9891, pp. 505–508. Springer, Cham (2016). https://doi.org/10.1007/978-3-319-45153-4_50
3. Burstein, J., Tetreault, J., Madnani, N.: The e-rater® automated essay scoring system. In: Handbook of Automated Essay Evaluation, pp. 77–89. Routledge (2013)
4. Cambre, J., Klemmer, S., Kulkarni, C.: Juxtapeer: comparative peer review yields higher quality feedback and promotes deeper reflection. In: Proceedings of the 2018 CHI Conference on Human Factors in Computing Systems, pp. 1–13 (2018)
5. Corbett, A.T., Anderson, J.R.: Knowledge tracing: modeling the acquisition of procedural knowledge. User Model. User-Adap. Inter. 4(4), 253–278 (1994)
6. Crawford, K.: The trouble with bias. In: Conference on Neural Information Processing Systems, invited speaker (2017)
7. Denny, P., Hamer, J., Luxton-Reilly, A., Purchase, H.: PeerWise: students sharing their multiple choice questions. In: Proceedings of the Fourth International Workshop on Computing Education Research, pp. 51–58 (2008)
8. Devlin, J., Chang, M.W., Lee, K., Toutanova, K.: BERT: pre-training of deep bidirectional transformers for language understanding. arXiv preprint arXiv:1810.04805 (2018)
9. Erickson, J.A., Botelho, A.F., McAteer, S., Varatharaj, A., Heffernan, N.T.: The automated grading of student open responses in mathematics. In: Proceedings of the Tenth International Conference on Learning Analytics & Knowledge, pp. 615–624 (2020)
10. Friedler, S.A., Scheidegger, C., Venkatasubramanian, S., Choudhary, S., Hamilton, E.P., Roth, D.: A comparative study of fairness-enhancing interventions in machine learning. In: Proceedings of the Conference on Fairness, Accountability, and Transparency, pp. 329–338 (2019)

11. Graham, S., Perin, D.: Writing next: effective strategies to improve writing of adolescents in middle and high schools. A report to Carnegie Corporation of New York. Alliance for Excellent Education (2007)
12. Heffernan, N.T., Heffernan, C.L.: The ASSISTments ecosystem: building a platform that brings scientists and teachers together for minimally invasive research on human learning and teaching. Int. J. Artif. Intell. Educ. **24**(4), 470–497 (2014). https://doi.org/10.1007/s40593-014-0024-x
13. Hill, H.C., Schilling, S.G., Ball, D.L.: Developing measures of teachers' mathematics knowledge for teaching. Elem. Sch. J. **105**(1), 11–30 (2004)
14. Jacob, R., Hill, H., Corey, D.: The impact of a professional development program on teachers' mathematical knowledge for teaching, instruction, and student achievement. J. Res. Educ. Effect. **10**(2), 379–407 (2017)
15. Kamishima, T., Akaho, S., Asoh, H., Sakuma, J.: Fairness-aware classifier with prejudice remover regularizer. In: Flach, P.A., De Bie, T., Cristianini, N. (eds.) ECML PKDD 2012. LNCS (LNAI), vol. 7524, pp. 35–50. Springer, Heidelberg (2012). https://doi.org/10.1007/978-3-642-33486-3_3
16. Kim, Y.S.G., Schatschneider, C., Wanzek, J., Gatlin, B., Al Otaiba, S.: Writing evaluation: rater and task effects on the reliability of writing scores for children in grades 3 and 4. Read. Writ. **30**(6), 1287–1310 (2017)
17. Livne, N.L., Livne, O.E., Wight, C.A.: Enhanching mathematical creativity through multiple solution to open-ended problems online (2008). http://www.iste.org/Content/NavigationMenu/Research/NECC_Research_Paper_Archives/NECC2008/Livne.pdf
18. Mayfield, E., et al.: Equity beyond bias in language technologies for education. In: Proceedings of the Fourteenth Workshop on Innovative Use of NLP for Building Educational Applications, pp. 444–460 (2019)
19. Mikolov, T., Chen, K., Corrado, G., Dean, J.: Efficient estimation of word representations in vector space. arXiv preprint arXiv:1301.3781 (2013)
20. Pennington, J., Socher, R., Manning, C.D.: GloVe: global vectors for word representation. In: Proceedings of the 2014 Conference on Empirical Methods in Natural Language Processing (EMNLP), pp. 1532–1543 (2014)
21. Price, T., Zhi, R., Barnes, T.: Evaluation of a data-driven feedback algorithm for open-ended programming. International Educational Data Mining Society (2017)
22. Roscoe, R.D., Allen, L.K., McNamara, D.S.: Contrasting writing practice formats in a writing strategy tutoring system. J. Educ. Comput. Res. **57**(3), 723–754 (2019)
23. Silver, E.A.: The nature and use of open problems in mathematics education: mathematical and pedagogical perspectives. Zentralblatt fur Didaktik der Mathematik/Int. Rev. Math. Educ. **27**(2), 67–72 (1995)
24. Sukkarieh, J.Z., Pulman, S.G., Raikes, N.: Automarking: using computational linguistics to score short, free- text responses (2003)
25. Sukkarieh, J.Z., Blackmore, J.: C-rater: automatic content scoring for short constructed responses. In: Twenty-Second International FLAIRS Conference (2009)
26. Walton, D.N.: Plausible Argument in Everyday Conversation. SUNY Press (1992)
27. Williams, J.J., et al.: AXIS: generating explanations at scale with learnersourcing and machine learning. In: Proceedings of the Third (2016) ACM Conference on Learning@ Scale, pp. 379–388 (2016)

Industry and Innovation Track

An Intelligent Multimodal Dictionary for Chinese Character Learning

Jinglei Yu, Jiachen Song, Penghe Chen, and Yu Lu[✉]

Faculty of Education, Advanced Innovation Center for Future Education,
Beijing Normal University, Beijing, China
luyu@bnu.edu.cn

Abstract. Chinese character learning is difficult, as the character's definitions in dictionary are simple but abstract. The image representations of Chinese character's definitions are easy to understand and helpful to remember. To assist learning Chinese character and understanding definitions, we design an intelligent dictionary which supports text and image of printed character as input and text, image and video as output modes. Particularly, users could query each definition in text to obtain the corresponding image definition via the designed cross-modal retrieval mechanism. Besides, we also build the image database of character evolving process as well as the video databases of micro-lectures for extended learning. A mobile version of the dictionary has been developed, which supports the multimodal query and output information for the individual Chinese character.

Keywords: Chinese character learning · Cross-modal retrieval · Multimodal dictionary

1 Introduction

Chinese character learning is a challenging task for learners, as it is hard to recognize single character, understand its multiple meanings, make appropriate phrases and form long-term memories. Chinese dictionary is an efficient and useful tool to learn Chinese characters, mainly containing pinyin, glyph information and multiple definitions. Pinyin refers to the character's pronunciation and glyph typically includes character's structure, radical (semantic or phonetic component), and number and sequence of strokes. Definition is the statement of character's meanings in different context using simple but abstract description.

By leveraging the current Chinese dictionary, learners could obtain the necessary information of the individual character as well as the commonly used phrases. However, it is still difficult for learners to get a quick understanding and form long-term retention of all the character's information, especially the character's multiple and ambiguous definitions. The previous studies show that images are appropriate for representing abstract scenario and unusual objects [5]. According to the dual coding theory [4], the verbal and visual dual representation

© Springer Nature Switzerland AG 2022
M. M. Rodrigo et al. (Eds.): AIED 2022, LNCS 13356, pp. 79–83, 2022.
https://doi.org/10.1007/978-3-031-11647-6_13

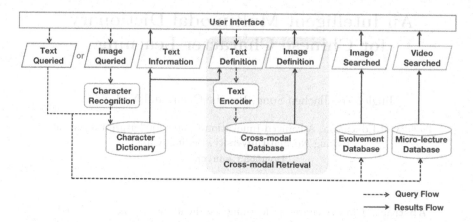

Fig. 1. The simplified block diagram of the designed dictionary.

could enhance learner's memory, especially when word and image are strongly associated with each other [7]. Hence, we design and implement an intelligent Chinese dictionary featured with multimodal input and output. Each definition of the character could be queried to show its image representation. Besides, since most Chinese characters have their unique historical evolving processes, where the original script is pictorial and may reflect the visual meaning from the perspective of character formation [8], we also provide characters' evolving process images and correspondent micro-lectures videos for extended learning.

2   Dictionary Design

Figure 1 illustrates the block diagram of the dictionary. The dictionary supports two types of retrieval functions, namely multimodal information search and cross-modal retrieval.

2.1   Multimodal Information Search

In Fig. 1, the whole framework except the cross-modal retrieval part is the multimodal information search part. The input supports either typing a character or uploading an image of the printed character, where the optical character recognition (OCR) service is used to extract the queried character from the image. Based on the queried character, the system requests the online API of Xinhua dictionary, which is the most popular Chinese dictionary, for the character's basic text information, including pinyin, glyph information and definitions. For the character evolving process, we build a dedicated database to store the image collections of characters in five chronologically formed scripts from calligraphy works. We also build another database for 15 exemplary characters by manually recording 1–2 min micro-lectures for each character to further explain their glyphs and definitions from the historical perspective.

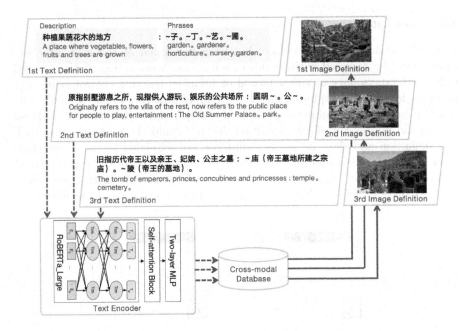

Fig. 2. The simplified block diagram and its query results of cross-model retrieval.

2.2   Cross-modal Retrieval

Based on the multimodal information search results, the system supports learner clicking on each text definition to query its image definition. As shown in Fig. 2, the queried Chinese character is "Yuan" and each definition could be split into a description and example phrases. Each of them is a query string. The cross-model retrieval mechanism firstly extracts each text's feature by text encoder and searches the image with maximum cosine similarity in the cross-modal database. After that, the image from text-image pairs with the maximum similarity would be selected as the image definition. The image features are extracted and stored in the database in advance.

In practice, the text and image features are extracted by encoders from large-scale multimodal pre-trained model BriVL [1,2]. Specifically, BriVL is a two-tower training framework consisting of two replaceable text and image encoders, which are connected by InfoNCE loss in the training process. After the pre-training, the two encoders could work independently and provide APIs that we utilized. The image encoder is based on Efficient-Net_B7 [6] and the text encoder is based on the Chinese pre-trained version of RoBERTa_Large [3]. Both of them are followed by self-attention block and multi-layer perception (MLP) block to project features to the same cross-modal space. The pre-trained model is learned from 650 million image-text pairs crawled from web. The wide coverage of topics and scenarios is rationally enough for our retrieval design.

2.3   User Interface

A mobile version of the dictionary has been developed as the user interface, which could be accessed without downloading, as shown in Fig. 3. Users could look up the dictionary in either formal or informal learning environment. The input could be either typing or simply uploading a picture of the printed character for both native and non-native speakers. To reduce the cognitive load and deepen the impression of the definition, learner could choose to click on each text definition and show its image definition. For extended learning, the character evolving process and micro-lectures are provided. From the perspective of character formation, micro-lectures analyze character's glyph, original meanings, shape changes during the historical process and its current usages in phrases.

Fig. 3. The user interface of the dictionary.

3   Conclusion and Future Work

By leveraging the multimodal pre-trained model, we design and implement the intelligent Chinese dictionary with interactive image representation for each definition to smooth the learners' path to acquire Chinese characters. Besides, 15 exemplary characters' evolving processes and micro-lectures are provided and implemented on the mobile version, which would be constantly enlarged to cover more basic Chinese characters. For the future work, the text-to-image retrieval recall of the encoders needs further improvement by updating state-of-art single-modal encoders for BriVL and fine-tuning the cross-modal framework. Besides, user's feedback mechanism would also be useful to correct inaccurate image definitions and collect bad cases for model fine-tuning. Additionally, considering cloud storage load, images should be collected from online resources and only image URL-feature pairs are stored locally with the regularly validation checking. We are currently deploying the dictionary to serve the school students and teachers, and the usability study is also in plan.

Acknowledgements. This research is supported by the Fundamental Research Funds for the Central Universities and the National Natural Science Foundation of China (No. 62077006, 62177009).

References

1. Fei, N., et al.: WenLan 2.0: make AI imagine via a multimodal foundation model. arXiv preprint arXiv:2110.14378 (2021)
2. Huo, Y., et al.: WenLan: bridging vision and language by large-scale multi-modal pre-training. arXiv preprint arXiv:2103.06561 (2021)
3. Liu, Y., et al.: RoBERTa: a robustly optimized BERT pretraining approach. arXiv preprint arXiv:1907.11692 (2019)
4. Paivio, A.: Mental Representations: A Dual Coding Approach. Oxford University Press, Oxford (1990)
5. Szczepaniak, R., Lew, R.: The role of imagery in dictionaries of idioms. Appl. Linguis. **32**(3), 323–347 (2011)
6. Tan, M., Le, Q.: EfficientNet: rethinking model scaling for convolutional neural networks. In: International Conference on Machine Learning, pp. 6105–6114. PMLR (2019)
7. Underwood, J.: HyperCard and interactive video. Calico J. 7–20 (1989)
8. Yu, J., Song, J., Lu, Yu., Yu, S.: Back to the origin: an intelligent system for learning chinese characters. In: Roll, I., McNamara, D., Sosnovsky, S., Luckin, R., Dimitrova, V. (eds.) AIED 2021. LNCS (LNAI), vol. 12749, pp. 457–461. Springer, Cham (2021). https://doi.org/10.1007/978-3-030-78270-2_81

Leveraging Natural Language Processing for Quality Assurance of a Situational Judgement Test

Okan Bulut[1]([⊠]) [iD], Alexander MacIntosh[2] [iD], and Cole Walsh[2] [iD]

[1] University of Alberta, Edmonton, AB T6G 2G5, Canada
bulut@ualberta.ca
[2] Altus Assessments Inc., Toronto, ON M5V 2Y1, Canada
{amacintosh,cwalsh}@altusassessments.com

Abstract. Situational judgement tests (SJTs) measure various non-cognitive skills based on examinees' actions for hypothetical real-life scenarios. To ensure the validity of scores obtained from SJTs, a quality assurance (QA) framework is essential. In this study, we leverage natural language processing (NLP) to build an efficient and effective QA framework for evaluating scores from an SJT focusing on different aspects of professionalism. Using 635,106 written responses from an operational SJT (Casper), we perform sentiment analysis to analyze if the tone of written responses affects scores assigned by human raters. Furthermore, we implement unsupervised text classification to evaluate the extent to which written responses reflect the theoretical aspects of professionalism underlying the test. Our findings suggest that NLP tools can help us build an efficient and effective QA process to evaluate human scoring and collect validity evidence supporting the inferences drawn from Casper scores.

Keywords: Quality assurance · Validity · Situational judgement · Natural language processing

1 Introduction

Most medical schools in Canada and the United States require situational judgement tests (SJTs) to evaluate applicants' knowledge and skills in non-academic areas such as professionalism [1]. In SJTs, applicants are asked to review a series of hypothetical real-life scenarios and then describe the course of action they are likely to take [2]. Altus Assessments' Casper [3] is an online SJT that consists of video-based and text-based scenarios focusing on ten aspects of professionalism based on the CanMEDS framework defining key competencies (e.g., collaboration and communication) for the health professions [4]. Examinees sitting the Casper test receive a set of questions associated with each scenario and write responses to each question. Therefore, unlike multiple-choice or rating questions in traditional SJTs, Casper's constructed-response format allows examinees to draw from their own experiences and use their own words as they describe what actions or decisions they would take for each scenario and their rationale for doing so.

M. M. Rodrigo et al. (Eds.): AIED 2022, LNCS 13356, pp. 84–88, 2022.
https://doi.org/10.1007/978-3-031-11647-6_14

Although the use of constructed-response questions has advantages, it also creates some challenges that need to be addressed to obtain reliable and valid test scores from Casper. For example, scoring responses involves human judgment [4]. Raters from various backgrounds and professions review underlying theory and guiding statements for each scenario and then rate answers to each scenario holistically. These scores are expected to vary based on the difficulty levels of questions and examinees' ability levels in the aspects being measured by Casper. However, other factors such as examinees' writing ability (or lack thereof) along with their interpretation of each scenario may also influence how they respond and subsequently how the human raters interpret and score those responses. Also, while answering the questions in Casper, examinees are not required to speak directly to any of the aspects underlying Casper. While this provides freer expression, how the responses from each applicant relate to the targeted aspects of professionalism remains in question.

To ensure the correct interpretation and use of scores, a quality assurance (QA) process evaluating the content, internal structure, and response processes is necessary [5]. However, research shows that psychometric procedures designed for multiple-choice and rating questions are not suitable for SJTs [6]. Casper involves long written responses from large numbers of examinees, which cannot be examined based on conventional psychometric methods. Thus, Altus Assessments aims to build a new QA framework for evaluating Casper scores efficiently using language processing (NLP) tools. The outcomes of this framework can help Altus increase the interpretive value of the Casper responses, while enhancing rater training and feedback process.

In this study, our objective is to demonstrate how state-of-the-art NLP methods can be leveraged to evaluate and validate written responses to the Casper test. We propose a new analysis framework to answer the following questions: 1) Do the subjectivity and tone of written responses affect scores assigned by human raters? and 2) Do written responses for each scenario accurately reflect the aspects of professionalism underlying Casper?

2   Related Work

To date, a variety of test-related tasks have been studied using NLP methods, such as automatic text summarization [7], automatic item generation [8], automated essay scoring [9], and topic modeling [10]. In this study, we employ an unsupervised text classification method known as Lbl2Vec [11] to categorize written responses in Casper into the aspects of professionalism. Unlike topic modeling that clusters documents based on word (or phrase) patterns and identifies latent topics, Lbl2Vec categorizes documents based on semantic similarities between the documents and predefined keywords representing a topic [12]. After transforming a document into word and document embeddings [13], the algorithm calculates the centroid of embeddings for each predefined topic and then calculates the cosine similarity to find the most relevant topic for each document. In this study, we use Lbl2Vec to categorize examinees' written responses into the ten aspects of professionalism underlying Casper. We also conducted sentiment analysis to check the subjectivity and sentiments of written responses in Casper. In lexicon-based approaches, sentiments within a document are calculated using the number of words

classified as either positive or negative based on a pre-defined dictionary. To calculate sentiment scores (ranging from 0 to 1 where 0 is negative and 1 is positive), we
utilized VADER [14] in Python, which is sensitive both the polarity and the intensity
of sentiments. For subjectivity, we used TextBlob [15], which calculates a *subjectivity* score based on the number of personal opinions and factual information shared in
the document. Subjectivity ranges from 0 to 1 where 0 is very objective and 1 is very
subjective.

3   Method

3.1   Corpus

The corpus consisted of anonymized 635,106 written responses to 311 unique scenarios
in Casper. Based on the content underlying the scenarios, subject matter experts (SMEs)
involved in developing Casper categorized the scenarios into one of the ten intended
professionalism aspects: collaboration, communication, empathy, equity, ethics, motivation, problem solving, professionalism, resilience, and self-awareness. During test
administration, examinees provided written responses to three questions related to each
scenario and received a score between 1 and 9. In this study, we combined examinees'
responses to three questions associated with each scenario.

3.2   Analysis Framework

Our analysis framework consisted of three steps. First, we performed text preprocessing to prepare the data. Second, we conducted sentiment analysis, saved polarity and
subjectivity scores, and evaluated the relationship between sentiment scores and scores
assigned by human raters. We expected that sentiments expressed through the responses
would not have any relationship with the scores. Finally, we used unsupervised text
classification to categorize written responses into one of the ten aspects of professionalism based on predefined keywords obtained from the CanMEDS framework. Then,
we compared the aspects assigned by Lbl2Vec with the intended aspects assigned
by the subject matter experts for each scenario. This allowed us to determine whether
written responses are related to the intended aspects underlying the scenarios. Figure 1
demonstrates the analysis framework involving these three steps.

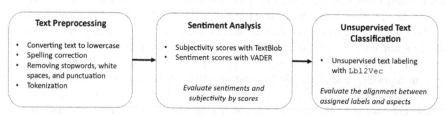

Fig. 1. Data analysis framework used to analyze written responses to Casper.

4   Results

Figure 2 shows the distribution of subjectivity and sentiment scores by scores assigned by human raters. Lower score categories (i.e., 1 to 3 points) have larger variation in subjectivity, indicating that examinees with very low or high subjectivity in their responses receive lower scores. Sentiment scores indicated high levels of negative skewness for all score categories, suggesting that nearly all examinees used a positive tone in their written responses, regardless of their scores.

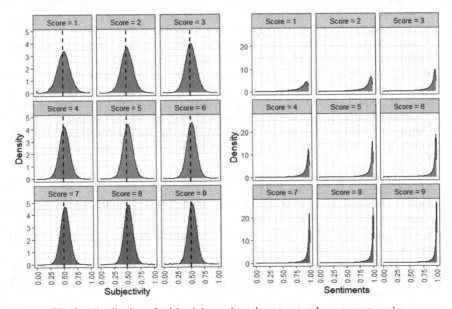

Fig. 2. Distribution of subjectivity and sentiment scores by score categories

To evaluate the alignment between the aspects of professionalism identified by SMEs and those assigned by Lbl2Vec, we calculated precision as TP/(TP + FP) and recall using TP/(TP + FN) where TP is true positive, FP is false positive, and FN is false negative. We found that collaboration, communication, motivation, and resilience had high precision (around .76) and recall (around .70), whereas professionalism, ethics, and self-awareness were much more difficult to detect in the written responses.

5   Conclusions and Future Research Directions

In this study, we demonstrated how to leverage NLP methods to build an QA framework for Casper. Instead of relying on human input on a large volume of written responses in Casper, we used NLP tools to examine the impact of the tone used in written responses on scores assigned by human raters and the alignment between theoretical aspects of professionalism and the aspects extracted from written responses. We will expand our QA process with new NLP-based applications, such as rater training procedures based on sentiment and subjectivity analyses and an automated scoring engine using transformers such as BERT [16] to evaluate scoring accuracy.

References

1. Webster, E., Paton, L., Crampton, P., Tiffin, P.: Situational judgement test validity for selection: a systematic review and meta-analysis. Med. Educ. **54**, 888–902 (2020)
2. Weekley, J., Ployhart, R.: Situational Judgment Tests. Taylor and Francis, Hoboken (2013)
3. Dore, K.L., Reiter, H.I., Kreuger, S., Norman, G.R.: CASPer, an online pre-interview screen for personal/professional characteristics: prediction of national licensure scores. Adv. Health Sci. Educ. **22**(2), 327–336 (2016). https://doi.org/10.1007/s10459-016-9739-9
4. Baldwin, D., Fowles, M., Livingston, S.: Guidelines for Constructed-Response and Other Performance Assessments. Educational Testing Service, Princeton (2005)
5. Standards for Educational and Psychological Testing. American Educational Research Association, Washington, USA (2014)
6. Sorrel, M., Olea, J., Abad, F., de la Torre, J., Aguado, D., Lievens, F.: Validity and reliability of situational judgement test scores. Organ. Res. Methods **19**, 506–532 (2016)
7. Christian, H., Agus, M., Suhartono, D.: Single document automatic text summarization using term frequency-inverse document frequency (TF-IDF). ComTech Comput. Math. Eng. Appl. **7**, 285 (2016)
8. Gierl, M., Lai, H., Turner, S.: Using automatic item generation to create multiple-choice test items. Med. Educ. **46**, 757–765 (2012)
9. Attali, Y., Burstein, J.: AUTOMATED ESSAY SCORING WITH E-RATER® V.2.0. ETS Research Report Series. 2004, i-21 (2004)
10. Shin, J., Guo, Q., Gierl, M.: Multiple-choice item distractor development using topic modeling approaches. Front. Psychol. **10**, 825 (2019)
11. Schopf, T., Braun, D., Matthes, F.: Lbl2Vec: an embedding-based approach for unsupervised document retrieval on predefined topics. In: Proceedings of the 17th International Conference on Web Information Systems and Technologies (2021)
12. Chang, M., Ratinov, L., Roth, D., Srikumar, V.: Importance of semantic representation: dataless classification. In: The Proceedings of the 23rd National Conference on Artificial Intelligence, pp. 830–835. AAAI Press (2008)
13. Dai, A., Olah, C., Le, Q.: Document embedding with paragraph vectors. arXiv preprint arXiv: 1507.07998 (2015)
14. Hutto, C., Gilbert, E.: VADER: a parsimonious rule-based model for sentiment analysis of social media text. In: The Proceedings of the International AAAI Conference on Web and Social Media, Ann Arbor, MI, pp. 216–225 (2014)
15. Loria, S.: TextBlob: Simplified Text Processing — TextBlob 0.16.0 documentation. https://textblob.readthedocs.io/en/dev/
16. Devlin, J., Chang, M.-W., Lee, K., Toutanova, K.: BERT: pre-training of deep bidirectional transformers for language understanding. arXiv preprint arXiv:1810.04805 (2018)

AI-Enabled Personalized Interview Coach in Rural India

Shriniwas Nayak[1(✉)], Satish Kumar[2], Dolly Agarwal[2(✉)], and Paras Parikh[1(✉)]

[1] UBS, Mumbai, India
{shriniwas.nayak,paras.parikh}@ubs.com
[2] Pratham Education Foundation, Mumbai, India
{satish.k,dolly.agarwal}@pratham.org

Abstract. Soft skills are one of the highly-valued skills integral in one's personal as well as professional life. Recent advancements in technology and the penetration of the internet enable new ways to build systems to identify and help learners who need assistance in improving their soft skills. This paper presents the design, development and evaluation of a prototype intelligent system for training youth in rural areas. The system assesses a personal interview of a learner using their audio/video captured from a mobile phone or web camera. Verbal and non-verbal cues are evaluated using speech processing, computer vision and natural language processing and a detailed feedback is generated using Natural language generation. The real-world data used for training the machine learning models for the system were collected from rural areas in India and were annotated by reliable experts. We validated the performance of the system on an out-of-sample data-set which included 100 interviews collected by piloting the application.

Keywords: Computer vision · NLP · AI in education

1 Introduction

The International Labor Organization defines core skills for employability into four categories: learning to learn, communication, teamwork and problem solving [1]. These skills are essential for gaining and retaining employment. Soft skills form an important part of the core skills and are highly valued in the job market. According to the ILO [2] 40% of the Indian population is aged between 13 to 35. A considerable portion of this population would be entering the labor market and would require a combination of the core skills to gain employment. One of the major problems faced by youth in rural areas of India is the lack of guidance and training support in the area of soft skills development. In order to address this issue, we have created a platform for youth to practise and enhance their personal interview skills. Automated interview assessment system may not be suited for hiring employees but it can be very useful for training candidates and preparing them for the workforce, which is not a high-stake situation.

M. M. Rodrigo et al. (Eds.): AIED 2022, LNCS 13356, pp. 89–93, 2022.
https://doi.org/10.1007/978-3-031-11647-6_15

In this paper we give an overview of our end to end system which automatically assesses an audio/video personal interview on the basis of verbal and non verbal cues which are extracted from the video using computer vision and natural language processing techniques. We have developed an online portal for rural youth where they can submit their interview videos and get an automated evaluation report based on various features like eye contact, smile, lip bite, face obstruction and sentiment. This automated report can be accessed by the candidate using their credentials and they can submit multiple videos in order to improve their interview skills. The paper presents literature survey, data collection method, our system architecture & implementation and also discusses results obtained from the first pilot.

2 Related Work

Work related to automated assessment of video interviews has been done in different capacities and use cases. An attempt to predict interpersonal skills and personality traits has been done by Suen [3] where an automatic personality recognition system was built and tested on 120 real job applicants. Another effort in the creation of a corpus of 1891 monologue job interviews data set collected from 260 online workers has been done by L. Chen [4]. Using text classification they were able to predict personality traits such as openness, conscientiousness, extraversion, agreeableness and emotional stability. An attempt to create a data set of 100 dual interviews (i.e. a person giving a face to face interview and the same person giving an interface based interview without the interviewer) and predict the communication of the candidate using the transcriptions has been done by Rasipuram [5]. Further, Hemamou [6] collected a corpus of more than 7000 candidates to predict the hireability of the candidate as evaluated by the recruiters. Most of the existing video interview assessment systems focus mainly on worker hireability in corporate settings. Our system is a one of a kind attempt to cater to rural youth in India in low resource settings.

3 Methodology

The data collection for model training was done in 2 states of India: Uttar Pradesh and Rajasthan. The training data includes videos and audios of interviews of youth in age group 14–18 years. The video interviews were conducted on zoom while the audio only interviews via voice calls. The test set videos/audios were collected via the proposed platform under the supervision of a field volunteer. Users used their smartphones to access the platform. System Architecture of the solution is as presented in Fig. 1. The system accepts input as a video/audio file from web application built using Python based Django framework (4.0.3), wherein the end-user can select the language of submission, a system generated unique id is associated with the file - which acts as a primary key in the database. A scheduler triggers the AI/ML based pipeline to process the videos/audios using multiple metrics based on video, audio and text analysis. After the evaluation

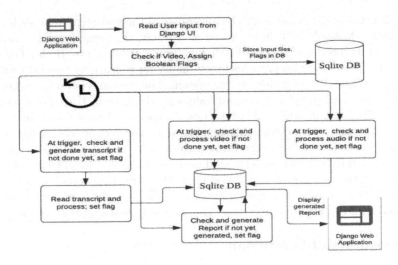

Fig. 1. System architecture

of the file is complete and results corresponding to each metric have been stored in the database, the report generation module is triggered; this module generates a feedback report for the end user in human readable natural language. For video analysis, the input video is divided into a series of frames. Using contour mapping [7] and face recognition [8,9], different attributes are identified from the video, including eye contact, smile index, face obstruction detection and lip bite index. The system draws contours around the eyes and lips to identify the eye contact and smile respectively. Open source libraries cv2 (4.4.0), tensorflow (2.3.1) and face_recognition (1.3.0) have been used for these metrics. All these models produce a numerical output that is converted into a categorical variable using thresholds learnt using grid search methods on the test data. Under text processing, sentiment analysis is performed on the transcript generated from audio/video; sentiment here is indicative of the kind of words(i.e. positive or negative), that were used during the course of the interview. The text analysis pipeline uses the Azure Speech to Text API [10] to obtain text data from the audio/video file. Natural Language pre-processing tasks: punctuation removal, stemming, case generalization and stop words removal are performed using NLTK (3.5) library, this data is then fed to three models for sentiment analysis which use TextBlob (0.15.3) [11], Vader (3.3.2) [12] and word cloud [13] approach respectively. These models generate a numerical output which is then converted to categorical values namely 'High', 'Medium' and 'Low' using predefined thresholds. The thresholds have been learned by the system using the grid search technique, the trained model was tested on a previously evaluated video/audio data set, and the system then performs a grid search in order to obtain the maximum accuracy for the models. The system converts the numerical output of the models to categorical output by dividing the output range using thresholds; for example, the output range of the sentiment analysis is classified using thresholds 0.1 and 0.3, and as a result, the range is categorized as, ≤ 0.0 & <0.1: 'Low', ≥ 0.1 & <0.3: 'Medium' and ≥ 0.3 & ≤ 1.0: 'High'. The system under discussion also provides functionality for re-learning these thresholds using supervised data that flows in the system. The system is fed with corpora of sentences corresponding to the result values for different metrics. For example, for the smile index metric result value 'Low', sentence - 'Please maintain a smile on your face on the interview, this helps to show your confidence' is fed to the system. Using the sentence corpora and the results in the database a natural language friendly report is generated. This report is stored in the database and rendered on the web portal for the end user. This report suggests improvements to the interviewee and acknowledges the strengths as well, thereby encouraging the candidate to perform better and simultaneously boosting their confidence.

4   Results and Discussion

After piloting the platform, we conducted a survey amongst the youth who submitted and viewed their results on the portal. We asked the following questions to them in order to understand their perception regarding the platform. 1. Would you like to practise more on the personal interview assessment portal?, 2. Did you find the feedback report useful?, 3. Did you face any issues while uploading the video?, 4. How did you like the design of the personal interview assessment portal?

86% respondents found the feedback report useful or very useful; 96% of the respondents wanted to practise more on the personal interview assessment portal indicating the efficacy and usability of the portal.

91% of the respondents found the design of the portal user friendly; 61% of the respondents did not face any issue in uploading their audio/video to the portal. One of the most frequently reported technical issue was that of multiple file extension support for audio and video files. Overall the youth in the sample set appreciated the feedback report, liked the design of the portal and would like to reuse the platform to improve their soft skills.

5   Conclusion

This paper showcases an impactful use case of AI to up-skill youths in rural India. The objective of this work is to provide interview coaching services with customized constructive feedback. While similar kind of systems do exist, this work is unique because it is custom build to cater to youths in rural areas by collecting relevant data set. We discussed our current system architecture and metrics evaluated to generate a feedback report. This is just the first prototype of such a system. In phase 2 of the project we plan to include other metrics like confidence detection, emotion recognition and multi-lingual support. With our youth net outreach, we can scale this system to reach up to 92, 000 youths in India. It is important to come up with innovative solutions to solve this very practical and real world problem.

References

1. Brewer, L.: Enhancing Youth Employability: What? Why? and How? Guide to Core Work Skills, 1st edn. International Labour Organization, Geneva (2013)
2. Decent work for youth in India. https://www.ilo.org/newdelhi/info/WCMS_175936/lang--en/index.htm4. Accessed 18 Mar 2022
3. Suen, H.-Y., Hung, K.-E., Lin, C.-L.: Intelligent video interview agent used to predict communication skill and perceived personality traits. HCIS **10**(1), 1–12 (2020). https://doi.org/10.1186/s13673-020-0208-3
4. Chen, L., Zhao, R., Leong, C.W., Lehman, B., Feng, G., Hoque, E.: Automated video interview judgment on a large-sized corpus collected online. In: 2017 Seventh International Conference on Affective Computing and Intelligent Interaction (ACII), pp. 504–509 (2017)
5. Rasipuram, S., Jayagopi, D.B.: Automatic assessment of communication skill in interview-based interactions. Multimedia Tools Appl. **14**(77), 18709–18739 (2018)
6. Hemamou, L., Felhi, G., Vandenbussche, V., Martin, J.-C., Clavel, C.: HireNet: a hierarchical attention model for the automatic analysis of asynchronous video job interviews. In: Proceedings of the AAAI Conference on Artificial Intelligence, vol. 1, no. 33, pp. 573–581 (2019)
7. Contour approximation method. https://docs.opencv.org/3.4/d4/d73/tutorial_py_contours_begin.html. Accessed 18 Mar 2022
8. Face-recognition. https://pypi.org/project/face-recognition/. Accessed 18 Mar 2022
9. OpenCV-Python. https://pypi.org/project/opencv-python/. Accessed 18 Mar 2022
10. Speech-to-text overview - speech service - azure cognitive services. https://docs.microsoft.com/en-us/azure/cognitive-services/speech-service/speech-to-text. Accessed 18 Mar 2022
11. textblob 0.16.0 documentation_2020. https://textblob.readthedocs.io/en/dev/quickstart.html. Accessed 18 Mar 2022
12. Hutto, C., Gilbert, E.: VADER: a parsimonious rule-based model for sentiment analysis of social media text. In: Proceedings of the International AAAI Conference on Web and Social Media, vol. 1, no. 8, pp. 216–225 (2014)
13. Bashri, M.F.A., Kusumaningrum, R.: A Sentiment analysis using Latent Dirichlet Allocation and topic polarity wordcloud visualization. In: 5th International Conference on Information and Communication Technology (ICoICT), pp. 1–5 (2017)

An Integrated Approach to Learning Solutions: UCD + LS&D + AIEd

Kristen S. Herrick[1]([⊠]), Larisa Nachman[1], Kinta D. Montilus[1],
K. Rebecca Marsh Runyon[1], Amy Adair[2], and Lisa Ferrara[1]

[1] Educational Testing Service (ETS) AI Research Labs ™, Princeton, USA
kherrick@ets.org
[2] Rutgers University, New Brunswick, USA

Abstract. User-centered design (UCD), learning science and instructional design (LS&D), and AI in education (AIEd) can be powerful, yet siloed practices when developing educational products. This paper describes our Agile ways of working in the ETS® AI Labs™ and how we are taking an integrated approach to educational product development. We discuss lessons learned and ways we can bridge the gap between learning theories and practices, artificial intelligence, and user experience/research to craft effective learning solutions.

Keywords: User-Centered Design (UCD) · Product Development Lifecycle (PDLC) · Agile

1 Challenges Applying AIEd

Leveraging artificial intelligence in educational contexts dates back to the 1970s with the emergence of Computer-Aided Instruction or what is now often called Intelligent Tutoring Systems [5]. Fifty years later, there are still many potential issues around AI applications to educational contexts. Given what we know or can surmise about the promises and pitfalls of AI in education (AIEd), are there approaches we can take to reduce the negative and enhance the positive impact on educational product design and development? We believe so!

The intersections of practice and theory—although trepidatious—are at the heart of the educational product design lifecycle. User-centered design (UCD), learning science and instructional design (LS&D), and AI in education (AIEd) can be powerful, yet siloed practices when developing educational products that facilitate learning (i.e., learning solutions). Thus, in the ETS® AI Labs™, we've spent the past two years applying these practices contemporaneously to develop learning solutions.

This paper aims to ignite discourse about our Agile ways of working in the ETS® AI Labs™ and how we are taking an integrated approach to educational product development. We discuss lessons learned and ways we can bridge the gap between learning theories and practices, artificial intelligence, and user experience/research to create effective learning solutions.

M. M. Rodrigo et al. (Eds.): AIED 2022, LNCS 13356, pp. 94–98, 2022.
https://doi.org/10.1007/978-3-031-11647-6_16

2   An Integrated Approach to Educational Product Development: UCD + LS&D + AIEd & an Applied Example

The learning solution development process is a nuanced recipe calling for equal parts lived, human experiences and scientific learnings. Moreover, creating high-quality, innovative learning solutions not only requires application of evidence based **AIEd** and **LS&D**, but also engagement in **UCD** to ensure we accurately and adequately address user wants, needs, and pain points. By taking this approach, we ensure that we expose the AIEd meaningfully within the learning context, leveraging capabilities as technological affordances to enhance the solution experience and best meet user needs.

- **User Centered Design (UCD)** is an iterative process that involves users throughout the ideation, design, prototyping, and testing phases of learning solution development to understand the whole user experience and subsequently create highly effective solutions for users [8]. UCD is important given research has emphasized the need for user engagement in the development and implementation of AIEd [10].
- **Learning sciences** is an interdisciplinary field focused on furthering our understanding of how learning happens in real-world settings and designing educational experiences that maximally support learning [4]. **Instructional Design** is a dynamic field focused on the design and management of various learning experiences, as well as supporting the pedagogical efforts of teachers and other educators [2]. Inputs from the fields of learning science and instructional design are referred to as **LS&D**.
- **Artificial Intelligence** has been described as "computing systems capable of engaging in human-like processes such as adapting, learning, synthesizing, correcting, and using various data required for processing complex tasks" [3]. AI has intersected with educational product development in many ways including use of NLP to interpret texts and conduct semantic analysis; use of NLP to enable speech to text and chat bot features; and use of AI in intelligent tutoring systems [7].

UCD, LS&D, and AIEd are uniquely poised to positively impact learning solution design and development. Within the ETS® AI Labs™, we develop learning solutions in cross-functional teams that work together to leverage expertise in each of these areas (i.e., UCD, LS&D, AIEd), apply it iteratively to learning solution design, and collaboratively address issues as they arise upon consistent engagement with users. Working together in this way not only helps ensure we're leveraging each practice appropriately to address user wants/needs/pain points, but also combining them to complement one another and optimize implementation within our learning solutions. The following provides an applied example of this integrated UCD + LS&D + AIEd approach and lessons learned from within the ETS® AI Labs™.

For example, we wanted to deliver an effective, delightful feedback experience for users within our solution. However, sometimes user, learning science, and AIEd perspectives don't lend themselves to the same solution design or development decisions:

- *User perspectives.* We heard from teachers that they want to be able to give students high quality feedback on their writing, but this is very time consuming so it can create a pain point for them. Students need clear and specific information about suggested revisions to their writing in order to improve it and earn high marks. Teachers and students appreciate and value immediate feedback. However, both understandably expressed mistrust of automated feedback coming from a computer system or online solution. Many educators share these same concerns [1].

- *LS&D research.* We found that LS&D best practices included use of specific types of feedback (i.e., mastery-oriented, explanatory, strengths-based). However, there is somewhat conflicting research about the timing of feedback delivery to learners. That is, the effectiveness of feedback delivery often depends on various factors including student ability level, type of feedback, and learning contexts [6, 10]. Moreover, if users do not value and trust feedback, their motivation to understand and apply feedback may diminish, which would likely impair their ability to revise and improve.

- *AIEd applications.* AI affords the opportunity to further personalize the feedback experiences for users, meeting them within their individual learning needs by leveraging factors such as user interests, proficiency levels, writing self-efficacy, or learning motivations. However, developing AI capabilities has historically required large amounts of data to build models, which can take months to years to acquire, clean, engineer, and test, delaying evaluating the usefulness and impact of an AI-driven experience with users.

To help mitigate these varying perspectives and develop an effective learning solution, we integrated best practices from UCD, LS&D, and AIEd.

- **UCD.** We interviewed teachers to more deeply understand how they feel about automated or solution-generated feedback on student work. We're also learning from teachers about things that would help increase their trust in automated feedback provided by AI in a learning solution. We've asked students how they feel about solution-generated feedback as well. We'll engage in user journey mapping to better understand student and teacher thoughts and feelings across each major segment in our learning solution. As we develop visual depictions of the automated feedback aspects of the learning solution, we'll share these with teachers to determine how we can best meet their needs and continue to improve our designs before developing the solution further. Additionally, we'll engage in co-design with teachers and students so that their needs/pain points related to automated feedback can be prioritized and addressed as we develop specific content and/or features of our solution.

- **LS&D.** Given LS&D research suggests that the effectiveness of feedback delivery depends on various factors, we have adopted learning frameworks and theories that help facilitate feedback that can be personalized to each student's learning behaviors, previous accomplishments, and sentiments. We're also designing feedback using a strengths-based methodology that factors in a student's unique talents, capabilities, and previous accomplishments to make feedback more personalized and trustworthy. We're aiming to provide feedback that is trustworthy by ensuring it is specific and actionable. The solution will also help students see how their effort to understand and use feedback helped them improve over time, which should engender more trust.

- **AIEd.** Because we're already engaging with subject matter experts (i.e., teachers) on solution design and experience, we leverage the signals they already trust and use for evaluating and generating feedback as a starting point for defining and designing our model, ensuring that system- or solution- generated feedback aligns with user expectations in the specific context of the solution experience. Next, we identify initial logic that combines the user response data with behavioral data, in this case identifying how to combine natural language processing (NLP) signals extracted from student writing with student usage of system tools like a graphic organizer with their identified writing strength to inform how to package those inputs into meaningful and actionable representation to the user. Finally, we build initial models leveraging existing data sets (from similar products, user profiles, etc.) to create solution prototypes that we can get in front of users in a more timely manner, leveraging user feedback to highlight and therefore prioritize the targeted development, training, or refinement a model needs in the specific context. By treating the AI development iteratively, much like many other assets and components within a user experience, the initial models can be used in the early prototypes, allowing for target user data to be collected, allowing further refinement of the model over time.

Taking an integrated UCD + LS&D + AIEd approach is novel, and achieving optimization is difficult. We're learning valuable lessons about how/when we can integrate UCD, LS&D, and AIEd throughout learning solution design and development:

- We must be agile. Thinking big, but building light, testing, and continuously iterating the solution will help us appropriately integrate users, learning science, and AI.
- We must work in cross-functional teams that sprint together and communicate effectively to leverage and weigh/balance perspectives from UCD, LS&D, and AIEd.
- We must ensure that designs are innovative enough to bring value to our users. Feedback helps our cross-functional team stay accountable to innovation.
- We must be user-obsessed. First versions of models may include incorrect judgments. UX design can account for potential errors, allowing for user-driven choice if they feel feedback is incorrect, or an algorithm to detect mistakes in the models and allow for correction early on, based on subsequent performance or behavior data.

References

1. Anson, C., Filkins, S., Hicks, T., O'Neill, P., Pierce, K.M., Winn, M.: Machine scoring fails the test. NCTE Position Statement on Machine Scoring (2013)
2. Beirne, E., Romanoski, M.P.: Instructional design in higher education: Defining an evolving field. OLC Research Center For Digital Learning & Leadership (2018)
3. Chatterjee, S., Bhattacharjee, K.K.: Adoption of artificial intelligence in higher education: a quantitative analysis using structural equation modelling. Educ. Inf. Technol. **25**(5), 3443–3463 (2020)
4. Fischer, F., Goldman, S.R., Hmelo-Silver, C.E., Reimann, P.: Introduction: evolution of research in the learning sciences. In: International Handbook of the Learning Sciences, pp. 1–8. Routledge (2018)
5. Guan, C., Mou, J., Jiang, Z.: Artificial intelligence innovation in education: a twenty-year data-driven historical analysis. Int. J. Innov. Stud. **4**(4), 134–147 (2020)

6. Mathan, S.A., Koedinger, K.R.: An empirical assessment of comprehension fostering features in an intelligent tutoring system. In: Cerri, S.A., Gouardères, G., Paraguaçu, F. (eds.) ITS 2002. LNCS, vol. 2363, pp. 330–343. Springer, Heidelberg (2002). https://doi.org/10.1007/3-540-47987-2_37
7. Miao, F., Holmes, W., Huang, R., Zhang, H.: AI and education: Guidance for policy-makers. UNESCO (2021)
8. Norman, D.A., Draper, S.W.: User Centered System Design: New Perspectives on Human-Computer Interaction. L. Erlbaum Associates (1986)
9. Shute, V.J.: Focus on formative feedback. Rev. Educ. Res. **78**(1), 153–189 (2008)
10. Zawacki-Richter, O., Marín, V.I., Bond, M., Gouverneu, F.: Systematic review of research on artificial intelligence applications in higher education–where are the educators? Int. J. Educ. Technol. High. Educ. **16**(1), 1–27 (2019)

Artificial Intelligence (AI), the Future of Work, and the Building of a National Talent Ecosystem

Linda Molnar[1] , Ranjana K. Mehta[2] , and Robby Robson[3] (✉)

[1] National Science Foundation, 2415 Eisenhower Avenue, Alexandria, VA 22314, USA
[2] Texas A&M University, College Station, TX, USA
[3] Eduworks Corporation, STE 110, Corvallis, OR 97333, USA
robby.robson@eduworks.com

Abstract. This article presents the background and vision of the *Skills-based Talent Ecosystem for Upskilling* (STEP UP) project. STEP UP is a collaboration among teams participating in the US National Science Foundation (NSF) Convergence Accelerator program, which supports translational use-inspired research. This article details the context for this work, describes the individual projects and the roles of AI in these projects, and explains how these projects are working synergistically towards the ambitious goals of increasing equity and efficiency in the US talent pipeline through skills-based training. The technologies that support this vision range in maturity from laboratory technologies to field-tested prototypes to production software and include applications of Natural Language Understanding and Machine Learning that have only become feasible over the past two to three years.

Keywords: Convergence Accelerator · National Science Foundation · AI · Future of Work · SkillSync · Talent Ecosystem · VR · AR · Fairness

1 Introduction

In 2022, science education remains at the forefront of discussions about national priorities in the United States, as is highlighted in the June 25, 2021, Executive Order on Diversity, Equity, Inclusion, and Accessibility (DEIA) in the Federal Workforce [1] and in recent symposia and reports from the National Academies of Sciences, Engineering, and Medicine [2, 3]. Artificial Intelligence (AI) is at the intersection of these needs, including AI as it influences work and how AI can be part of the solution in ways that are fair and equitable. The US National Science Foundation's *Convergence Accelerator* [4] seeks to develop truly convergent areas of use-inspired research where AI is converged with other disciplines – including education and behavioral and cognitive science – to achieve a just and fair workplace.

The NSF Convergence Accelerator Tracks B1 (Artificial Intelligence and Future Jobs) and B2 (National Talent Ecosystem) – together, *Track B* – were created in 2019 with

© Springer Nature Switzerland AG 2022
M. M. Rodrigo et al. (Eds.): AIED 2022, LNCS 13356, pp. 99–103, 2022.
https://doi.org/10.1007/978-3-031-11647-6_17

the overarching goal of providing accessible and inclusive opportunities for everyone and are rooted in STEM (science, technology, engineering, and mathematics) education. An intentional approach to working towards a fair future of work has been central to their work and aligns well with the priorities cited above and with the Biden-Harris priority of "equipping the American middle class to succeed in a global economy" [5]. All Convergence Accelerator tracks were asked to work towards a shared vision of "Track Integration." This paper describes two of the Track B projects, the roles of AI in these projects, and the shared vision they are pursuing, called *STEP UP*.

2 The SkillSync Project

The SkillSync project, led by the last author, seeks to improve workforce upskilling by addressing the connection between employers and college non-degree programs. These programs are underutilized for upskilling due to many factors, including misalignment with desired skills, lack of communication between colleges and company HR and training departments, and mismatched business processes [6]. As stated in a report on a 2021 MIT Open Learning conference on this theme, "The labor market information chain is broken: Workers don't know what skills they need, educators don't know what skills to educate for, and employers don't know what skills workers have" [7].

To address this, the SkillSync project has developed a set of AI services for extracting knowledge, skills, and abilities (KSAs) from unstructured text (and specifically from job descriptions), for de-biasing and anonymizing job descriptions and associated KSA by replacing "company identifiable information" and biased language with generic and unbiased terminology, and for computing the alignment between a set of courses and a prioritized set of skills. These services apply large pre-trained language models such as BERT [8] and GPT-2 [9]. They then use transfer learning to fine-tune these models to perform specific tasks in specific domains and multiple techniques to reduce gender, racial, ethnic and other biases that have been shown to affect job searches [10]. Ensuring fair, ethical, and equitable treatment of workers is a key guiding principle in the creation of tangible deliverables for Track B.

These AI services support a SkillSync web app that digitizes the connection between companies and colleges, enabling companies to identify and communicate skills needs and enabling colleges to identify and respond with training opportunities that are aligned to these needs. The SkillSync app also incorporates an intelligent virtual agent called *AskJill*, which is being developed by Dr. Ashok Goel's Design & Intelligence Lab at the Georgia Institute of Technology. AskJill serves as a text-based dialog agent that answers user questions about the SkillSync app. Its goal is to provide contextual help and, more importantly, to increase user understanding and trust in the AI. It is based on technology developed for the Jill Watson virtual teaching assistant [11] and uses a two-dimensional hybrid machine learning and semantic processing model.

3 The LEARNER Project

LEARNER (Learning Environments with Augmentation and Robotics for Next-gen Emergency Responders), led by the second author, is engaged in developing an intelligent training platform for first responders who use new technologies such as augmented reality (AR) and exoskeletons. The platform employs a unique human-centered adaptive training framework that incorporates physiological, neural, and behavioral markers of learning, together with user preference and training history, into real-time exercises that use AR, virtual reality (VR), and force feedback. Two specific emergency response curricula are being developed: Triage and Patient Handling. A core tenet of LEARNER is supporting multiple access levels and modalities, ranging from web-based training delivery to physical live trainings and spanning the spectrum of virtual to physical realities. This is critical for providing access to all first responders and has implications for the design and development of adaptive training algorithms. AI is currently used to analyze pre-training exercises for the purpose of differentiating learners and will be used in algorithms that adapt learning at micro and macro levels.

The LEARNER team is creating adaptation models for each exercise based on the transferability of learning markers across different access levels, explainability of the machine learning models, and associated computation cost for near real-time adaptation. These learning markers are biometric markers that have been correlated with learning speed and effectiveness. In web-based desktop learning, markers are limited to mouse clicks and dwell time, whereas VR and AR-based training modalities at higher access levels can offer rich insights into gaze behavior [12] and can capture markers such as heart rate variability and neural activity through integrated neurophysiological sensors. The LEARNER architecture supports both performance-based and state-based evaluation, which enable macro and micro adaptation to optimize training effectiveness. Performance-based adaptations are guided by heuristics developed with input from subject-matter experts (e.g., trainers). Affect- and behavior-based markers gathered during learning using neurophysiological data will also inform state-based adaptation to ensure effective encoding of information from training materials.

4 Collaborations

The first significant collaboration among Track B projects was the formation of a *National Talent Ecosystem Council* (NTEC). This council is intended to guide, disseminate, and increase the impact of translational research in the AI and Future of Work domain regardless of its origin. It is also intended to serve as a mechanism that helps identify vetted research, which is important as more and more workforce and training applications use AI without identifying its limitations and potential biases and without exposing the underlying techniques they use. This council was formally launched in October 2021 and is intended to be a sustainable council that continues past the end of NSF funding for the Track B projects.

The second planned collaboration among Track B projects is a project that investigates the tradeoffs among various training modalities with different levels of fidelity and sophistication, ranging from non-adaptive desktop training to AI-supported virtual, augmented, and extended reality (VR, AR, XR) environments and adaptive instructional systems. This collaboration will produce guidelines for evaluating the pros and cons of using different approaches and training modalities in research and real world settings and will disseminate these through NTEC.

The third and most significant collaboration – Track Integration – is research and development of the *Skills-based Talent Ecosystem Platform for Upskilling* (STEP UP), envisioned as supporting a complete skills-based talent pipeline ecosystem. STEP UP will (a) allow employers to create job profiles that identify the skills needed for in-demand jobs; (b) enable individuals to create (and validate) skills profiles that identify the skills they have; (c) use AI to match skills profiles to job profiles and to identify skill gaps; (d) enable individuals to find and enroll in training to fill skills gaps; (e) issue skills-based credentials that validate newly acquired skills; and (f) enable individuals to provide existing or potential employers with those credentials. All these functions will be supported by extensions of SkillSync AI services and by freely available infrastructure, including the Credential Engine and the Open Competency Framework Collaborative [13]. These will be used to (a) extract skills profiles from work history and other data provided by workers (subject to human editing and validation), (b) to determine the skills addressed by specific training opportunities, (c) to recommend training pathways, (d) to select training and training modalities in accordance with the guidelines mentioned above, and (e) to match worker skills with employer needs.

5   The Long Term Vision

STEP UP is creating infrastructure to help advance a fair and efficient transition to the future of work. It is envisioned that new workforce technologies and associated skills can be rapidly added to the STEP UP platform with the assistance of AI that mines job postings and other sources; that providing skills-based credentials, training profiles, worker profiles and job profiles will create more opportunities to support different types of learning and career pathways; and that new AI-assisted training technologies and modalities can accelerate the speed and effectiveness with which skills can be acquired. STEP UP is also envisioned as a public good that can support organizations who have similar visions of skills-based talent management and that is integrated into emerging infrastructure such as *Learner and Employment Records* (LERs), currently being piloted in universities, government agencies, and US fortune 50 companies and with efforts such as the US Chamber of Commerce T3 Innovation network, which now includes more than 500 organizations [14]. The contribution of STEP UP lies in the translation of basic AI and learning science research into practice and in the ability to continue to advance this research. This long-term vision can only be achieved by careful integration of multiple disparate fields and by establishing a platform that supports continued integration of future advances in all of them.

Acknowledgement. This work reported here by the second and third authors was supported by National Science Foundation awards #2033578 and #2033592 respectively.

References

1. Biden, J.R.: Executive order on diversity, equity, inclusion, and accessibility in the federal workforce, The White House. 25 Jun 2021

2. National Academies of Sciences, Engineering, and Medicine, Imagining the Future of Undergraduate STEM Education. In: Proceedings of a Virtual Symposium. The National Academies Press, Washington DC (2022)
3. Devlin, J., Chang, M.-W., Lee, K., Toutanova, K.: BERT: Pre-training of Deep Bidirectional Transformers for Language Understanding, arXiv [cs.CL] 11 Oct 2018
4. National Science Foundation, Convergence Accelerator. https://beta.nsf.gov/funding/initiatives/convergence-accelerator. Accessed 18 Mar 2022
5. The White House, The Biden-Harris Administration Immediate Priorities, The White House, 10 Jan 2021. https://www.whitehouse.gov/priorities/. Accessed 18 Mar 2022
6. Modern Campus, State of Continuing Education survey highlights growing engagement gap, 09 Mar 2021. https://moderncampus.com/blog/2021-state-of-continuing-education-survey-highlights-growing-engagement-gap.html. Accessed 22 Feb 2022
7. MIT, Bridging the gap between education and employment: Community college and beyond, MIT News | Massachusetts Institute of Technology, 17 Aug 2021. https://news.mit.edu/2021/bridging-education-workforce-gap-community-college-beyond-0817. Accessed 22 Feb 2022
8. Devlin, J., Chang, M.-W., Lee, K., Toutanova, K.: BERT: Pre-training of Deep Bidirectional Transformers for Language Understanding, arXiv [cs. CL] 11 Oct 2018
9. Radford, A., et al.: Language models are unsupervised multitask learners. OpenAI blog 1(8), 9 (2019)
10. Kirk, H.R., et al.: Bias out-of-the-box: an empirical analysis of intersectional occupational biases in popular generative language models, Adv. Neural Inf. Process. Syst. 34, 2611–2624 (2021)
11. Goel, A.K., Polepeddi, L.: Jill Watson. In: Learning Engineering for Online Education, pp. 120–143 (2018)
12. Shi, Y., Zhu, Y., Mehta, R.K., Du, J.: A neurophysiological approach to assess training outcome under stress: a virtual reality experiment of industrial shutdown maintenance using functional Near-Infrared Spectroscopy (fNIRS). Adv. Eng. Inform. 46, 101153 (2020)
13. US Chamber of Commerce Foundation, Open competency framework collaborative, Open Competency Framework Collaborative. https://www.ocf-collab.org/. Accessed 22 Feb 2022
14. US Chamber: The T3 Innovation Network, U.S. Chamber of Commerce Foundation (2022). https://www.uschamberfoundation.org/t3-innovation. Accessed 06 Feb 2022

Introducing EIDU's Solver Platform: Facilitating Open Collaboration in AI to Help Solve the Global Learning Crisis

Aidan Friedberg[✉]

Head of Data, EIDU, Berlin, Germany
aidan.friedberg@eidu.com

Abstract. EIDU provides child-focused learning content along with digitally supported structured pedagogy programmes for teachers and is being rolled out to all public pre-primary schools across four counties in Kenya. EIDU is content-agnostic, allowing any provider to integrate new content into the platform, with the choice and order of content for each individual child optimised through personalisation algorithms. Autonomous digitised assessment tools enable real-time learning measurement which can be fed back to content providers and researchers, facilitating continuous improvement cycles. EIDU is providing open access to its platform to facilitate collaboration in the personalisation space with 'the Solver Platform'. Researchers will have access to a vast, anonymised learning dataset to train and develop personalisation algorithms. These algorithms can be deployed onto the platform using a plug-in system and will automatically be evaluated and selected based on their measured learning impact, always ensuring the safety of learners comes first.

1 Introduction

Although becoming ever more prevalent in high-income countries, digital educational interventions are still mostly non-existent at scale in low-middle-income countries (LMICs), especially within sub-Saharan Africa [1]. This means much of the insight into what and how children learn, as well as how effective digital educational interventions are, often depends on research which may not be relevant for these environments [2]. Where LMIC-specific studies have been conducted, they have demonstrated considerable potential for Digital Personalised Learning (DPL) tools to improve learning outcomes. For example, in a meta-analysis, Major et al., 2021 found that DPL tools can generate significant learning gains for students in the LMIC context [3]. The largest gains were seen where the DPL tool contained adaptive elements which afforded a 'teach at the right level' approach. How to build personalisation algorithms which optimally apply this kind of adaptivity to different age groups, subjects and learning environments, or in cases of conflicting objectives, is an area of research which is still very much in its infancy [4]. EIDU is seeking to add momentum to this area of research, especially in the LMIC context.

EIDU has created a technology-supported programme designed for LMIC education systems. The programme consists of educational support such as structured pedagogy

© Springer Nature Switzerland AG 2022
M. M. Rodrigo et al. (Eds.): AIED 2022, LNCS 13356, pp. 104–108, 2022.
https://doi.org/10.1007/978-3-031-11647-6_18

training, government capacity building, and digital content for teachers and learners. In partnership with the Kenyan government and the German Federal Ministry for Economic Cooperation and Development (BMZ), the programme is being implemented in all public Early Childhood Development (ECD) centres across four counties in Kenya and will have more than 100,000 active learners by 2023. EIDU aims to reach all two million pre-primary children in Kenya by 2025.

An integral part of the programme is the content-agnostic digital platform delivered through low-cost smartphones. This combines structured pedagogy content for teachers, such as lesson plans, with digital learning exercises and validated autonomous assessments for children. Various content creators like onebillion and Solocode integrate their content with the platform. These digital exercises are then mapped to the relevant subject, domain, learning objective, and skill; facilitating an easy alignment between learner exercises, teacher lesson plans, and the relevant national curricula. At the time of writing, approximately 40,000 pre-primary children complete an average of 100 learning exercises and one assessment exercise every week on the platform in Kenya.

2 Personalisation at EIDU

A fundamental optimisation problem that needs to be solved in education is selecting the optimal learning exercise for a given learner at a certain point in time. Since our platform combines content from various creators – that were not initially designed to co-exist on a single platform – an effective personalisation or ordering algorithm is crucial to ensuring meaningful learning experiences for children. Currently, EIDU is testing two algorithms, detailed below, against a manually ordered static curriculum in an A/B/C test, with each partition containing approximately 15,000 learners. This was done using the Solver Platform described in Sect. 3 as a proof of concept of the system. Final results are expected in June 2022.

2.1 Elo Based Adaptivity

The first partition treats the personalisation problem as a matchmaking problem. The idea behind this is that optimal learning can be achieved when learners are in a state of flow [5]. Within each learning objective, the difficulty of exercises and the skill level of children are calculated using an Elo-type rating system – specifically the Glicko-2 algorithm [6]. In this context children and exercises are treated as players in a zero-sum game. If a child solves an exercise the child 'wins' and gains rating points while the exercise loses points. If the child fails to solve an exercise the converse holds true. Ratings aren't just nominal but can be used to make probabilistic predictions about exercise outcomes (solved/not solved), which can be tested for their accuracy (AUC 0.81). Our first implementation aims to always provide children with exercises they have an 85% chance of solving. This specific target is based on the work of Wilson et al. 2019 [7]. Ultimately, this targeted difficulty level is arbitrary and could be refined through further rounds of A/B testing.

2.2 Deep Knowledge Tracing

The second partition is an implementation similar to the Deep Knowledge Tracing (DKT) algorithm described by Piech et al. 2015 [8]. This Long Short-Term Memory neural network takes a child's performance history as a binary input sequence (solved/not solved) and outputs a vector of probabilities for solving a given set of work units. While DKT only moderately outperforms 2.1 in terms of predicting exercise outcomes (AUC: 0.87), the major advantage of a neural network is the ability to run inference many timesteps into the future. Our implementation takes the play history of a child and runs inference for multiple timesteps in the future, selecting the next exercise by maximising the expected average gain of future probability vectors.

3 EIDU Solver Platform

EIDU hopes to create a new ecosystem in the AI research community with 'the Solver Platform', enabling researchers to build, deploy and iterate on learning algorithms, like those described above, in the LMIC context. The Solver Platform broadly consists of three components: providing researchers with access to our data, providing a plug-in system for researchers to deploy personalisation algorithms, and finally monitoring and evaluation tools to facilitate the publishing of findings and further iterations.

3.1 Data

EIDU has built up an extensive historical learning database while scaling our programme over the last year, with almost 100 million numeracy and literacy exercises done by 150,000 pre-primary and early primary children in Kenya and from our pilots in Ghana and Nigeria. Opening up this historical data as well as our real-time data to researchers creates the largest research database of learning in sub-Saharan Africa. Specifically, for each learning exercise done, this data will include:

Usage Data: the anonymous learner ID; the learner-specific sequence number; the exercise ID; the grade, subject, domain, learning objective, and skill that the exercise belongs to; the outcome of the exercise; the final score of the learner; the time taken to finish the exercise; ID of the most recent lesson taught by the teacher.

Assessment Data: the anonymous learner ID; the learner-specific sequence number; the assessment unit ID; the assessment battery the assessment belongs to; the skill being assessed; the assessment question; the learner's answer; the outcome of the assessment; the time taken to finish the assessment; Lesson ID of the most recent lesson taught by the teacher.

3.2 Plug-In System

Personalisation modules developed by researchers should be able to take the above data for a single learner's play history and recommend at runtime which exercise the child should attempt next from a given subset of exercises. This subset is based on the learning objective being taught in the class at that time. The algorithm can recommend a single work unit or a selection of work units that the child can then choose from.

A Software Development Kit (SDK) with extensive documentation will be provided for researchers (an example of how this works can be seen in the SDK we provide to our content partners at https://dev.eidu.com/). The EIDU platform runs on Android, and the personalisation plug-in system requires researchers to build modules with this compatibility in mind. A test environment will also be provided to ensure researchers can identify compatibility or latency issues early on. Personalisation modules have access to a native TensorFlow Lite interpreter, allowing researchers to load any saved TensorFlow models.

Our total user base is randomly partitioned, with each partition being assigned a module. Each research partner is provided with a dedicated AWS S3 bucket where they can upload and update versions of their module and trained models. Every time a device syncs with our servers, it will download the latest version of the module which the device is associated with.

3.3 Learning Analytics Platform

As soon as a module has gone live, a research partner will be given access to EIDU's Learning Analytics Platform (LAP) to facilitate monitoring and evaluation. For learning outcomes, this includes various time series of aggregated assessment data that can be broken down by assessment type, as well as longitudinal learning gains where learners have been retested on the same assessment unit. To track engagement, completion rates, scores and time spent per exercise and learning objective, allow researchers to understand how personalisation is shaping the learner's experience.

EIDU recognises the role aggregated data can play in the marginalisation of particular groups in education [9]. To help mitigate this, we disaggregate data by age, gender, and geography. We are expanding the demographic information we collect to ensure we can capture further characteristics such as neurodiversity in our data. Not only does this allow us to monitor whether groups of children are being left behind by interventions, but it also facilitates a better understanding of learning differences and the possibility to implement targeted interventions for specific subpopulations.

3.4 Algorithm Selection

As an increasing number of researchers utilise the Solver Platform, classical A/B/n testing methodologies become inefficient. Statistically detecting small differences is a hard problem, which is further compounded when there are numerous treatment arms to compare. Moreover, classical testing means learners are kept in sub-optimal modules for needlessly long periods of time. Instead, we employ a multi-armed bandit algorithm, in order to prioritise high-impact modules over low-impact modules on the Solver Platform. Multi-armed bandits have been used extensively in website optimisation and have been shown to be dramatically more efficient at finding the best arm than traditional statistical experiments, with an increasing advantage as the number of arms grows [10]. To further safeguard learning and child safety, algorithms are vetted and continuously monitored and evaluated. Where they are found to be damaging or nefarious, they will be completely removed from the system.

4 Next Steps

This paper has discussed the potential for collaboration in the personalisation and AI space using the Solver Platform. The platform aims to dramatically speed up innovation cycles in the education sector, especially for LMIC environments. In 2022 and 2023, EIDU will provide early access to selected researchers in this space in partnership with the Futures Forum on Learning, the Bill and Melinda Gates Foundation, and EdTech Hub. Outcomes and learnings derived from this process will be published, in the hope of seeding a community and body of research which can demonstrate how collaboration in AI in education can contribute to solving the global learning crisis.

References

1. Krönke, M.: Africa's digital divide and the promise of E-learning (2020)
2. Rao, N., et al.: Early childhood development and cognitive development in developing countries: a rigorous literature review. Department for International Development (2014)
3. Major, L., Francis, G.A., Tsapali, M.: The effectiveness of technology-supported personalised learning in low-and middle-income countries: a meta-analysis. Br. J. Edu. Technol. **52**(5), 1935–1964 (2021)
4. Maghsudi, S., Lan, A., Xu, J., van Der Schaar, M.: Personalized education in the artificial intelligence era: what to expect next. IEEE Signal Process. Mag. **38**(3), 37–50 (2021)
5. Nakamura, J., Csikszentmihalyi, M.: The concept of flow. In: Csikszentmihalyi, M. (ed.) Flow and the Foundations of Positive Psychology, pp. 239–263 Springer, Dordrecht (2014). https://doi.org/10.1007/978-94-017-9088-8_16
6. Glickman, M.E.: Example of the Glicko-2 system, pp.1–6. Boston University (2012)
7. Wilson, R.C., Shenhav, A., Straccia, M., Cohen, J.D.: The eighty five percent rule for optimal learning. Nat. Commun. **10**(1), 1–9 (2019)
8. Piech, C., et al.: Deep knowledge tracing. In: Advances in Neural Information Processing Systems, 28 (2015)
9. DFID Girls Education Challenge: Thematic review understanding and addressing educational marginalisation: Part 2: educational marginalisation in the GEC (2018)
10. Scott, S.L.: Multi-armed bandit experiments in the online service economy. Appl. Stoch. Model. Bus. Ind. **31**(1), 37–45 (2015)

Workshops and Tutorials

Intelligent Textbooks: Themes and Topics

Sergey Sosnovsky[1]([✉]), Peter Brusilovsky[2], and Andrew Lan[3]

[1] Utrecht University, Princetonplein 5, 3584 CC Utrecht, The Netherlands
s.a.sosnovsky@uu.nl
[2] University of Pittsburgh, 135 North Bellefield Avenue, Pittsburgh, PA 15260, USA
[3] University of Massachusetts Amherst, Amherst, MA 01003, USA

Abstract. The transition of textbooks from printed copies to digital and online formats has facilitated numerous attempts to enrich them with various kinds of interactive functionalities, link them with external resources or extract valuable information from them. As a result, new research challenges and opportunities emerge that call for the application of artificial intelligence methods to enhance digital textbooks and students' interaction with them. In this report, we summarize our workshop series on Intelligent Textbooks from 2019 to 2022. We focus on the evolution of the topics covered in the workshops' programs and identify the main themes that have been proposed by the intelligent textbooks community.

Keyword: Intelligent textbooks

1 Historic Overview

This year, the workshop on Intelligent Textbooks is organized for the fourth time. It builds on the success of the three previous workshops conducted as a part of the satellite program of the International Conference on Artificial Intelligence in Education in 2019, 2020 and 2021. Overall, 36 (20 full and 16 short) contributions were published in the workshop proceedings over these years; 14 intelligent textbooks prototypes and technologies were showcased as demo presentations.

At the first workshop in 2019, a majority of accepted submissions have focused on various aspects of making textbooks adaptive through navigation, recommendation, or problem solving support. Other popular topics were integration of interactive content, orchestration of learning around digital textbooks and automated analysis of the textbook content for various purposes.

In 2020, adaptivity and interactivity remained important aspects of intelligent textbooks. However the trend on leveraging machine learning, natural language processing and semantic technologies to automate processing or construction of textbook content became much more prevalent. Several papers and demos have presented approaches for textbook generation, transformation, linking to external content and extraction of knowledge from textbooks.

© Springer Nature Switzerland AG 2022
M. M. Rodrigo et al. (Eds.): AIED 2022, LNCS 13356, pp. 111–114, 2022.
https://doi.org/10.1007/978-3-031-11647-6_19

The third workshop explored a variety of topics. Two new trends that separated it from its predecessors were: demonstration-based papers presenting prototypes of domain-oriented textbook applications and projects exploring automated approaches to extract from textbooks different kinds of learning objects.

In 2022, the workshop has attracted ten submissions. And while the review process is still ongoing at the moment of writing this report, we can already provide a preliminary overview of the topics covered by the potential program of the Intelligent Textbooks 2022 workshop and relate them to the main themes explored by the workshop participants over the last 4 years. Table 1 presents the summary of these topics together with counts of corresponding papers presented at each of the workshops from 2019 to 2022[1].

2   Workshop Themes

Intelligent interfaces for online textbooks could be considered as an "end product" of several other research directions. This is where the "intelligence" reaches the readers augmenting their textbook experience with a range of functionalities not available in traditional textbooks. While these new functionalities are typically based on modeling and knowledge processing technologies, the main concern of papers focused on intelligent interfaces is not these foundation technologies, but how to present the new functionality to the readers. Among interface-focused papers presented at the past workshops, several paper focused on open student models - a visual presentation of learned knowledge and progress (computed by AI student modeling algorithms) to the learners. Other papers focused on such technologies as personalized guidance within the textbook, run-time recommendation of external learning content, and augmenting user interactions with interactive tools such as concept maps and chatbots.

One of the appealing ways to extend online textbooks with additional functionalities not available in traditional textbooks is adding so-called "smart content" items - interactive activities, which engage students into learning by doing (rather than just by reading). Since most of these "smart content" items are in fact learning exercises supporting automatic assessment, working with smart content also enables the students to check their content understanding and receive feedback on their performance. While only a fraction of "smart content" activities could be intelligent by themselves (i.e., ITS problems with intelligent scaffolding) the work on smart content is important for intelligent textbooks as a whole. Most importantly, "smart content" activities produce a much richer volume of learning data that is crucial for both student modeling and knowledge extraction. The work on smart content has been well-represented at the workshop, especially in papers focused on computer science textbooks, the domain where smart content is becoming increasingly popular. The key research issues examined in the workshop papers focused on smart content are the infrastructure (how to connect an interactive item to a textbook while maintaining authentication and data collection) and content matching (how to assign a smart content

---

[1] The numbers for 2022 are projections.

Table 1. Number of papers under each topic over the years.

Topic/Year	2019	2020	2021	2022
Intelligent interfaces	5	2	1	0
Smart content	1	1	2	4
Knowledge extraction	1	2	0	1
Learning content construction	0	1	3	2
Intelligent textbook generation	1	2	0	1
Interaction mining and crowdsourcing	1	1	3	1
Domain-focused textbooks and prototypes	1	0	2	0
Miscellaneous	4	0	2	1

item to its proper place within a textbook). Smart content grouped together with the intelligent interfaces forms the first large research theme that is focused on textbook enrichment with extra features and functionality.

A textbooks can be seen not only as an object of enrichment, but also as a source of domain knowledge whereas intelligent textbooks are often written in formats that enable automated extraction of this knowledge. Works published in our workshop have covered many different subtopics related to knowledge extraction. Several papers discussed the construction of knowledge representations from textbooks, using a wide range of approaches, from human-labeling to automated construction via both traditional feature-based natural language processing (NLP) techniques and modern embedding-based NLP techniques. Several works focused on extracting prerequisite relations and studied the noisy nature of this process. The extracted knowledge representations can be useful in many downstream tasks, including entity relationship visualization, matching across textbooks and between textbooks and external learning content, monitoring ideas in student discourse, and personalized learning support.

In recent years, with rapid development in neural language models in NLP research, especially generative ones such as GPT, our workshop has seen an uptick in the number of works on automated content construction for intelligent textbooks using these models. These models are highly capable of effectively transferring what they learn from pre-training on web-scale text to different contexts, resulting in high levels of fluency and consistency of the generated text. The vast majority of these works use generative language models to automatically produce assessment questions for intelligent textbooks. Works have focused on generating both different formats of questions, from multiple-choice to short-answer, and different types of questions, from factual to reasoning. Additionally, several works have also considered generating other types of learning content, such as textbook indices and concept definitions. Learning content construction and knowledge extraction constitute the second theme of research on intelligent textbooks aimed at utilizing the textbooks themselves as a source of an added value, be it (elements of) domain knowledge or (elements of) learning material.

Another stream of research that has been explored by the workshop participants is textbook generation and assembly. The works on this topic are characterized neither by a uniform methodology, nor by the common attributes of the final product. Yet, they have had a common objective - propose a technology that facilitates creation of digital textbooks from external resources. These resources range from Wikipedia content to specially-formatted material such as Jupyter notebooks to existing digital textbooks in PDF format. The proposed technologies ranged from a community-oriented authoring platform for digital textbook assembly to a framework for automated generation of intelligent textbooks enriched with semantic and adaptive services. Generally speaking, textbooks generation represents the third workshop theme that focuses on a textbook itself as the final product.

Most technologies presented at the workshop are aimed at developing and/or supporting complex applications built around textbooks. These applications provide their users with various methods of interaction with the actual content of textbooks, integrated smart content items and or value-adding services enriching textbook functionality. Evaluation of these interactions to data-mine typical patterns, model parameters or characteristics of students has been a common topic for several workshop contributions. Another related line of research has become an organization of interaction between users and textbooks in such a way that the textbook application could crowdsource execution of challenging tasks to its users. The outcomes of these interactions would provide the textbook application with the elements of crowdsourced "intelligence" (e.g., concept map-based exercises helping extract types of relations between domain terms, or text highlighting behavior helping to train a student modelling approach). Interaction mining and crowdsourcing constitute the fourth large theme of research on intelligent textbooks that is concerned with the link between the textbook and the user and aims at extracting "intelligence" from it.

Intelligent textbooks are broadly applicable in many subject domains. Therefore, tools that are customized to the specifics of each subject domain are necessary. For example, in the domain of reading education, intelligent textbooks can benefit from embedded tools for authoring support, student modeling, personalized activity, and mini-games. In the domain of quantum cryptography, intelligent textbooks can benefit from built-in coding environments, interactive visualizations, and self-graded quizzes, driven by learning styles and student objectives. In medical domains such as dentistry and cardiovascular anatomy, intelligent textbooks can benefit from chatbots built into the textbook to ask learners questions and interact with them or take the form of a mobile application helping students learn the logical connections behind key technical terms.

Finally, over the years, the workshop has attracted several contributions that are hard to categorize under a single label. Some of these papers proposed technologies that are too unique (e.g., a prototype using paper-based workbooks and a mobile application scanning and grading hand-written solutions). Others focused on very particular tasks (e.g., a new format facilitating representation and retrieval of math formulae). In addition, we have had several position papers envisioning new way to organise and orchestrate lessons around intelligent textbooks of the future.

Interdisciplinary Approaches to Getting AI Experts and Education Stakeholders Talking

Rachel Dickler[✉], Shiran Dudy, Areej Mawasi, Jacob Whitehill, Alayne Benson,
and Amy Corbitt

NSF AI Institute for Student-AI Teaming (iSAT), Boulder, CO 80302, USA
rachel.dickler@colorado.edu

Abstract. The present workshop aims to facilitate conversations within the Artificial Intelligence in Education (AIED) community around bridging the gap between AI efforts and educational stakeholders' (teachers, students, parents, administrators, etc.) needs. In particular, the workshop will address existing barriers to collaboration between researchers and stakeholders, approaches envisioned and taken to address these challenges, and corresponding insights to help move the field forward in this area. The workshop invites papers (with a paper format that is aligned to the AIED submission instructions) that present approaches to cross-disciplinary research, exemplify novel interdisciplinary research approaches in the broad field of AIED, etc. Papers from international research communities are particularly welcomed. The structure of the workshop will include presentations of accepted papers, followed by an in-depth small group and whole group discussion regarding common themes across presentations. An overview of the lessons learned and key take-aways presented will be published in a 1-page summary in the conference proceedings.

Keywords: Interdisciplinary · Stakeholders · Collaboration

1 Workshop Motivation, Expected Outcomes, and Impact

1.1 Workshop Motivation

While the field of Artificial Intelligence in Education (AIED) continues to innovate and push the boundaries of AI technologies, it is critical to collaborate with educators and other core stakeholders (e.g., students, parents, administrators) to ensure that technologies are beneficial rather than harmful when implemented in actual classrooms. With real-world deployment of AIED systems, there may arise subtle and important concerns about privacy, equity, bias, etc. [1, 3]. It is essential that researchers attune to sociocultural, ethical, and political aspects of technology implementations in the real-world (e.g., [4]). Collaboration with education stakeholders is one essential approach to developing AI systems that support learning while also attending to stakeholders' key values and concerns [1].

Given the importance of collaboration with stakeholders, this workshop aims to bring researchers of different disciplines together, to share and learn about colleagues'

© Springer Nature Switzerland AG 2022
M. M. Rodrigo et al. (Eds.): AIED 2022, LNCS 13356, pp. 115–118, 2022.
https://doi.org/10.1007/978-3-031-11647-6_20

challenges, experiences, and lessons learned in developing AI innovations for k-12 settings. Participants will reflect on approaches that have been implemented to promote collaboration with stakeholders and corresponding outcomes. Specifically, we propose the present workshop to: 1) learn about different efforts to promote collaboration with stakeholders, including the development of interdisciplinary hubs such as the NSF AI institutes and 2) visit approaches, such as participatory design (e.g., [2, 5]), that promote more equitable collaboration.

1.2   Expected Outcomes

Rich discussions will be fostered between interdisciplinary experts and, correspondingly, there will be opportunities for initiating new collaborations. In particular, a portion of the workshop will be dedicated towards group discussions where participants can make connections across disciplines and share about varying experiences. The workshop participants will also synthesize major practices conducted in interdisciplinary teams with educators and stakeholders that attend to equity in the design, implementation, and analysis of AIED research and curricula. Finally, the synthesis of practices based on presentations and discussion will be summarized by the workshop organizers in the form of lessons learned to share with the community.

1.3   Impact

The workshop provides a unique space with opportunities for discussion that will result in ideas integrated into future research efforts beyond the workshop. Specifically, researchers will be able to connect with fellow colleagues (nationally and internationally) to build connections and innovate on new approaches to interdisciplinary research. The workshop outcomes will be presented as recommendations towards supporting collaboration with stakeholders.

2   Call for Papers from the AIED Community and Review

For one component of our proposed workshop, we will solicit papers (2–4 pages formatted using the AIED 2022 conference proceeding guidelines; not including references) from the AIED community. The event and the call-for-papers will be advertised through a Google Site (created by the workshop organizers) and global email-lists. Co-organizers of the workshop are from the communication team at the AI Institute for Student-AI Teaming (iSAT), who will advertise via their website, twitter, and other media outlets. Papers will be submitted to a google email address created for the workshop. These papers can include case studies, experimental studies, and narratives about (note that the following list is not exhaustive): specific ideas or approaches for interdisciplinary AIED research that connects researchers and stakeholders, research questions about how interdisciplinary research teams operate, barriers and specific challenges not encountered in more siloed teams, unique benefits that arise due to interdisciplinarity across collaborations, and "lessons learned" from existing interdisciplinary teams that bridge the gap between AI and stakeholders.

In terms of the review process, review will be single-blind (reviewers are anonymous, authors are not). Authors will be sent the anonymous reviews along with the paper decisions via email. Reviewers will be asked to recommend the paper for acceptance or rejection, provide a rationale for their recommendation, and rate the papers on a scale of -1, 0, or 1 for the following metrics: relevance to workshop, interest to AIED community, and paper quality. We roughly estimate 10–15 submissions and a 80–100% acceptance rate. If there are fewer papers submitted, then each paper will be allotted a slightly longer presentation window and additional time will be spent on discussion. The schedule for the call for papers and review is as follows:

- Call for papers: March 28, 2022
- Paper submission deadline: June 27, 2022
- Paper review period: June 28, 2022 - July 8, 2022
- Final paper decisions: July 9, 2022 - July 12, 2022
- Notification of acceptance: July 13, 2022
- Camera-ready deadline: July 20, 2022
- Workshop day: July 27th, 2022 *or* July 31st, 2022

3 Tentative Workshop Schedule and Format

We propose a half-day online workshop in Zoom consisting of paper presentations from members of the community followed by structured discussion around themes that emerge in approaches to cross-disciplinary research. The schedule is as follows: Introductions, Opening remarks (from workshop organizers), Paper presentations with time for questions after each paper, Small group and whole group discussion on themes across papers. Questions can be asked via the Zoom chat throughout the workshop, as well as via a Slack channel that will be accessible following the conclusion of the workshop.

4 Organizers (Affiliation, Email Address, and Biography)

- **Rachel Dickler** (Research Associate, iSAT, University of Colorado Boulder, rachel. dickler@colorado.edu). Rachel's background is in the field of the Learning Sciences with a focus on the development and implementation of AI in K-12 settings. Her research examines interactions between students and AI agents, as well as the design of dashboard technologies.
- **Shiran Dudy** (Research Associate, Institute of Cognitive Science, University of Colorado Boulder, shiran.dudy@colorado.edu). Shiran's background is in Natural Language Processing, and development of communication systems. Shiran is working on the conversational system at iSAT where she is designing a dialogue system that supports students' collaboration and learning.
- **Areej Mawasi** (Research Associate, Institute of Cognitive Science, University of Colorado Boulder, armw7353@colorado.edu). Areej draws on learning sciences, design-based methods, and critical STEM education to study the sociocultural and political aspects of learning and technology-rich learning environments. Within iSAT, Areej studies co-design processes towards building AI-curriculum with nondominant youth and teachers.

- **Jacob Whitehill** (Assistant professor of Computer Science, Worcester Polytechnic Institute, jrwhitehill@wpi.edu). Jake's research is in multimodal machine learning and its intersection with education. His group's work is published at AIED, EDM, as well as speech recognition and machine learning venues such as ICASSP, AAAI and ICML.
- **Alayne Benson** (AI Communications & Outreach, iSAT, University of Colorado Boulder, Alayne.Benson@colorado.edu). Alayne is an experienced communications professional and former educator who is driven to ignite the passion and purpose audiences need to create a brighter future. She's also an experienced content creator of digital and print Social Studies and English Language Arts products for teachers and students.
- **Amy Corbitt** (Communications Professional, iSAT, University of Colorado Boulder, Amy.Corbitt@colorado.edu). Amy is a communications professional with more than two decades of experience writing and editing. She is supporting the team at iSAT by sharing their important work with the broader community.

5   Program Committee

The program committee will consist of the workshop organizers as well as members of iSAT including: Peter Foltz (Research Professor, University of Colorado), Leanne Hirshfield (Research Professor, University of Colorado), Jamie Gorman (Professor, Georgia Institute of Technology), Sadhana Puntambekar (Professor, University of Wisconsin), Michael Tissenbaum (University of Illinois), Tamara Sumner (Professor, University of Colorado), Sidney D'mello (Professor, University of Colorado), Michael Chang (Research Associate, CIRCLS and iSAT).

References

1. Adams, C., Pente, P., Lemermeyer, G., Rockwell, G.: Artificial intelligence ethics guidelines for K-12 education: a review of the global landscape. In: Roll, I., McNamara, D., Sosnovsky, S., Luckin, R., Dimitrova, V. (eds.) AIED 2021. LNCS, vol. 12749, pp. 24–28. Springer, Cham (2021). https://doi.org/10.1007/978-3-030-78270-2_4
2. Cober, R., Tan, E., Slotta, J., So, H.J., Könings, K.D.: Teachers as participatory designers: two case studies with technology-enhanced learning environments. Instr. Sci. **43**(2), 203–228 (2015)
3. Holstein, K., Doroudi, S.: Equity and artificial intelligence in education: will "AIEd" amplify or alleviate inequities in education? arXiv preprint arXiv:2104.12920 (2021)
4. Ogbonnaya-Ogburu, I.F., Smith, A.D., To, A., Toyama, K.: Critical race theory for HCI. In: Proceedings of the 2020 CHI Conference on Human Factors in Computing Systems, pp. 1–16 (2020)
5. Luckin, R., Puntambekar, S., Goodyear, P., Grabowski, B.L., Underwood, J., Winters, N.: Sketch-Ins: a method for participatory design in technology-enhanced learning. In: Handbook of Design in Educational Technology, pp. 104–113. Routledge (2013)

ETS® AI Labs™ Ways of Working Tutorial: How to Build Evidence-Based, User-Obsessed, AI-Enabled Learning Solutions in an Agile Framework

K. Rebecca Marsh Runyon[✉], Kinta D. Montilus[✉], Larisa Nachman[✉], Kristen Smith Herrick[✉], and Lisa Ferrara[✉]

Educational Testing Service (ETS) AI Labs™, Princeton, US
{krunyon,kmontilus,lnachman,kherrick,laferrara}@ets.org

Abstract. How do you advance the science and engineering of digital learning solutions at your institution, business, or organization? Bring your current ways of working to this tutorial and get ready to innovate them alongside your fellow researchers, practitioners, business owners, and policy makers. As we work together to share our knowledge and lived experiences, presenters will assist participants in co-creating action plans for how they can utilize best practices from user-centered design (UCD), Design thinking, and Agile to deliver user-obsessed, AI-enabled, efficacious learning solutions.

Keywords: Agile · Design Thinking · UCD

1 Themes

- Closing the gap between research and application (i.e., referencing, synthesizing, and applying the appropriate Learning Science & Design and AIED research within learning experience design and iteration)
- Working collaboratively, cross-functionally, effectively, and efficiently to deliver user-obsessed, AI-enabled, and efficacious digital learning solutions
- How to apply Agile, UCD, or Design Thinking to your own context to address stakeholder needs in technology-enhanced, AI-driven solutions

2 Target Audience and Participation Capacity

Learning experience design researchers, practitioners, business associates or owners, and/or policy makers interested in gaining insight into how various companies, organizations, and/or institutions and their teams work to create evidence-based, innovative, and efficacious, AI-enabled learning solutions at scale. 50 participants maximum.

© Springer Nature Switzerland AG 2022
M. M. Rodrigo et al. (Eds.): AIED 2022, LNCS 13356, pp. 119–122, 2022.
https://doi.org/10.1007/978-3-031-11647-6_21

3 Relevance to AIED Community

Creating innovative and efficacious digital learning solutions requires a combination of varied expertise, research methodology, cross-functional collaboration, and software development. To deliver these solutions at scale and an appropriate velocity, many educational technology companies or initiatives have begun to implement Lean Startup and/or Agile methodologies, which are new to many other contexts or companies. This tutorial will provide an overview of how the ETS® AI Labs™ cross-functional teams work in Agile and go on to discuss how we also leverage UCD and Design Thinking in product conceptualization, development, and testing to create research-based, user-obsessed, and AI-enabled learning solutions.

Participants will have the opportunity to learn about these methodologies, their applications within the ETS® AI Labs™, and determine how these approaches facilitate their own work. Participants will engage with the presenters and each other to create an action plan for how they can use UCD, Design Thinking, and/or Agile methods to work in cross-functional ways that enable AI to enhance their solution development and implementation. Beyond learning about these methodologies and their prospective applications within their own context, participants will also discuss what they have learned and how it can be applied more broadly to improve AIEd research and application by and large.

4 Content, Format, and Activities

This three-part tutorial session will include:

- Tutorial welcome, introduction, and overview to describe and illustrate how the ETS® AI Labs™ use Agile methods in traditionally "non-agile" contexts, reference and engage in the Product Development Life Cycle (PDLC) and implement UCD processes to create effective AI-enabled learning solutions at scale (150 min, Interactive Lecture and Discussion with interspersed short breaks). Presenters will provide an overview of:

  - Agile and the Product Development Life Cycle (PDLC): Brief description of what it is, how it can be leveraged generally/across contexts to deliver effective solutions at scale, and examples of how we apply it in the ETS® AI Labs™ (e.g., project team cross-functional roles/responsibilities and their ways of working)
  - UCD: Brief description of what it is, how it can be leveraged more generally/across contexts to deliver user-centered solutions, and examples of how we apply it in the ETS® AI Labs™
  - Design Thinking: Brief description of what it is, how it can be leveraged more generally/across contexts to deliver creative, innovative solutions, and examples of how we apply it in the ETS® AI Labs™
  - Lessons Learned: Provide a working list of key takeaways or "lessons learned" about these methodologies, our ways of working, and how they intersect with AIEd research and practice

- 60-min Lunch Break

- Breakout session to identify specific problems to be solved for users, learners, or stakeholders in participants' unique contexts and identify how Agile/cross-functional ways of working, UCD, and/or Design Thinking address these problems within AI-enabled learning solutions (100 min, Hands-on).

  - Participants break out into small groups (consisting of 10 or fewer) with one presenter dedicated to each group to facilitate activities and discussion. Participants will:

    - Engage in mock problem statement interviews with their fellow participants (facilitated by presenters) to apply aspects of UCD and Design Thinking and establish one core, highly prioritized problem to be solved w/ an AI-enabled learning solution appropriate for their context.
    - Participants will then create an Action Planning activity which will facilitate planning the 'Who, What, Where, and When' of how they will incorporate specific aspects of UCD, Design Thinking, and/or Agile methodologies into their ways of working at their institution, business, or organization to solve this problem.
    - Participants will discuss their plans with presenters and fellow participants to further refine and solicit additional ideas and inspiration (e.g., identifying one potential gain, risk, and/or an area of AI innovation and engage with fellow participants to brainstorm mitigation or iteration leveraging aspects of UCD, Design Thinking, and Agile methods, other).

- Collective discussion and discourse (50 min, Discussion)

  - Participants will share insights garnered from their small group engagements that may be relevant to others in attendance and/or AI/AIEd research or application, in general (e.g., how to iterate collective processes/ways of working and facilitate closing the gap between AIEd research and practice)
  - Q&A with participants and presenters

5   Presenters and Facilitators

K. Rebecca Marsh Runyon, krunyon@ets.org: Becca is Director of Learning Science & Design within the ETS® AI Labs™. She earned her Master's in Cognitive Science (2010) and PhD in Assessment and Measurement from James Madison University (2013). For the past 10 years, she has worked primarily in higher education technology digital strategy development and product design as a learning researcher, scientist/designer, and assessment specialist.

Kinta D. Montilus, kmontilus@ets.org: Kinta is a Senior Instructional Designer within the ETS® AI Labs™. She has over six years of leading and contributing to highly engaging and effective learning designs. Her favorite thing is to co-design with users to showcase where relevant scientific literature can be applied in innovative ways to create solutions for real learning problems. Kinta graduated from Seton Hall University (2022) with a PhD in Education Research, Assessment, and Program Evaluation.

Larisa Nachman, lnachman@ets.org: Larisa is a User Researcher with the ETS® AI Labs™. Prior to her work at ETS, she worked for 8 years as an educator and academic advisor in K-12 settings and for two years as an impact researcher in higher education publishing. Larisa earned her Master's in Secondary English Education (2018) from Teachers College, Columbia University.

Kristen Smith Herrick, kherrick@ets.org, serves as a Learning Scientist within the ETS® AI Labs™. Since 2012, Kristen has worked in various assessment, learning improvement, and learning science-based roles - in higher education and industry settings. This includes 8 + years leading assessment and learning improvement efforts at higher education institutions.

Lisa Ferrara, laferrara@ets.org: Lisa is Senior Director of Product Development within the ETS® AI Labs™. She earned her Master's in Instructional Design and Educational Technology (2009) and PhD in Learning and Cognitive Science (2017) from the University of Utah. She leads product strategy by ensuring that learning science principles underpin the research and development frameworks used to fuel product innovation.

Practitioner Track

Synchronization Ratio of Time-Series Cross-Section and Teaching Material Clickstream for Visualization of Student Engagement

Konomu Dobashi[1]([✉]), Curtis P. Ho[2], Catherine P. Fulford[2], Meng-Fen Grace Lin[2], and Christina Higa[2]

[1] Aichi University, Nagoya, Aichi 453-8777, Japan
dobashi@vega.aichi-u.ac.jp
[2] University of Hawai'i at Mānoa, Honolulu, HI 96822, USA

Abstract. This paper describes a study that analyzed the synchronization ratio of learners' teaching material browsing behavior during lessons using a learning management system and online teaching materials. In lessons attended by many learners, teachers often instruct the learners to open the material, and in such cases, it is important that the browsing behavior of as many learners as possible is synchronized. Therefore, data mining technology and time-series cross-section analysis were applied to the learning log, and a method was developed to calculate the synchronization ratio of teaching material browsing and to visualize it in tables and graphs. It is possible to calculate the average synchronization ratio and the average material synchronization ratio for the section of teaching materials and quizzes used during lessons.

Keywords: Synchronization ratio · Learning management system · Class improvement · Clickstream · Student engagement · Time-series · Cross-section

1 Introduction

In many computer lessons, the teacher presents the prepared materials to the learner, and the teacher often gives examples of computer operations as the learning progresses. In such a lesson in which browsing of teaching materials and operation of a personal computer proceed almost at the same time, it is desirable that the behavior of all the learners is synchronized with the instructions of the teacher (Fujii et al., 2018).

Therefore, if the learning behavior can be grasped in real-time, a support effect useful for class implementation can be expected. In general, if teachers unilaterally proceed with any lesson, the number of learners who fall behind on the learning will increase. In order to avoid such a situation, it is indispensable to accumulate learning logs using the latest information technology such as learning management system (LMS) and to develop an analysis method utilizing data mining. Additionally, when it is necessary to analyze and list a large number of learners at the same time, the time-series cross-section

© Springer Nature Switzerland AG 2022
M. M. Rodrigo et al. (Eds.): AIED 2022, LNCS 13356, pp. 125–131, 2022.
https://doi.org/10.1007/978-3-031-11647-6_22

(TSCS) analysis method used in the field of statistics is effective (Beck and Katz, 2011). In this study, based on TSCS analysis, the synchronization ratio and the average material synchronization ratio when the learner opens the teaching materials were obtained, and the teaching material browsing behavior was analyzed and examined.

2 Related Research

Research on accumulated learning logs has become more important and more active (Bradley, 2021). Many researchers are engaged in research called educational data mining (EDM) or learning analytics (LA). These two areas are primarily focused on various data related to education and learning, such as learner test scores, data such as GPAs, and data related to LMS learning logs (Blikstein and Worsley, 2016).

In addition, research is being conducted on dashboards that track and analyze learner behavior from several perspectives and develop an easy-to-understand graphical user interface for learning situations in school lessons (Bodily et al., 2018). The purpose of developing these systems are to grasp the state of the learner with numerical values and graphs and to support the lesson by conducting the lesson while looking at the analysis result displayed on the dashboard (Vieira et al., 2018).

Research on synchronization and synchronization ratios has been conducted in various fields. In the field of education, research is being conducted on the synchronization of movements in music learning for young children and the synchronization of browsing in e-book teaching material. A system is being developed that uses Moodle LMS and an e-book system to conduct lessons and analyze learner learning logs in real-time. The purpose is to provide feedback and analysis to teachers and learners to support lectures in real-time (Shimada et al., 2018). In the system, the heat map and graphs displayed on the dashboards of teachers and learners showed the synchronization ratio of teaching materials. Their research also showed that teachers can adjust the lecture speed and that learners can easily understand the situation of other learners.

3 Experimental Method

Since the faculty to which the first author belongs is conducting research related to modern China at Aichi University, the class covered in this paper is named "China Data Analysis". This class has been held continuously in recent years at Aichi University, utilizing the data of the china data published on the Internet. The purpose the class is to acquire the basics of statistics using Excel.

The teaching materials are equivalent to 15 lessons, and the overall structure contains 13 chapters, 95 sections, and 183 pages. The average number of sections per chapter is 7.3, and the average number of pages per section is 1.9. In Moodle, 13 chapters and 95 sections were created in topic mode. PDF files were created to upload the teaching materials to Moodle for each section. Links to the materials are provided on the gateway page of the Moodle course, linking to the top page of each section, which is a transition section for moving through the materials section by section.

In the first half of the lesson, all the learners went along with the teacher's instruction and learned the data processing with Excel. In the second half, exercises related to the lecture for the day were prepared and the learners decided to work on their own like self-regulated learning. In order to collect learning logs in Moodle, the teacher required learners to open materials at all times during class followed by the instructions.

4 Experimental Result

In Fig. 1, the space limitation of this paper has reduced the screenshot of the entire Excel file, snipping out and enlarging the important parts. In column A of Fig. 1, the anonymized learner IDs are displayed. From column B–AG is a summary of the quizzes and clickstreams of the teaching materials. The first row from the top of the screen shows the time interval every minute. The cells that display a numerical value on the sheet show the frequency of opening the teaching material for each learner. The rightmost AG column is the total number of clicks by a learner up to 13:27, and the 52nd row at the bottom of the table is the total number of times the material was opened. Furthermore, the synchronization ratio is shown in row 53, which is the original data for the bar graph. The synchronization ratio was calculated by dividing the number of people who opened the teaching materials by the number of attendees (49 people) on the day. Both the number of attendees on the day and the number of people who opened the teaching materials were calculated from the TSCS in Fig. 1.

The graph in Fig. 1 shows that many learners synchronized at the beginning of the lesson, but most are clickstreams of quizzes. The time limit is 5 min, and the reason why the last value of the quiz is high at 13:05 is because this is when the quiz score confirmation page could be opened. After that, the teacher instructed the learners to open the teaching materials at 13:06 and the class proceeded. The three colored columns (N, P, AD) on the sheet in Fig. 1 are the times when the teacher opened the teaching materials on the computer on the teacher's desk. The class started at 13:00, but Fig. 1 shows the data and graphs from 3 min before the class started (12:57) to 13:27. The file in Fig. 1 was generated on a laptop computer in the classroom at around 13:28 during the class. Figure 1 also includes multiple material clicks in addition to the quiz. It is necessary to calculate the synchronization ratio for each teaching material.

The teacher opened the entry page in column N of Fig. 1 and opened the teaching materials in each section in columns P and AD. The teacher opened the teaching materials and instructed the learners at 13:11 (column P) and 13:25 (column AD), and it can be seen that the synchronization ratio is slightly higher at these times. On the day of the class, the first half of the lesson was held using the teaching materials in the two sections "07.01 Deviation/Variance/Standard Deviation" and "07.02 Finding the Variance". From Fig. 2, it can be seen that at 13:11, 32.7% of learners (16 people) opened and synchronized the former material. The latter material was opened by the teacher at 13:25, but only 18.4% of learners (nine people) were in synchronization with these instructions. It can be observed that many learners had already opened these two materials before the teacher directed them to. It can be seen that the former teaching material was used until around 13:25 in class, and then, the latter teaching material was used (Fig. 2).

As shown above, the synchronization ratio calculated in this paper will change in various ways depending on the lesson type, the method of presenting the teaching materials, etc. Since the clickstreams for opening the teaching materials are analyzed, it changes from moment to moment depending on the browsing behavior of the learner, as is shown by the synchronization ratio graph in Fig. 1. Therefore, it is possible to see the synchronization ratio at a certain time, but it is difficult to use as it is as an index for the synchronization ratio of the entire lesson. Therefore, when the lesson was completed, the average synchronization ratio of the learners over the course of the entire lesson was calculated. The ASR was defined as the synchronization ratio during a certain period of time. The ASR of the above two teaching materials is shown in Fig. 2 with the maximum synchronization ratio.

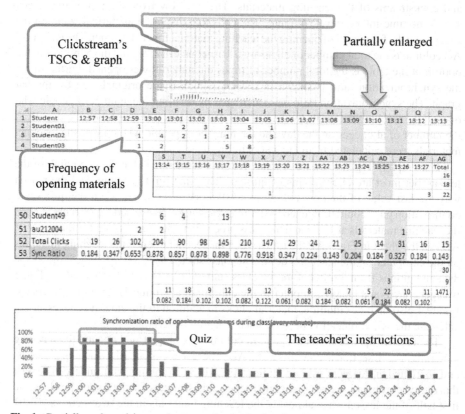

Fig. 1. Partially enlarged image for example graph showing TSCS and synchronization ratio for transition section (minute-by-minute; 2/11/2021; China Data Analysis; 49 attendees; Excel screenshot).

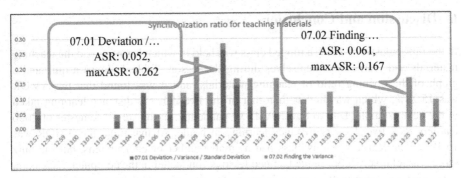

Fig. 2. Example of synchronization ratio of transition section of teaching materials.

5 Average Material Opening Ratio (AMOR)

In the lessons of this paper, learners can open multiple teaching materials at the same time. That includes teaching materials that are not used in the class on the day. Therefore, the AMOR was defined separately by targeting multiple teaching materials opened by the learner at once, including the teaching materials that are synchronized. Multiple teaching material logs used in the lesson on the day were extracted, and the ratio was calculated by dividing the number of learners who opened the teaching materials during the lesson by the number of attendees on the day. The following is an example of finding the average opening ratio of teaching materials from the lessons that have been conducted so far. The example of Table 1 analyzed lessons using the same section of materials as Fig. 1. The target classes are flipped classroom (Nov. 26, 2019), live delivery type distance lesson (Jul. 7, 2020), and blend type class (Jun. 1, 2021; Nov. 22, 2021). Preprocessing was performed based on the Moodle logs of these lessons, and the AMOR was calculated based on the synchronization ratio every minute to five minutes during the lesson time. All of the examples given here were taught by the teacher using the same teaching materials. Table 1 shows a list of lesson types and the times when the teaching materials were opened by the teacher. For AMOR and Quiz, the test results of the mean values by ANOVA were not significantly different ($p > 0.05$).

Table 1. Average teaching material opening ratio by class type.

Date/Year	Lesson type	Attendees		Clicks	Quiz	AMOR				
		Ma	Fe			1 min	2 min	3 min	4 min	5 min
11/26/2019	Flipped	19	30	1016	4.27	0.1528	0.1958	0.2347	0.2706	0.3129
7/7/2020	Live	17	29	824	5.66	0.1519	0.1698	0.2099	0.2480	0.2916
6/1/2021	Blended	11	36	1054	4.98	0.1630	0.1968	0.2568	0.2886	0.3579
11/2/2021	Blended	27	22	1100	4.43	0.1514	0.2364	0.3000	0.3522	0.4044

6   Discussion and Conclusion

In the fall semester of 2019, a flipped class was held in which learners read the teaching materials and learned by themselves during the lesson. Table 1 shows a quiz and a graph of the opening ratio of teaching materials, and it was found that the AMOR was 15.28%, which was higher than that of the live delivery distance learning taken in the next example. Due to the COVID-19 pandemic, in the spring semester of 2020 learners took live delivery distance learning at home. At that time, the AMOR was 15.19%. The learners in this class had a small number of clickstreams, but the quiz score was relatively high. In the blended lessons conducted in the computer classroom, the teacher can observe the learners' screens from the back of the classroom during the quiz. Additionally, during the exercise time in the second half of the class, the teacher patrolled the classroom and accepted questions. The AMOR was 16.30%, which was the highest of the three classes.

Since the time series in this paper is divided by time, even a difference of 10 s or less from the teacher's instruction may be included in the time zone before or after. Therefore, when looking at the synchronization ratio, it is necessary to take into consideration the time zones before and after the teacher gives the instructions. The synchronization ratio and AMOR in this paper are generated in real-time in the classroom on-site, so teachers can observe the learner's teaching materials browsing behavior and concentration during class. When finding that not many learners have the teaching material open, the teacher should require the learner to open the teaching material. In addition, the teacher can wait until most learners open the material.

It was clarified that the synchronization ratio and the teaching material opening ratio change depending on various factors such as the lesson format, contents, how the lesson proceeds, and how the teaching materials are presented. However, these can still be regarded as indicators of the degree of concentration in a lesson and may be used as one of the indicators of the evaluation of a lesson.

Acknowledgments. This work was supported by JSPS KAKENHI Grant Numbers 18K11588 and 21K12183.

References

Blikstein, P., Worsley, M.: Multimodal learning analytics and education data mining: using computational technologies to measure complex learning tasks. J. Learn. Anal. 3(2), 220–238 (2016)

Beck, N., Katz, J.N.: Modeling dynamics in time-series–cross-section political economy data. Annu. Rev. Polit. Sci. 14, 331–352 (2011)

Bodily, R., Kay, J., Aleven, V., Jivet, I., Davis, D., Xhakaj, F., Verbert, K.: Open learner models and learning analytics dashboards: a systematic review. In: Proceedings of the 8th International Conference on Learning Analytics and Knowledge, pp. 41–50 (2018)

Bradley, V.M.: Learning management system (LMS) use with online instruction. Int. J. Technol. Educ. 4(1), 68–92 (2021)

Fujii, K., Marian, P., Clark, D., Okamoto, Y., Rekimoto, J.: Sync class: Visualization system for in-class student synchronization. In: Proceedings of the 9th Augmented Human International Conference, pp. 1–8 (2018)

Shimada, A., Konomi, S.I., Ogata, H.: Real-time learning analytics system for improvement of on-site lectures. Interact Technol. Smart Educ. 15(4), 315–331 (2018)

Vieira, C., Parsons, P., Byrd, V.: Visual learning analytics of educational data: a system-atic literature review and research agenda. Comput. Educ. 122, 119–135 (2018)

Interpretable Knowledge Gain Prediction for Vocational Preparatory E-Learnings

Benjamin Paaßen[1]($\boxtimes$) (iD), Malwina Dywel[2], Melanie Fleckenstein[2], and Niels Pinkwart[1,3] (iD)

[1] German Research Center for Artificial Intelligence, Berlin, Germany
{benjamin.paassen,niels.pinkwart}@dfki.de
[2] Provadis Partner für Bildung und Beratung GmbH, Frankfurt, Germany
[3] Humboldt-University of Berlin, Berlin, Germany

Abstract. Vocational further education typically builds upon prior knowledge. For learners who lack this prior knowledge, preparatory e-learnings may help. Therefore, we wish to identify students who would profit from such an e-learning. We consider the example of a math e-learning for the Bachelor Professional of Chemical Production and Management (CCI). To estimate whether the e-learning would help, we employ a predictive model. Developing such a model in a real-world scenario confronted us with a range of challenges, such as small sample sizes, overfitting, or implausible model parameters. We describe how we addressed these challenges such that other practitioners can learn from our case study of employing data mining in vocational training.

Keywords: Multi-dimensional item response theory · Performance modeling · Knowledge gain · Vocational education · Further education

1 AIEd Implementation

Vocational further education aims to teach certain, job-related skills to a wide variety of learners. For example, the Provadis trade school offers a two-year course for the Bachelor Professional of Chemical Production and Management (BP-CPM; German: "Industriemeister Chemie", as defined by the chamber of commerce and industry, IHK) to acquire the skills necessary to supervise workers and apprentices in chemical plants[1]. A particular challenge for such further education courses is that incoming students have very different levels of prior knowledge, depending on their prior school education, their amount of job experience, and further individual factors. For example, some may enter the program right after high school education and apprenticeship, while others may have left school after lower secondary education, followed by an apprenticeship and multiple years of work experience before entering the program. Accordingly, students may lack or have forgotten pre-requisite knowledge, which severely impacts their chances for a successful qualification [2,3,7].

---

[1] https://www.provadis.de/weiterbildung/fuer-berufstaetige/chemische-produktion/industriemeister/-in-chemie-ihk/.

© Springer Nature Switzerland AG 2022
M. M. Rodrigo et al. (Eds.): AIED 2022, LNCS 13356, pp. 132–137, 2022.
https://doi.org/10.1007/978-3-031-11647-6_23

Our ultimate educational goal is to provide personalized support in order to maximize learners' chances at succeeding in the program, irrespective of their individual starting point. In this paper, we focus on one specific strategy, namely recommending a preparatory mathematics e-learning to prospective students in the BP-CPM course who lack some pre-requisite math knowledge and would profit from the e-learning. Our recommendation scheme is fully automatic and has the following steps.

First, before entering the course, students perform a (voluntary) pre-test with twenty-one math questions regarding six skills, namely basic algebra, fractions, equation solving with a single variable, text tasks with two variables, powers, and linear functions[2]. Second, we diagnose gaps in math knowledge via classic test theory, that is, counting the rate of correct responses for each skill. Third, we predict how much the rate could improve if the student would visit the preparatory e-learning. Finally, we recommend the e-learning if the model predicts a sufficiently high gain (at least 5%, averaged over all skills) and we communicate the prediction to the student.

1.1 Predictive Model

pre-test responses   prior knowledge   post-test knowledge   post-test responses

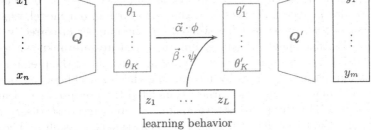

learning behavior

Fig. 1. The proposed prediction pipeline. For prediction, we only use pre-test responses (black). For the training data, we also record learning behavior and post-test responses (blue). The predictive model itself is shown in orange. (Color figure online)

To predict knowledge gain, given prior knowledge as measured by a pre-test, we use a predictive model. Developing such a model posed a challenge, particularly due to the small sample size in our educational setting. Given that the BP-CPM course is highly specialized, only very few students sign up per year (often less than ten). Accordingly, it is impossible to accumulate enough data from this population to train a very data-hungry model. On the other hand, a very simple,

---

[2] https://projekte.provadis.de/showroom/provadis/Mathematik_Orientierungstest/online/#p=32.

linear model is likely insufficient because we expect nonlinearities. In particular, we expect a bell-shaped relation between prior knowledge and knowledge gain, because very low prior knowledge means that a preparatory e-learning is insufficient (the skills would need to be learned from scratch) and high prior knowledge means that nothing is left to be learned.

Accordingly, we need a nonlinear model which requires as little training data as possible. To achieve such a model, we applied three assumptions: First, we assume that the relation between prior knowledge and knowledge gain can be modeled well by a linear combination of (few) bell curves. Second, we assume that knowledge gain is always non-negative, that is, there is no forgetting during the e-learning. This makes sense because the e-learning is fairly short (roughly five hours), such that forgetting is unlikely. Third, we assume that there is no interaction between skills, that is, prior knowledge in one skills does not help with acquiring another skill. This last assumption is most likely false but crucial to reduce the number of free parameters and, thus, the need for data.

In more detail, our model has the following structure. Let θ_k be the prior knowledge of a student for skill k and θ'_k be the knowledge *after* visiting the e-learning. Further, let $\phi_1, \ldots, \phi_L$ be bell-curves, centered at different knowledge values. Then, we assume that the following, non-linear relationship holds:

$$\theta'_k = \theta_k + \sum_{l=1}^{L} \alpha_{k,l} \cdot \phi_l(\theta_k) + \text{auxiliary feature influences} \qquad (1)$$

where $\alpha_{k,l}$ is a model parameter that weighs the influence of the lth bell-curve for skill k. We also include influences of auxiliary features in our model, especially the number of tasks a student has worked on, the number of correctly solved tasks, and the time spent on skill k, but these features are unknown for new students and, hence, we treat these influences as constant in the prediction.

We fit our parameters $\alpha_{k,l}$ by maximizing the log-likelihood on observed data. In particular, a sample of learners completed the pre-test, the e-learning, and a post-test. For the post-test, we assume an item response theory model [1,4,5], where Eq. 1 describes the ability of each student for skill k. The item-to-skill assignments for pre- and post-test Q and Q' where manually designed by experts. Using this model, we performed a maximum likelihood estimation of the difficulty parameters for each post-test question and the parameters $\alpha_{k,l}$. As such, our approach bears some resemblance to performance factors analysis, which also replaces student-specific ability parameters with an expression of prior knowledge plus learning behavior [6]. Our model is visualized in Fig. 1.

1.2 Experiments and Results

For model training and evaluation, we recorded the data of $N = 30$ learners in the final year of their Chemical Production Technician or Chemical Laboratory Assistant apprenticeship (ages 16–19). This represents the population of learners who may later become BP-CPM. While 30 is a small sample size, it reflects the overall small population: A full cohort at the Provadis trade school consists of roughly 120–140 learners. We recruited all learners who were currently preparing for their final exam and were not sick or otherwise excused. Our sample included students both with prior high school education as well as students with lower secondary education. The study was performed as part of in-classroom teaching and lasted for four to five hours, including the pre-test, the e-learning, and the post-test (time varied depending on each learner's individual speed). This time span might appear short for a full e-learning, but bear in mind that the e-learning is meant to refresh knowledge, not necessarily teach from scratch.

To gauge the accuracy of our proposed model (refer to Fig. 1), we compared it to a direct logistic regression which predicts post-test answers from pre-test answers (logreg), a deep learning model which learns Q and Q' (deep), and a variation of our proposed model which uses a different encoder for pre-test abilities which is based on ranking the difficulties of items (rank). As hyper-parameters, we used $L = 8$ bell-curves, regularization strength $C = 1$ for the difficulty parameters in the post-test, and regularization strength $C = 10^{-3}$ for the parameters α. For deep learning, we used 10,000 epochs of Adam optimization with a learning rate of 10^{-3}.

Table 1. Mean accuracy ($\pm$ std.) in leave-one-out crossvalidation over $N = 30$ students.

	Logreg	Deep	Rank	Proposed
Train	**0.99 $\pm$ 0.00**	0.82 $\pm$ 0.01	0.83 $\pm$ 0.00	0.83 $\pm$ 0.00
Test	0.81 $\pm$ 0.14	0.80 $\pm$ 0.18	0.81 $\pm$ 0.17	**0.83 $\pm$ 0.16**

The results of our experiments are shown in Table 1. As we can see, our proposed pipeline performs best on the test set, although logreg is far superior on the training set. This highlights the very real risk of overfitting for a small data set. The deep model does worst, indicating that learning the encoder does not offer a big advantage for this data.

Finally, we re-trained our proposed model on the data of all $N = 30$ students to obtain a model for practical application. Figure 2 shows the learned curves for four example skills. Note that the curves differ for different skills, which may indicate that the e-learning is—on average—more effective in teaching some skills (such as basic algebra) compared to other skills (such as functions) in our sample. Nonetheless, given the small sample size and the interaction with other factors, we perform such interpretations with care.

2 Reflection of the Challenges and Opportunities

Developing artificial intelligence applications for vocational education is a challenging endeavor: The subjects are, typically, highly practical and thus difficult to translate to a virtual setting. Skillsets are, typically, highly specialized which

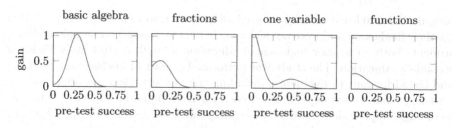

Fig. 2. The predicted gain $\sum_{l=1}^{L} \alpha_{k,l} \cdot \phi(\theta_k)$ (y-axis) versus pre-test success rate θ_k (x-axis) for the skills 'basic algebra', 'fractions', 'equation solving with a single variable', and 'functions' as learned by our model.

means that cohorts are small and data is scarce. Finally, vocational education topics, typically, evolve quickly such that any AI teaching tool for a specific topic bears the risk of being out-of-date soon. Such practical issues might explain why artificial intelligence has is still much less common in vocational education compared to university or school education [8].

In this paper, we considered math knowledge, which is relevant across a broad range of vocational topics and less subject to change. Nonetheless, we need to develop a model which works specifically for the population of chemistry (lab) technicians who might be interesting in becoming BP-CPM—which is a small population. Thus, we only had a small sample size. We tried to address this challenge by a combination of expert knowledge and model design. In particular, experts manually assigned each pre- and post-test question as well as each part of the e-learning to one of six skills which enabled us to make dedicated predictions for each skill. Regarding model design, we assumed that the relationship between prior knowledge and knowledge gain is bell-shaped, that knowledge gain is never negative, and that knowledge gain only depends on factors related on the same skill, not other skills.

Importantly, while our model is nonlinear, it can still be visualized and interpreted by human experts in the form of knowledge gain curves (refer to Fig. 2), which enables our domain experts to judge whether the learned model is plausible or whether further changes need to be applied. In fact, we used such visualizations repeatedly during model development to arrive at our current version. Prior versions of our model did not use the bell-curve assumption or the non-negativity assumption and yielded visibly implausible models.

Our work provides some insight into the challenges and opportunities of applying educational data mining and artificial intelligence for vocational education settings. In particular, we conclude that it is necessary to record data and expertise that is applicable to the specific educational context and population at hand, for example, chemistry technicians who may be interested in becoming BP-CPM. In many cases, this means small sample sizes and highly specialized knowledge, which means that very data-efficient models with strong inductive bias have to be applied. We hope that our own case provides an example how to develop such a model for a practically relevant task—namely recommending a math preparatory e-learning.

3   Description of Future Steps

In future work, we intend to apply the learned model in practice for recommending the preparatory math e-learning to potential learners in the BP-CPM course. We will also monitor learners' feedback. We will also generalize the approach to other e-learnings, concerning topics such as chemistry, pharmacy, or process engineering, for courses such as the Bachelor Professional of Pharmaceutics (CCI).

Acknowledgements. Funding by the German Ministry for Education and Research (BMBF) under grant number 21INVI1403 (project KIPerWeb) is gratefully acknowledged.

References

1. Baker, F.: The basics of item response theory. In: ERIC Clearinghouse on Assessment and Evaluation, College Park, MD, USA (2001). https://eric.ed.gov/?id=ED458219
2. Fulano, C., Magalhães, P., Núñez, J.C., Marcuzzo, S., Rosário, P.: As the twig is bent, so is the tree inclined: lack of prior knowledge as a driver of academic procrastination. Int. J. Sch. Educ. Psychol. 9(sup1), S21–S33 (2021). https://doi.org/10.1080/21683603.2020.1719945
3. Hailikari, T., Katajavuori, N., Lindblom-Ylanne, S.: The relevance of prior knowledge in learning and instructional design. Am. J. Pharm. Educ. 72(5) (2008). https://doi.org/10.5688/aj7205113
4. Hambleton, R., Swaminathan, H.: Item Response Theory: Principles and Applications. Springer, New York (1985). https://doi.org/10.1007/978-94-017-1988-9
5. Hambleton, R.K., Jones, R.W.: Comparison of classical test theory and item response theory and their applications to test development. Educ. Meas. Issues Pract. 12(3), 38–47 (1993). https://doi.org/10.1111/j.1745-3992.1993.tb00543.x
6. Pavlik, P.I., Cen, H., Koedinger, K.R.: Performance factors analysis - a new alternative to knowledge tracing. In: Proceedings of the International Conference on Artificial Intelligence in Education (AIED), pp. 531–538 (2009). https://eric.ed.gov/?id=ED506305
7. Schaap, H., Baartman, L., De Bruijn, E.: Students' learning processes during school-based learning and workplace learning in vocational education: a review. Vocat. Learn. 5(2), 99–117 (2012). https://doi.org/10.1007/s12186-011-9069-2
8. Seifried, J., Ertl, H.: Forschungsrichtungen zur künstlichen intelligenz in der beruflichen bildung. In: Seufert, S., Guggemos, J., Ifenthaler, D., Ertl, H., Seifried, J. (eds.) Künstliche Intelligenz in der beruflichen Bildung: Zukunft der Arbeit und Bildung mit intelligenten Maschinen?!, pp. 341–347. Franz Steiner Verlag, Stuttgart (2021). (in German)

Adaptive Cross-Platform Learning for Teachers in Adult and Continuing Education

Thorsten Krause[1]([✉]), Henning Gösling[1], Sabine Digel[2], Carmen Biel[3], Sabine Kolvenbach[4], and Oliver Thomas[1]

[1] German Research Center for Artificial Intelligence, Parkstr. 40, 49080 Osnabrück, Germany
{thorsten.krause,henning.goesling,oliver.thomas}@dfki.de
[2] Universität Tübingen, Münzgasse 11, 72070 Tübingen, Germany
sabine.digel@uni-tuebingen.de
[3] German Institute for Adult Education, Heinemannstr. 12-14, 53175 Bonn, Germany
biel@die-bonn.de
[4] Schloss Birlinghoven, Fraunhofer FIT, 53757 Sankt Augustin, Germany
sabine.kolvenbach@fit.fraunhofer.de

Abstract. Available online trainings and learning contents for teachers in adult and continuing education (ACE) are scattered across many learning platforms. We presume great synergies by enabling teachers to combine learning content of multiple ACE platforms in their individual learning paths and receive verifiable credentials as proof of competency afterwards. In our three-year consortium research project KUPPEL, we investigate how to enable personalized and adaptive cross-platform learning for teachers in ACE. KUPPEL will connect existing ACE platforms using a multi-agent system in a cloud. Each learner will be represented by a learner agent that will continuously give personalized recommendations for content and learning peers. In this paper, we describe the main components of the KUPPEL cloud, i.e., the multi-agent system, the recommender system and self-sovereign identity and verifiable credentials with blockchain technology. With our work, we seek to ease access to online learning resources, improve learning outcomes and make online learning more attractive to learners.

Keywords: Cross-platform learning · Education · Adult and continuing education (ACE) · Multi-agent system · Recommender system · Self-sovereign identity (SSI) · Verifiable credential (VC) · Distributed ledger

1 Introduction

Online learning platforms for teachers in adult and continuing education (ACE) provide learning opportunities to teachers looking to pick up new skills or refine their teaching competencies. These teachers have little experience with online-based learning and using learning management systems in a pedagogically sensible way. The group of ACE teachers consists of approximately 530,000 people in Germany and shows a high willingness for own further training [1]. However, available trainings and learning contents are scattered across many online learning platforms. To the best of our knowledge, no

M. M. Rodrigo et al. (Eds.): AIED 2022, LNCS 13356, pp. 138–143, 2022.
https://doi.org/10.1007/978-3-031-11647-6_24

technical solution exists to enable learners to execute cross-platform trainings composed of learning content from various platforms. Our three-year consortium research project KUPPEL investigates how to execute personalized and adaptive trainings for teachers across different ACE platforms. Specifically, we pursue basic scientific questions regarding the demand-oriented and learning-promoting design of (1) cross-platform selection and sequencing of learning content, (2) cross-platform recommendations of learning content and peers, and (3) cross-platform tamper-proof certifications of competencies gained in trainings. We seek to answer these questions by developing a competency-based framework curriculum (:DTrain) for our target group, which we operationalize according to DQR[1] level 5 upwards and which provides the basis for content development and adaptation. This curriculum will consist of learning content from two existing ACE platforms. We want to connect these two ACE platforms through a multi-agent-based hybrid cloud (KUPPEL cloud) which will contain one software agent per learner. This so-called learner agent will use hybrid recommender systems to suggest suitable content and peers for collaborative learning activities based on a learner's behavior, rating, interests, or professional background. The platform will also issue blockchain-based verifiable credentials based on the respective learner's progress and competencies.

2 The Multi Agent-Based KUPPEL Cloud for Cross-Platform Learning, Recommendations, and Certification

2.1 Requirements for Adult and Continuing Education

During the application phase of our project, we identified three main requirements for improving the online-based ACE of trainers: (1) cross-platform execution of (collaborative) trainings, (2) adaptive and personalized recommendations of learning content and peers from different platforms, and (3) unforgeable certificates issued for successful completion of cross-platform trainings.

Cross-Platform Selection and Sequencing of Learning Content. The available learning content for trainers in ACE is scattered all over existing learning platforms. Moreover, these teachers have little experience with online-based learning [2]. Connecting available learning content to cross-platform sequences might increase its accessibility and reduce switching costs.

Cross-Platform Recommendations of Learning Content and Peers. ACE teachers are very heterogeneous in terms of their expertise and competencies and often work under precarious circumstances. Their own professional development often competes with employment. For that reason, adaptive and personalized recommendations should take the learner's activities, feedback, interests, workload, and professional background into account. For certain learning content, peers should receive recommendations for learning partners to benefit from collaborative learning activities.

---

[1] https://www.dqr.de/dqr/en/the-dqr/dqr-levels/dqr-levels_node.html.

Certifications for Cross-Platform Trainings. With the increasing digitalization in all areas of life, lifelong learning becomes essential for professional careers. The standardized documentation and registration of individual learning records from cross-platform trainings in long-term tamper-proof distributed ledgers will enhance transparency and trust in certification processes.

2.2 Trial ACE Platforms

The competency-based framework curriculum :DTrain for our target group will consist of learning content from the two ACE platforms vhs.cloud and EULE. With vhs.cloud, the German Adult Education Association (DVV) operates its own communication and learning platform. Among other features, vhs.cloud provides learning units and trainings for continuing education courses as well as self-directed learning. EULE is a cross-institutional open educational resource (OER), located in the DIE's portal wb-web. It reinforces the professional development of teachers in further education. EULE holds self-learning offers that serve the development of teaching competency. Both platforms are established in the field of ACE and already provide learning content that we can adapt according to the :DTrain modules. Moreover, both contain typical elements of learning management systems yet differ significantly regarding their underlying concepts and structures for building and displaying learning sequences. Therefore, they act as ideal prototypes for testing the interface-based innovations within our project to ensure broad transferability to other platform types used in ACE. The medium-term goal is to expand the hybrid offering to include additional partner platforms in line with the content of the framework curriculum.

2.3 Components of the KUPPEL Cloud

The core components of the KUPPEL cloud will be a multi-agent system [3], a hybrid recommender system, an SSI infrastructure, and verifiable credentials with blockchain technology. In this section, we describe their main functions.

Multi-agent System. The multi-agent system in the KUPPEL cloud supports cross-platform execution of ACE trainings. Each learner is represented by a learner agent. The learner agent guides the learner during training across the associated platforms. Additionally, the multi-agent system includes a pedagogical agent, a certificate agent, and an ID agent. The pedagogical agent supports the modelling process, configuration, and release of learning sequences. The certificate agent is connected to a blockchain network and issues certificates based on a learner's learning paths and assessments. The ID agent assigns unique digital identities to agents and learners. The learner agent activates whenever a learner operates on an associated ACE platform. A hybrid recommender system suggests the learner's next task based on the learner's activities, ratings, interests, or professional background. The agent sends the recommendation to the platform that shows it to the learner. If the learner accepts the recommendation and the corresponding task is on the current platform, the learner agent notifies the platform to show the selected task. If the task is not on the current platform, the learner agent redirects the learner to the corresponding platform. Then, the learner agent notifies this platform to show the selected task. Figure 1 describes the interaction between learners and the corresponding learner agents in the KUPPEL cloud. Learner A stays on the current ACE platform and Learner B switches to a task on another ACE platform. Alternatively, as illustrated for learner C, the agent could transmit learning content between platforms instead of redirecting the learner. The colored arrows mark the sequences of events triggered by the respective learner's actions.

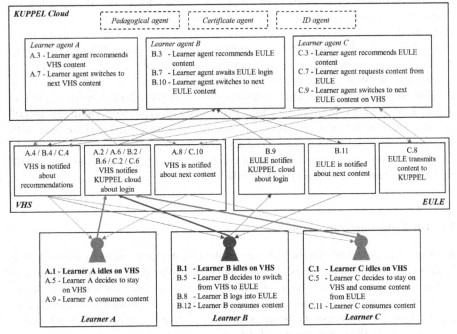

Fig. 1. Interactions between learners, ACE platforms, and the KUPPEL cloud

The learner can request a learning partnership while performing tasks. In that case, multi-agent coordination is necessary to assign the learning partner requests to another learner. Those learners picked in the multi-agent coordination mechanism as potential learning partners can select the learning partner request out of the list of recommended tasks. Besides consuming (collaborative) learning content, learners can perform assessments to prove their competencies.

Recommender System. We employ several AI-based recommender systems to adapt and personalize the learning experience. First, learners will receive recommendations for learning content. The target system will not only recommend closed sequences from different platforms, but also rearrange their components into new sequences. Second, the system will suggest peers for collaborative learning activities. Recommendations will be based on users' behavior, ratings for completed assignments, competency level, interests, and professional background. Accordingly, we will employ two hybrid recommender systems. These will include collaborative, content-based and knowledge-based components [4]. The collaborative component is described in [5]. We will further enable the learner or a responsible pedagogic person to select which data (user-behavior, rating, interests, or professional background) affects recommendations. The recommendations are not supposed to substitute the learner's decision process. Our goal is to support the learner make better decisions, enhancing their learning activity in what may be called a "Human-AI"-partnership [6].

SSI and Verifiable Credentials with Blockchain Technology. The unforgeable documentation of certificates using a blockchain has been described and tested [7]. In Germany, the first test network for educational institutions has been set up with digicerts.de. This solution already ensures proof of authenticity and integrity of documents. Still, it lacks cross-platform identity management and issuance and validation of data based on international W3C standards. SSI with blockchain technology and concepts using the W3C standards Decentralized Identifiers (DIDs) and Verifiable Credentials (VCs) creates new opportunities to use learning platforms with cross-platform digital identities, to control identity data, to obtain and manage credentials, and to share them with third parties in a generally accepted standard that cryptographic methods can reliably verify. Moreover, in our approach special emphasis is also placed on compliance with the General Data Protection Regulation (GDPR).

2.4   Challenges and Opportunities

We develop the technical solution following an iterative design science research approach [8] that contains multiple phases of requirement elicitation, design, demonstration, and evaluation. During the requirement elicitation phase, we define user stories and derive meta and functional requirements in weekly focus groups with pedagogical and technical experts. During this process, we identified several challenges. The protection and control of personal data, the secure storage of data, and data integrity are major challenges to gaining users' trust in the digitization of education processes. Other challenges relate to acquiring data for the machine learning-based methods such as collaborative filtering, designing enough learning content for the competency-based framework curriculum :DTrain, and playing sequences of learning content from various platforms on a single ACE platform to avoid redirecting the user between each step. We will consider these challenges through the upcoming design phase.

However, several opportunities come with the KUPPEL cloud. Learning will become more effective by linking existing ACE platforms and personalizing cross-platform recommendations for adaptive content. Users will find useful platforms they did previously not know about, and providers will attract new users who joined their platform through the KUPPEL cloud. Collaborative learning experiences will potentially increase learning success. Finally, the KUPPEL cloud offers self-sovereign identity management and decentralized registration of tamper-proof certificates based on blockchain technology.

3   Conclusion and Outlook

In this contribution, we propose a technical solution, the KUPPEL cloud, for linking the two existing German ACE platforms vhs.cloud and EULE using a multi-agent system. Each learner will be represented by a learner agent. The learner agent will guide the learner through learning content of associated ACE platforms. A hybrid recommender system will support the learner by constructing personalized and adaptive cross-platform learning paths. We will use blockchain technology to create tamper-proof records of the learner's learning path and competencies.

We are currently specifying the meta requirements and functions of the KUPPEL cloud, and will soon start its implementation. We have examined various blockchain infrastructures from third-party providers as well as national and international initiatives and selected Hyperledger Indy with the Hyperledger Aries Agent Framework for the implementation of the identity and certificate agent. We intend to implement our multi-agent system in Java using the Java Agent Development (JADE) framework and develop the agents' routines as Finite-State-Machines. The Contract-Net-Protocol will be used to identify suitable learning partners. We intend to implement the recommender system using the TensorFlow-framework and combine it with content- and knowledge-based approaches. Further, we will investigate how recommendations' design affects user perception. By connecting learning platforms, we will offer a unified catalogue of online learning content. This way, we seek to remove barriers and ease access for learners. We expect the recommender system to improve learning outcomes and our corresponding research will unveil design principles that can enhance existing recommender systems in the domain. The Verifiable Credentials will enable learners to verify their achievements and thus make online learning more attractive.

Acknowledgments. This contribution originates from the research project KUPPEL (KI- unterstützte plattformübergreifende Professionalisierung erwachsenen-pädagogischer Lehrkräfte), funded by the German Federal Ministry of Education and Research (BMBF), ref. no. 21INVI0802, 21INVI0803, 21INVI0805.

References

1. Koscheck, S., et al.: Das Personal in der Weiterbildung. wbv Media (2016)
2. Bildungsberichterstattung, A.G.: Bildung in Deutschland 2020. wbv Media (2020)
3. Dorri, A., Kanhere, S.S., Jurdak, R.: Multi-agent systems: a survey. IEEE Access **6**, 28573–28593 (2018). https://doi.org/10.1109/ACCESS.2018.2831228
4. Aggarwal, C.C.: Recommender Systems. Springer International Publishing, Cham (2016). https://doi.org/10.1007/978-3-319-29659-3
5. Krause, T., Stattkus, D., Deriyeva, A., Beinke J.H., Thomas, O.: Beyond the rating matrix: debiasing implicit feedback loops in collaborative filtering. In: Proceedings of the 17th International Conference on Wirtschaftsinformatik (2022)
6. Borgwardt, A.: Bit für bit in die Zukunft : Künstliche Intelligenz in Wissenschaft und Forschung. Friedrich-Ebert-Stiftung, Abt. Studienförderung, Berlin (2020)
7. Gräther, W., Kolvenbach, S., Ruland, R., Schütte, J., Torres, C., Wendland, F.: Blockchain for Education: Lifelong Learning Passport (2018)
8. Peffers, K., Tuunanen, T., Rothenberger, M.A., Chatterjee, S.: A design science research methodology for information systems research. J. Manag. Inf. Syst. **24**, 45–77 (2007)

Complex Learning Environments: Tensions in Student Perspectives that Indicate Competing Values

Minghao Cai[1(✉)], Carrie Demmans Epp[1,2], and Tahereh Firoozi[2]

[1] EdTeKLA Research Group, University of Alberta, Edmonton, Canada
{minghaocai,cdemmansepp}@ualberta.ca
[2] Center for Research in Applied Measurement and Evaluation, University of Alberta, Edmonton, Canada
tahereh.firoozi@ualberta.ca

Abstract. Advances in software have enabled an increase in the complexity of online learning environments (OLE). How students perceive learning when these environments are more complex is not yet understood. We conducted a focus group with students who were taking a pilot-training course to understand their experiences with a complex OLE. This online course integrated Moodle, a virtual world, and a simulator among other technologies to facilitate student learning. Student perceptions and experiences highlight tensions that indicate nuanced concerns about the learning environment. Many of these concerns are similar to those previously identified. Student concerns over features like leader boards suggest a need for updating online-learning theories to explicitly address issues around data privacy and information sharing.

Keywords: Online learning · Virtual worlds · Pilot training · Simulation

1 Introduction

A considerable body of literature has reported on explorations of student perceptions of online learning environments (OLE), where an OLE is any technology that is used to facilitate learning via the Internet. OLEs include learning management systems, learning content management systems, massive open online courses [1, 2], video-conferencing tools (e.g., zoom, Google Meet), virtual worlds, and social media tools. In most cases, this literature explores the use of a single online learning environment that delivers multimedia content and has features that are meant to support student interaction [3]. Very few of the studied OLEs have provided complex environments that integrate multiple technologies to go beyond more traditional OLEs by incorporating simulations and virtual worlds.

The integration of the varied capabilities of these different types of OLE is both a technical and pedagogical challenge. Guidance for how to use these tools to support learning tends to focus on a single system, and we have recent evidence that small differences in OLE feature design and use are associated with reasonably large differences in student behaviours and experiences [1, 3, 4].

© Springer Nature Switzerland AG 2022
M. M. Rodrigo et al. (Eds.): AIED 2022, LNCS 13356, pp. 144–149, 2022.
https://doi.org/10.1007/978-3-031-11647-6_25

Given this recent evidence, it is important to explore how students respond when different learning technologies are integrated to enable learning experiences that are meant to support a range of activities from knowledge acquisition to its application in simulated environments. We take a first step in this direction by exploring student responses to an OLE that integrates four separate learning technologies into a Moodle-based online learning environment so that pilot trainees (tertiary educational students) can learn what they need to know to pass their licensing exam. The learning environment was designed to simulate a real flight-school setting using a virtual world. We ask, **"How do students experience an online learning environment that integrates multiple technologies and approaches?"**.

2 Theoretical Framework

In this study, online learning refers to a type of teaching and learning environment which leads to effective learning by considering learners, content, assessment, and community. These key components of online learning should be structured and designed to ensure that appropriate interactions among student, teacher, and content will be encouraged so that each learning objective is supported [5].

Taking a constructivist view, learning objectives are constructed by diagnosing and considering the unique cognitive structures and the personalities that students bring to the learning context. Further, like in theories of general learning, effective online learning is achieved through defining pedagogy and content and then effectively communicating those with teachers and learners. In addition, high quantity and quality assessment of the content knowledge gives students an opportunity to reflect on their learning experience and construct their understanding through the use of online computer-marked assessments that can extend beyond quizzes to simulation exercises, virtual labs, and other automated assessments of active student learning [6].

As a critical component of online learning, the interaction within the community of inquiry framework [7] should be included. This interaction can be used to support and challenge learners to construct the intended knowledge and reach the planned learning objectives. A variety of web-based activities can support the types of communication and interaction that enable students to construct knowledge. Because OLEs vary in the extent to which they support specific interactions [3] and pedagogies [8], it is important to investigate how different OLEs support learning.

Consistent with the above socio-constructivist view, we adopted the Community of Inquiry (CoI) [7] framework for data analysis. CoI highlights social presence, cognitive presence, and teaching presence as three core elements for student success in online learning environments. While the presence of the three elements is essential, creating an effective OLE depends on the interface among these elements.

3 Methods

To explore how pilot trainees (students) respond to a complex online learning environment, we conducted a qualitative case study of their experiences.

3.1 Learning Context and System

The course students took aims to equip them with enough knowledge and comprehension to pass the Transport Canada governed Private Pilot - Aeroplane (PPAER) exam. As part of this course, students were exposed to four integrated systems within their OLE: VR City, LM Flight, Flight Forge, and ManeuvAIR.

VR City is a 3D virtual campus where students were able to create their own avatars - these were used to interact with VR City and other students. Students were encouraged to attend synchronous lectures hosted in private classrooms by licensed flight instructors (Fig. 1.a). Likewise, private study rooms were provided to enable small-group collaborative learning, and students could access a Moodle-based LMS (called LM Flight) through computers in the virtual world (Fig. 1.b).

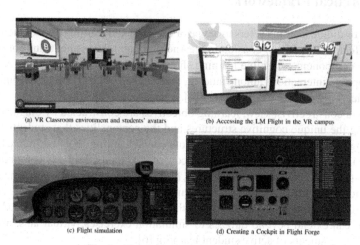

(a) VR Classroom environment and students' avatars   (b) Accessing the LM Flight in the VR campus

(c) Flight simulation   (d) Creating a Cockpit in Flight Forge

Fig. 1. The online learning environment

Within LM Flight, students had access to content to help prepare them for the PPAER exam. Content was delivered through interactive multimedia notes that included animations. Support for vocabulary was provided through a chatbot, and assessment was performed through quizzes. These quizzes were aligned with the licensing exam: they assess student knowledge and comprehension of key rules and concepts. Discussion forums, private messaging, and chatrooms were provided to support student interaction.

Within the OLE, students also had access to a tool (Flight Forge – Fig. 1.d) that would allow them to create virtual aircraft, which they could later fly in a simulator (ManeuvAIR – Fig. 1.c). In this case, students had the opportunity to modify and fly a small plane (i.e., Cessna Skyhawk 172 SP).

3.2 Data Collection

We identified key areas of interest in coordination with the lead instructor of the school where the above course is taught. Based on these discussions, a focus group protocol was designed and conducted following student use of the existing OLE. All work was reviewed and approved by our university research ethics office.

Six students (3 female and 3 male) of varying levels consented and participated in this focus group. The focus group lasted for approximately 1 h, was recorded, and was transcribed verbatim. The main questions centered on why students were trying to learn about flying; what they had previously tried to learn in this area; and their experiences with the OLE. This included discussion of the learning approaches that were relatively new to learners (e.g., the flight simulator and virtual world).

3.3 Data Analysis

We analysed the transcripts using ATLAS.ti. In the initial step, the coding system was developed by one team member based on a literature review. We coded and structured the content of the focus group sentence by sentence, relating the text to the key questions of our study. We then grouped the extracted quotations by the technology involved and student preferences. In the second step, the code system was refined based on the Community of Inquiry (CoI) framework [7]. Extracted data were then analyzed to reveal student perspectives as they relate to social presence, teaching presence, and cognitive presence. Two critical peer debriefers (authors 2 and 3) checked the coding separately to increase the reliability of analyses [9].

4 Findings

Students facilitated their learning using external platforms (e.g., zoom) because the OLE provided inadequate support for collaborative learning activities, for navigating course activities across platforms, or for completing those activities. When reflecting on the course materials and the learning process, students reported that the assessments helped them better remember key information.

VR City provided students with a virtual-reality campus environment that they found comfortable. The integration of this environment supported the formation of social connections and an enjoyable classroom atmosphere. However, they expressed concern over the unrestricted inclusion of too many entertainment-based features that are unaligned with educational goals. Students suggested limiting access to these features so they do not become a distraction.

When reflecting on their learning practices, students affirmed a need for integrated support materials so they do not have to leave the primary learning environment. They also suggested adding features that better support collaborative practices (e.g., screen sharing and shared simulation experiences) and self-directed learning (e.g., virtual study hall) to promote engagement and enable more authentic learning.

Students expressed concerns that sharing information through gamification and progress tracking features would lead to unproductive competition or pressure. This concern shows a tension with their requests for additional gamification features to support motivation.

5 Discussion and Conclusion

Interest in virtual-reality (VR) environments and virtual representations as tools to increase learning engagement has been expanding [10, 11]. In our study, the multi-dimensional online learning environment that was used appears to have helped students overcome geographical barriers and enhanced their online-learning experience by introducing new virtual presence and interaction techniques. However, the balance of entertainment elements and educational materials needs to be handled carefully and should be adjusted to enable a sense of community to be developed and maintained while keeping students focused on learning. This tension can be partly addressed through the timing of access to such features. VR and virtual world elements will help enhance student engagement and community development when students are new to a course. As the course progresses, maintaining a focus on purely social and entertainment activities may distract students from their learning.

The current use of online learning environments still tends towards instructivist approaches [3]. In contrast to this common approach, our study supports the use of learning designs that facilitate collaborative activities, which several learning theories argue are important, e.g., CoI and knowledge building [7, 12].

If we consider the theoretical frameworks that are often relied upon in online learning, it is important to note that they fail to explicitly account for issues of privacy and learner control over their data despite these theories encouraging the inclusion of social elements or sharing as part of learning. This is true of the CoI framework [7], communities of practice [13], connectivism [14], and the more recent integrated model [15]. This lack of coverage is in contrast to concerns expressed by participating students who were worried about the effects of sharing their information through gamification features, such as leader boards. They were specifically concerned about the pressure and competition that these types of features may place on students. This is in explicit opposition to prior studies of gamified learning that argue introducing competition can motivate students to work towards achieving assigned learning tasks [16]. Since social features are being integrated into online learning to motivate learners through mechanisms like leader boards and social comparison [17–19], this issue of privacy and comparison against others will need to be explicitly addressed by online learning theories.

Based on these findings, the virtual world has been redesigned and the integrations between systems are being adapted. As part of this, agents are being developed to provide guidance to learners while they are flying a plane in the simulator. Moving forward, we need to work to better understand the impact of student concerns on their learning performance and improve system design accordingly.

Acknowledgements. This work received financial support from Mitacs and the Natural Sciences and Engineering Research Council of Canada (NSERC), [RGPIN-2018-03834].

References

1. Baikadi, A., Demmans Epp, C., Schunn, C.D.: Participating by activity or by week in MOOCs. Inf. Learn. Sci. **119**, 572–585 (2018)

2. Kizilcec, R.F., Kambhampaty, A.: Identifying course characteristics associated with sociodemographic variation in enrollments across 159 online courses from 20 institutions. PLoS One **15**, e0239766 (2020). https://doi.org/10.1371/journal.pone.0239766
3. Demmans Epp, C., Phirangee, K., Hewitt, J., Perfetti, C.A.: Learning management system and course influences on student actions and learning experiences. Educ. Technol. Res. Dev. **68**(6), 3263–3297 (2020). https://doi.org/10.1007/s11423-020-09821-1
4. Baikadi, A., Schunn, C.D., Long, Y., Demmans Epp, C.: Redefining "What" in analyses of who does what in MOOCs. In: 9th International Conference on Educational Data Mining (EDM 2016). pp. 569–570. International Educational Data Mining Society (IEDMS), Raleigh, NC, USA (2016)
5. Anderson, T.: Toward a theory of online learning. Theory and Practice of Online Learning (2004)
6. Koohang, A., Riley, L., Smith, T., Schreurs, J.: E-Learning and constructivism: from theory to application. Interdisciplinary Journal of E-Learning and Learning Objects **5**, 91–109 (2009)
7. Garrison, D.R., Anderson, T., Archer, W.: Critical inquiry in a text-based environment: computer conferencing in higher education. Internet High. Educ. **2**, 87–105 (1999). https://doi.org/10.1016/S1096-7516(00)00016-6
8. Scardamalia, M., Bereiter, C.: Pedagogical biases in educational technologies. Educ. Technol. **48**, 3–11 (2008)
9. Creswell, J.W.: Research design: qualitative, quantitative, and mixed methods approaches **342** (2017)
10. Chang, C.-W., Lee, J.-H., Wang, C.-Y., Chen, G.-D.: Improving the authentic learning experience by integrating robots into the mixed-reality environment. Comput. Educ. **55**, 1572–1578 (2010). https://doi.org/10.1016/j.compedu.2010.06.023
11. Akcayir, G., Demmans Epp, C.: Designing, Deploying, and Evaluating Virtual and Augmented Reality in Education. IGI Global (2020)
12. Scardamalia, M., Bereiter, C., Bereiter, C.: A brief history of knowledge building. Canadian Journal of Learning and Technology/La Revue Canadienne de L'apprentissage et de la Technologie **36** (2010)
13. Wenger, E.: Communities of Practice: Learning, Meaning, and Identity. Cambridge University Press (1999)
14. Goldie, J.G.S.: Connectivism: a learning theory for the digital age. Int. J. Instr. Technol. Distance Learn. **38**(10), 1064–1069 (2004)
15. Picciano, A.G.: Theories and frameworks for online education: seeking an integrated model. OLJ **21**, (2017). https://doi.org/10.24059/olj.v21i3.1225
16. Cagiltay, N.E., Ozcelik, E., Ozcelik, N.S.: The effect of competition on learning in games. Comput. Educ. **87**, 35–41 (2015)
17. Brooks, C., Panesar, R., Greer, J.: Awareness and collaboration in the iHelp courses content management system. In: Nejdl, W., Tochtermann, K. (eds.) Innovative Approaches for Learning and Knowledge Sharing, pp. 34–44. Springer Berlin Heidelberg, Berlin, Heidelberg (2006). https://doi.org/10.1007/11876663_5
18. Davis, D., Jivet, I., Kizilcec, R.F., Chen, G., Hauff, C., Houben, G.-J.: Follow the successful crowd: raising MOOC completion rates through social comparison at scale. In: Proceedings of the Seventh International Learning Analytics & Knowledge Conference, pp. 454–463. Association for Computing Machinery, New York, NY, USA (2017)
19. Loboda, T.D., Guerra, J., Hosseini, R., Brusilovsky, P.: Mastery grids: an open source social educational progress visualization. In: Rensing, C., de Freitas, S., Ley, T., Muñoz-Merino, P.J. (eds.) EC-TEL 2014. LNCS, vol. 8719, pp. 235–248. Springer, Cham (2014). https://doi.org/10.1007/978-3-319-11200-8_18

Embodied Agents to Scaffold Data Science Education

Tanmay Sinha[1]([✉]) [iD] and Shivam Malhotra[2] [iD]

[1] Professorship for Learning Sciences and Higher Education, ETH Zürich, Zürich, Switzerland
`tanmay.sinha@gess.ethz.ch`
[2] Department of Electrical Engineering, Indian Institute of Technology Kanpur, Kanpur, India
`mshivam@iitk.ac.in`

Abstract. Arguing and working with data has become commonplace in several study domains. One way to immerse students in hands-on exploration with data is to provide them with problem-solving environments, for example jupyter notebooks, which can scaffold students' reasoning and bring them closer to disciplinary ways of thinking. Although the intrinsic affordances of jupyter notebooks (e.g., interaction with multiple data representations, automation of procedural task aspects) allow students to engage in rich learning experiences, students lack crucial social scaffolding that directly targets the process of learning. We are developing an AIED infrastructure *EASEx* for use in higher education contexts that brings in the affordances of embodied pedagogical agents to significantly advance educational practice by scaffolding students in a personalized manner as they work through problems using jupyter notebooks.

Keywords: Data science education · Failure · Pedagogical agents · Scaffolding

1 Introduction and Motivation

Research has shown that the best learning emerges in the context of supportive relationships that make it challenging, engaging and meaningful (e.g., [1, 2]). Learning from failure-driven problem-solving in data-rich environments (e.g., jupyter notebooks), which can be emotionally taxing [3] and negatively affect students' task motivation, reflects an important learning context where students often lack crucial social support. Embedded social scaffolding based on continuous monitoring of students' activity in problems with high failure-likelihood, can (i) alleviate cognitive and affective demands, and (ii) grease the wheels of the task interaction by developing rapport and empathizing with students' frustration and motivational insecurities (e.g., low self-esteem). Scaffolding failure-driven problem-solving, however, can take many forms, for example, that which is directed towards the task to help students attend to critical features (cognitive support) or engage them in time management and reflection on the progress made (metacognitive support), that which nudges students to regulate their emotions (affective support), that which nudges persistent behavior via directed praise or error encouragement (motivational support), and finally that which builds rapport with the student via

M. M. Rodrigo et al. (Eds.): AIED 2022, LNCS 13356, pp. 150–155, 2022.
https://doi.org/10.1007/978-3-031-11647-6_26

face management and coordination strategies like self-disclosure, referring to shared experience (relational support). Despite stemming from seemingly disparate lines of research (e.g., [4, 5]), invariably, instantiating these myriad forms of social support for improving learning outcomes is a complex endeavor.

Embodied pedagogical agents [6], which can be designed to provide dynamic scaffolds using both verbal and nonverbal behaviors (e.g., smile, eye gaze, gestures), hold high potential to improve the social context of learning through problem-solving. Seminal studies in human-computer interaction have found that humans apply social rules and heuristics (e.g., trust, friendliness, cooperation) from the domain of people to the domain of machines, provided that the technology (i) resembles human-like shape and form while avoiding the uncanny valley of appearance realism [7], and more crucially, (ii) behaves in a socially-competent manner to evoke natural/genuine responses from people via rich multimodal interaction [8]. 2D/3D computer characters in the form of embodied pedagogical agents, which are one instantiation of such a technology, have been shown to enhance STEM learning [9, 10] because of key capabilities such as using conversational signals (e.g., head nod), gaze and gesture as attentional guides, conveying/eliciting emotion, and adaptively responding to turn-taking and task actions.

Research on scaffolding learning through problem-solving (especially, within data-rich environments) and technology-focused work on embodied pedagogical agents, however, has developed predominantly independently, with the former typically limited to technology-poor traditional STEM settings (e.g., math). However, with the proliferation of data-driven decision-making, classroom learning today is increasingly moving beyond the understanding of mere mathematical formalisms (e.g., correlation) to equipping students with the ability to read, work with, analyze, and argue with data [11]. To provide accessible opportunities for students to engage and persist in ill-defined problems, follow paths that are ultimately not productive, and create new visual and numerical representations, we are implementing a content-agnostic AIED infrastructure called *EASEx* (Embodied Agents to Scaffold Education, x = any data-rich domain, e.g., physics, medicine, psychology, astronomy) for university-wide dissemination. See Fig. 1 for an overview. *EASEx* can offer in-situ social support via embodied pedagogical agents during data-driven problem-solving in jupyter notebooks. The implementation can be found at https://github.com/EASEx/Deployment.

2   Design Affordances of *EASEx*

2.1   Natural Language Understanding

The primary purpose of this module is answer matching. We perform agglomerative clustering using sentence transformers [12] on previously submitted student answers and automatically assign available topics to each cluster, wherein only a single topic corresponds to the correct answer. Low scoring clusters can be reported to an instructor for manual assignment to a topic. Upon submitting, students' answer is matched to the nearest cluster to check for correctness, and a corresponding scaffold can be triggered. This module also consists of a code parser responsible for analyzing students' programming code for any errors and reporting the usage of modules and functions in the code to the scheduler (cf. section 2.3).

2.2 Natural Language Rendering and Generation

Most embodied agents rely on manual gesture authoring, but that leaves little room for personalization and real-time modification of parameters. To circumvent such issues, we draw on automatic gesture authoring toolkits, for example, speech visemes for lip-syncing and SG Toolkit [13] for intuitive gesture authoring. We overlay the gestures generated by the toolkits over a custom mesh (body) and generate the videos using Blender's python API. Several predesigned character textures (male, female, gender-ambiguous) are available for usage. There is no standby animation, i.e., the agent is only displayed when there is some interaction required to ensure minimal intrusiveness to students' problem-solving activities. Speech synthesis forms an integral part of our rendering process, and inspired by the literature on teacher emotions [14], our architecture draws on frameworks for expressive styles (e.g., [15], where we convert emotion markup language (EML) to valid speech synthesis markup language and generate speech using compatible text-to-speech) to allow course designers to employ EML tags for producing naturalistic speech (e.g., with pauses, discrete/dimensional emotions). To further provide personalization to the dialog, filler tags (e.g., favorite subject) can be specified for replacement with actual values during the rendering.

2.3 Scheduler

The scheduler comprises the core of *EASEx* and is responsible for the backend server and routing requests to other services via ZeroMQ sockets. The scheduler has several loops running concurrently and is responsible for scheduling problem-solving sessions, metrics, and various kinds of scaffolds. It uses an SQLite database (long-term storage) and a BadgerDB cluster (temporary storage) to provide concurrency and responsive read/write capabilities. Traditional knowledge-tracing based learner modeling approaches [16] rely on students' longstanding history of past performance and are thus limited in their ability to provide scaffolds within isolated instances of novel data-driven programming tasks. Recent advances too (e.g., [17]) focus primarily on algorithmic programming, and cannot be readily modified to data science education.

Our implementation of *EASEx* therefore uses a rich set of metrics from several data streams – (i) responsiveness (determined using mouse, keyboard and cell execution activity), (ii) stagnancy (determined using the pattern of code execution and errors), (iii) code errors, functions and module usage (determined using the NLU). Each stream consists of a set of indicators (e.g., last keypress time for responsiveness stream) connected via Boolean expressions, which determine whether and what scaffold to trigger. The scheduler can analyze real-time student metrics with rules set by a course designer and trigger scaffolds that are allowed. Each scaffold is assigned a type (e.g., cognitive, metacognitive, affective, motivational, relational), and a course designer can (i) choose when to trigger a scaffold, and (ii) assign either a priority order or a fixed scaffold type to each critical feature (subgoal) involved in the problem-solving task. Our infrastructure also allows configuring custom rules to set a minimum/maximum of each scaffold type to be triggered within a problem-solving session. Based on these aforementioned inputs, we use a backtracking algorithm to determine all suitable scaffold sequences and an optimal sequence to be delivered during runtime using the data streams.

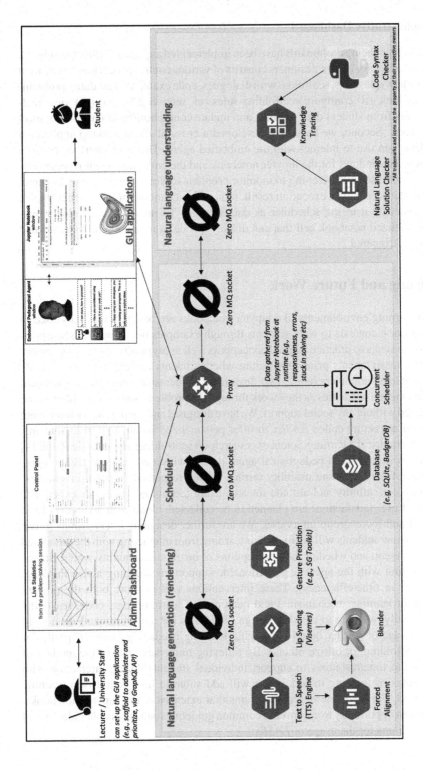

Fig. 1. Architecture of *EASEx* (Embodied Agents to Scaffold Education, x = any data-rich domain)

2.4 Administrative Dashboard

Two stakeholder-facing dashboards have been implemented as part of *EASEx* (see Fig. 1). First, the interface for course designers consists of various control panels that can be used to (i) create new questions, scaffolds (with dialogues, code examples, text data), problem-solving sessions, (ii) configure scaffolding rules (cf. section 2.3), and (iii) view real-time statistics from student data streams, and update configurations if they do not meet desired effects. Second, we have also developed a cross-platform desktop application that students can use to interact with the embodied agent. This user interface consists of two browser windows for the jupyter notebook and the embodied agent respectively. The application uses web scraping to monitor problem-solving activity, parse the input, output and status (e.g., last execution result, total cell executions) of the notebook cells, and communicate it to the scheduler as data streams. To provide a scaffold in real-time, a specialized notebook cell that can display a code example and/or an associated text/image is triggered.

3 Ongoing and Future Work

Data-rich learning environments like jupyter notebooks are being increasingly used in higher education contexts to walk students through examples of a concept, create post-lecture assignments to practice learned concepts as well as atypically, create preparatory sensemaking assignments prior to a lecture where students can explore and generate relevant problem parameters (e.g., [18]). Invariably, students are often left to their own devices and mental capacities as they work through the problem, and they are likely to be overwhelmed without any social support. We have designed and are iteratively improving an AIED infrastructure called *EASEx* to offer personalized social support to students, drawing on empirical Learning Sciences research on scaffolding students' learning [4] and advances in embodied pedagogical agent design within educational technologies [6]. As part of the upcoming usability testing, survey responses focused on students' background (e.g., affinity and attitude towards using technology) and user experience (e.g., ease of use, satisfaction) are planned to be collected, along with audio data from think-aloud and focus-group interviews. We are further designing interventions to test whether and how students will profit in their learning from interacting with the embodied pedagogical agent, and whether their perceptions of instructional quality would change after interacting with the agent (e.g., actionable support encouraging active thinking, improvement in time-efficiency). These interventions will measure both the process (e.g., relevant learning mechanisms) and outcomes of learning (e.g., code, reasoning quality). Finally, we plan to develop design guidelines for university staff on how to customize usage of *EASEx* based on specific teaching needs (e.g., turn off particular kinds of scaffolding, prioritize others). By relieving university staff of the anxiety of not having the time/resources to support individual students during interaction with a vast syllabus of learning materials, we will add value to how data-driven teaching is approached. For complete degree programs that extensively use jupyter notebooks, there is also an opportunity to co-develop common guidelines and accessible video-based training programs for the usage of *EASEx*.

References

1. Roschelle, J., Teasley, S.D.: The construction of shared knowledge in collaborative problem solving. In: Computer supported collaborative learning, pp. 69–97. Springer, Berlin, Heidelberg (1995)
2. Barron, B.: When smart groups fail. Journal of the Learning Sciences 12(3), 307–359 (2003)
3. Sinha, T.: Enriching problem-solving followed by instruction with explanatory accounts of emotions. Journal of the Learning Sciences 31(2), 151–198 (2022)
4. Quintana, C., et al.: A scaffolding design framework for software to support science inquiry. Journal of the Learning Sciences 13(3), 337–386 (2004)
5. Zhao, R., Sinha, T., Black, A.W., Cassell, J.: Socially-aware virtual agents: automatically assessing dyadic rapport from temporal patterns of behavior. In: International conference on intelligent virtual agents, pp. 218–233. Springer, Cham (2016)
6. Lugrin, B., Pelachaud, C., Traum, D. (eds.) The Handbook on Socially Interactive Agents: 20 years of Research on Embodied Conversational Agents, Intelligent Virtual Agents, and Social Robotics Volume 1: Methods, Behavior, Cognition, 1st. ed. ACM Books, vol. 37. Association for Computing Machinery, New York, NY, USA (2021)
7. von der Pütten, A.M., Krämer, N.C., Gratch, J., Kang, S.-H.: "It doesn't matter what you are!" Explaining social effects of agents and avatars. Comput. Hum. Behav. 26(6), 1641–1650 (2010)
8. Cassell, J., Tartaro, A.: Intersubjectivity in human–agent interaction. Interact. Stud. 8(3), 391–410 (2007)
9. Johnson, W.L., Lester, J.C.: Face-to-face interaction with pedagogical agents, twenty years later. Int. J. Artif. Intell. Educ. 26(1), 25–36 (2016)
10. Sinatra, A.M., Pollard, K.A., Files, B.T., Oiknine, A.H., Ericson, M., Khooshabeh, P.: Social fidelity in virtual agents: Impacts on presence and learning. Comput. Hum. Behav. 114, 106562 (2021)
11. Vahey, P., Finzer, W., Yarnall, L., Schank, P.: CIRCL primer: data science education. In CIRCL Primer Series (2017). http://circlcenter.org/data-science-education
12. Reimers, N., Gurevych, I.: Sentence-bert: Sentence embeddings using siamese bert-networks (2019). arXiv preprint arXiv:1908.10084
13. Yoon, Y., Park, K., Jang, M., Kim, J., Lee, G.: Sgtoolkit: An interactive gesture authoring toolkit for embodied conversational agents. In: The 34th Annual ACM Symposium on User Interface Software and Technology, pp. 826–840 (2021)
14. Frenzel, A.C., Daniels, L., Burić, I.: Teacher emotions in the classroom and their implications for students. Educational Psychologist 56(4), 250–264 (2021)
15. Charfuelan, M., Steiner, I.: Expressive speech synthesis in MARY TTS using audiobook data and emotionML. In: Interspeech, pp. 1564–1568 (2013)
16. Pelánek, R.: Bayesian knowledge tracing, logistic models, and beyond: an overview of learner modeling techniques. User Model. User-Adap. Inter. 27(3–5), 313–350 (2017). https://doi.org/10.1007/s11257-017-9193-2
17. Jiang, B., Wu, S., Yin, C., Zhang, H.: Knowledge tracing within single programming practice using problem-solving process data. IEEE Trans. Learn. Technol. 13(4), 822–832 (2020)
18. Sinha, T., Kapur, M., West, R., Catasta, M., Hauswirth, M., Trninic, D.: Differential benefits of explicit failure-driven and success-driven scaffolding in problem-solving prior to instruction. J. Educ. Psychol. 113(3), 530–555 (2021)

Democratizing Emotion Research in Learning Sciences

Tanmay Sinha[1]([⊠]) [iD] and Sunidhi Dhandhania[2] [iD]

[1] Professorship for Learning Sciences and Higher Education, ETH Zürich, Zürich, Switzerland
tanmay.sinha@gess.ethz.ch
[2] Department of Computer Science, Indian Institute of Technology Kanpur, Kanpur, India
sunidhi@iitk.ac.in

Abstract. We describe the design of an AIED tool *EmoInfer* to accelerate process-based research on emotions in Learning Sciences and help educational stakeholders understand the interplay of cognition and affect in ecologically-valid learning situations. Through an iterative implementation pipeline, we have developed a user interface to streamline automatic annotation and analysis of videos with facial expressions of emotion. *EmoInfer* can be applied to quantify and visualize the frequency and the temporal dynamics of naturally occurring or induced emotions (both on-the-fly and posthoc after data collection). By offering an accessible toolkit with "low floor, high ceiling and wide walls", we aim to initiate democratizing emotion research in Learning Sciences.

Keywords: Emotions · Graphical user interface · Learning analytics

1 Introduction and Motivation

Emotions lie at the heart of human learning. Although there has been a steady surge in research on emotions across psychological contexts [1], advances in automated computer-based analyses for emotion inference have the potential to further accelerate this trend. Learning Sciences researchers, who typically rely on manually annotated process-focused descriptions of affective processes (along with other tools such as self-reports), however, often may not possess the methodological/programming toolkit to incorporate these quantitative advances in their studies. This missed opportunity may constrain the scope of research questions that can be asked of the data, and weaken the validity of the presented evidence. To bridge this gap, we are iteratively building an open-source tool *EmoInfer* to simplify the annotation and analysis of video data with discrete emotion labels automatically inferred from facial expressions. *EmoInfer* builds on state-of-the-art research in the Learning Sciences [2] and can currently be used in offline contexts (posthoc after the data collection), with the development of an online version (for on-the-fly usage during data collection) underway. The development of *EmoInfer* reflects our broader efforts into creating intuitive graphical user interfaces for use by educational stakeholders with a non-programming background (e.g., researchers, practitioners and instructors). For instance, data-driven emotion inference can augment an

© Springer Nature Switzerland AG 2022
M. M. Rodrigo et al. (Eds.): AIED 2022, LNCS 13356, pp. 156–162, 2022.
https://doi.org/10.1007/978-3-031-11647-6_27

instructor's contextual assessment of students' observable cues in a (virtual) classroom, in turn supporting more targeted scaffold delivery. *EmoInfer* can be freely downloaded via https://github.com/EmoInfer.

2 Design Affordances of *EmoInfer*

2.1 Facial Action Unit Prediction

Human annotations of observable changes in facial movements (action units), best captured by the facial action coding system [3], form the starting point for classification via machine learning models. Powered by OpenFace [4], *EmoInfer* can take high-quality frames from single or multiple input video files (frames that have a face successfully tracked and facial landmarks detected with confidence greater than 80%), predict and visualize the presence/continuous intensities of up to 18 facial action units. Action unit intensities are binarized using a stakeholder-defined threshold (between 1 and 5) for further analyses. Thresholds for head movements (pitch, roll) can also be user-specified. See Fig. 1 for illustration of the interface.

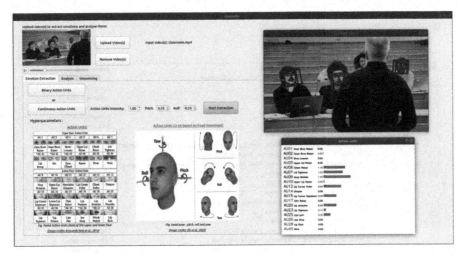

Fig. 1. Interface for action unit prediction with emotion inference computation in the backend

2.2 Emotion Inference

Drawing on [2], *EmoInfer* then uses three cross-cultural coding schemes[1] [5–7] to map the automatically annotated facial movements into 28 discrete emotion categories on a frame-by-frame basis (see Table 1), with the rationale that an individual's experience of

---

[1] For example, in one of the emotion inference coding schemes [5], facial expressions of emotions appeared at above chance rates in five cultures with varying societal characteristics (China, India, Japan, Korea, United States), with no gender differences in the frequency of occurrence of these patterns across the cultures.

emotion can be influenced by sensorimotor feedback from these facial action units [8]. Several non-prototypical emotions (e.g., shame, contempt, confusion) and compound emotions[2] (e.g., happily surprised, angrily disgusted) mimicking real-life displays, especially in ecologically-valid learning situations, can be inferred and visualized further. See Fig. 2 for illustration of a saved datafile with inferred emotions.

Table 1. List of emotions annotated by *EmoInfer*. Please refer to [2] for physical descriptions

Emotion categories	Exemplars
Self-conscious	Shame, Pride, Embarrassment
Knowledge	Interest, Surprise, Confusion
Hostile	Fear, Anger, Disgust, Contempt
Pleasurable	Happiness, Amusement, Awe
Compound	Happily surprised, Happily disgusted, Sadly fearful, Sadly angry, Sadly surprised, Sadly disgusted, Fearfully angry, Fearfully surprised, Fearfully disgusted, Angrily surprised, Angrily disgusted, Disgustedly surprised, Appalled/hatred
Other	Pain, Sadness

Frame	Face_id	Cordaro et al. [5]	Du et al. [6]	Keltner et al. [7]
1	1	Surprise	Awe, Surprise, Fearfully Surprised	Surprise
7	0	Anger, Confused	Anger	Confused
18	0	Pride	Disgust	Embarrassment
20	0	Pride	Happiness	Embarrassment
21	0	Pride	Happiness	Embarrassment
22	0	Pride	Happiness	Embarrassment
23	0	Pride	Happiness	Embarrassment
30	0	Awe, Surprise	Happiness, Awe, Surprise, Happily Surprised, Fearfully Surprised	Surprise, Interest
31	0	Surprise	Awe, Surprise, Fearfully Surprised	Surprise
35	1	Awe, Surprise	Happiness, Awe, Surprise, Happily Surprised, Fearfully Surprised	Surprise, Interest
41	0	Pride	Happiness	Embarrassment
42	0	Pride	Happiness	Disgust, Embarrassment
43	0	Pride	Happiness	Disgust, Embarrassment
44	0	Pride, Embarrassment	Happiness, Happily Disgusted	Embarrassment
45	0	Pride	Happiness	Embarrassment
46	0	Pride	Happiness	Embarrassment
47	0	Pride, Embarrassment	Happiness, Happily Disgusted	Embarrassment
48	0	Pride, Embarrassment	Happiness, Happily Disgusted	Embarrassment
49	0	Pride, Embarrassment	Happiness, Happily Disgusted	Embarrassment
50	0	Pride, Embarrassment	Happiness, Happily Disgusted	Embarrassment
51	0	Pride, Embarrassment	Happiness, Happily Disgusted	Embarrassment
52	0	Pride, Embarrassment	Happiness, Happily Disgusted	Embarrassment
53	0	Pride, Embarrassment	Happiness, Happily Disgusted	Embarrassment
54	0	Pride, Embarrassment	Happiness, Happily Disgusted	Embarrassment
55	0	Pride, Embarrassment	Happiness, Happily Disgusted	Embarrassment
55	1	Awe, Surprise	Happiness, Happily Disgusted	Surprise, Interest

Fig. 2. Saved datafile after emotion inference

---

[2] Compound emotions [6] are those that can be distinctively expressed because of overlap in action unit patterns as well as unambiguously discriminated by observers.

2.3 Data Aggregation

For all subsequent analyses post emotion inference, *EmoInfer* provides choices to stakeholders for (i) setting a time granularity to downsample the temporal resolution of the data (e.g., ranging from 1/30th of a second to 300 s) and obtain the relevant descriptive statistics and visualizations (after application of majority voting to the video frames), as well as for (ii) specifying whether the calculation of descriptive statistics and visualizations should proceed independently (e.g., videos of two participants, one from an experimental and the other from a control condition) or together (e.g., videos of two participants from the experimental condition). Specifically, for a selected coding scheme, the emotion with the highest incidence (occurring in the highest percentage of video frames) is displayed, along with violin plots and corresponding parametric and nonparametric statistics for each inferred emotion. This can allow stakeholders to have a consolidated look at the emotion data both numerically and graphically. Since the underlying datafiles are downloadable, stakeholders are free to use other statistical programs to perform more complex analyses. See Fig. 3 for illustration of the interface.

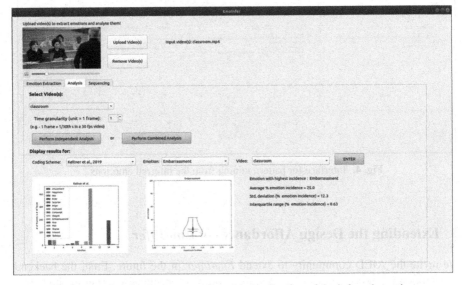

Fig. 3. Interface for data aggregation and visualization of the inferred emotions

2.4 Pattern Mining

In addition to the frequency of occurrence of emotions, *EmoInfer* also allows probing into frequently co-occurring and temporally contingent sequences of emotions. Because human perception of facial expressions is heavily influenced by dynamics [9], pattern mining capabilities within *EmoInfer* can allow stakeholders to decipher the meaning of frequently subtle facial expressions that may not be identifiable in static presentations. Powered by SPMF [10], *EmoInfer* summarizes emotion dynamics obtained by using a closed sequential pattern mining algorithm with time constraints in three objective ways that can be corroborated with theoretical lenses a stakeholder brings into the analyses – (i) sequences comprising each unique emotion inferred, (ii) sequences of multiple lengths, (iii) sequences comprising more than one emotion. Parameters like the minimum percentage of sequences comprising a particular pattern (support, e.g., 75% or more), and the time interval between itemsets in a sequence can be chosen (e.g., 0.5–1.5 s). The rationale for our recommendations can be viewed at https://github.com/EmoInfer. See Fig. 4 for illustration of the interface.

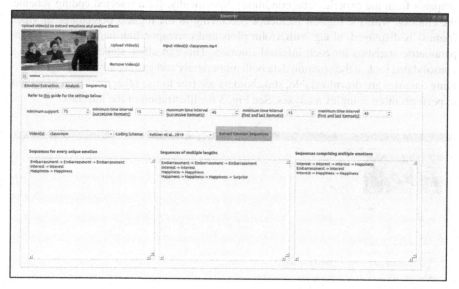

Fig. 4. Interface for pattern mining from the inferred emotions

3   Extending the Design Affordances of *EmoInfer*

We invite the AIED community to extend *EmoInfer* in the future. First, the backend for emotion inference can be expanded to incorporate machine learning methods like vision transformers [11] that are inspired by transfer learning approaches. The underlying intuition would be to use models pretrained on abundant data available for the six basic emotions (e.g., happiness, surprise), and use that to fine-tune classification for rarer data on non-prototypical emotions relevant to learning contexts (e.g., shame, confusion). Second, to allow for a comparative view of emotion dynamics in the data, the pattern mining interface (Fig. 4) can be strengthened to incorporate alternatives to pattern mining (e.g., epistemic network analysis [12]). Taken together, we envision that *EmoInfer* can be used by stakeholders to understand the interplay of emotion and cognition in at least three ways – (i) examining the relationship between naturally occurring emotions and learning (e.g., see [2] for a case study based on the *EmoInfer* pipeline), (ii) experimentally manipulating emotions [13] to assess their differential effect on learning [14], (iii) triangulating process data from facial expressions of emotions with other self-reported or physiological measures [15] to nudge participants in meaningful ways (e.g., scaffold them towards regulating emotions [16, 17]). Invariably, *EmoInfer* can allow stakeholders to carry out manipulation checks for observable correlates of emotion and cut down the time-taking process of using video-based data streams to create rich, explanatory accounts of learning. With increasing efforts to improve multimodal data collection in the field (e.g., [18]) and the criticality of implementation fidelity in educational interventions, disseminating accessible evidence-based tools such as *EmoInfer* to facilitate effective practice holds high significance.

References

1. Dukes, D., et al.: The rise of affectivism. Nature Human Behaviour **5**(7), 816–820 (2021)
2. Sinha, T.: Enriching problem-solving followed by instruction with explanatory accounts of emotions. Journal of the Learning Sciences **31**(2), 151–198 (2022)
3. Ekman, P., Friesen, W.V.: Measuring facial movement. Environmental Psychology and Nonverbal Behavior **1**(1), 56–75 (1976)
4. Baltrusaitis, T., Zadeh, A., Lim, Y.C., Morency, L.P.: Openface 2.0: facial behavior analysis toolkit. In: 13th IEEE international conference on automatic face & gesture recognition 2018, pp. 59–66. IEEE (2018)
5. Cordaro, D.T., Sun, R., Keltner, D., Kamble, S., Huddar, N., McNeil, G.: Universals and cultural variations in 22 emotional expressions across five cultures. Emotion **18**(1), 75 (2018)
6. Du, S., Tao, Y., Martinez, A.M.: Compound facial expressions of emotion. Proc. Natl. Acad. Sci. **111**(15), E1454–E1462 (2014)
7. Keltner, D., Sauter, D., Tracy, J., Cowen, A.: Emotional expression: Advances in basic emotion theory. J. Nonverbal Behav. **43**(2), 133–160 (2019)
8. Buck, R.: Nonverbal behavior and the theory of emotion: The facial feedback hypothesis. J. Pers. Soc. Psychol. **38**(5), 811 (1980)
9. Ambadar, Z., Schooler, J.W., Cohn, J.F.: Deciphering the enigmatic face: the importance of facial dynamics in interpreting subtle facial expressions. Psychol. Sci. **16**(5), 403–410 (2005)
10. Fournier-Viger, P., et al.: The spmf open-source data mining library version 2. In: Joint European Conference on Machine Learning and Knowledge Discovery in Databases. Riva del Garda. Springer, Italy (2016)
11. Dosovitskiy, A., et al.: An image is worth 16x16 words: Transformers for image recognition at scale. arXiv preprint arXiv:2010.11929 (2020)
12. Shaffer, D.W.: Epistemic network analysis: understanding learning by using big data for thick description. In: International handbook of the learning sciences, 1st edn, pp. 520–531. Routledge (2018)
13. Joseph, D.L., Chan, M.Y., Heintzelman, S.J., Tay, L., Diener, E., Scotney, V.S.: The manipulation of affect: a meta-analysis of affect induction procedures. Psychol. Bull. **146**(4), 355 (2020)
14. Wong, R.M., Adesope, O.O.: Meta-analysis of emotional designs in multimedia learning: A replication and extension study. Educ. Psychol. Rev. **33**(2), 357–385 (2021)
15. Harley, J.M., Lajoie, S.P., Frasson, C., Hall, N.C.: Developing emotion-aware, advanced learning technologies: A taxonomy of approaches and features. Int. J. Artif. Intell. Educ. **27**(2), 268–297 (2017)
16. Tamir, M.: Why do people regulate their emotions? a taxonomy of motives in emotion regulation. Pers. Soc. Psychol. Rev. **20**(3), 199–222 (2016)

17. Quoidbach, J., Mikolajczak, M., Gross, J.J.: Positive interventions: an emotion regulation perspective. Psychol. Bull. **141**(3), 655 (2015)
18. Schneider, B., Hassan, J., Sung, G.: Augmenting social science research with multimodal data collection: the EZ-MMLA Toolkit. Sensors **22**(2), 568 (2022)

Sensing Human Signals of Motivation Processes During STEM Tasks

Richard DiNinni[✉] and Albert Rizzo

Institute for Creative Technologies, University of Southern California, 12015 Waterfront Drive, Los Angeles, CA 90094, USA
dininni@usc.edu

Abstract. This paper outlines the linking of a multi-modal sensing platform with an Intelligent Tutoring System to perceive the motivational state of the learner during STEM tasks. Motivation is a critical element to learning but receives little attention in comparison to strategies related to cognitive processes. The EMPOWER project has developed a novel platform that offers researchers an opportunity to capture a learner's multi-modal behavioral signals to develop models of motivation problems that can be used to develop best practice strategies for instructional systems.

Keywords: Motivation · Sensing · EMPOWER

1 Introduction

The Enhancing Mental Performance and Optimizing Warfighter Effectiveness and Resilience (EMPOWER) project is a U.S. Defense Health Program in the Education and Training Medical Readiness category. This effort aims to advance state-of-the-art AI techniques and machine-based human perception to support and sharpen the acquisition of critical cognitive and emotional skills that service members need to achieve and sustain optimal performance under diverse conditions. Towards that objective, the EMPOWER project has developed OmniSense, a multi-modal sensing system deploying novel algorithms for facial expression analysis, voice activity detection, automatic speech recognition, and body tracking to accurately detect changes in human physiological state and performance while predicting affect (arousal and valence). Thus, OmniSense represents a research platform that provides a synchronized, distributed solution for realtime human behavior and physiology acquisition and analysis, offering an appealing machine learning and pattern recognition tool for research across a wide range of human performance, training, and educational contexts.

2 Stem

The National Academies of Sciences, Engineering, and Medicine (2017) determined there is a widely expanding need across industries and professions to build a broadly

© Springer Nature Switzerland AG 2022
M. M. Rodrigo et al. (Eds.): AIED 2022, LNCS 13356, pp. 163–167, 2022.
https://doi.org/10.1007/978-3-031-11647-6_28

diversified workforce with the necessary skills and knowledge in science, technology, engineering, and mathematics (STEM). Yet that objective is difficult to achieve through standardized instruction in classrooms filled with students who have differing interests, capabilities, and experiences. Studies show a 1:6 time to learn ratio in the classroom, meaning slower students need six times longer than the fastest students to learn a single lesson to the required level of mastery (Gettinger, 1983). The individual differences and variability among learners results in barriers to effective instruction in a group setting. These issues point to the potential for individualization in education and training, particularly when considering the complex, abstract concepts and advanced knowledge needed to satisfactorily complete STEM instructional programs. Hence, the wide application of computer-based training systems for STEM requirements.

3 Intelligent Tutoring Systems

A meta-analysis of Intelligent Tutoring Systems (ITS) by Kulik and Fletcher (2016) found that almost all of those systems address STEM topics. The deep STEM-ITS connection can be seen largely as the result of the Department of Defense's research funding, as many occupational specialties involve the need for service members to consistently complete tasks that are highly technical and linked to other critical mission requirements (Fletcher, 2009). Yet, the knowledge, skills, and cognitive dexterity needed to progress through STEM programs have broader applicability than strictly STEM disciplines (Carnevale and Smith, 2013). Moreover, ITS provide a cost effective approach to one-on-one tutoring and deliver training and education on demand and at the point of need. A next step to consider is whether an ITS that can perceive the motivational state of the user offers a path to further improve learning outcomes.

4 Motivation

In the realms of education and training, motivation and self-regulatory strategies are key elements for achieving desired learning outcomes. Though often overlooked, motivational issues are more complex to determine and address than knowledge gaps (Clark and Estes, 2008). Most studies and considerably more strategies target cognitive processes rather than providing the necessary attention to the impact of motivation on learning, transfer, and performance improvement (Clark and Saxberg, 2018).

The principal indicators of motivated action are the choice to initiate a task, persistence when encountering distractions or difficulties, and applying the necessary mental effort to complete the task successfully (Mayer, 2011). Active choice refers to a stakeholder embarking on the initial steps needed to achieve performance goals (Clark and Estes, 2008). Persistence involves the capacity and resilience to overcome barriers, constraints, and other environmental challenges to maintain momentum toward performance goals. The third aspect of motivation focuses on whether an individual allocates and maintains the amount of deliberate mental effort required to solve problems, implement solutions, and ultimately reach the stated performance goals. Critical across these steps is the integration of positive feedback so that any challenges or difficulties an individual encounters are attributed to controllable rather than uncontrollable factors, such as the ability to self-regulate an increase in focused effort (Clark and Saxberg, 2018).

Wigfield and Eccles (2000) provide a model of motivation to help explain how an individual's internal expectation for success in coordination with an assessment of underlying task values can influence his or her choices, persistence, and performance. Expectancy-value theory suggests that expectancies and values interact to predict an individual's persistence, mental effort, and performance once he or she chooses a goal or task. Expectancies refers to an individual's belief in his or her ability to succeed. Wigfield and Eccles (2000) offer that task values point to the four elements that motivate an individual's choice behavior: Attainment Value (tied to identity or self), Intrinsic Value (interest in the task, goal, or something related), Utility Value (its usefulness or relevance), and Cost (the tradeoff or loss of time).

Bandura's (2005) description of human agency indicates that people regulate and control their own actions and develop self-efficacy to motivate personal behavior. Self-efficacy theory focuses on how empowering individuals with a sense of human agency can serve to motivate attainment of performance goals (Bandura, 2000). Self-efficacy manifests itself through an individual's beliefs, expectations, and perceptions of his or her capabilities for producing successful outcomes. Individuals have a higher tendency to persist when encountering distractions and obstacles if they feel confident in their ability to succeed (Bandura, 2000; Eccles, 2006). Of critical importance, self-efficacy theory holds that individuals must have opportunities to build mastery experiences and observe or experience reinforcing models to promote development of positive personal beliefs (Bandura, 2000).

5 Modeling Motivation Problems

As noted earlier, the EMPOWER program's OmniSense platform enables the dynamic capture and quantification of behavioral signals across the following components: body tracking, visual attention, facial expression, head gesture, speech recognition, voice activity, and acoustic analysis (See Fig. 1). These informative behavioral signals serve two purposes. First, they are broadcast to other software components of the system to inform the state and actions of the participant. Second, they produce the capability of analyzing the occurrence and quantity of behaviors, over the course of the interaction/exercise, to inform detection of cognitive and emotional/motivational state. This paper proposes the integration of OmniSense capabilities with an ITS to model the motivational state of a learner while working through a set of STEM tasks.

A research prototype that delivers on the necessary requirements would build on existing OmniSense features. For example, a robust and accurate head gesture recognition component developed through a recurrent neural network can classify behavioral cues and train a machine learning model that identifies interruptions in motivational processes, such as delays in starting on a lesson unit, lack of persistence, or trouble investing necessary mental effort to successfully complete the unit. Visual attention and facial expression components similarly provide actionable data from the learner's behavioral markers, supporting the development of deep learning models that correlate with other indicators and further automatic accurate assessment of motivational state.

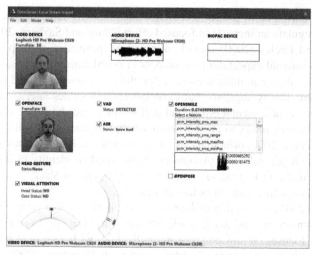

Fig. 1. OmniSense facial expression analysis, head gesture recognition, and visual attention components.

As an end-to-end multimodal system, the OmniSense pipeline synchronizes and produces data streams as output from recognition and analysis of component signals. Applications of OmniSense use deep learning models to send outputs to a virtual human agent for generation of rapport building behaviors and empathetic responses based on imitating human-human interaction. Similarly, automatic assessment of motivational state can trigger an agent or other ITS feedback method to assess the learner's task value for working on the assigned lesson and the belief they hold about their self-efficacy to complete the lesson. Further the system architecture can activate positive messages to support the learner's belief about the value of the task and attribute success to factors within the learner's control.

The combination of machine learning techniques and OmniSense sensing capabilities across multiple modalities also provides the ability to develop models of the individual learner. It can build a unique profile of each learner based on distinctive behavioral signals during their time on task. This opens the door to more personalized instruction, adding to assessment of prior knowledge and task completion.

6 Conclusion

There are benefits to identifying and separating motivation challenges from learning strategy issues (Clark and Saxberg, 2018). A data-driven, evidence-based path to understanding motivational issues through the capture of multimodal indicators during a learner's time on a STEM task offers an opportunity to provide behavioral metrics that can enhance the development of strategies to improve motivation. Prior work leveraging automatic analysis of behavioral signals to support the effective assessment of post-traumatic stress disorder (Rizzo et al., 2016; Scherer et al., 2013; Stratou et al., 2013) demonstrates the promise of detecting verbal and nonverbal behavioral cues and training machine learning models in an educational context. Thus, we suggest the approach to multimodal sensing and modeling of a learner's motivational processes outlined here presents a rich and potentially fruitful area of research to study, inform, and optimize performance.

References

Bandura, A.: Exercise of human agency through collective efficacy. Curr. Dir. Psychol. Sci. **9**(3), 75–78 (2000)

Bandura, A.: The evolution of social cognitive theory. In: Smith, K.G., Hitt, M.A. (eds.) Great minds in management, pp. 9–35. Oxford University Press (2005)

Building America's Skilled Technical Workforce: In Policy File. National Academy of Sciences (2017)

Carnevale, A., Smith, N.: Workplace basics: the skills employees need and employers want. Hum. Resour. Dev. Int. **16**(5), 491–501 (2013)

Clark, R., Estes, F.: Turning research into results: A guide to selecting the right performance solutions. Information Age (2008)

Clark, R., Saxberg, B.: Engineering motivation using the belief-expectancy-control framework. Interdisciplinary Education and Psychology **2**(1), 4–32 (2018)

Eccles, J.: Expectancy value motivational theory. Education.com (2006)

Fletcher, J.: Education and training technology in the military. Science **323**(5910), 72–75 (2009)

Gettinger, M.: Effects of learner ability and instructional modifications on time needed for learning and retention. J. Educ. Res. **76**(6), 362–369 (1983)

Kulik, J., Fletcher, J.: Effectiveness of intelligent tutoring systems: a meta-analytic review. Rev. Educ. Res. **86**(1), 42–78 (2016)

Mayer, R.: Applying the science of learning. Pearson (2011)

Rizzo, A., et al.: Detection and computational analysis of psychological signals using a virtual human interviewing agent. Journal of Pain Management **9**(3), 311–322 (2016)

Scherer, S., et al.: Automatic behavior descriptors for psychological disorder analysis. In: 2013 10th IEEE International Conference and Workshops on Automatic Face and Gesture Recognition (FG), pp. 1–8 (2013)

Stratou, G., Scherer, S., Gratch, J., Morency, L.-P.: Automatic nonverbal behavior indicators of depression and PTSD: the effect of gender. Journal on Multimodal User Interfaces **9**(1), 17–29 (2014). https://doi.org/10.1007/s12193-014-0161-4

Wigfield, A., Eccles, J.: Expectancy-value theory of achievement motivation. Contemp. Educ. Psychol. **25**(1), 68–81 (2000)

Implementation of a Mathematics Formative Assessment Online Tool Before and During Remote Learning

Jamie Gillespie(✉) , Kevin Winn(✉) , Malinda Faber(✉), and Jessica Hunt(✉)

The Friday Institute for Educational Innovation, North Carolina State University,
1890 Main Campus Dr., Raleigh, NC 27606, USA
{jggilles,kwinn,mmfaber,jhunt5}@ncsu.edu

Abstract. ASSISTments is a free online learning tool for improving students' mathematics achievement by providing immediate feedback and hints to students, detailed information on how students performed to teachers, and instructional suggestions for teachers to use. Researchers at the Friday Institute for Educational Innovation conducted an intrinsic, longitudinal multiple-case study of 7th-grade mathematics teachers' implementation of ASSISTments and its impact on their instruction before and during the COVID-19 pandemic. The study examined teachers' use of ASSISTments in three instructional contexts: in- person only, remote only, and both in-person and remote. Our findings indicate that teachers in all contexts changed their instructional practices for homework review and for determining whether their students had understood lessons. Teachers used the ASSISTments auto-generated reports to focus their homework reviews, based on their students' performance, and to provide instructional interventions and/or re-teaching. They also used the instructional suggestions provided by the ASSISTments platform to plan lessons to re-teach concepts or to review prior instruction with their students.

Keywords: ASSISTments · Mathematics education · Educational technology · Teaching support · Feedback · Formative assessment · Data-based decisions

1 Introduction

The COVID-19 pandemic introduced unprecedented disruption to education in the U.S. Students' achievement in mathematics was more negatively affected than other subjects by the effects of closing schools and turning to remote instruction [1]. Many schools used educational technologies, such as ASSISTments, an online mathematics instructional platform, to maintain learning during school closures. ASSISTments saw significantly increased use during the pandemic, going from supporting 800 teachers to supporting 20,000 teachers and their 500,000 students.

To gain a better understanding of teachers' practices when using educational technologies, and how the learning analytics they provide affect instruction, researchers at the Friday Institute for Educational Innovation (Friday Institute) completed a case study

© Springer Nature Switzerland AG 2022
M. M. Rodrigo et al. (Eds.): AIED 2022, LNCS 13356, pp. 168–173, 2022.
https://doi.org/10.1007/978-3-031-11647-6_29

exploring how ASSISTments was used during in-person and remote instruction and how, if at all, the use of ASSISTments changed teachers' instructional practices. Researchers were interested in teachers' use of ASSISTments for homework and/or classwork and the impact of ASSISTments on teachers' instructional practices for homework and/or classwork, and in instructional decision-making, in general.

The study's research questions were: (1) *How did teachers implement ASSISTments during in-person instruction and during remote instruction?* and (2) *How did the use of ASSISTments affect teachers' instructional practices during in-person instruction and remote instruction?*

1.1 ASSISTments

The case study is part of a large-scale randomized controlled trial in North Carolina that seeks to replicate the findings of an earlier study, completed in Maine, that examined the efficacy of ASSISTments. The Maine study found that ASSISTments significantly increased students' achievement and changed teachers' instructional practices [2].

The ASSISTments platform contains mathematics questions/problems that teachers may assign to a class of students, to groups of students, or to individual students. The questions/problems are drawn from open educational resources (e.g., Illustrative Math, Engage NY/Eureka Math, Open Up Resources) and there was also the option during the study for teachers to enter their own questions/problems. While completing the assignments, students are given immediate feedback on their accuracy and some problem types also provide hints on how to improve their answers or help separate multistep problems into parts. Once they have completed the assignments, their teachers receive automated reports which provide data regarding how long each student worked, whether they needed multiple chances to answer any question/problem, and whether they asked for hints. Teachers also receive class-level reports showing the accuracy rates of a class per question/problem and whether there were any common wrong answers.

Teachers use ASSISTments to assign online mathematics tasks and to see students' results easily. Teachers can use ASSISTments to assign tasks in class or as homework. By providing individual and class-level reports of students' responses to the tasks and data analysis, ASSISTments allows teachers to quickly assess students' learning. In this way, ASSISTments creates opportunities to use classwork and homework as formative assessments. ASSISTments also provides instructional suggestions to teachers that they may use in re-teaching the whole class, in small group instruction, or in one-on-one instruction.

The ASSISTments theory of change posits that the use of ASSISTments increases the likelihood that teachers will make instructional changes in response to homework results. This process, which is a form of formative assessment, would be described by Duckor and Holmberg [3] as "a dynamic pedagogical process between students and teachers" (p. 336). Research suggests that the use of formative assessment results to make instructional decisions increases students' achievement ([2, 4–7]). The ASSISTments theory of change argues that the use of ASSISTments leads to the use of formative assessment, resulting in teachers making instructional changes based on students' performance. This, then, leads to increased student achievement.

2 Methods and Analysis

This study investigated the implementation of ASSISTments and its impact by gathering data pertaining to teachers' use of ASSISTments and pertaining to their instructional practices both before using ASSISTments and with ASSISTments. Researchers focused on ASSISTments use among teachers of 7th-grade mathematics during in-person instruction and during remote instruction, although some teachers of other grades were also included.

2.1 Survey

In November of 2020 – the second semester of pandemic-response instruction for North Carolina's teachers – the research team invited 544 ASSISTments users in North Carolina to take a Qualtrics survey to share their experiences using ASSISTments both in person and online. Users included those who began using ASSISTments prior to, during, and after the pandemic. The survey asked participants if they had used ASSISTments during in-person instruction, remote instruction, or both, and how they had used ASSISTments (classwork, homework, assessments, other), as well as their plans for future use. Participants were also asked to reflect on factors that made using ASSISTments difficult or easy, as well as their opinions and practices regarding formative assessments and homework. Ninety-seven teachers completed the survey. Closed survey items were analyzed using descriptive statistics. Open-ended survey items were analyzed for themes using an open coding approach [8].

2.2 Interviews

Three researchers reviewed the open-ended responses on the survey to narrow down a pool of potential interviewees. While the researchers looked at all open-ended responses to determine whom to interview, they looked most closely at responses to a question that asked participants to share their definitions of formative assessments. This sampling strategy was used because of the wide variety of individual definitions of the term. The team wanted to understand a wide array of perspectives. However, due to the low response rate to the initial email invitations, the team eventually reached out to all survey respondents who had provided an email address and completed 31 interviews (n = 17 both in-person and remote; n = 13 remote only; n = 1 in-person only). Interview questions encouraged teachers to reflect on their experiences using ASSISTments and asked about: (1) how they used ASSISTments in their instruction; (2) how, if at all, using ASSISTments changed their teaching practices; (3) differences in their use of ASSISTments between in-person and remote instruction; and (4) their perceptions of homework and formative assessments, in general. Interviews lasted between 14 min and 37 min, and audio recordings were transcribed using Rev.com. The research team used Atlas.ti to analyze the transcripts, determining interrater reliability by coding three transcripts together. The team established a set of codes based on the interview questions and also used open coding and eclectic coding in the analysis [9]. Multiple rounds of coding narrowed down the findings to themes explained in the next section. To aid in analysis, participants were labeled by the instructional environments in which they had used ASSISTments: in person only, remote only, and both remotely and in person.

3 Findings

3.1 Implementation of ASSISTments

In response to the first study question regarding how teachers implemented ASSIST-ments during in-person and remote instruction, researchers found that use of ASSIST-ments remained consistent across instructional environments. The survey indicated that teachers used the program for homework and classwork the most, and the purpose was largely to practice new skills, with reviewing old skills the second-most common purpose. This lack of variability between in-person and remote instruction suggests that ASSISTments is a flexible program which is easy for teachers to access across modes of instruction. This finding was affirmed through an examination of ASSISTments log data and through interviews with teachers.

Researchers noted several changes in the ways that teachers viewed homework more broadly. Teachers indicated that assigning homework through ASSISTments helped them provide more focused assignments for their students. For example, teachers were more selective in which tasks they assigned, choosing to give fewer items each night, but ensuring that each item was aligned with the day's lesson. Interviews also illuminated the tension that existed when students knew their homework was being used as a formative assessment rather than as a graded assignment. When homework "didn't count," some students did not complete the work.

Additionally, remote instructional environments changed teachers' conceptions of homework and classwork. Some teachers said that every assignment during remote learning was homework because they were trying to limit the use of synchronous screen time to direct instruction rather than individual practice. Many teachers noted that they no longer assigned homework during remote learning because they did not want to overburden students with additional screen time, because their students struggled to concentrate in a remote learning environment.

3.2 Impact of ASSISTments

In response to the second study question, results showed that using ASSISTments changed most teachers' instructional practices, regardless of the learning environment (in-person or remote). The survey indicated that 73% of teachers found that using ASSISTments changed how they knew whether their students had understood a lesson. They used the ASSISTments reports to understand where students had struggled and what their errors were. The most common change for teachers was in how they reviewed homework with students. Teachers overwhelmingly agreed that ASSISTments helped them choose items to review in class that were more targeted to their students' needs. Using the ASSISTments-generated reports, teachers saw which items the students got correct, which they struggled with, and whether there were any common wrong answers.

A few teachers shared that, before using ASSISTments, they were inconsistent in how they reviewed homework and in how they understood students' confusion. Some waited for students to ask questions during class, and others made guesses as to which items they thought were the hardest for their students to answer. Others simply reviewed all the homework items with their students, without knowing how they had performed on them. In implementing ASSISTments, however, teachers analyzed the reports, which showed (1) how many times the students attempted to answer each item, (2) whether they needed to ask for hints, and (3) how long they spent answering each item. These report features gave teachers data to understand which items most students needed to review. The reports also showed teachers if there were any common wrong answers, which helped teachers see if there were misconceptions among the students. This helped teachers determine whether they needed to re-teach a concept to the entire class or to small groups of students.

Most teachers reported using the data from ASSISTments to differentiate instruction and place students into small groups based on their performance. During remote instruction, teachers placed students in virtual breakout rooms and gave them short lessons based on their needs identified via ASSISTments. One teacher provided virtual one-on-one instruction "after school."

Although there were similarities between teachers who used ASSISTments both in person and remotely and teachers who used it only remotely, it was notable that remote- only users (n = 13) found that ASSISTments improved their organization and efficiency. These teachers shared that ASSISTments made reviewing students' work easier and timelier, so that students had feedback within 24 h. Their classes became more efficient and more targeted to students' needs when they used ASSISTments.

4   Significance and Implications

Findings from this study add to the body of knowledge on the use of computer-based platforms to assess students, provide formative feedback to students, and provide usable data for teachers to make instructional decisions. It also adds to the growing body of knowledge on teachers' instructional practices during remote instruction and on instructional practices when moving from in-person to remote instruction. As the pandemic lingers, understanding which instructional practices are used, and how, during remote instruction will assist schools in improving remote instruction. Further, this research is particularly timely as many schools across the U.S. plan to retain virtual "schools" and remote learning options beyond the pandemic.

Our findings also support prior research demonstrating the positive impact of immediate feedback on students' mathematics achievement and the positive impact of the use of formative assessment data to adapt instruction for student achievement. Although this study does not provide achievement results due to the canceled End-of- Grade Exams, it does indicate how teachers changed their instructional practices when provided with easy-to-access and clear formative assessment results that identify specific areas of student struggle. Schools can keep this in mind when planning professional development and creating data teams. Knowing which practices are likely to occur during teachers' use of specific tools, like ASSISTments, will make it easier for schools to prepare teachers to use them effectively.

This study is also significant in that it demonstrates that a specific instructional tool can change how teachers determine whether their students have understood a lesson. On the survey, 73% of teachers said that ASSISTments had changed how they gauged their students' learning. These findings were reinforced in our interviews, where teachers shared that they were using the reports to determine what they needed to re- teach, which students needed reinforcement on which skills, and whether there were any common wrong answers. This level of understanding is formative assessment at its most effective, allowing teachers to differentiate and personalize students' learning in order to achieve student growth. Teachers also reported using the instructional suggestions provided in ASSISTments, helping them create those targeted instructional activities more efficiently.

References

1. Kuhfeld, M., Tarasawa, B., Johnson, A., Ruzek, E., Lewis, K.: Learning during COVID-19: initial findings on students' reading and math achievement and growth. NWEA (2020)
2. Roschelle, J., Feng, M., Murphy, R.F., Mason, C.A.: Online mathematics homework increases student achievement. AERA Open **2**(4) (2016)
3. Duckor, B., Holmberg, C.: Mastering Formative Assessment Moves: 7 High-Leverage Practices to Advance Student Learning. ASCD, Alexandria, VA (2017)
4. Andersson, C., Palm, T.: The impact of formative assessment on student achievement: a study of the effects of changes to classroom practice after a comprehensive professional development programme. Learn. Instr. **49**, 92–102 (2017)
5. Chen, F., Lui, A.M., Andrade, H., Valle, C., Mir, H.: Criteria-referenced formative assessment in the arts. Educ. Assess. Eval. Account. **29**(3), 297–314 (2017). https://doi.org/10.1007/s11 092-017-9259-z
6. Yin, Y., Tomita, M.K., Shavelson, R.J.: Using formal embedded formative assessments aligned with a short-term learning progression to promote conceptual change and achievement in science. Int. J. Sci. Educ. **36**(4), 531–552 (2014)
7. Polly, D., Wang, C., Martin, C., Lambert, R.G., Pugalee, D.K., Middleton, C.W.: The influence of an internet-based formative assessment tool on primary grades students' number sense achievement: influence of an internet-based formative assessment tool. Sch. Sci. Math. **117**(3–4), 127–136 (2017)
8. Miles, M.B., Huberman, A.M., Saldaña, J.: Qualitative Data Analysis: A Methods Sourcebook, 4th edn. SAGE, Los Angeles (2020)
9. Saldaña, J.: The Coding Manual for Qualitative Researchers, 4th edn. SAGE, Los Angeles (2021)

Using Genetic Programming and Linear Regression for Academic Performance Analysis

Guilherme Esmeraldo[1], Robson Feitosa[1,4(✉)], Cicero Samuel Mendes[2],
Cícero Carlos Oliveira[1], Esdras Bispo Junior[2,3], Allan Carlos de Sousa[1],
and Gustavo Campos[4]

[1] Federal Institute of Ceará, Crato, Ceará, Brazil
{guilhermealvaro,robsonfeitosa,cicerocarlos,allancarlos}@ifce.edu.br
[2] Federal University of Pernambuco, Recife, Pernambuco, Brazil
csrm@cin.ufpe.br
[3] Federal University of Jataí, Jataí, Goiás, Brazil
bispojr@ufj.edu.br
[4] State University of Ceará, Fortaleza, Ceará, Brazil
gustavo.campos@uece.br

Abstract. The academic evaluation process, even today, is the subject of much discussion. This process can use quantitative analysis to indicate the level of learning of students to support the decision about whether the student can attend the next curriculum phase. From this context, this paper analyzes the history of students' grades in the 1st year of a technical course in informatics integrated to high school, for the years 2020 and 2021, through the linear regression method, supported by genetic programming, to find out the influence of the grades of the first two bimesters concerning the final grade. The main results show that the genetic programming algorithm favored the search for linear regression models with a good fit to the datasets with students' data. The resultant models proved accurate and explained more than 74% of the datasets.

Keywords: Academic performance analysis · Linear regression · Genetic programming

1 Introduction

One of the missions of academic institutions is to enhance the education quality of its students. However, many obstacles arise during the teaching-learning process. Aiming to improve this process, professionals in the educational area have concentrated efforts on analyzing the factors that affect the learning, to propose pedagogical interventions that can assist in this process [1].

There are several initiatives to improve the learning process evaluation [2–4]. However, quantitative assessment is still the most used means as a base parameter to compute the academic performance of the students, in different subjects,

M. M. Rodrigo et al. (Eds.): AIED 2022, LNCS 13356, pp. 174–179, 2022.
https://doi.org/10.1007/978-3-031-11647-6_30

in the most diverse academic institutions in Brazil [5]. For the students of the Brazilian federal technical programs at the Federal Institute of Ceará, the evaluation process takes place every two months, allowing them to improve or maintain their performance in the course over a year. After this period, students receive a weighted average (final grade) computed from the bi-monthly evaluations. If this final grade is greater than a minimum limit, the student is considered able to attend the next curriculum phase.

Knowing the importance of an adequate quantitative evaluation by course for analyzing the students' performance, the present work aims to use genetic programming and linear regression to establish a systematic academic monitoring approach based on the evaluation scores of the first two bimesters. Therefore, this proposal evaluates the probability of the student not reaching a satisfactory final grade by predicting it with sufficient time to take preventive actions (diagnostic assessment) aiming to support the student learning, such as (i) referral to the academic reinforcement with support of monitors, (ii) parallel supplementary lessons, (iii) study groups, (iv) psychological and/or social assistance, and (v) other kinds of formative assessment. As a result of these initiatives, it hopes the percentage decrease of retained students.

2 Proposed Approach for Academic Performance Evaluation

This study uses the collected data used at the academic records coordination of the Federal Institute of Ceará - campus Crato. It anonymized all records from the four bimesters of the subjects "Logic and Programming Language" and "Web Development Language I". These records refer to the 1st year of the Technical Course in Informatics integrated into High School, 2020 and 2021 (excluding transfer and dropout records). One of these records is the final grade which is computed according to the following equation:

$$M_f = (n_1 + 2 * n_2 + 3 * n_3 + 4 * n_4)/10 \qquad (1)$$

where M_f consists of the final grade, and n_1, n_2, n_3 and n_4 refer to the grades obtained in the four bimesters.

Linear Regression (LR) is a method used to model a linear relationship between a dependent variable and one (or more) independent variables. The process of statistical analysis by linear regression involves the following steps: (i) formulating the models, which consists of choosing the model variables (independent variables, also called systematic or predictive, and the dependent variable, also known as response), (ii) fitting the model, which aims to estimate the linear parameters of the models and determine the functions of the estimates of these parameters; and (iii) performing inference, which verifies the model fitness and carries out the analysis of local discrepancies, which when significant may lead to the choice of another model or to accept the existence of discrepant data [6].

In this work, the independent variables consisted of the grades obtained by the students in the first two academic bimesters (n_1 and n_2), and the response variable consisted of the final grade (M_f). For automation of the steps of the statistical analysis method by linear regression, it opted to use GP4LR, a new regression analysis tool supported by genetic programming (as described in the following section).

To validate the proposed approach, it used a case study. It chose two subjects, "Logic and Programming Language" and "Web Development Language I", to result in two statistical models. It generated a Linear Regression Model (LRM) for each subject from the academic grade register data analysis, taking into account only the grades from the first two bimesters to infer the final grade. With the models obtained, the idea is to estimate the student's final grade two bimesters in advance and, thus, support the planning and execution of preventive initiatives against student retention. In addition to these two LRM, it generated a third model with the data from all the subjects of the mentioned course for the years 2020 and 2021. Thus, it compared these three models to investigate their generalization ability under datasets with different subjects, as detailed in Sect. 4.

3   GP4LR Tool

Genetic Programming and Linear Regression have been used together in different applications, such as software/hardware projects [7], weather forecasting [8], food quality assessment [9], among others. This combination, in contrast to Symbolic Regression, addresses a new class of problems, making it necessary to explore it in order to establish its main characteristics and demands with support of error analysis.

GP4LR is a tool that aims to support statistical analysis and has as its main objective the selection of an LRM with the support of the Genetic Programming (GP) technique. The GP parameters used in this tool follow those commonly adopted in the Genetic Algorithms literature [10]. The sequence of steps for a statistical analysis using the tool presented here is given in five stages: 1) initially, it is necessary to select the dataset which will be analyzed in the study; 2) configure the parameters of the GP algorithm; 3) set up the LR parameters; 4) the dataset is processed, according to the parameters configured in the previous steps; and, finally, 5) a report is generated containing statistical analysis results related to the LRM obtained, as the best solution for the dataset under analysis. This process is iterative and can be repeated countless times.

4   Experimental Results

In order to conduct the experiments, it has been necessary to randomly split the original dataset into training and testing sets, with proportions of 70% and 30% respectively. The datasets for the subjects "Logic and Programming Language" (dataset 1) and "Web Development Language I" (dataset 2) contain 110 and 79 records, respectively; as well as the dataset with all the subjects of the mentioned course, for the years 2020 and 2021 (dataset 3), containing 5.732 records.

The following parameters of the GP technique have been used to generate the LRMs: 1) 100 individuals for the initial population size; 2) The removal of duplicate individuals has been considered; 3) Elitism; 4) Mutation rate of 5%; 5) Best individual for 30 generations and 100 as the maximum number of generations, both for stopping criterion; 6) 3 individuals to constitute the tournaments; and 7) For the evaluation criteria of genetic individuals, the RMSE metric has been used to fit the LRMs to the data from the datasets. The RMSE has been chosen because it accentuates the magnitude of the errors, thus highlighting the presence of outliers. The other parameters have been chosen after an empirical and arbitrary analysis.

After executing the GP4LR tool 30 times to generate each of the models (1, 2 and 3), the same LRM has been obtained for the three datasets. The resulting simplified model is given in Eq. (2):

$$M_f \sim n_1 + n_2 + n_1 * n_2 \tag{2}$$

Table 1 illustrates the results of checking the fitness of the LRMs to the training datasets and to verify the assumptions about the nature of their presented errors, in which: Mann-Whitney-Wilcoxon and Coefficient of Determination (R^2) have been used to evaluate the fitness of the LRMs to the training datasets; Kolmogorov-Smirnov for normality tests; Breusch-Pagan for homoscedasticity tests; and Durbin-Watson for tests of independence of residuals [11].

Table 1. Quality checking results of the LRMs fitted to the training dataset

Metrics for training datasets	Datasets		
	1	2	3
Mann-Whitney-Wilcoxon Hypothesis Test (Fitness, p-value)	0.64	0.83	0.49
Coefficient of Determination (R^2) (Fitness)	0.76	0.71	0.76
Kolmogorov-Smirnov Hypothesis Test (Residuals Normality, p-value)	0.0	0.0	0.0
Breusch-Pagan Hypothesis Test (Residuals Homoscedasticity, p-value)	0.0	0.0	0.0
Durbin-Watson Hypothesis Test (Residuals Independence, p-value)	0.77	0.64	0.97

According to Table 1, it can be seen that the LRMs vary from medium to good for the model fitness to the training dataset data, according to the Mann-Whitney-Wilcoxon and R^2, and the independence of errors (Durbin-Watson Hypothesis Test). The tests for normality (Kolmogorov-Smirnov) and homoscedasticity (Breusch-Pagan) failed (probably due to the existence of outliers, which respective records have not removed from the datasets). However, analyzing the plots in Fig. 1, it can be concluded that the distribution of errors tends to follow a Normal distribution (histograms charts at Fig. 1 (a), (b) and (c)) and the variance of the errors tends to be constant (scatter plots charts at Fig. 1 (d), (e) and (f)), thus validating the LRMs obtained for the training datasets.

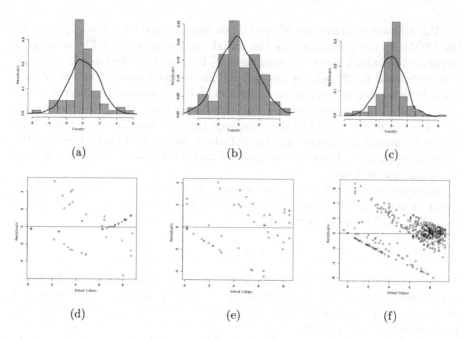

Fig. 1. Graphical residual analysis: histograms and scatter plots of the residuals

To evaluate the prediction performance of the obtained LRMs, the metrics RMSE (1.81, 1.72 and 1.75), MAE (1.25, 1.44 and 1.15) and R^2 (0.74, 0.82 and 0.74) have been computed considering the observed values in the test sets of the datasets 1, 2 and 3 and the predicted values by using the LRMs, respectively. Comparing the values obtained for the RMSE and MAE metrics, it is observed that the RMSE values have larger magnitudes than the MAE values. However, by jointly analyzing the RMSE and MAE results, it is verified that they contain approximate values, thus it can conclude a low variability of the errors [12]. This can also be verified from the calculation of the Variance Account For (VAF) metric, which values for datasets 1, 2 and 3, consisted of 74.5%, 82.7% and 81.9% (results closer to 100% are better [13]). Finally, the R^2 metric, which is used as a measure of the quality of fitness of the LRMs, presents the percentages of the total variation in which the LRMs explain the data in the test sets. Thus, the obtained models 1, 2 and 3 explain 74%, 82% and 74% of the datasets 1, 2 and 3, respectively.

5 Conclusions

The obtained LRMs can be employed for modeling and predicting the students' final grades from the grades of the first two bimesters. Comparing the case study scenarios, it is possible to conclude that the LRMs showed similar behaviors/trends and, consequently, the technique presented in this work can be employed for data analysis of different subjects.

Both quantitative and qualitative evaluation approaches are very relevant for monitoring the students' performance [1]. But the qualitative aspects evaluation is a task that demands a lot of time, human and financial resources, and the result can arrive too late, that is, after the student has been retained or given up on the course. Thus, as future work, new variables can be explored to enrich the work presented here. Since the automation of LRM parameter checking, contemplated by the GP4LR tool, makes it possible to investigate how the incorporation of new qualitative or quantitative variables (e.g. related to socioeconomic data) can be explored, thus contributing to the investigation of new hypotheses related to other pedagogical processes, such as dropout. In addition, the approach proposed in this work analyzed a dataset that only included students' grades from a technical school. Therefore, it may be interesting to investigate the performance of this approach in a context with more subjective evaluation metrics, such as in arts or social science classes.

References

1. Schoeffel, P., Wazlawick, R.S., Ramos, V.: Impact of pre-university factors on the motivation and performance of undergraduate students in software engineering. In: 2017 IEEE 30th Conference on Software Engineering Education and Training (CSEE&T), pp. 266–275 (2017)
2. Gibbs, G., Simpson, C.: Conditions under which assessment supports students' learning. Learn. Teach. High. Educ. 3–31 (2005)
3. Black, P., Harrison, C., Lee, C.: Assessment for Learning: Putting It Into Practice. McGraw-Hill Education (UK), London (2003)
4. Entwistle, N.: Concepts and conceptual frameworks underpinning the ETL project. Occas. Rep. **3**, 3–4 (2003)
5. Luckesi, C.C.: Avaliação da aprendizagem escolar: estudos e proposições. Cortez editora (2014)
6. Nelder, J.A., Wedderburn, R.W.M.: Generalized linear models. J. Roy. Stat. Soc. Ser. A (Gen.) **135**(3), 370–384 (1972)
7. Esmeraldo, G., Barros, E.: A genetic programming based approach for efficiently exploring architectural communication design space of MPSOCs. In: 2010 VI Southern Programmable Logic Conference (SPL), pp. 29–34 (2010)
8. Babovic, V., Keijzer, M.: Rainfall runoff modelling based on genetic programming. Hydrol. Res. **33**(5), 331–346 (2002)
9. Arnaldo, I., Krawiec, K., O'Reilly, U.: Multiple regression genetic programming. In: Proceedings of the 2014 Annual Conference on Genetic and Evolutionary Computation, GECCO 2014, pp. 879–886 (2014)
10. Goldberg, D.E.: Genetic Algorithms. Pearson Education India, Noida (2006)
11. Weisberg, S.: Applied Linear Regression, vol. 528. Wiley, Hoboken (2005)
12. Willmott, C.J., Matsuura, K.: Advantages of the mean absolute error (MAE) over the root mean square error (RMSE) in assessing average model performance. Climate Res. **30**(1), 79–82 (2005)
13. Faradonbeh, R.S., Jahed Armaghani, D., Monjezi, M.: Development of a new model for predicting flyrock distance in quarry blasting: a genetic programming technique. Bull. Eng. Geol. Env. **75**(3), 993–1006 (2016). https://doi.org/10.1007/s10064-016-0872-8

Where Is the AI? AI Literacy for Educators

Lesley Wilton(✉) ⓘ, Stephen Ip, Meera Sharma, and Frank Fan

University of Toronto, Toronto, ON, Canada
`Lesley.Wilton@utoronto.ca`

Abstract. This paper responds to the emerging call from researchers of many disciplines (computer science, engineering, learning sciences, HCI community, education) to address the need for fostering AI literacy in those with or without technical backgrounds. There is an urgent need for research to support educators' understandings of the potential challenges and opportunities surrounding the appropriate and responsible use of AI tools in formal education spaces. This contribution to the scholarly literature is based on three years of reflective data gathered from an author-instructor's experiences of working with graduate students who identify and analyze AI applications in an introductory AIED course. The course was designed by the author-instructor to critically examine ethics, bias, privacy, inclusion, data collection and explainability in popular AIED tools. The emerging scholarship on AIED is reviewed to identify common understandings and justifications of AI literacy. Reflective data is shared to highlight the need for educators to better understand the implications of integrating AI applications into teaching. This article is intended to inspire the promotion of AI Literacy for educators (AILE) and to contribute to the development of meaningful AI literacy frameworks and guidelines.

Keywords: Artificial intelligence in education (AIED) · AI literacy for educators (AILE) · AI and ethics · Teacher education · Higher education · K-12 education · AI literacy guidelines

1 Introduction to Artificial Intelligence in Education (AIED) for Educators

1.1 Why Educators Need an Informed Understanding of AI

Advances in artificial intelligence (AI) have led to a prevalence of AI tools in everyday lives (smartphone applications, Google, Alexa, Siri, toys, games, and more). It is expected that educators are teaching learners who commonly engage with AI-supported applications. ISTE (2019) explains that while only 33% of people think they are using AI, nearly 77% already are [8]. Teachers and their students may recognize that the underlying technology involves AI of some kind but may not necessarily understand its implications. Dignum explains that the concept of AI is multifaceted and broad as it "deals not only with how to represent and use complex and incomplete information logically but also with questions of how to see (vision), move (robotics), communicate

© Springer Nature Switzerland AG 2022
M. M. Rodrigo et al. (Eds.): AIED 2022, LNCS 13356, pp. 180–188, 2022.
https://doi.org/10.1007/978-3-031-11647-6_31

(natural language, speech) and learn (memory, reasoning, classification)" [6]. Dignum further explains that the responsible use of AI means that everyone "should be able to get proper information about what AI is and what it can mean for them, and also to have access to education about AI and related technologies" [6]. The need for responsible and trustworthy AI is called for worldwide [6, 24]. Responsible AI "is about ensuring that AI systems are ethical, legal, beneficial and robust, that these properties are verifiable, and that organizations that deploy or use these systems are held accountable" [23]. Given the rapid pace of AI application development and the need for common understandings of what AI applications are and what they do, we are looking at how to address these issues within the field of education. This paper looks at the author-instructor's reflective experiences in the context of understanding the implications of AI in education (AIED) and the need for AI literacy for educators (AILE).

1.2 Context

The author-instructor draws on reflections gathered from three years of experience of developing and teaching an introduction to AIED course to higher education students (pre-service, in-service and graduate level). With self-study as a methodological frame for inquiry-guided research [2, 21], the AIED researchers and author-instructor are examining AI applications in educational contexts. This study is a response to the call for educators to engage in important AIED and AILE conversations. This examination begins by exploring the need to identify common understandings of AI concepts, particularly for those without technical backgrounds [1, 6, 11]. This is followed by a review of the importance of identifying responsible and ethical concerns in AIED and the need to foster AILE through guidelines, defined competencies and learning opportunities to prepare educators and their students for the everyday integration of AI technologies now and in the future [11, 15, 22, 23].

1.3 AI as a Starting Point

As a starting point, Luckin (2018) [13] provides a broad definition of AI as "technology capable of actions and behaviors requiring intelligence when done by humans." While AIED research has been taking place for over three decades, primarily in computer science, AIED must be recognized as more than the implementation of AI technology in education. We must also consider the integration of pedagogical, social and economic dimensions when incorporating AIED [16, 17, 19, 20]. There is a need for developing ethical principles and practices for AI applications in education [7]. The following views of AI in the context of education guide the challenges we aim to address:

- AI literacy for educators (AILE) encompasses basic competencies and fundamental skills for educators to understand, create, use, apply, and evaluate AI applications in an educational context. There is a need for AI literacy to develop appropriate technological and pedagogical strategies [15, 18];
- Educators need to be supported to update their AI knowledge to enable them to address contemporary teaching opportunities and challenges in responsible ways. Examples

of this are current considerations of student safety and privacy, and explainable under-standings of AI-supported personalized learning applications that determine learning pathways and recommend grades [15]; and

- It is important to educate teachers on human-centered considerations around AI that foster social responsibility and ethical awareness of AI for societal work. Beyond enhancing one's AI abilities and interests, we must also consider inclusiveness, fairness, accountability, transparency, bias and ethics [15, 25].

1.4 AI Literacy

There are valid concerns about the uncertainties of identifying the underlying AI tech-nology in educational tools. Common misconceptions can limit educators' abilities to effectively use and understand the risks of engaging with AI. One promising dimen-sion of AIED focuses on fostering AI literacy [3, 16]. AI Literacy is an emerging term describing the competencies necessary to critically evaluate AI technologies; to com-municate and collaborate effectively with AI; and to use AI as a tool in the classroom [12]. The Council of Birmingham City School of Education and Social Work reports that the term AI Literacy has evolved to mean the development of skills and technical awareness around the application, practical considerations, and transformative thinking about AI education [4].

AI literacy is defined in emerging literature as an individual's ability to access and use AI-related knowledge and skills to build sound understandings about the principles of AI and its applications [5, 9, 10]. Looking at the K-12 context, for example, scholars note that AI literacy in K-12 education is at its starting point [15]. Ng et al. (2021) [15] identify four common goals of fostering AI literacy through learning curricula: to support the building of fundamental AI concepts; to facilitate the application of AI concepts in applicable contexts; to critically evaluate and engage with AI technologies within those contexts; and to better understand the existing and emerging ethical implications of AI applications. Researchers identify the importance of accounting for those with no prior knowledge and the need for learning artifacts when educating people about AI [15]. They emphasize that the reliable, trustworthy and fair use of AI must be addressed by broadening common understandings of ethics and responsible use not only from within the technical AI field, but also from the field of education. While many recognize that forms of AI technology are increasingly infused in our everyday lives, AI's role in formal education is less clear. Some are predicting that AI can enhance teaching and learning by complimenting instructional and assessment practices through the uses of big data, machine learning and sophisticated prediction. Some see the promise of AI in the fulfillment of supporting roles, such as chat-bots and intelligent tutors. Others are concerned about the impact of AI on educators and learners, particularly related to security, privacy, data collection, unexplained decision-making, inherent bias, job loss and loss of control.

1.5 Responding to the Need for AIED Understandings and AILE Supports

Aligned with the needs highlighted above, the author-instructor designed the introduction to AIED course for educators to address the key overarching questions: What definitions, terminology and core concepts of AI are important to understand as they relate to education? How do those in the field of education plan for and understand the complexities and implications of the integration of AI tools? How do educators stay current in the context of rapid and complex AI application developments [14]? What are the impacts and implications of AI integration in education today and in the future?

From the author-instructor's experiences, the following section highlights a sample of some of the AI tools examined for the type of AI, intended effects on learning, bias, ethics, privacy, data collection, inclusion, explainability, and other implementation considerations.

2 Challenges and Opportunities Associated with Fostering AI Literacy

To support educators' development of current understandings of the responsible use of AI in teaching, a fundamental level of AI literacy is required. Through examination of the collected data, it stood out that conceptions of AI greatly influenced assumptions of responsible use. Both the author-instructor and the graduate students were challenged to understand exactly what underlying AI technology was in use in some educational applications claiming to use AI. As a course project, the graduate students could choose any AI-related tool that applied to their context. The author-instructor reflected that despite marketing claims, it was often challenging to identify applications currently used in an educational context that truly featured AI. More than sixty AIED applications of interest have been identified in our study. Table 1, below, highlights a few examples of AI tools examined by the instructor and the learners.

Table 1. Sample of AI applications explored by educators.

AI tool	Description provided from developer/marketer	More information at
ALEKS	ALEKS is a research-based, adaptive learning platform designed to offer quality online tutoring and assessment programs for math, chemistry, statistics and business. This application has been trained to efficiently identify the exact topics each individual student has mastered, and which ones they are ready to learn, to accurately place them moving forward	www.aleks.com/

(*continued*)

Table 1. (*continued*)

AI tool	Description provided from developer/marketer	More information at
ELSA SPEAK	ELSA Speak is an artificial intelligence-powered language application designed to democratize English learning for learners anywhere. It uses speech technology with AI and deep learning to detect users' mispronunciations and provide real-time feedback with specific improvement suggestions	elsaspeak.com/en/
MATHia	MATHia is an adaptive learning math software that personalizes instruction for middle- and high-school students based on how they learn over time. It uses artificial intelligence and cognitive science to mirror a human tutor with more complexity and precision than any other math software	carnegielearning.com/solutions/math/mathia/
MuseNet	MuseNet is a deep neural network that uses unsupervised multi-purpose technology such the GPT-2 Sparse Transformer to create musical compositions with different instruments and musical styles. The transformer allows MuseNet to predict the next note based on a given set of notes and input positions	openai.com/blog/musenet/

(*continued*)

Table 1. (*continued*)

AI tool	Description provided from developer/marketer	More information at
Packback	Packback is an inquiry-driven online discussion platform designed to improve student motivation and critical thinking through discussions. It incorporates machine learning to scale instant 1-to-1 student feedback, automatically moderate the discussion community, and support grading for any class sizes	www.packback.co/
SnatchBot	SnatchBot is a chatbot platform that processes human conversation, allowing students to communicate with the digital device as if they were interacting with a real person. From answering questions about registrations to providing information on student groups, SnatchBot is praised for its capability to provide logistical support for student inquiries in a timely manner, alleviating staff workloads and increasing student satisfaction. This application attempts to move the capabilities of intelligent virtual assistants towards supporting teachers by identifying learning gaps, driving efficiencies, and streamlining educational tasks	snatchbot.me/

Table 1 provides the name of the tool, a short description of the functions and features of the tool supplied by the manufacturer/developer, and a link to the website for more information.

In addition to the data in Table 1, the following observations highlight some of the challenges encountered in examining AI applications in educational contexts.

1. The underlying AI technology can be difficult to identify/understand. Some tools that are promoted as featuring AI, may simply be referring to AI as a marketing strategy. Of those tools that truly feature AI, many AI features are difficult to understand.

2. Many of the product's explanations of the tool's use of AI were complex or limited. While this may have been a result of patented technology, these complicated or incomplete descriptions obscure educators' understandings of the tool.

3. Technology transience [14] creates surmounting pressure for educators. Free AI applications appeal to educators who face financial constraints. Some tools researched within the past few years can no longer be found online. An example of this is the chatbot feature of Duolingo.

4. Identifying those factors affecting issues of bias, ethics, explainability, and so on was very difficult. Some applications do not provide the specific detail about data collection by the AI tool. This is particularly concerning for free AI-based educational tools which may silently collect data. Undisclosed data collection is concerning given educators' obligations to protect student privacy. Privacy of student data is often regulated through established laws and policies. Educators must remain mindful of professional obligations. Nonetheless, who is responsible when educators do not understand that student privacy may be compromised? How do educators make sense of terms of service jargon that is often written in language that is difficult to understand? What happens when explainability is beyond the scope of an educators' understanding?

5. Although AI tools can display smart computation in an educational domain, they can fail to enhance learning. Some of AI features in educational applications were intended for a general situation that could not address the needs of a particular domain, specific learning activities, or teaching goals.

3 Conclusion and Future Steps

It is essential to continue to address the need for deepening educators' understandings of AIED and for supporting the development of AILE opportunities. The following suggestions are proposed for consideration.

1. AILE course development. It is recommended that AI Literacy for Educators (AILE) course development be expanded within the education discipline. An introductory course should foster educators' understandings of AI concepts and terminologies, and support the identification of appropriate AI tools for safe and responsible use in an educational context.

2. Interdisciplinary collaborations. Opportunities for those in the field of education to collaborate with other disciplines such as Computer Science (CS), Engineering or Human Computer Interaction (HCI) communities should be fostered. This collaboration is twofold. Teacher education (pre-service and in-service) offers an important opportunity for developing AI competencies [12, 18]. CS, Engineering or

HCI expertise could support the development of these competencies while benefiting from educator views. Scholars in other AI research communities would also benefit from pedagogical, curriculum, and learner perspectives.

3. Sharing the findings and community building. We encourage the dissemination of findings like these, and the fostering of learning communities. We are a small research group seeking to continue our research and share our findings [23]. We hope to bridge the gap between academia, business, and non-profit as we seek to prepare future generations for safe, responsible, and ubiquitous AI integration in education. We wish to initiate a SIG to include those from many fields, including education, as we address the need for AILE.

We end this paper by extending an open invitation to engage in discussions about best practices for advancing opportunities for those in education to engage with AIED and AI literacy.

References

1. Adams, C., Pente, P., Lemermeyer, G., Rockwell, G.: Artificial intelligence ethics guidelines for K-12 education: a review of the global landscape. In: Roll, I., McNamara D., Sosnovsky, S., Luckin, R., Dimitrova, V. (eds.) AIED 2021, LNAI, vol. 12749, pp. 24-28. Springer, Cham (2021). https://doi.org/10.1007/978-3-030-78270-2
2. Berry, A., Crowe, A.R.: Many miles and many emails: using electronic technologies in self-study to think about, refine and reframe practice. In: Fitzgerald, L., Heston, M., Tidwell, D. (eds.) Research Methods for the Self-study of Practice, pp. 83–96. Springer, Netherlands (2009). https://books.scholarsportal.info/en/read?id=/ebooks/ebooks0/springer/2010-02-11/2/9781402095146
3. Cetindamar, K., Wu, M., Zhang, Y., Abedin, B., Knight, S.: Explicating AI literacy of employees at digital workplaces. IEEE Trans. Eng. Manag. 1–14 (2022). https://doi.org/10.1109/TEM.2021.3138503
4. Cui, V., Wheatcroft, L.: AI Literacy – the role of primary education. Birmingham City School of Education and Social Work (2021). https://www.bcu.ac.uk/education-and-social-work/research/news-and-events/cspace-conference-2021/blog/ai-literacy-the-role-primary-education
5. Dai, Y., Chai, C.S., Lin, P.Y., Jong, M., Guo, Y., Qin, J.: Promoting students' well-being by developing their readiness for the artificial intelligence age. Sustainability 12(16), 6597 (2020). https://doi.org/10.3390/su12166597
6. Dignum, V.: Responsible Artificial Intelligence: How to Develop and Use AI in Responsible Way. Springer, Switzerland (2019)
7. Hwang, G.J., Xie, H., Wah, B.W., Gasevic, D.: Vision, challenges, roles and research issues of artificial intelligence in education. Comput. Educ. Artif. Intell. (1) (2020). https://doi.org/10.1016/j.caeai.2020.100001
8. ISTE: Teach AI: Prepare our Students for the Future (22 Jun 2019). Available at https://youtu.be/YmZJmmys7Ps
9. Kandlhofer, M., Steinbauer, G.: A driving license for intelligent systems. In: Proceedings of AAAI Conference on Artificial Intelligence, vol. 32, no. 1, pp. 7954–7955. AAAI Press (2018). https://ojs.aaai.org/index.php/AAAI/article/view/11399
10. Lee, I., Ali, S., Zhang, H., DiPaola, D., Breazeal, C.: Developing middle school students' AI literacy. In: Proceedings of the 52nd ACM Technical Symposium on Computer Science Education (SIGCSE '21). Association for Computing Machinery, pp. 191–197 (2021). https://doi.org/10.1145/3408877.3432513

11. Long, D., Magerko, B.: What is AI literacy? Competencies and design considerations. In: Proceedings of the 2020 CHI Conference on Human Factors in Computing Systems. Association for Computing Machinery, pp. 1–16 (2020). https://doi.org/10.1145/3313831.3376727
12. Luckin, R., Holmes, W., Griffiths, M., Forcier, L.B.: Intelligence Unleashed: An argument for AI in Education. Pearson, London (2016)
13. Luckin, R.: Machine Learning and Human Intelligence: The Future of Education for the 21st Century. UCL Press, London (2018)
14. Muilenberg, L.Y., Berge, Z.L.: Revisiting teacher preparation: responding to technology transience in the educational setting. Q. Rev. Dist. Learn. 16(2), 93–105 (2015)
15. Ng, D.T.K., Leung, J.K.L., Chu, K.W.S., Qiao, M.S.: AI literacy: definition, teaching, evaluation and ethical issues. In: Proceedings of the Association for Information Science and Technology, vol. 58, no. 1, pp. 504–509 (2021). https://doi.org/10.1002/pra2.487
16. NSF AI Institute for Engaged Learning: The National Science Foundation AI Institute for Engaged Learning. https://sites.google.com/ncsu.edu/ai-engage/home
17. Ouyang, F., Jiao, P.: Artificial intelligence in education: the three paradigms. Comput. Educ. Artif. Intell. (2) (2021). https://doi.org/10.1016/j.caeai.2021.100020
18. Pedro, F., Subosa, M., Rivas, A., Valverde, P.: Artificial Intelligence in Education: Challenges and Opportunities for Sustainable Development. UNESCO (2019)
19. Perrotta, C., Selwyn, N.: Deep learning goes to school: toward a relational understanding of AI in education. Learn. Media Technol. 45(3), 251–269 (2020). https://doi.org/10.1080/17439884.2020.1686017
20. Selwyn, N., et al.: What's next for Ed-Tech? Critical hopes and concerns for the 2020s. Learn. Media Technol. 45(1), 1–6 (2020). https://doi.org/10.1080/17439884.2020.1694945
21. Tidwell, D., Heston, M., Fitzgerald, L.: Introduction. In: Fitzgerald, L., Heston, M., Tidwell, D. (eds.) Research Methods for the Self-study of Practice, pp. 1–16. Springer, Netherlands (2009). https://books.scholarsportal.info/en/read?id=/ebooks/ebooks0/springer/2010-02-11/2/9781402095146
22. Touretzky, D., Gardner-McCune, C., Martin, F., Seehorn, D.: Envisioning AI for K-12: what should every child know about AI? In: Proceedings of the AAAI Conference on Artificial Intelligence, vol. 33, no. 1, pp. 9795–9799. AAAI Press (2019). https://doi.org/10.1609/aaai.v33i01.33019795
23. UNICEF: Policy guidance on AI for children 2.0 (2021). https://www.unicef.org/globalinsight/media/2356/file/UNICEF-Global-Insight-policy-guidance-AI-children-2.0-2021.pdf
24. Verdegem, P.: AI for Everyone: Critical Perspectives. University of Westminster Press, London (2021)
25. Vincent-Lancrin, S., van der Viles, R.: Trustworthy artificial intelligence (AI) in education: promises and challenges. OECD Education Working Papers. OECD Publishing, Paris (2020). https://doi.org/10.1787/a6c90fa9-en

Posters and Late-Breaking Results

Increasing Teachers' Trust in Automatic Text Assessment Through Named-Entity Recognition

Candy Walter[✉][iD]

University of Hildesheim, Hildesheim 31141, Germany
candy.walter@uni-hildeheim.de

Abstract. In this study, we investigate teachers' trust in automatic text assessments and explore whether teachers' trust increases when we show them, in addition to the rating classification score, contextual didactic keywords (e.g., important didactic terms or persons, so-called entity) in the text through named-entity recognition (NER). The results of the present study make it clear that for the majority of the participating teachers (N = 34), trust in automatic text assessment increases as soon as they are shown important task-related keywords in the text. In order to increase teachers' acceptance and trust in the assessment result of an automatic text response, we recommend that when developing future correction-supporting assessment tools, the automatic extraction of important task-related didactic keywords from the text should be considered in addition to text score.

Keywords: Named-entity recognition (NER) · Natural language processing (NLP) · Text assessment systems · Trust

1 Introduction

The processing of human languages, such as English or German, is known in the context of artificial intelligence (AI) and machine learning (ML) as *natural language processing* (NLP). There are numerous studies that not only demonstrate the usefulness of trained NLP systems in reducing the assessment burden in teaching, but also confirm the reliability of fully automated text assessment over human correction [2]. A major challenge is still to increase teachers' trust in NLP-based applications, for automatic text assessment in education. To counteract this, we have developed an NLP system that is capable of automatic text assessment [3] as well as extraction of task-related didactic keywords [4]. In an exploratory case study, we pursued the following question: "Does teachers' trust in an automatic assessment system increase when we highlight contextual didactic terms in the response text in addition to the text scoring?"

2 An NLP System for the Analysis of Mathematics Didactic Free Text Answers

To answer the question, we have developed an NLP system that consists of the combination of artificial neural networks [3,4], which we have trained with the

ⓒ Springer Nature Switzerland AG 2022
M. M. Rodrigo et al. (Eds.): AIED 2022, LNCS 13356, pp. 191–194, 2022.
https://doi.org/10.1007/978-3-031-11647-6_32

help of German-language student responses from various mathematics didactic lectures. For this purpose, student teachers from different universities were asked to respond to various mathematics didactic tasks in writing – e.g., "Explain the EIS principle known according to J. Bruner." We classified all students' responses with "correct" and "incorrect". Our classifications were checked by independent peers. Subsequently, all student responses were artificially augmented via the online language API *DeepL* with the back-translation method, resulting in a corpus of about 3600 individual responses on average for the training of the ML models for each task. The underlying algorithm classifies the responses based on a trained recurrent neural network (RNN) with LSTM layers [3]. For the extraction of didactic keywords and their visualization in the text, we used the open source library *spaCy* and the annotation tool *doccano*. A language model that can filter mathematics didactic entities from German texts was trained. For example, the used language model recognizes the words enactive, iconic, and symbolic as "forms of representation", or it extracts from texts the word EIS principle as a "teaching concept". Finally, the trained models were merged, into a web application using the framework *flask*, resulting in a ready-to-use NLP system for the browser.

3   Case Study: Objectives, Implementation and Results

Objectives and Test Persons: The aim of this study is to evaluate teachers' confidence in automatic text analysis by means of a psychometric scale measurement (Likert scale [1]) and qualitative interviews in a comparative way, in order to get clues for future developments of innovative assessment systems in education. The sample consisted of 34 teachers aged 27 to 54 years who teach mathematics and its didactics at different German educational institutions. In order to minimize person-related confounding variables, such as the influence of "prior knowledge" and "attitudes toward digital media", care was taken when selecting the subjects to ensure that they had a certain degree of willingness and openness to use innovative digital teaching support systems. However, they had no prior knowledge regarding automated text assessments or other ML-based assessment tools.

Implementation: In order to avoid external influences, the interviews took place in a quiet environment of the respective teaching institute. Each interview lasted about 15 min, during which the subjects were asked to assess an automatically assessed free-text response in two different test scenarios. First, teachers were shown automatic assessment of student responses without NER analysis and afterwards including NER analysis. Study participants indicated their trust in the automatic text assesment on a 4-point Likert scale for the item in both situations: "I trust the result of the automatic text assessment and agree with the system assessment". The response scale for the level of agreement with the item ranged from complete rejection to complete approval and included the four points: (1) *Strongly disagree*; (2) *Disagree little*; (3) *agree quite a bit* and (4) *Strongly agree*. Since we wanted the subjects to take a concrete position in the tests, we decided against a middle category with the response option *Neither agree nor disagree* in the scale construction. During the tests, the teachers were interviewed and their statements were documented. To ensure that the study participants did not read the student responses in the text field and thereby influence their statements by their own evaluations, the responses were concealed by us after the analysis – approximately 2 sec after text entry.

Results: For the outcome evaluation, the responses from the tests were compared with the interview statements of the subjects. For each test, the item provided different intensities for the measured attribute: "Trust". In each test, all participating teachers convincingly positioned themselves on one of the four scale responses offered. We compared both the absolute values and the average score of the test responses to identify response tendencies regarding teachers' trust in the NLP system. The average score can be between 1 and 4 according to the Likert scale. Where values less than or equal to 2 can be interpreted with a rather disagreeing position and values greater than 2 with a rather agreeing position.

In the *first test*, 32.35% study participants do not trust the automatic assessment by the system at all and 38.24% have a rather low level of trust. In the interview, 18 teachers (approximately 52.94%) indicated that they have little trust in the NLP system because they do not receive additional information on text analysis and their concerns are high that correct student responses will be classified as incorrect by the system. These concerns are independent of the age and gender of the interviewees. However, 29.41% of the teachers interviewed, trust the system quite a bit. In absolute numbers it can be recognised that 24 out of 34 teachers interviewed do not trust or only slightly trust the automatic text assessment without NER analysis. Only 10 study participants trust the NLP system without NER analysis, not a single test person trusts the assessment tool completely.

In the *second test*, 11.76% of study participants have no trust at all and 26.47% have bit of trust in the system. In contrast, 61.76% of teachers now have quite a bit of trust in the automatic text assessment due to the additional keywords displayed. In other words, 13 of the 34 teachers interviewed also show slight or no trust in automatic text assessment in the second test, while 21 teachers now fairly trust the NLP system with named-entity recognition. Also, in the second test, not a single teacher fully trusts the assessment tool. The interview statements of the teachers indicate that the gain in trust is accompanied by increased assessment transparency through the NLP system. The additional automatic indication of didactic technical terms makes the result of the text assessment more trustworthy for teachers. Six out of 34 teachers note that the extracted technical terms can also be used as an opportunity for feedback to the students. In addition, for almost all teachers interviewed, the points of system security, equal opportunities, data protection, and privacy are significant basic requirements that significantly influence and determine trust and acceptance in automatic assessment systems.

Comparison of Test Results: The additional visualization of important didactic keywords increases teachers' trust in automatic text assessment by 32.35%. On the other hand, the complete rejection of the automatic text assessment decreases by 20.59%. The proportion of teachers' who have little trust in the assessment tool also decreases by 11.77% when NER analysis is included in addition to text assessment. Interestingly, roughly 11.8% of study participants which didn't trust the NLP system in the first test, had quite a bit of trust in the second test, due to additional NER analysis. In the first test, the teachers tended to select the second response on the Likert scale (average score: 1.97) and thus have recognizably little trust in the result of the automatic text assessment. The trust of the teachers' in the NLP system increases as soon as they are offered, in addition to the text assessment, the automatic extraction of didactic keywords from the text. In this case, teachers seem to have quite a bit of trust in automatic text assessment. In the second test, the interviewed teachers, with an average score of 2.5, the majority tends toward the third response of the Likert scale. The fact that no study participant fully trusts the automated assessment system is positive in the context of the study. It demonstrates responsible and reflective teacher behavior when using digital technology in educational context.

4   Conclusions

Innovative assessment systems can massively reduce the correction effort assessing written work and demonstrably relieve teachers. The willingness to use trained NLP systems for teaching and learning in everyday life increases and decreases with the trust of teachers in the functionality of such systems. The results of the present study show that the majority of participating teachers have a higher trust in automatic text assessment when task-related entities are visibly highlighted in the text. We therefore recommend that in the future development of NLP systems for correcting text responses, care must be taken to ensure that teachers are not only informed about the scoring result of the analyzed response, but that they are also shown important didactic keywords from the response text. The additional text-related information increases teachers' trust in the analyzing assessment tool.

References

1. Batterton, K. A., Hale, K. N.: The likert scale what it is and how to use it. Phalanx **50**(2), 32–39 (2017). www.jstor.org/stable/26296382
2. Condor, A.: Exploring automatic short answer grading as a tool to assist in human rating. In: Bittencourt, I.I., Cukurova, M., Muldner, K., Luckin, R., Millán, E. (eds.) AIED 2020. LNCS (LNAI), vol. 12164, pp. 74–79. Springer, Cham (2020). https://doi.org/10.1007/978-3-030-52240-7_14
3. Hochreiter, S., Schmidhuber, J.: Long short-term memory. Neural Comput. **9**(8), 1735–1780 (1997). https://doi.org/10.1162/neco.1997.9.8.1735
4. Jiang, J.: Information extraction from text. In: Aggarwal, C., Zhai, C. (eds.) Mining Text Data, pp. 11–41. Springer, Boston, MA. (2012). https://doi.org/10.1007/978-1-4614-3223-4_2

A Transparency Index Framework for AI in Education

Muhammad Ali Chaudhry[✉], Mutlu Cukurova, and Rose Luckin

University College London, London, UK
dtnvma1@ucl.ac.uk

Abstract. Numerous AI ethics checklists and frameworks have been proposed focusing on different dimensions of ethical AI such as fairness, explainability, and safety. Yet, no such work has been done on developing transparent AI systems for real-world educational scenarios. This paper presents a Transparency Index framework that has been iteratively co-designed with different stakeholders of AI in education, including educators, ed-tech experts, and AI practitioners. We map the requirements of transparency for different categories of stakeholders of AI in education. The main contribution of this study is that it highlights the importance of transparency in developing AI-powered educational technologies and proposes an index framework for its conceptualization for AI in education.

Keywords: AI in education · Transparency in AI · Algorithmic transparency · AI development pipelines · Bias in AI · Human-centred AI

1 Introduction and Literature

Ethical AI is a rapidly developing field with a number of tools, frameworks and research papers coming out at a high frequency. Considering the blurred boundaries between different dimensions of ethical AI like fairness, accountability, transparency and explainability, it is important to first define what we mean by transparency in the context of AI. Both, AI and transparency can be interpreted in many ways [1]. In the context of this study, transparency in AI refers to a process with which all the information, decisions, decision-making processes and assumptions are made available to be shared with the stakeholders and this shared information has the potential to enhance the understanding of these stakeholders.

While there has been significant work on ethical AI [2], the focus on transparency as a necessary construct to enable ethical AI is limited. More specifically in education, it is even fewer [11]. The research presented in this paper aims to fill in this gap by presenting and evaluating a framework that tackles the challenges of ethical AI through transparency in the entire product lifecycle of machine learning powered AI products in educational contexts, from the initial planning of an AI system till it is deployed and iteratively improved in the real world.

Transparency of machine learning powered AI products in education is a relatively new concept compared to other sectors like healthcare, recruitment or the justice system

© Springer Nature Switzerland AG 2022
M. M. Rodrigo et al. (Eds.): AIED 2022, LNCS 13356, pp. 195–198, 2022.
https://doi.org/10.1007/978-3-031-11647-6_33

where AI-powered products are being widely used. Within AIED, the transparency of AI-powered products needs to be considered within the context of ethics of education like what is the purpose of learning (to pass exams or help students achieve self-actualization), what is the role of AI (to empower teachers or replace them) and will AI be creating equal access to education or widen the gap between affluent and disadvantaged communities [3]. Recently, there has been a growing interest among AI researchers and practitioners within education in the ethical implications of AI products on learners and teachers. Holmes et al. [4] have presented a community-wide framework for developing ethical AI for education. The Institute for Ethical AI in Education [5] collaborated with over 200 experts in AIED through interviews, roundtables and The Global Summit on the Ethics of AI in Education to prepare an ethical framework for AI in education that provides detailed criteria and a checklist to educators for evaluating AI-powered ed-tech products. They also identify 'transparency and accountability' as one of the objectives that educators need to look for when selecting AI products for their educational institutions.

2 Methodology

A mixed-methods approach was used to first develop, then evaluate and improve the framework developed in this study. The framework was created in three steps: firstly, based on the literature review of different stages of the AI development process: data processing stage, machine learning modelling stage and deployment and iterative improvements stage. In this step, popular frameworks for the documentation, robustness and reproducibility for each of these stages were identified.

In the second step, the selected framework for each stage of the AI development process was applied in the real world during the development of an AI tool in an educational context for a training organization that envisioned to become a leader in educational technology. This application of domain agnostic frameworks for AI tool development in an educational context enabled us to identify any gaps in these frameworks when applied in educational settings.

In the third step, the framework was iteratively evaluated in two phases with three groups of educational stakeholders including educators, ed-tech experts and AI practitioners. 40 candidates were recruited in two phases to participate in this research. 18 candidates responded to this request and participated in interviews.

3 Results and Discussion

In this section, we present the results of this research in the form of the final version of the Transparency Index framework that was built based on the literature review, the co-design of an AI-powered ed-tech tool with trainers and after the interviews with educators, ed-tech experts and AI practitioners. The Transparency Index framework for AI in Education and further details of these processes can be accessed here: https://eda rxiv.org/bstcf.

For the data processing stage, datasheets from Gebru et al. [6] were chosen as the benchmark for documenting different components of the data processing stage in the AI development process. For the Machine Learning modelling stage, model cards by

Mitchell et al. [7] were chosen as a baseline to document the details of the ML model used. Factsheets by Arnold et al. [8] were added as a basic requirement for documenting the usability of AI tools. Some additional requirements were also added for each stage to record the various decisions and assumptions made specifically for the AI-powered products for educational contexts. These requirements were derived from our experience of applying AI to enhance the understanding of domain experts [9] and then improved based on the feedback from educators, ed-tech experts and AI practitioners.

In the interviews, some themes appeared across all the stakeholders from different groups, irrespective of their backgrounds. In contrast, some local themes appeared across all stakeholders from a particular group, with a specific background.

Across all the groups, stakeholders thought that the framework was useful in enhancing their understanding of AI products and where these products can go wrong. Educators stated that the framework could help them get a better understanding of AI-powered ed-tech products and the contexts in which they work best. Some educators who were currently evaluating ed-tech tools were very interested in using the framework as an auditing tool to get a better understanding of these products.

For 17 out of the 18 stakeholders interviewed for this research, the conversation on transparency regarding ethical considerations in AI products was a relatively new phenomenon. Seven educators were using some form of ed-tech and AI products in schools, but ethical AI was a relatively new conversation for them. It seemed they were not aware of the adverse consequences of AI going wrong.

All eighteen interviewees, including most of the AI practitioners who build AI products, did not perceive transparency in machine learning development pipelines the way it was being addressed by the proposed framework in this research. For all nine educators interviewed, this was the first time they were having conversations on ethical AI and what kind of documentation or precautionary measures to expect from companies applying AI in education.

One of the themes that emerged across all three stakeholders of AI in education, including educators, ed-tech experts and AI practitioners focused on how all the groups found transparency through the Transparency Index framework useful. Educators as the users of AI-powered ed-tech products and AI practitioners as developers and providers of AI systems in education believed transparency helped them in understanding their products better. But some researchers have also argued against complete transparency similar to the arguments presented in [10]. Complete transparency like making the code of an AI tool public can hinder innovation as companies will be sharing the secret sauce that makes AI work in certain contexts and provides them with a competitive edge over others. It can be argued that many times users may not even know what information they need. What is useful for them, what kind of impact lack of transparency can have on them or what is too much transparency for them that leads to cognitive overload. This is especially the case for tier 3 users who are not tech experts and do not know exactly what kind of information from the entire AI tool's development pipeline will be useful for them. Our results show that educators, in comparison to other stakeholders such as ed-tech experts, tech enthusiasts and researchers have mostly never had conversations on ethical AI and/or transparency in AI before. AI practitioners also confirmed this by saying that they have never received requests for transparency or concerns about ethical aspects of AI in product development from their clients (educators).

4 Conclusion

The Transparency Index framework proposed in this paper integrates the popular frameworks of ethical AI like Datasheets, Model Cards and Factsheets into one coherent framework that addresses the whole AI product development timeline for Educational Technology interventions. We contextualize these tools in a single framework for their applicability in educational contexts and validate these modifications through interviews with various stakeholders of AI in education. In future, the Transparency Index framework for education can be further developed into a practical tool to evaluate transparency levels of AI-powered ed-tech products for practitioners.

References

1. Weller, A.: Challenges for transparency. arXiv preprint arXiv:1708.01870 (2017)
2. Jobin, A., Ienca, M., Vayena, E.: The global landscape of AI ethics guidelines. Nat. Mach. Intell. **1**(9), 389–399 (2019)
3. Holmes, W., Bialik, M., Fadel, C.: Artificial intelligence in education. Promises and Implications for Teaching and Learning. Center for Curriculum Redesign (2019)
4. Holmes, W., et al.: Ethics of AI in education: towards a community-wide framework. Int. J. Artif. Intell. Educ. (2021)
5. Turilli, M., Floridi, L.: The ethics of information transparency. Ethics Inf. Technol. **11**(2), 105–112 (2009)
6. Gebru, T., et al.: Datasheets for datasets. arXiv preprint arXiv:1803.09010 (2018)
7. Mitchell, M., et al.: Model cards for model reporting. In: Proceedings of the conference on fairness, accountability, and transparency, pp. 220–229 (Jan 2019)
8. Arnold, M., et al.: FactSheets: increasing trust in AI services through supplier's declarations of conformity. IBM J. Res. Dev. **63**(4/5), 6-1 (2019)
9. Kent, C., et al.: Machine learning models and their development process as learning affordances for humans. Lecture Notes in Computer Science, pp. 228–240 (2021)
10. Carabantes, M.: Black-box artificial intelligence: an epistemological and critical analysis. AI Soc. (2019). https://doi.org/10.1007/s00146-019-00888-w
11. Khosravi, H., et al.: Explainable artificial intelligence in education. Comput. Educ.: Artif. Intell. **3**, 100074 (2022). https://doi.org/10.1016/j.caeai.2022.100074

Benchmarking Partial Credit Grading Algorithms for Proof Blocks Problems

Seth Poulsen(✉), Shubhang Kulkarni, Geoffrey Herman, and Matthew West

University of Illinois at Urbana-Champaign, Urbana, IL 61801, USA
sethp3@illinois.edu

Abstract. Proof Blocks (proofblocks.org) is a software tool that allows students to practice writing mathematical proofs by dragging and dropping lines instead of writing proofs from scratch. Because of the large solution space, it is computationally expensive to calculate the difference between an incorrect student solution and some correct solution, restricting the ability to automatically assign students partial credit. We benchmark a novel algorithm for finding the edit distance from an arbitrary student submission to some correct solution of a Proof Blocks problem on thousands of student submissions, showing that our novel algorithm can perform over 100 times better than the naïve algorithm on real data. Our new algorithm has further applications in grading Parson's Problems, task planning problems, and any other kind of homework or exam problem where the solution space may be modeled as a directed acyclic graph.

Keywords: Mathematical proofs · Automated feedback · Scaffolding

1 Introduction

Traditionally, classes which cover mathematical proofs expect students to read proofs in a book, watch their instructor write proofs, and then write proofs on their own. Students often find it difficult to write proofs on their own, even when they have the required content knowledge [6]. Because proofs need to be graded manually, it often takes a while for students to receive feedback.

Proof Blocks (proofblocks.org) is a software tool that allows students to construct a mathematical proof by dragging and dropping instructor-provided lines of a proof instead of writing from scratch (similar to Parson's Problems for writing code). This tool scaffolds students' learning as they transition from reading proofs to writing proofs while also providing instant machine-graded feedback. To write a Proof Blocks problem, an instructor specifies lines of a proof and their logical dependencies. The autograder accepts any ordering of the lines that satisfies all logical dependencies (see [5] implementation details).

The initial version of Proof Blocks lacked a way for students to receive partial credit. Receiving partial credit for their work is a significant concern for students taking exams in a computerized environment [1]. A simple partial credit scheme

© Springer Nature Switzerland AG 2022
M. M. Rodrigo et al. (Eds.): AIED 2022, LNCS 13356, pp. 199–203, 2022.
https://doi.org/10.1007/978-3-031-11647-6_34

would be based on the location of the first incorrect line of the proof. For example, if the first 3 lines of a student's proof were correct, and the correct proof had 10 lines, the student would receive 3/10. This does not always work well. For example, what if a student has the first line of the proof wrong, but every other line was correct? Under this simple partial credit scheme, this answer would receive no partial credit, despite being almost completely correct.

A better way to assign partial credit is based on the minimum edit distance from the student submission to some correct solution. This gives a more holistic measure of the similarity of the student submission to some correct solution. However, because of the large solution space, it is computationally expensive to exhaustively check all possible solutions. This naïve approach will not scale to many students needing feedback at the same time (as in an active learning setting in a large classroom), or to problems of longer than 8 lines. To solve these problems in scaling, we propose a novel algorithm which calculates the minimum edit distance by directly manipulating the student submission into a correct solution and runs asymptotically faster than the naïve solution.

2   The Partial Credit Algorithms

A *proof blocks problem* $P = (C, G)$ is a set of *proof lines* C together with a directed acyclic graph (DAG) $G = (V, E)$, which defines the logical structure of the proof. The proof lines and DAG are provided by the instructor who writes the question (see [5] for question authoring details). The set of vertices V of the graph G is a subset of the set of proof lines C. A *submission* $S = s_1, s_2, ...s_n$ is a sequence of distinct proof lines, constructed by a student who is attempting to solve a proof blocks problem. If a submission S is a topological ordering of the graph G, we say that S is a *solution* to the proof blocks problem P.

If a student makes a submission S to a proof blocks problem $P = (C, G)$, we want to assign partial credit such that: (1) students get 100% only if the submission is a solution (2) partial credit declines linearly with the number of edits needed to convert the submission into a solution (3) credit is guaranteed to be in the range $0 - 100$. To satisfy these properties, we assign partial credit as follows: score $= 100 \times \frac{\max(0, |V| - d^*)}{|V|}$, where d^* is the minimum edit distance from the student submission to some correct solution of P, that is: $d^* = \min\{d(S, O) \mid O \in$ ALLTOPOLOGICALORDERINGS$(G)\}$, where $d(S, O)$ is the least common subsequence edit distance between the sequence S and the topological ordering O. This means, for example, that if a student's solution is 2 deletions and 1 insertion (3 edits) away from a correct solution, and the correct solution is 10 lines long, the student will receive 70%.

Baseline Algorithm. The most straightforward approach to calculating partial credit as we define it is to iterate over all topological orderings of G and for each one, calculate the edit distance to the student submission S. While this is effective, this algorithm is computationally expensive.

MVC Algorithm. Our more efficient algorithm operates by reducing the problem to the *minimum vertex cover* (MVC) problem over a subset of the student's

Table 1. Performance of baseline vs. MVC algorithm. (*) denotes a statistically significant difference in algorithm time, $p < 0.001$ in all cases. Algorithm run times are reports in mean milliseconds, followed by their standard deviations. For all problems with more than 3 possible solutions, the MVC algorithm was significantly faster, with a speedup of up to almost 200 times. This speedup gap will continue to grow as instructors write more complex problems.

Question number	Proof length	Possible solutions	Submissions	Submission size (mean)	Prob. subgraph size (mean)	MVC size (mean)	Baseline Alg. time	MVC Alg. time	Speedup factor
1	4	1	291	4.0	0.5	0.3	0.1 (0.03)	0.16 (0.04)	0.6*
2	5	1	104	5.0	2.1	1.0	0.12 (0.05)	0.21 (0.05)	0.6*
3	6	1	551	5.8	3.7	2.3	0.14 (0.04)	0.31 (0.09)	0.5*
4	6	3	19	5.0	0.2	0.1	0.25 (0.07)	0.26 (0.09)	1.0
5	6	6	674	6.2	0.1	0.0	0.41 (0.11)	0.25 (0.06)	1.6*
6	7	20	488	6.5	0.0	0.0	1.46 (0.49)	0.26 (0.07)	5.6*
7	9	24	28	9.0	3.2	1.5	2.23 (0.31)	0.47 (0.09)	4.7*
8	9	24	529	8.9	3.5	1.6	2.22 (0.53)	0.47 (0.10)	4.7*
9	9	35	376	8.5	1.1	0.4	3.03 (0.97)	0.41 (0.12)	7.4*
10	9	42	13	8.8	2.2	1.0	3.49 (0.46)	0.44 (0.08)	8.0*
11	10	21	623	9.8	4.6	1.8	2.36 (0.42)	0.57 (0.15)	4.1*
12	10	96	145	8.3	3.0	1.2	8.92 (2.38)	0.78 (0.64)	11.4*
13	10	1100	616	9.4	3.1	1.3	92.04 (20.78)	0.47 (0.13)	194.9*

submission. Rather than iterate over all topological orderings, this algorithm works by directly manipulating the student's submission until it becomes a correct solution. In order to do this, we define a few more terms. We call a pair of proof lines (s_i, s_j) in a submission a *problematic pair* if line s_j comes before line s_i in the student submission, but there is a path from s_i to s_j in G, meaning that s_j must come *after* s_i in any correct solution. We define the *problematic subgraph* to be the graph where the nodes are the set of all proof lines in a student submission that appear in some problematic pair, and the edges are the problematic pairs. Then the minimum edit sequence from the student submission is given by deleting each line in an MVC of the problematic subgraph, then adding lines to reach the correct solution. More details of the algorithm and a proof of its correctness are in [3].

3   Benchmarking Algorithms on Student Data

We collected data from homework, practice tests, and exams from the Discrete Mathematics course in the Computer Science department at the University of Illinois during the Fall 2020 semester [4]. In total, we collected 9,610 submissions to Proof Blocks problems. After discarding correct solutions and problems with too few data points, left us with a total of 4,457 submissions for our benchmarking data set. We used Python with NetworkX [2] to implement the algorithms and ran benchmarks on an Intel i5-8530U CPU with 16GB of RAM. Table 1 shows the benchmarks and statistical comparisons of our novel MVC-based algorithm and the baseline algorithm. The baseline algorithm performed about 2 times as fast as the MVC algorithm when there was one solution—more trivial cases when both algorithms took less than a third of a millisecond. The MVC algorithm performed about 5 to 200 times faster than the baseline algorithm. when there were 20 or more possible solutions. In Fig. 1 (A), we show that the run time of the baseline algorithm scales exponentially with the number of topological orderings of the proof graph, while in (B) we show that the run time of the MVC algorithm scales exponentially with the length of the proof. Thus, a relatively short Proof Blocks problem could have a very long grading time with the baseline algorithm, while with the MVC algorithm, we know that we can guarantee a bound on grading time given the proof length.

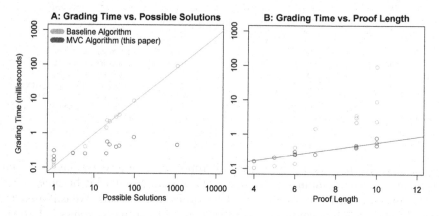

Fig. 1. Comparison of grading time for the two grading algorithms. Subplot (A) is a log-log plot showing that we see that the baseline algorithm scales with the number of possible solutions, Subplot (B) is a log-linear plot showing that the MVC Algorithm runtime scales with the length of the proof. This is a critical difference, because the number of topological orderings of a DAG can be $n!$ for a graph with n nodes.

References

1. Darrah, M., Fuller, E., Miller, D.: A comparative study of partial credit assessment and computer-based testing for mathematics. J. Comput. Math. Sci. Teach. **29**(4), 373–398 (2010)
2. Hagberg, A.A., Schult, D.A., Swart, P.J.: Exploring network structure, dynamics, and function using networkx. In: Varoquaux, G., Vaught, T., Millman, J. (eds.) Proceedings of the 7th Python in Science Conference, pp. 11–15. Pasadena, CA USA (2008)
3. Poulsen, S., Kulkarni, S., Herman, G., West, M.: Efficient partial credit grading of proof blocks problems. arXiv preprint arXiv:2204.04196 (2022)
4. Poulsen, S., Viswanathan, M., Herman, G.L., West, M.: Evaluating proof blocks problems as exam questions. In: Proceedings of the 17th ACM Conference on International Computing Education Research, pp. 157–168 (2021)

5. Poulsen, S., Viswanathan, M., Herman, G.L., West, M.: Proof blocks: autogradable scaffolding activities for learning to write proofs. In: Proceedings of the 27th ACM Conference on Innovation and Technology in Computer Science Education V. 1. Association for Computing Machinery, New York, NY, USA (2022). https://doi.org/10.1145/3502718.3524774
6. Weber, K.: Student difficulty in constructing proofs: the need for strategic knowledge. Educ. Stud. Math. **48**(1), 101–119 (2001). https://doi.org/10.1023/A:1015535614355

Robust Adaptive Scaffolding with Inverse Reinforcement Learning-Based Reward Design

Fahmid Morshed Fahid[1]([✉]) [iD], Jonathan P. Rowe[1], Randall D. Spain[1],
Benjamin S. Goldberg[2], Robert Pokorny[3], and James Lester[1]

[1] North Carolina State University, Raleigh, NC 27695, USA
{ffahid,jprowe,rdspain,lester}@ncsu.edu
[2] U.S. Army CCDC - STTC, Orlando, FL 32826, USA
benjamin.s.goldberg.civ@mail.mil
[3] Intelligent Automation Inc, Rockville, MD 20855, USA

Abstract. Reinforcement learning (RL) has shown significant potential for inducing data-driven scaffolding policies but designing reward functions that lead to effective policies is challenging. A promising solution is to use inverse RL to learn a reward function from effective demonstrations. This paper presents an inverse reward deep RL framework for inducing scaffolding policies in an adaptive learning environment. The framework centers on generating a data-driven model of immediate rewards by sampling high learning-gain episodes from previous student interactions and applying inverse RL. The resulting reward model is used to induce an adaptive scaffolding policy using batch constrained deep Q-learning. We evaluate this framework on data from 487 learners who completed an adaptive trianing course that provided direct instruction on principles of leading stability operations. Results show that the framework yields significantly better scaffolding policies more quickly compared to several RL baselines.

Keywords: Inverse reinforcement learning · Reward modeling · Adaptive learning environments · Adaptive scaffolding · ICAP

1 Introduction

Adaptive learning environments provide scaffolding to support students in their learning activities. Reinforcement learning (RL), more specifically, deep RL has shown significant promise in providing effective scaffolding in adaptive learning environments [2, 4]. A key factor in inducing effective scaffolding policies is designing reward functions. Researchers commonly use outcomes such as learning gains, post-test scores, and ratings of effort as rewards for inducing scaffolding policies [2, 4]. However, none of these can fully represent the effectiveness of individual scaffolds for a learning task. A promising solution is to use inverse RL, a branch of RL that learns reward models from demonstrations by finding maximum margin or maximum entropy between expert demonstrations and non-expert demonstrations. Inverse RL has been utilized to learn hidden rewards in a variety of domains [1], but limited work has explored its benefits in adaptive learning environments. In this paper we propose an inverse RL-based reward design framework for inducing adaptive scaffolding polices using deep RL.

© Springer Nature Switzerland AG 2022
M. M. Rodrigo et al. (Eds.): AIED 2022, LNCS 13356, pp. 204–207, 2022.
https://doi.org/10.1007/978-3-031-11647-6_35

2   Method

2.1   Dataset

To investigate the efficacy of using inverse RL for reward design we used log data from 487 learners (54% male) who completed a 90-min training course designed to teach principles of leading stability operations [4]. The course contained a series of short videos that provided direct instruction on leadership principles and multiple-choice questions to check students' learning. If learners incorrectly answered a multiple-choice question, the learning environment randomly provided one of four types of ICAP-inspired remediation: (1) read a feedback statement indicating the learner's answer was incorrect, (2) passively reread a transcription of the concept covered in the question, (3) reread the transcript and actively highlight the portion of the transcription that is related to the missed question, or (4) reread the transcript and constructively summarize the answer to the missed question in the learner's own words [3]. The final dataset includes a total of 4,998 instances of ICAP-inspired remediation. Each learner received on average 10 instances of remediation ($SD = 12.7$, $min = 1$, $max = 113$). Further analysis showed that there was a significant difference ($p < 0.001$) between pretest ($M = 4.18$, $SD = 2.30$,) and posttest scores ($M = 8.32$, $SD = 2.96$) among the learners.

2.2   Inverse Reward Deep Reinforcement Learning Framework

Our inverse reward-based deep RL framework involves a two-step procedure that combines inverse reward learning and deep Q networks to devise data-driven adaptive scaffolding policies to improve learning gains. In the first step, the framework uses prior data generated from a behavioral (i.e., scaffolding) policy π_b to learn a reward function using the Offline Reinforced Imitation Learning (ORIL) algorithm. ORIL is an inverse reinforcement learning algorithm that leverages deep learning techniques on prior data [7]. The approach uses a discriminator network as a reward function $R_\psi(s)$ that distinguishes between the expert demonstrations from unlabeled demonstrations.

For the second step, the framework calculates the reward-components for all the state-action pairs in the dataset and distributes the normalized learning gain (NLG) of an episode according to the reward-components of that episode to calculate a series of immediate rewards R_{imm}. Finally, we use Batch-Constrained Deep-Q Networks (BCQ) [5] to learn the optimum scaffolding policy π^* based on the immediate rewards to improve the overall expected student learning gain.

2.3   States, Actions, Rewards and Evaluations

We formalize our Markov decision process as follows. To represent states ($\mathcal{S}$), we created 31 features from learners' log data divided into 3 groups: (1) survey features, (2) video features, and (3) remediation engagement features. See our earlier work for a detailed description of each state [6]. Our action set ($\mathcal{A}$) contains the following four ICAP-inspired actions: $\mathcal{A} = \{none, passive, active, constructive\}$. We used pretest and posttest scores to calculate the normalized learning gains (NLG) and used that as delayed reward $R_{del} \in [-100, 100]$. Immediate rewards $R_{imm} \in [-100, 100]$ were calculated using our inverse reward deep RL framework. We used Expected Cumulative Reward (ECR) and Weighted Doubly Robust (WDR) to evaluate the policy [6].

3 Results and Discussion

In our experiment, we used an η of 0.5. Our discriminator used three fully connected neural network layers (64 neurons each) with hyperbolic tangent function in all but output layer, which uses the sigmoid function. For BCQ, we also used three layers of fully connected neural networks with 128 neurons each and the ReLU activation functions. We used τ as 0.3 and discount factor γ as 0.95.

For all models, we used a target network (θ') and an online network (θ) to improve stability and copied the value of the online network to the target network after every 100 epochs. We have also used the Adam optimizer (learning rate of $1e^{-3}$). For robustness, we ran all our experiments 10 times with different seeds. We did not use test-train splits as this is not necessary [6].

We explored three different sampling heuristics: sampled based on linear weight of NLG (*irl_linear*), sampled based on exponential weight of NLG (*irl_exp*) and sampled from episodes with at least 75th percentile NLG (*irl_q3*). For the exponential heuristic, we used an exponential time constant of 100 to keep the final weight between $[-e, e]$. In all cases, we sampled only 10% of the demonstrations as experts.

We compare the induced policies against three baselines. The *base_delayed* is the baseline where NLG is directly used as a delayed reward. The *base_q3* and *base_q2* assume all the reward-components in episodes with at least 75th and 50th percentile NLG are 1 and the rest are 0, respectively.

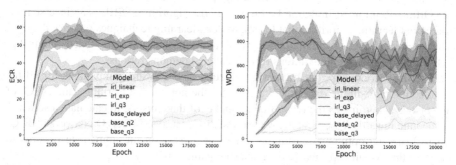

Fig. 1. The left figure shows ECR, and the right figure shows WDR for all models with 10 random repeats for 20,000 iterations.

As depicted in Fig. 1, all inverse RL-based reward policies perform significantly better (*p-value* < 0.001) than all the other baselines, including the *base_delayed* for both metrices. The inverse RL models converged significantly faster (around 1,500 epochs) than the delayed reward baseline (12,500 epochs).

4   Conclusion

Devising effective strategies for adaptive scaffolding is challenging and creating RL-based scaffolding policies raises important questions about RL reward design. We used an inverse RL-based reward design framework to guide deep RL models for adaptive scaffolding. The framework learns the rewards of different scaffolding demonstrations by sampling higher learning gain episodes using inverse RL and using batch constrained deep Q networks to optimize scaffolding policies to increase learning gains. Empirical analysis shows that it outperforms policies that directly use learning gains as rewards along with other baselines. The results suggest that the framework shows significant promise, and future work investigating the integration of data-driven scaffolding policies induced using this framework is an important direction.

Acknowledgements. This work is supported the U.S. Army Research Laboratory under cooperative agreement W911NF-15-2-0030. The statements and opinions expressed in this article do not necessarily reflect the position or the policy of the United States Government, and no official endorsement should be inferred.

References

1. Arora, S., Doshi, P.: A survey of inverse reinforcement learning: challenges, methods and progress. Artif. Intell. **297**, 1–28 (2021)
2. Sanz Ausin, M., Maniktala, M., Barnes, T., Chi, M.: Exploring the impact of simple explanations and agency on batch deep reinforcement learning induced pedagogical policies. In: Bittencourt, I.I., Cukurova, M., Muldner, K., Luckin, R., Millán, E. (eds.) AIED 2020. LNCS (LNAI), vol. 12163, pp. 472–485. Springer, Cham (2020). https://doi.org/10.1007/978-3-030-52237-7_38
3. Chi, M.T.H., Wylie, R.: The ICAP framework: linking cognitive engagement to active learning outcomes. Educ. Psychol. **49**(4), 219–243 (2014)
4. Fahid, F.M., Rowe, J.P., Spain, R.D., Goldberg, B.S., Pokorny, R., Lester, J.: Adaptively scaffolding cognitive engagement with batch constrained deep Q-networks. In: Roll, I., McNamara, D., Sosnovsky, S., Luckin, R., Dimitrova, V. (eds.) AIED 2021. LNCS (LNAI), vol. 12748, pp. 113–124. Springer, Cham (2021). https://doi.org/10.1007/978-3-030-78292-4_10
5. Fujimoto, S., Meger, D., Precup, D.: Off-policy deep reinforcement learning without exploration. In: Proceedings of the 36th International Conference on Machine Learning, pp. 2052–2062. PMLR (2019)
6. Thomas, P.S., Brunskill, E.: Data-efficient off-policy policy evaluation for reinforcement learning. In: Proceeding of the 33rd International Conference on Machine Learning, pp. 2139–2148. PMLR (2016)
7. Zolna, K., et al.: Offline learning from demonstrations and unlabeled experience. In: arXiv preprint arXiv:2011.13885 (2020)

A Design of a Simple Yet Effective Exercise Recommendation System in K-12 Online Learning

Shuyan Huang[1] , Qiongqiong Liu[1] , Jiahao Chen[1] , Xiangen Hu[2] ,
Zitao Liu[1,3](✉) , and Weiqi Luo[3]

[1] TAL Education Group, Beijing, China
{huangshuyan,liuqiongqiong1,chenjiahao,liuzitao}@tal.com
[2] The University of Memphis, Memphis, TN, USA
xhu@memphis.edu
[3] Guangdong Institute of Smart Education, Jinan University, Guangzhou, China
lwq@jnu.edu

Abstract. We propose a simple but effective method to recommend exercises with high quality and diversity for students. Our method is made up of three key components: (1) candidate generation module; (2) diversity-promoting module; and (3) scope restriction module. The proposed method improves the overall recommendation performance in terms of recall, and increases the diversity of the recommended candidates by 0.81% compared to the baselines.

Keywords: Exercise recommendation · K-12 online education

1 Introduction

The recommender systems (RSs) are widely adopted in internet applications to provide relevant items according to users' preference history [1,6,10,15]. In the online learning scenario, users are learners and items are exercises. As a key component in online learning, exercise recommendation systems provide personalized exercises to every student according to his or her individual requirements [4,8,13]. Despite the promising results achieved by the existing exercise recommendation methods [2,9,12,14,16], building an effective exercise RS still presents many challenges that come from special characteristics of real-life student learning scenarios. First, classic RS approaches are likely to recommend lexically similar exercises while a desired RS is supposed to generate exercises with different content but the same knowledge concepts (KCs) and help students avoid rote memorization. Second, compared to the data scale in internet companies, student-exercise interaction datasets are much smaller and even more sparse. Third, the exercise RSs accompany students for the entire semester and the recommended results should follow the learning progress of the in-school syllabus. Hence, we propose a simple yet effective exercise recommendation system to recommend KC-relevant exercises within the scope of the in-school syllabus.

© Springer Nature Switzerland AG 2022
M. M. Rodrigo et al. (Eds.): AIED 2022, LNCS 13356, pp. 208–212, 2022.
https://doi.org/10.1007/978-3-031-11647-6_36

2   System Design

The entire exercise recommendation workflow is illustrated in Fig. 1.

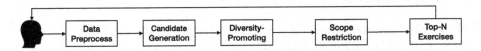

Fig. 1. The overview of our exercise recommendation system.

Candidate Generation Module. We conduct the candidate generation by using an enhanced random walk algorithm on an educational tripartite graph that is composed of a set of KCs, exercises and class materials. By incorporating KCs and class materials in the graph, we are able to capture the underlying knowledge similarities of exercises while focusing on the in-class teaching content. For each student, we create a query set Q that may contain the most recently incorrectly answered exercises as well as exercises practiced in the class materials. Then, we generate recommendations using the enhanced random walk algorithm given the query set. Let E_1 and E_2 be the edges between KCs and exercises and edges between exercises and class materials. Each random walk produces a sequence of steps. Each step is composed of three operations. First, given the current exercise q, i.e., $q \in Q$, we select an edge e from E, i.e., $E = E_1 \cup E_2$. If e comes from E_1, it means q is connected with a KC k. Otherwise, e connects q with a class material m. Then, we select exercise e' by sampling an edge e' from either $E_1(k)$ that connects k and e' or $E_2(m)$ that connects m and e'. Third, the current exercise is updated to e' and the step repeats. We use $E_1(k)$ and $E_2(m)$ to denote exercises connect to KC k and class material m respectively. The length of each short random walk and the total number of steps across all such short random walks are determined by two hyper-parameters α and T respectively. We record the visit count for each candidate exercise q in each step and maintain a counter V that maps the candidate to the visit count. Finally, the N exercises with the highest visit count are selected as the candidate recommendations.

Diversity-Promoting Module. In this module, we remove candidates with high lexical similarities. First, we further pre-train the widely used large-scale language model, i.e., *RoBERTa*, using the exercise data [7]. Then we obtain the exercise semantic representations by averaging all the word-level representations of its textual content. After that, similarities between all the exercises are calculated via their exercise embedding and the recommended candidates with high similarities, i.e., over a certain threshold, are filtered out. In this way, we provide a set of diverse exercises to help students avoid rote memorization.

Scope Restriction Module. In K-12 education, the recommended exercises should align with the in-school learning progress of the students. To achieve this property, we first roughly sort all the KCs in sequential order according to the class syllabus arrangement in each semester. Then, we only select and recommend candidates whose KCs are never beyond the scope of KCs covered by the up-to-date course progress.

3 Experiments

We conduct our method on a real-world K-12 online learning dataset with 1679 KCs, 1,153,690 exercises and 135,637 class materials. These class materials are created by teaching professionals from a third-party online educational platform[1]. We randomly withhold test data of 244 class materials, which contain 494 KCs and 8,340 exercises. Our exercise recommendation evaluation task is *finding a set of exercises which are relevant to the set of KCs from the student's incorrectly answer exercises of the handcrafted class material.* Similar to [3], the recommendation performance is evaluated by (1) recall, which measures the percentages of recommended exercises that the handcrafted class material contains as well; and (2) Distinct-2 , which measures the diversity of selected exercises by computing the ratios of the number of different bi-grams to the total number of bi-grams [5]. We compute recall scores from both micro and macro perspectives [11]. We set the α is 0.04 and the total steps T of a random walk procedure is 100,000. For each KC, we select the top $N \in \{10, 25, 100\}$ recommended exercise candidates.

Table 1. Results (%) of different recommendation methods on the online learning dataset. TG, DP, and SR denote the tripartite graph based candidate generation module, the diversity-promoting module and the scope restriction module respectively.

Model	Top-10		Top-25		Top-100		Distinct-2
	Macro recall	Micro recall	Macro recall	Micro recall	Macro recall	Micro recall	
TG	13.52	9.63	25.73	19.48	44.70	36.22	13.05
TG+DP	13.52	9.64	25.69	19.48	44.41	36.06	13.21
TG+SR	13.75	9.80	**26.25**	**19.90**	**45.71**	**36.93**	13.69
TG+DP+SR	**13.80**	**9.82**	26.20	19.86	45.42	36.80	**13.86**

The recommendation results are shown in Table 1. From Table 1, we make the following observations: First, when comparing TG+DP+SR to TG+DP, we can see that SR in general improves the overall recommendation performance. We believe this is because the SR module is able to remove the out-of-scope recommended candidates generated from the random walk based candidate generation module. Second, comparing TG and TG+DP, TG+SR and TG+DP+SR, we can see, the DP module increases the diversity of the recommended exercises up to 0.17%.

---

[1] https://www.xesvip.com/.

4   Conclusion

In this work, we propose a simple but effective method utilizing exercises, KCs, and class materials to recommend exercises for students. Our recommendation framework relies on the candidate generation module, the diversity-promoting module and the scope restriction module. We evaluate our approach on an offline educational data set and the results show that the proposed method is able to give accurate and more diverse recommended exercises.

Acknowledgements. This work was supported in part by National Key R&D Program of China, under Grant No. 2020AAA0104500; in part by Beijing Nova Program (Z201100006820068) from Beijing Municipal Science & Technology Commission and in part by NFSC under Grant No. 61877029.

References

1. Barria-Pineda, J., Akhuseyinoglu, K., Želem-Ćelap, S., Brusilovsky, P., Milicevic, A.K., Ivanovic, M.: Explainable recommendations in a personalized programming practice system. In: Roll, I., McNamara, D., Sosnovsky, S., Luckin, R., Dimitrova, V. (eds.) AIED 2021. LNCS (LNAI), vol. 12748, pp. 64–76. Springer, Cham (2021). https://doi.org/10.1007/978-3-030-78292-4_6
2. Chen, J., Li, H., Ding, W., Liu, Z.: An educational system for personalized teacher recommendation in K-12 online classrooms. In: Roll, I., McNamara, D., Sosnovsky, S., Luckin, R., Dimitrova, V. (eds.) AIED 2021. LNCS (LNAI), vol. 12749, pp. 104–108. Springer, Cham (2021). https://doi.org/10.1007/978-3-030-78270-2_18
3. Eksombatchai, C., et al.: Pixie: a system for recommending 3+ billion items to 200+ million users in real-time. In: WWW, pp. 1775–1784 (2018)
4. Huang, Z., Liu, Q., Zhai, C., Yin, Y., Chen, E., Gao, W., Hu, G.: Exploring multi-objective exercise recommendations in online education systems. In: CIKM, pp. 1261–1270 (2019)
5. Li, J., Galley, M., Brockett, C., Gao, J., Dolan, W.B.: A diversity-promoting objective function for neural conversation models. In: NAACL, pp. 110–119 (2016)
6. Liu, T., Wang, Z., Tang, J., Yang, S., Huang, G.Y., Liu, Z.: Recommender systems with heterogeneous side information. In: WWW, pp. 3027–3033 (2019)
7. Liu, Y., et al.: RoBERTa: a robustly optimized BERT pretraining approach. arXiv preprint arXiv:1907.11692 (2019)
8. Mohseni, M., Maher, M.L., Grace, K., Najjar, N., Abbas, F., Eltayeby, O.: Pique: recommending a personalized sequence of research papers to engage student curiosity. In: Isotani, S., Millán, E., Ogan, A., Hastings, P., McLaren, B., Luckin, R. (eds.) AIED 2019. LNCS (LNAI), vol. 11626, pp. 201–205. Springer, Cham (2019). https://doi.org/10.1007/978-3-030-23207-8_38
9. Pang, Y., Jin, Y., Zhang, Y., Zhu, T.: Collaborative filtering recommendation for MOOC application. Comput. App. Eng. Educ. 25(1), 120–128 (2017)
10. Schafer, J.B., Konstan, J.A., Riedl, J.: E-commerce recommendation applications. Data Mining Knowl. Disc. 5(1), 115–153 (2001)
11. Van Asch, V.: Macro-and micro-averaged evaluation measures. CLiPS 49, Belgium (2013)
12. Wan, S., Niu, Z.: An e-learning recommendation approach based on the self-organization of learning resource. Knowl. Based Syst. 160, 71–87 (2018)

13. Wang, S., Wu, H., Kim, J.H., Andersen, E.: Adaptive learning material recommendation in online language education. In: Isotani, S., Millán, E., Ogan, A., Hastings, P., McLaren, B., Luckin, R. (eds.) AIED 2019. LNCS (LNAI), vol. 11626, pp. 298–302. Springer, Cham (2019). https://doi.org/10.1007/978-3-030-23207-8_55

14. Xu, G., Jia, G., Shi, L., Zhang, Z.: Personalized course recommendation system fusing with knowledge graph and collaborative filtering. Comput. Intell. Neurosci. **2021**, 9590502 (2021)

15. Yan, Y., Liu, Z., Zhao, M., Guo, W., Yan, W.P., Bao, Y.: A practical deep online ranking system in e-commerce recommendation. In: Brefeld, U.U., et al. (eds.) ECML PKDD 2018. LNCS (LNAI), vol. 11053, pp. 186–201. Springer, Cham (2019). https://doi.org/10.1007/978-3-030-10997-4_12

16. Yang, D., Piergallini, M., Howley, I., Rose, C.: Forum thread recommendation for massive open online courses. In: EDM 2014. Citeseer (2014)

Wide & Deep Learning for Judging Student Performance in Online One-on-One Math Classes

Jiahao Chen[1], Zitao Liu[1,2](✉), and Weiqi Luo[2]

[1] TAL Education Group, Beijing, China
{chenjiahao,liuzitao}@tal.com
[2] Guangdong Institute of Smart Education, Jinan University, Guangzhou, China
lwq@jnu.edu

Abstract. In this paper, we investigate the opportunities of automating the judgment process in online one-on-one math classes. We build a Wide & Deep framework to learn fine-grained predictive representations from a limited amount of noisy classroom conversation data that perform better student judgments. We conducted experiments on the task of predicting students' levels of mastery of example questions and the results demonstrate the superiority and availability of our model in terms of various evaluation metrics.

Keywords: Performance judgement · Deep learning · Online education

1 Introduction

Accurate student performance judgments provide essential information for teachers deciding what and how to teach and form the fundamental prerequisites for adaptive instructions [4,7,9,13]. When inaccurate diagnoses and judgments are made of students' abilities, prior knowledge, learning and achievement motivation, or other student characteristics, teaching becomes less effective in terms of learning and achievement gains [8,12]. Our work is different from many existing studies along this direction [4,6,10,11]. First, previous approaches try to manually conduct user studies to verify the accuracy of teacher judgment and the corresponding effects and influence. Second, the majority of existing works use the results of assignments or exams to represent student performance, which is a very lagged and coarse indicator.

In this paper, we aim to explore the opportunities of building an AI-driven model to perform automatic fine-grained student performance assessment by analyzing the in-class conversational data at the question level and aim to predict whether the student understands each example question on slides in the online one-on-one class. Our work focuses on utilizing the deep learning and neural network techniques to learn effective representations from a limited amount of

M. M. Rodrigo et al. (Eds.): AIED 2022, LNCS 13356, pp. 213–217, 2022.
https://doi.org/10.1007/978-3-031-11647-6_37

noisy classroom conversation data that perform better student judgments. Our framework builds upon the Wide & Deep learning architecture [3] to achieve both memorization and generalization in one model by training a multi-layer perceptron (MLP) on representations from both wide and deep components. The wide component effectively memorizes sparse feature interactions between teachers and students, while deep neural networks can generalize and handle the linguistic variations in classroom conversations. Experiments show that the Wide & Deep framework performs better than alternative methods when predicting student performance on a real-world educational dataset.

2   The Wide & Deep Framework

In this work, we design the Wide & Deep learning framework to achieve both memorization and generalization in one model, by training a MLP model with representations from both the wide component and the deep component.

The Wide Component. In the wide component, we handcraft 25 features to capture the teacher-student interactions on each example question in the online one-on-one math class. Our manually-engineered features include both continuous-valued features, i.e., conversation duration and Jaccard similarity between teacher's and student's sentences, and discrete-valued features, i.e., the number of words per conversation from the teacher, the number of student-spoken sentences with less than 3 words. We use one-hot encoding for discrete features and project them to dense embedding space with a linear projection matrix $\mathbf{W}$, i.e., $\mathbf{x}_d^* = \mathbf{W}\mathbf{x}_d$, where $\mathbf{x}_d$ represents the one-hot representations of discrete features and $\mathbf{x}_d^*$ represents the projected dense representations for discrete features. The overall features from the wide component $\mathbf{x}_{wide}$ is the concatenation of the original continuous features $\mathbf{x}_c$ and projected dense features $\mathbf{x}_d^*$, i.e., $\mathbf{x}_{wide} = \mathbf{x}_c \oplus \mathbf{x}_d^*$ where $\oplus$ denotes the concatenation operation. A two-layer fully connected deep neural network is trained on features from wide component $\mathbf{x}_{wide}$.

The Deep Component. We design a deep component for learning effective low-dimensional representations of the rich in-class conversational information and teaching styles and instructions variations. Firstly, we use the "EduRoBERTa" [1] to obtain the sentence embedding $\mathbf{e}_i$ of each spoken transcription sentence $\mathbf{s}_i$, i.e., $\mathbf{e}_i = \text{EduRoBERTa}(\mathbf{s}_i)$. Then, we use a deep sequential model to capture the relation of long-range dependencies among the sentences in the original conversational texts. In this work, we use bidirectional LSTM as our sentence encoder, i.e., $\{\mathbf{h}_1, \mathbf{h}_2, \cdots, \mathbf{h}_n\} = \text{BiLSTM}(\{\mathbf{e}_1, \mathbf{e}_2, \cdots, \mathbf{e}_n\})$. Furthermore, we apply the structured self-attention mechanism to each learned representation from BiLSTM, i.e., $\{\mathbf{q}_1, \mathbf{q}_2, \cdots, \mathbf{q}_n\} = \text{Attention}(\{\mathbf{h}_1, \mathbf{h}_2, \cdots, \mathbf{h}_n\})$. The structured self-attention is an attention mechanism relating different positions of a single sequence in order to compute a representation of the sequence [15]. At the end, we use the average pooling operation to get the low-dimensional dense representation $\mathbf{x}_{deep}$ from the deep component, i.e., $\mathbf{x}_{deep} = \text{MeanPooling}(\{\mathbf{q}_1, \mathbf{q}_2, \cdots, \mathbf{q}_n\})$. A four-layer fully connected deep neural network is trained on features from wide component $\mathbf{x}_{wide}$.

Joint Learning. The wide component and deep component are combined using an element-wise sum of their output log odds as the prediction, which is then fed to a Softmax function to output the probability score of the student's level of mastery of each example question. Our Wide & Deep Model is learned by minimizing the standard cross-entropy loss function.

3 Experiments

In this work, we collect 5226 unique samples of teacher-student conversations of solving example question from 497 recordings of online one-on-one math classes in grade 8 from a third-party K-12 online education platform. We use a third-party publicly available ASR service for the conversation transcriptions. We evaluate the performance of our proposed approach (denoted as **W&D**) by comparing it with the following baselines: (1) **GBDT**: Gradient boosted decision tree [5] on handcrafted wide features; (2) **LSTM w. EduRoBERTa**: Similar to the proposed approach but it eliminates the contribution of features from the wide component. We utilize the XGBoost library [2] for the GBDT training where we grid search the max_depth over $[1, 2, \cdots, 9]$ and n_estimators over $[10, 20, \cdots, 90]$ and with a fixed subsample of 0.9. For the BiLSTM model, the size of hidden states is grid searched over $[64, 128, 256]$. We set batch size to 256 and conduct the hyper-parameter tuning by selecting the one with the highest accuracy on the validation set and we report the model performance on the test set.

Table 1. Overall performance in terms of accuracy, micro-F1 and macro-F1.

Model name	Accuracy	micro-F1	macro-F1
GBDT	0.705	0.676	0.606
LSTM w. EduRoBERTa	**0.728**	0.704	0.643
W&D(Ours)	**0.728**	**0.709**	**0.649**

We list the test results in terms of accuracy and F1 scores from both micro and macro perspectives [14], shown in Table 1. We make the following two observations: (1) The LSTM w. EduRoBERTa led to a macro F1 of 0.643, which is a notable improvement over the performance of GBDT. It showed that text information is important for judging student performance; and (2) our W&D model achieves the best performance in all metrics by combining wide and deep features. This suggests that although the deep model has pretty good than GBDT, the wide features can improve the performance of LSTM. Thus, it is essential to incorporate both wide and deep features.

4 Conclusion

In this paper, we present a Wide & Deep framework to predict students' level of mastery of questions. Combining the handcraft features with a deep learning-based model can improve the performance. As part of future work, we plan to find more efficient method to encode the conservation text of students and teachers. Furthermore, we will do a more profound analysis of handcraft features, which can help us understand which features are essential for judging students' performance.

Acknowledgements. This work was supported in part by National Key R&D Program of China, under Grant No. 2020AAA0104500; in part by Beijing Nova Program (Z201100006820068) from Beijing Municipal Science & Technology Commission and in part by NFSC under Grant No. 61877029.

References

1. Eduroberta. https://github.com/tal-tech/edu-bert. Accessed 07 Feb 2022
2. Chen, T., Guestrin, C.: XGBoost: a scalable tree boosting system. In: SIGKDD, pp. 785–794 (2016)
3. Cheng, H.T., Koc, L., Harmsen, J., Shaked, T., Chandra, T., Aradhye, H., Anderson, G., Corrado, G., Chai, W., Ispir, M., et al.: Wide & deep learning for recommender systems. In: Proceedings of the 1st Workshop on Deep Learning for Recommender Systems. pp. 7–10 (2016)
4. Dhamecha, T.I., Marvaniya, S., Saha, S., Sindhgatta, R., Sengupta, B.: Balancing human efforts and performance of student response analyzer in dialog-based tutors. In: Penstein Rosé, C., Martínez-Maldonado, R., Hoppe, H.U., Luckin, R., Mavrikis, M., Porayska-Pomsta, K., McLaren, B., du Boulay, B. (eds.) AIED 2018. LNCS (LNAI), vol. 10947, pp. 70–85. Springer, Cham (2018). https://doi.org/10.1007/978-3-319-93843-1_6
5. Friedman, J.H.: Stochastic gradient boosting. Comput. Stat. Data Anal. **38**(4), 367–378 (2002)
6. Haataja, E., Moreno-Esteva, E.G., Salonen, V., Laine, A., Toivanen, M., Hannula, M.S.: Teacher's visual attention when scaffolding collaborative mathematical problem solving. Teach. Teach. Educ. **86**, 102877 (2019)
7. Hao, Y., Li, H., Ding, W., Wu, Z., Tang, J., Luckin, R., Liu, Z.: Multi-task learning based online dialogic instruction detection with pre-trained language models. In: Roll, I., McNamara, D., Sosnovsky, S., Luckin, R., Dimitrova, V. (eds.) AIED 2021. LNCS (LNAI), vol. 12749, pp. 183–189. Springer, Cham (2021). https://doi.org/10.1007/978-3-030-78270-2_33
8. Huang, G.Y., et al.: Neural multi-task learning for teacher question detection in online classrooms. In: Bittencourt, I.I., Cukurova, M., Muldner, K., Luckin, R., Millán, E. (eds.) AIED 2020. LNCS (LNAI), vol. 12163, pp. 269–281. Springer, Cham (2020). https://doi.org/10.1007/978-3-030-52237-7_22
9. Liu, H., Liu, Z., Wu, Z., Tang, J.: Personalized multimodal feedback generation in education. In: COLING, pp. 1826–1840 (2020)
10. McIntyre, N.A., Jarodzka, H., Klassen, R.M.: Capturing teacher priorities: Using real-world eye-tracking to investigate expert teacher priorities across two cultures. Learn. Instr. **60**, 215–224 (2019)

11. Praetorius, A.K., Drexler, K., Rösch, L., Christophel, E., Heyne, N., Scheunpflug, A., Zeinz, H., Dresel, M.: Judging students' self-concepts within 30 s? investigating judgement accuracy in a zero-acquaintance situation. Learn. Individ. Differ. **37**, 231–236 (2015)
12. Praetorius, A.K., Koch, T., Scheunpflug, A., Zeinz, H., Dresel, M.: Identifying determinants of teachers' judgment (in) accuracy regarding students' school-related motivations using a bayesian cross-classified multi-level model. Learn. Instr. **52**, 148–160 (2017)
13. Stürmer, K., Könings, K.D., Seidel, T.: Declarative knowledge and professional vision in teacher education: Effect of courses in teaching and learning. Br. J. Educ. Psychol. **83**(3), 467–483 (2013)
14. Van Asch, V.: Macro-and micro-averaged evaluation measures. Belgium: CLiPS 49 (2013)
15. Vaswani, A., et al.: Attention is all you need. NIPS 30 (2017)

Multimodal Behavioral Disengagement Detection for Collaborative Game-Based Learning

Fahmid Morshed Fahid[1]($\boxtimes$) [iD], Halim Acosta[1], Seung Lee[1], Dan Carpenter[1] [iD],
Bradford Mott[1] [iD], Haesol Bae[2], Asmalina Saleh[2], Thomas Brush[2],
Krista Glazewski[2], Cindy E. Hmelo-Silver[2], and James Lester[1]

[1] North Carolina State University, Raleigh, NC 27695, USA
{ffahid,hacosta,sylee,dcarpen2,bwmott,lester}@ncsu.edu
[2] Indiana University, Bloomington, IN 47405, USA
{haebae,asmsaleh,tbrush,glaze,chmelosi}@indiana.edu

Abstract. Collaborative game-based learning environments offer significant promise for creating effective and engaging group learning experiences. These environments enable small groups of students to work together toward a common goal by sharing information, asking questions, and constructing explanations. However, students periodically disengage from the learning process, which negatively affects their learning, and the impacts are more severe in collaborative learning environments as disengagement can propagate, affecting participation across the group. Here, we introduce a multimodal behavioral disengagement detection framework that uses facial expression analysis in conjunction with natural language analyses of group chat. We evaluate the framework with students interacting with a collaborative game-based learning environment for middle school science education. The multimodal behavioral disengagement detection framework integrating both facial expression and group chat modalities achieves higher levels of predictive accuracy than those of baseline unimodal models.

Keywords: Multimodal learning · Collaborative game-based learning · Behavioral disengagement

1 Introduction

Game-based learning environments are designed to enhance positive cognitive and affective outcomes among learners by embedding curricular content in gameplay [4]. Collaborative game-based learning environments integrate collaborative elements such as group chat so that students can interact with each other, which can potentially increase engagement [3, 4]. However, disengagement may appear throughout the learning process in game-based learning environments and the impact is even higher in a collaborative learning space as students' behavioral disengagement can distract other students, impeding the learning process and engendering negative attitudes [6]. Prior work has

M. M. Rodrigo et al. (Eds.): AIED 2022, LNCS 13356, pp. 218–221, 2022.
https://doi.org/10.1007/978-3-031-11647-6_38

characterized different types of disengagement behaviors and their impacts [5], but identifying disengagement behaviors is challenging, as what constitutes disengagement is often dependent on the context and modality of the learning environment [2].

This paper introduces a multimodal behavioral disengagement detection framework leveraging facial video recordings and group chat logs from middle school students and investigates the effectiveness of different modalities when identifying disengagement behaviors in a collaborative game-based learning environment, CRYSTAL ISLAND: ECOJOURNEYS. We labeled the two modalities for disengagement behaviors and investigated the impacts of features from both modalities for predicting disengagement. Results show that multimodal models can achieve higher predictive accuracy when automatically detecting disengagement behaviors compared to unimodal baselines.

2 Multimodal Disengagement Modeling

A classroom study was conducted with 26 middle school students (8 females; 18 males) as they interacted in a collaborative game-based learning environment for ecosystem science, CRYSTAL ISLAND: ECOJOURNEYS. In the game, the students were divided into groups of three or four (total 7 groups) to determine the cause of the sudden sickness of fish on a remote island using in-game chat and a collaborative whiteboard system. A total of 2,560 chat messages and 44 h of facial video recordings were collected.

We first annotated both modalities (group chat and video recording) as *engaged* or *disengaged*. Two raters marked all the chat messages that are either content or task related as *engaged* (35.35%) and the rest as *disengaged* ($\kappa = 0.925$). Next, for tagging the video recordings, two raters used the HELP coding framework [1] with a window size of 10 s [7] and tagged each window as *disengaged* whenever a student was disengaged for more than 4 s ($\kappa = 0.763$). The final labels contain 4,863 disengaged segments (30.85%) and 10,898 engaged segments (69.15%). To combine the chat-based labels with video-based labels, we considered the "window of a chat" to be a window of 10 s after a chat message was sent. We combined the labels using three different heuristics: *chat_first* prefers *chat-based* labels whenever there is a chat message; *engaged_first* prefers engaged behavior whenever there is disagreement between *chat-based* and *video-based* labels; and, *disengaged_first* prioritizes disengaged behavior whenever there is disagreement.

For chat-based features, we transformed the tokenized in-game chat messages into distributed vector representations using ELMo. For each chat message, we averaged a 1024-dimension vector over all tokens to generate a single word embedding. We refer to these as *chat_only* features. For video-based features, we utilized the action unit features, pose features, and gaze features (total 49 features) from the OpenFace[1] behavior analysis toolkit. We refer to these as *video_only* features. We also created a multimodal feature set called *all* features that combines *chat_only* and *video_only* features.

3 Results and Discussion

To compare our feature groups across different labels based on different modalities, we utilized three off-the-shelf classifiers, namely, Random Forest (RF), Decision Tree (DT),

---

[1] https://openface-api.readthedocs.io.

and Logistic Regression (LR). All classifiers were trained to predict engagement (1) and disengagement (0) behaviors among students on individual 10-s time segments. We repeated all experiments with three random seeds, each with five-fold cross-validation, and only report the mean scores (Table 1).

Table 1. Precision, Recall, and F1 scores (in percent) for unimodal and multimodal features across multiple labels using different classifiers. Highest F1 scores are bolded.

Labels	Classifier	*chat_only* unimodal features			*video_only* unimodal features			*all* multimodal features		
		Prec.	Rec.	F1	Prec.	Rec.	F1	Prec.	Rec.	F1
chat-based	LR	76	69	73	62	33	43	77	71	**74**
	DT	60	62	**61**	45	42	43	58	61	59
	RF	78	63	**70**	64	34	44	79	61	69
video-based	LR	16	1	1	74	44	55	72	47	**57**
	DT	20	2	3	60	56	58	61	57	**59**
	RF	34	1	1	79	60	**68**	80	56	66
chat_first (combined)	LR	77	12	21	72	43	54	75	52	**61**
	DT	61	11	18	58	54	56	60	56	**58**
	RF	77	10	18	78	58	66	79	59	**68**
engaged_first (combined)	LR	25	1	1	74	43	54	73	47	**57**
	DT	20	1	2	59	54	57	60	56	**58**
	RF	39	0	1	79	59	**68**	79	57	66
disengaged_first (combined)	LR	74	13	22	73	45	55	74	53	**62**
	DT	59	11	19	59	55	57	61	57	**59**
	RF	75	13	21	78	60	68	79	60	**69**

For unimodal features, we can see that *video_only* unimodal features outperform *chat_only* unimodal features in all cases except for *chat-based* labels. Only for *chat-based* labels, *chat_only* unimodal features are significantly better at predicting disengagement behaviors than *video_only* unimodal features. This shows that *video_only* features are dominant in predicting disengagement behaviors when disengagements are defined either using *video-based* annotations or using a combination of *video-based* annotation and *chat-based* annotations. But *video_only* features are ineffective in predicting *chat-based* disengagement behaviors. When *all* features are used, most of the classifiers perform better than the unimodal features, indicating that combined labels achieve better predictive scores when both modalities of features are present.

A potential explanation for *chat_only* features performing worse in most cases maybe the low number of segments that contain chat messages (2,560) when compared against the total number of segments (15,761) that contain *video_only* features. As for other limitations, we defined disengagement behaviors using 10-s segments for *video-based* annotations. Previous work also used different sizes of segmented windows to define student behavior [7], but in reality, such behaviors are continuous.

4 Conclusion

Collaborative game-based learning environments offer engaging learning opportunities for groups of students. However, students may become disengaged from the learning process, and these disengagement behaviors can be detrimental to the learning process of individuals as well as their groups. Automatically detecting disengagement behaviors is challenging as they are often difficult to identify with a single modality. We have introduced a multimodal behavioral disengagement detection framework that leverages in-game chat messages in conjunction with facial video recordings of individual learners to detect disengagement behaviors among students. The results show that the multimodal behavior disengagement detection framework outperforms unimodal models for detecting student disengagement. These findings suggest that the multimodal behavior disengagement detection framework can inform the design of adaptive scaffolding for collaborative game-based learning environments.

Acknowledgements. This work is supported by the National Science Foundation through grants IIS-1839966, and SES-1840120. Any opinions, findings, and conclusions or recommendations expressed in this material are those of the authors and do not necessarily reflect the views of the National Science Foundation.

References

1. Aslan, S., et al.: Human expert labeling process (HELP): towards a reliable higher-order user state labeling process and tool to assess student engagement. Educ. Technol. **57**, 53–59 (2017)
2. Giannakos, M.N., Sharma, K., Pappas, I.O., Kostakos, V., Velloso, E.: Multimodal data as a means to understand the learning experience. Int. J. Inf. Manag. **48**, 108–119 (2019)
3. Jeong, H., Hmelo-Silver, C.E., Jo, K.: Ten years of computer-supported collaborative learning: a meta-analysis of CSCL in STEM education during 2005–2014. Educ. Res. Rev. **28**, 100284 (2019)
4. de Jesus, Â.M., Silveira, I.F.: A collaborative game-based learning framework to improve computational thinking skills. In: 2019 International Conference on Virtual Reality and Visualization (ICVRV), pp. 161–166. IEEE (2019)
5. Langer-Osuna, J.M.: Productive disruptions: rethinking the role of off-task interactions in collaborative mathematics learning. Educ. Sci. **8**(2), 87 (2018)
6. Park, K., et al.: Detecting disruptive talk in student chat-based discussion within collaborative game-based learning environments. In: Proceedings of the 11th International Learning Analytics and Knowledge Conference, pp. 405–415 (2021)
7. Thomas, C., Jayagopi, D.B.: Predicting student engagement in classrooms using facial behavioral cues. In: Proceedings of the 1st ACM SIGCHI International Workshop on Multimodal Interaction for Education, pp. 33–40 (2017)

Using Metacognitive Information and Objective Features to Predict Word Pair Learning Success

Bledar Fazlija[1]([✉]) and Mohamed Ibrahim[2]

[1] ZHAW Zurich University of Applied Sciences, Winterthur, Switzerland
bledar.fazlija@zhaw.ch
[2] DijaVu Digital Incubator, Cairo, Egypt

Abstract. There is a variety of metacognitive information that can be used to model learning success, such as judgements of learning (JOLs) and feeling of difficulty (FOD). The latter is not widely used, and part of this study is to collect FODs through crowed annotation and demonstrate its potential as a predictor of learning success. While objective features related to task difficulty provide valuable information for the modeling task, we show evidence that FOD can provide similar insight. We examine and compare the use of objective and subjective features as predictors for learning success in second language word pair learning. The results indicate that metacognitive information is transferable across different groups of subjects. They also show that crowed annotation is a useful method for enriching datasets with FODs and potentially other metacognitive information.

Keywords: Metacognition · Machine learning · Transfer learning · Word difficulty · Crowed annotation · Feeling of difficulty · Judgments of learning

1 Introduction

Education is being transformed in many directions [1–7], particularly through the use of digital technologies. There are several successful approaches [7–10] that have been used to track student performance and development and to model students' cognitive states, abilities, and problem-solving skills. Other models are concerned with trying to predict students' learning success in solving tasks with high accuracy. This raises the question of what features should be included to predict students' task-solving abilities or to model their cognitive states. One important type of "pedagogical" feature is feedback, both from teacher to student and from student to teacher. Regarding the former, there is a large body of research discussed in Hattie's synthesis of meta-analyses [11], that demonstrates the effectiveness of teacher to student feedback. It is particularly effective when the feedback is immediate and frequent. Less researched is the way student feedback can be used by teachers. In the cognitive sciences, subjective feedback and reporting of various kinds have been studied. Metacognitive notions that have been explored and found useful include judgments of learning (JOLs) [12] and feeling of difficulty (FOD) [13]. It is shown in [12] that JOLs are effective in predicting outcome in the case of word pair

© Springer Nature Switzerland AG 2022
M. M. Rodrigo et al. (Eds.): AIED 2022, LNCS 13356, pp. 222–226, 2022.
https://doi.org/10.1007/978-3-031-11647-6_39

learning tasks such as those used in the present work. Unlike JOLs, FOD has not been extensively studied for modeling learning outcome.

The work in [13] examined this notion in the context of mathematics exercises for high school students. In this study, we attempt to show the potential of using FOD as a predictor for modeling word pair learning success. We incorporate objective features, as studied in [14–16], into the modeling of word pair learning success. We investigate the effects of using objective features (such as word length, frequency of occurrence in large English text corpora, word familiarity, number of hyponyms and hypernyms[1]) in combination with the subjective features FOD and JOL on predictive performance. We use transfer learning [17, 19] as an approach to gain knowledge from one scenario and context and transfer it to another. For our research, we used the OMNI[2] dataset [18], which contains JOLs of students learning word pairs in a second language. However, this dataset does not contain FODs. To overcome this problem and conduct our study with combined enrichment of JOLs and FODs, we employed crowded annotation and used a naïve form of transfer learning. For this purpose, online annotation services provided by the platform "www.figure-eight.com" were used. To predict students' performance in learning word pairs, we used different machine learning models.

2 Methodology

We used both the OMNI [18] dataset and data collected in an own study for which crowd annotation was utilized, using the annotation platform www.figure-eight.com.

OMNI Dataset. 45 Lithuanian-English word pairs were shown to the subjects at five different time points for about four seconds, first the English word and then, after a short delay, the Lithuanian word. After studying, subjects had to report JOLs, seeing the word pairs and being asked how likely they were to remember the pair a week later.

Crowed Annotation for FOD. Participants were shown each of the 45 English words from the OMNI dataset. They were then asked to rate how difficult they perceived each of the words on a scale from 0 (not at all difficult) to 100 (very difficult).

Objective Features. The dataset was additionally enriched by some objective features indicative of word difficulty.

Models. We tested support vector machines (SVM), AdaBoost, and random forests as classifiers. These model types generally perform well on small datasets. Transfer learning techniques was used to increase the prediction performance [17, 19].

3 Results

Table 1 shows that describing measurable features for difficulty of a task can play an positive role in the prediction of learning success.

Table 1. Results for objective features in terms of ROC AUC

Features	AdaBoost	SVM	Random forest
All objective features	0.609	0.599	0.608

Note: No subject-specific features (e.g., user id) or time lags were used as features

Table 2. AUC of JOLs and FODs and their combinations

Features	AdaBoost	SVM	Random forest
JOLs	0.671	0.671	0.668
Naïvely transferred FODs	0.561	0.512	0.569
JOLs + naïvely transferred FODs	0.648	0.674	0.673

Table 2 presents the AUC of the subjective features of JOLs and FODs.

The results also show that the use of JOLs alone can improve predictive power more than using FOD alone. However, it should be noted that the FODs used here are collected using crowed annotation and only enter in averaged form (for each word across all subjects) and are transferred equally to all subjects. Comparing the results in Table 1 with Table 2, we see that subjective metacognitive feedback is a better predictor than objective features alone.

Table 3 shows the results for combining objective features with JOLs and FODs individually and all together.

Table 3 AUC of combined objective-subjective features

Features	AdaBoost	SVM	Random forest
JOLs + objective	0.653	0.681	0682
FODs + objective	0.609	0.608	0.608
FODs + JOLs + objective	0.652	0.684	0.683

4　Discussion

We have demonstrated the usefulness of the metacognitive information of FOD, and objective features related to word difficulty in improving the predictive power of classifiers in the context of second language word pair learning. Despite the challenges

---

[1] https://dictionary.cambridge.org/de/.

[2] http://gureckislab.org/omni.

associated with using crowd annotation, studies involving metacognitive information such as FOD can also benefit from it. It is worth noting that the FOD is included in our analyses only as an average over a whole group of subjects. Nevertheless, the results are promising and suggest that a more sophisticated application of subject-related transfer learning will lead to even better results. Future studies could take advantage of the similarities between subjects to increase positive transfer. How is all this useful for practical applications? On the one hand, modeling students' cognitive state should interfere as little as possible with instructional activities in educational institutions. On the other hand, student feedback contains important information about the learning process and should not be neglected. FOD should therefore be further investigated because it has the advantage of being easy to collect and yet reflects important aspects of human metacognitive monitoring processes and thus human learning.

References

1. Fazlija, B.: Intelligent tutoring systems in higher education – towards enhanced dimensions. Zeitschrift für Hochschulentwicklung **14**(3), 217–233 (2019)
2. Li, K.C., Wong, B.Y.Y.: Revisiting the definitions and implementation of flexible learning. In: Li, K.C., Yuen, K.S., Wong, B.T.M. (eds.) Innovations in Open and Flexible Education. EIS, pp. 3–13. Springer, Singapore (2018). https://doi.org/10.1007/978-981-10-7995-5_1
3. Collis, B., Moonen, J.: Flexible learning in a digital world. Open Learn. J. Open Distance e-Learning **17**(3), 217–230 (2002)
4. De Boer, W., Collis, B.: Becoming more systematic about flexible learning: beyond time and distance. ALT-J Assoc. Learn. Technol. J. **13**(1), 33–48 (2005)
5. Anderson, J., Boyle, C., Reiser, B.: Intelligent tutoring systems. Science **228**(4698), 456–462 (1985)
6. Koedinger, K., Anderson, J., Hadley, W., Mark, M.: Intelligent tutoring goes to school in the big city. Int. J. Artif. Intell. Educ. (IJAIED) **8**, 30–43 (1997)
7. Ma, W., Adesope, O.O., Nesbit, J.C., Liu, Q.: Intelligent tutoring systems and learning outcomes: a meta-analysis. J. Educ. Psychol. **106**, 939–2176 (2014)
8. Corbett, A., Anderson, J.: Knowledge tracing: modeling the acquisition of procedural knowledge. User Model. User-Adap. Inter. **4**(4), 253–278 (1994). https://doi.org/10.1007/BF0109 9821
9. Piech, C., et al.: Deep knowledge tracing. In: Neural Information Processing Systems (NIPS) (2015)
10. Pardos, Z.A., Heffernan, N.T.: KT-IDEM: introducing item difficulty to the knowledge tracing model. In: Konstan, J.A., Conejo, R., Marzo, J.L., Oliver, N. (eds.) UMAP 2011. LNCS, vol. 6787, pp. 243–254. Springer, Heidelberg (2011). https://doi.org/10.1007/978-3-642-22362-4_21
11. Hattie, J.: Visible Learning: A Synthesis of Over 800 Meta-analyses Relating to Achievement. Routledge, Abingdon (2008)
12. Nelson, T., Dunlosky, J.: When People's Judgments of Learning (JOLs) are extremely accurate at predicting subsequent recall: the "delayed-JOL effect." Psychol. Sci. **2**(4), 267–270 (1991)
13. Efklides, A., Samara, A., Petropoulou, M.: Feeling of difficulty: an aspect of monitoring that influences control. Eur. J. Psychol. Educ. **14**, 461–476 (1999). https://doi.org/10.1007/BF0 3172973
14. Leroy, G., Kauchak, D.: The effect of word familiarity on actual and perceived. J. Am. Med. Inf. Assoc. **21**, e169–e172 (2014)

15. Beinborn, L.M.: Predicting and manipulating the difficulty of text-completion exercises for language learning. Technische Universität Darmstadt (2016)
16. Mukherjee, N., Patra, B. G., Das, D., Bandyopadhyay, S.: Ju_nlp at semeval-2016 task 11: identifying complex words in a sentence. In: Proceedings of the 10th International Workshop on Semantic Evaluation (SemEval-2016) (2016)
17. Pan, S.J., Yang, Q.: A survey on transfer learning. IEEE Trans. Knowl. Data Eng. **22**, 1345–1359 (2009)
18. Halpern, D., et al.: Knowledge tracing using the brain. In: Educational Data Mining (EDM) (2018)
19. Raina, R., Battle, A., Lee, H., Packer, B., Ng, A.Y.: Self-taught learning: transfer learning from unlabeled data. In: Proceedings of the 24th International Conference on Machine Learning (2007)

Classifying Different Types of Talk During Collaboration

Solomon Ubani$^{(\boxtimes)}$ (iD) and Rodney Nielsen

University of North Texas, Denton, TX 76203, USA
{solomon.ubani,rodney.nielsen}@unt.edu

Abstract. Pair programmers utilizing more Exploratory (critical, constructive) talk has been shown to help students achieve a better mutual understanding of problems they are solving. In this paper, we investigate the promise of fine tuning a pretrained transformer-based machine learning model to classify utterances into Exploratory, Cumulative, and Disputational talk. The task of classifying utterances into different types of *collaborative* talk was approached as a multi-label text classification problem. This is the first successful automatic classification of utterances into the different types of collaborative talk.

Keywords: Pair programming · CITS · CSCL · Collaborative learning

1 Introduction

Pair programming is a practice in software engineering where two programmers develop software together on the same physical or virtual workstation. During pair programming one of the programmers writes the code (the driver) while the other programmer watches that coding to catch any errors or find ways to improve it (the navigator). It is good practice for both pair programmers to verbalize their thoughts and to exchange questions and ideas [1]. Effective Pair programming is known for improving code quality, technical knowledge, and team morale but has its own obstacles, one of which is poor communication [2]. The research introduced here is a step toward improving such communication.

Neil Mercer [3] developed a framework for analyzing students' dialogue, which distinguishes more productive conversations from less productive ones. During collaboration and other joint activities among students, conversations can be grouped into three categories – Cumulative, Disputational, and Exploratory talk. "In Cumulative talk, speakers build positively but uncritically on what the other has said" [4]; whereas, "Disputational talk is characterized by disagreement and individualized decision making" [4]. "In Exploratory talk, participants engage critically but constructively with each other's ideas by offering alternative ideas" [4]. Fernández et al. [5] showed that Exploratory talk expands the joint Zone of Proximal Development [6] by aiding partners to achieve a better mutual understanding of the problem they are collaborating to solve. Previous studies that analyzed students conversations during pair programming have found that

© Springer Nature Switzerland AG 2022
M. M. Rodrigo et al. (Eds.): AIED 2022, LNCS 13356, pp. 227–230, 2022.
https://doi.org/10.1007/978-3-031-11647-6_40

without any intervention students use Cumulative talk more than Exploratory or Disputational talk [4] but more Exploratory talk can be achieved through human (teacher) interventions [7]. Ubani and Nielsen [8] highlighted the need for more intelligent collaborative systems that analyze dialogue and nudge positive collaborative behavior. The longer-term goal of this research is to develop tools that can provide such interventions automatically, rather than requiring substantial human intervention as seen in the work of Zakaria et al. [7]. This is the first work to successfully automate the classification of different types of collaborative talk – Exploratory, Cumulative, and Disputational talk.

An automatic classification model as investigated here can be integrated into a Collaborative Intelligent Tutoring System (CITS) that encourages improvements in pair programming by nudging programmers to improve communication, for example, by engaging in more Exploratory and less Disputational talk.

2 Dataset

Our approach to classifying collaborative talk utterances involves supervised machine learning which in this case requires a corpus of pair programming dialogue labeled at an appropriate level of detail to indicate the type of talk. We used the corpus of labeled pair programming dialogue collected and annotated in research by Zakaria et al. [7]. The corpus consisted of four activity sessions from 12 students (6 pairs) in a programming class at a school in the southeastern United States resulting in 24 total collaboration sessions. The dataset consists of a total of 7027 utterances. Zakaria et al. manually labeled each utterance for all relevant specific fine-grained subtypes of Disputational, Cumulative, Exploratory or Other talk (off-task dialogue) achieving a Cohen's kappa [9] average of 0.795 across the categories of collaborative talk. This is characterized as substantial by Landis and Koch [9].

3 Experiments

During our experiments, we performed a leave-one-(pair)-out evaluation based on the six programmer pairs (the same pairs were maintained for all four activities) resulting in a total of $6 \times 4 = 24$ distinct *testing* datasets. To act as a baseline, we first created a TF-IDF model that achieved a macro-average F_1-score over the 24 *testing* datasets of 0.31 for classifying Cumulative Talk, 0.03 for Exploratory, 0.02 for Disputational, and 0.20 for Other Talk. For all other experiments, we used our *training* data to finetune a pre-trained transformer model, DeBERTa [10], to classify collaborative talk. To address data imbalance across the classes, we augmented only the *training* set using two data augmentation techniques: Easy Data Augmentation [11] (the percent of the words in a sentence to be modified is set to 0.1) and Back translation [12] (English and German were the source and intermediary languages respectively). For Back (and Forward) translation, we used the Microsoft translator[1]. We used EDA and Back translation to augment the data in the training sets at ratios that led to balanced classes (each of the four classes ranged from 20% to 25% in each of the final 24 training subsets).

---

[1] https://docs.microsoft.com/en-us/azure/cognitive-services/translator/.

All experiment hyperparameter selection was performed using cross-validation *tuning* folds, then we retrained using both the *training* and *tuning* dataset and reported results obtained on the held-out testing dataset. Neither the training nor tuning data included utterances from programmers or activities represented in the corresponding testing dataset. We determined the hyperparameters based on performance on the *tuning* dataset, resulting in a training time of three epochs, batch size of 64, AdamW optimizer with a learning rate of 5×10^{-5}, categorical cross-entropy as the loss function, and a maximum sequence (utterance) length (MSL) of 200 tokens. We froze the first 12 blocks of the DeBERTa-Large pretrained model and finetuned the remaining 12 blocks.

4 Results

Results were computed for the fine-grained subcategories of talk and averaged over the 24 testing sets. The model achieved an F_1-score of 0.65 for detecting Cumulative Talk, 0.53 for Exploratory Talk, 0.52 for Disputational Talk, and 0.43 for Other Talk.

5 Discussion, Conclusion and Future Work

Our collaborative talk classification model achieved an F_1-score of 0.65 for detecting Cumulative talk, 0.53 for Exploratory, 0.52 for Disputational, and 0.43 for detecting Other talk. This is substantially better than the baseline, which only reached F_1-scores of 0.31, 0.03, 0.02, and 0.20, respectively. Since we partitioned the data by pair programming activities and pairs to ensure dataset independence, the model did not overfit to the idiosyncrasies of a specific activity or programmer. Therefore, our reported results should be indicative of what to expect for new programming pairs and activities in similar course settings. The results show that it is feasible to develop a system that significantly outperforms the baseline which is promising, but there is still a lot of room for improvement. In future work, we will systematically probe [13] our finetuned transformer models and augment the models with the linguistic and contextual features obtained from further error and computational linguistic analysis not already encoded in our models.

Since the primary goal of this model is to nudge students to engage in more Exploratory talk, a high F_1-score for detecting Exploratory talk is more important than for other types of talk. Higher recall will help ensure the system does not nudge users at a time when they just used Exploratory talk and higher precision will help ensure an accurate assessment of how often it occurs. Similarly, high precision is more important than high recall in detecting Disputational talk, since if the system erroneously classifies an utterance as Disputational, any resultant system feedback could be counterproductive, and it is not necessary to detect every instance of Disputational talk to provide occasional support with corrective nudges. This practical perspective on the results should guide further research into the most important model improvements. In future, we plan to deploy our model in a CITS to assess how effectively it can foster improved collaboration by encouraging students to engage in more Exploratory talk.

Acknowledgements. We would like to thank Zarifa Zakaria, Jessica Vandenberg, Jennifer Tsan, Danielle Cadieux Boulden, Collin F. Lynch, and Eric N. Wiebe from North Carolina State University, Raleigh, NC, USA and Kristy Elizabeth Boyer from University of Florida, Gainesville, FL, USA for granting us access to the dataset used in this paper.

References

1. Hughes, J., Walshe, A., Law, B., Murphy, B.: Remote pair programming. In: CSEDU 2020 - Proceedings of the 12th International Conference on Computer Supported Education, vol. 2, pp. 476–483 (2020). https://doi.org/10.5220/0009582904760483

2. Begel, A., Nagappan, N.: Pair programming: what's in it for me? In: Proceedings of the Second ACM-IEEE International Symposium on Empirical Software Engineering and Measurement, pp. 120–128, October 2008

3. Mercer, N.: Words and Minds: How We Use Language to Think Together. Routledge, London (2002)

4. Zakaria, Z., et al.: Collaborative talk across two pair-programming configurations (2019)

5. Fernández, M., Wegerif, R., Mercer, N., Rojas-Drummond, S.: Re-conceptualizing "scaffolding" and the zone of proximal development in the context of symmetrical collaborative learning. J. Classr. Interact. **50**(1), 54–72 (2015)

6. Vygotsky, L.: Interaction between learning and development. Read. Dev. Child. **23**(3), 34–41 (1978)

7. Zakaria, Z.: Two-computer pair programming: exploring a feedback intervention to improve collaborative talk in elementary students. Comput. Sci. Educ. **32**(1), 1–28 (2021)

8. Ubani, S., Nielsen, R.: Review of Collaborative Intelligent Tutoring Systems (CITS) 2009–2021. In: 2022 11th International Conference on Educational and Information Technology (ICEIT), pp. 67–75 (2022). https://doi.org/10.1109/ICEIT54416.2022.9690733

9. Landis, J.R., Koch, G.G.: The measurement of observer agreement for categorical data. Biometrics **33**(1), 159–174 (1977). https://doi.org/10.2307/2529310.JSTOR2529310.PMID843571

10. He, P., Liu, X., Gao, J., Chen, W.: DeBERTa: Decoding enhanced BERT with disentangled attention. arXiv preprint arXiv:2006.03654 (2020)

11. Wei, J., Zou, K.: EDA: Easy data augmentation techniques for boosting performance on text classification tasks. arXiv preprint arXiv:1901.11196 (2019)

12. Sennrich, R., Haddow, B., Birch, A.: Improving neural machine translation models with monolingual data. arXiv preprint arXiv:1511.06709 (2015)

13. Conneau, A., Kruszewski, G., Lample, G., Barrault, L., Baroni, M.: What you can cram into a single vector: Probing sentence embeddings for linguistic properties. arXiv preprint arXiv:1805.01070 (2018)

Nonverbal Collaboration on Perceptual Learning Activities with Chemistry Visualizations

Martina A. Rau[✉] and Miranda Zahn

Department of Educational Psychology, University of Wisconsin - Madison, Madison, USA
marau@wisc.edu

Abstract. Classroom orchestration tools rely on research showing when individual or collaborative activities are effective. No prior research has investigated effects of collaboration on perceptual learning, which is nonverbal and inductive. An experiment compared individual learning to nonverbal, gesture-based collaboration on a perceptual training. Results show an advantage of gesture-based collaboration on pre-to-posttest gains, especially for low-spatial skills students.

Keywords: Perceptual training · Collaboration · Gestures · Spatial skills

1 Introduction

Classroom orchestration tools may help teachers decide when to transition between individual and collaborative instruction. Designing such tools requires research that determines for which types of activities and for which collaboration is effective. For example, following the call to understand under what circumstances collaboration is beneficial [1], research showed that collaboration has differential effects on conceptual vs. procedural learning [2]. We expand this research by testing whether collaboration enhances students' benefit from activities that focus on perceptual learning. Further, we investigate effects on students with low spatial skills who tend to be underrepresented in science, technology, engineering, and mathematics (STEM) [3].

Visual representations are often used in STEM instruction to illustrate abstract concepts [4]. Students need to be able to fluently perceive information shown in visual representations [4, 5]. For example, when chemistry students see a Lewis structure (Fig. 1A), they need to be able to perceive the molecule's 3D geometry (Fig. 1B). Students acquire such perceptual fluency via nonverbal processes involved in inductive pattern learning [5]. These processes are considered nonverbal because verbalization interferes with perceptual learning [6]. To support perceptual fluency, perceptual trainings present students with short nonverbal tasks that ask students to quickly recognize or classify visual representations [4, 7].

A limitation of prior research on perceptual trainings is that it has exclusively focused on individual learning, possibly because of their focus on nonverbal learning. On the one hand, collaboration involves nonverbal communication via gesturing, which can

© Springer Nature Switzerland AG 2022
M. M. Rodrigo et al. (Eds.): AIED 2022, LNCS 13356, pp. 231–235, 2022.
https://doi.org/10.1007/978-3-031-11647-6_41

support perceptual learning [8]. This may be particularly helpful for students with low spatial skills [9]. On the other hand, collaboration enhances learning from complex tasks but not from simple tasks [10], whereas perceptual trainings typically involve simple, one-step tasks. To address these limitations, we test if nonverbal, gesture-based collaboration enhances perceptual learning, if spatial skills moderate the effect of gesture-based collaboration, and if problem-solving behaviors mediate this effect.

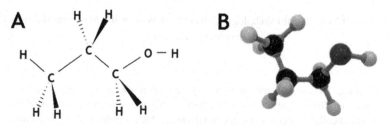

Fig. 1. A: 2D wedge-dash Lewis structure; **B:** 3D ball-and-stick model.

2 Methods

The study was conducted in an introductory chemistry course with 45 undergraduate students, taught by one of the authors. The course lasted 15 weeks; the study occurred in weeks 14–15. Three students were excluded from the analysis: two were absent in week 15, and one did not consent to participate in the study, yielding $N = 42$.

Individual students were randomly assigned to conditions. Students in the *collaborative condition* ($n = 22$) worked in dyads on the perceptual learning activities. Dyads were instructed not to talk but to communicate only via gestures, such as pointing and showing shapes. The *individual condition* ($n = 20$) worked alone on the activities.

The perceptual training was provided electronically and was designed based on prior research [4, 5]. To encourage nonverbal, inductive processes, the perceptual training asked students to "solve tasks fast and intuitively without overthinking it." Students received 20 short tasks asked them to select one of four visuals that showed the same molecule as a given visual. The choice options varied a range of visual features that are relevant to chemical isomerism and irrelevant features.

Students took a content knowledge pretest at the start of week 14, an immediate posttest after finishing the perceptual learning activities, and a delayed posttest at the start of week 15. Further, students took a mental rotations test in week 1. We also used logs from the perceptual training to compute Z scores for time on task and error rates. Students in the collaborative condition were assigned the group score.

To analyze the data, we first calculated the intraclass correlation (ICC) to estimate the degree of clustering due to students being nested in dyads. While the ICC for students' immediate posttest scores was nonsignificant ($p = .693$), it was significant for students' delayed posttest scores (ICC $= .506$, $p = .041$). This rejects the assumption of independence of students within dyads. Hence, we used a hierarchical linear model (HLM). Level 1 modeled repeated test-time within students (immediate and delayed posttest).

Level 2 modeled student characteristics (spatial skills, pretest scores). Level 3 modeled student-group variables (condition, interaction between condition and spatial skills). Levels 2 and 3 included random effect for students and student dyads, respectively. To conduct the mediation analysis, we constructed a causal path model.

3 Results

To test if collaboration is effective, we examined the main effect of condition, which was significant, $F(1, 30.7) = 9.89, p = .004$. Students in the collaborative condition had higher learning gains than students in the individual condition. To test if spatial skills moderate the effect of collaboration, we examined the interaction between condition and the continuous spatial skills variable, which was significant, $F(1, 34.4) = 6.29$, $p = .017$. Students with lower spatial skills benefitted more from collaboration (see Fig. 3A). To test if problem-solving behaviors mediate the effect of collaboration, we first verified whether collaboration affected problem-solving behaviors. The HLM showed a significant effect on error rates, $F(1, 33.3) = 5.48, p = .025$, but not on time on task ($F < 1$). There was no significant interaction between condition and spatial skills ($F < 1$). Thus, we used only error rates for the mediation analysis. The causal path model fits the data, $\chi^2 (9, N = 42) = 13.0074, p = .163$) (Fig. 3B). Error rates fully mediate the effect of condition on immediate posttest: collaborating students made fewer errors, which increased performance on the immediate posttest.

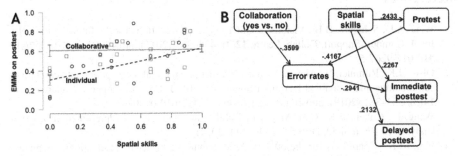

Fig. 3. A: Interaction of condition with spatial skills. The y-axis shows estimated marginal means (EMMs, averaged across immediate and delayed posttests). Error bars show standard errors of the means. **B:** Summary of causal path analysis model, unstandardized coefficients.

4 Discussion

This study tested whether nonverbal, gesture-based collaboration enhances perceptual learning. We found a large benefit of gesture-based collaboration, especially for low-spatial skills students, which was fully mediated by a reduction of error rates.

Our results extend prior research in several ways. First, adding to research showing that perceptual learning can occur in social settings [8], we found that gesture-based enhances perceptual learning. Second, in contrast to research suggesting that verbal collaboration interferes with perceptual learning [6] our findings suggest that this assertion does not extend to gesture-based collaboration.

Further, our findings expand prior research by providing pathways for addressing an equity issue for students with low spatial skills. Our study suggests that gesture-based collaboration may help address this issue. Specifically, students with low spatial skills may have difficulties in mentally rotating visual features so that they can translate between the visuals. A partner who is more proficient at this task may help them see mappings between the visuals and thereby help them recognize perceptual patterns.

Limitations of our study include that we did not observe students' nonverbal communication and that it had a relatively small sample size.

To conclude, our study is the first to establish that nonverbal collaboration via gesturing enhances students' benefit from activities focused on perceptual learning with visual representations. These findings can inform the design of classroom orchestration systems that can help instructors decide which activities lend themselves to collaborative work, and which students might benefit from them.

Acknowledgements. This work was supported by a 2019 National Academy of Education (NAEd)/Spencer Postdoctoral Fellowship. We thank Judith Burstyn, Edward Misback, and the Chemistry Department.

References

1. Wise, A.F., Schwarz, B.B.: Visions of CSCL: eight provocations for the future of the field. Int. J. Comput.-Support. Collab. Learn. **12**(4), 423–467 (2017). https://doi.org/10.1007/s11 412-017-9267-5
2. Olsen, J.K., Rummel, N., Aleven, V.: It is not either or: an initial investigation into combining collaborative and individual learning using an ITS. Int. J. Comput.-Support. Collab. Learn. **14**(3), 353–381 (2019). https://doi.org/10.1007/s11412-019-09307-0
3. Wang, L., Cohen, A.S., Carr, M.: Spatial ability at two scales of representation: a meta-analysis. Learn. Individ. Differ. **36**, 140–144 (2014)
4. Rau, M.A.: Conditions for the effectiveness of multiple visual representations in enhancing STEM learning. Educ. Psychol. Rev. , 1–45 (2016). https://doi.org/10.1007/s10648-016-9365-3
5. Kellman, P., Massey, C.: Perceptual learning, cognition, and expertise. In: Ross, B. (ed.) The psychology of Learning and Motivation, vol. 558, pp. 117–165. Elsevier, New York, NY (2013)
6. Schooler, J.W., Fiore, S., Brandimonte, M.: At a loss from words: verbal overshadowing of perceptual memories. Psychol. Learn. Motiv. **37**, 291–340 (1997)
7. Kellman, P.J., Massey, C.M., Son, J.Y.: Perceptual learning modules in mathematics. Top. Cogn. Sci. **2**, 285–305 (2010)
8. Singer, M.: The function of gesture in mathematical and scientific discourse in the classroom. In: Church, B., Kelly, S.A., M.W. (eds.) Why Gesture? How the Hands Function in Speaking, Thinking and Communicating, pp. 317–329. John Benjamins Publishing, Amsterdam (2017)

9. Wu, S.P., Corr, J., Rau, M.A.: How instructors frame students' interactions with educational technologies can enhance or reduce learning with multiple representations. Comput. Educ. **128**, 199–213 (2019)
10. Koedinger, K.R., Corbett, A.T., Perfetti, C.: The knowledge-learning-instruction framework. Cogn. Sci. **36**, 757–798 (2012)

Mining for STEM Interest Behaviors in Minecraft

Matt Gadbury(✉) and H. Chad Lane

University of Illinois Urbana-Champaign, Urbana, IL 61801, USA
gadbury2@illinois.edu

Abstract. We consider how pre-existing STEM interest influences the way in which adolescents engage an astronomy-themed Minecraft environment. Participants in an after-school program met for five sessions over the course of five weeks and explored a variety of hypothetical versions of Earth, such as Earth with no moon, in Minecraft. An association rule mining approach was taken to understand how differing levels of STEM interest influence in-game science tool usage and observations across worlds. Highest science tool use was observed among participants with moderate STEM interest, suggesting high engagement and desire to learn compared with the low and high STEM interest groups. High recorded observations among the high STEM interest group suggests confidence or high prior knowledge, while moderate tool use among low STEM interest learners might suggest development of interest in content.

Keywords: Association rule mining · STEM interest · Motivation · Minecraft

1 Introduction

Digital games offer extensive opportunities for learning [1] in a platform that has displayed evidence of being more effective at motivating learning and engagement than conventional methods [4]. Adolescents tend to shed or adopt STEM identities during their middle school and early high school years, because of feelings of incompetence or disinterest [2]. Interest develops through consistent opportunities to engage domain-specific content or topics, and once learners can formulate their own questions and make personal connections to content their interest tends to remain steady [3]. Using an open world digital sandbox game, Minecraft, we explore how adolescents of varying interest levels in STEM engage the environment through self-directed play and use of science tools. Association Rule Mining (ARM) is used to extract rules for how these groups interact with STEM content in the game and pinpoint differences between them.

2 Methods

2.1 Participants

Participants were all students at a middle school in the midwestern United States participating in an after-school program (n = 14, 7% female, 8 identified as White/Caucasian,

© Springer Nature Switzerland AG 2022
M. M. Rodrigo et al. (Eds.): AIED 2022, LNCS 13356, pp. 236–239, 2022.
https://doi.org/10.1007/978-3-031-11647-6_42

1 as American Indian or Alaskan Native, and 5 preferred not to answer). The median age for participants was 12. Consent was obtained virtually by at least one parent/guardian of each participant prior to the program start. The program was conducted on-site in the middle school library.

2.2 Materials

The primary mode for delivering content throughout the after-school program was a customized Minecraft server using Minecraft JAVA Edition (MC Server). The MC Server hosts 13 unique worlds to explore, including two orientation worlds to introduce players to science tools and observations, six "What-if" worlds, such as Earth as a moon (Mynoa), as well as five known Exoplanets. On each world players can use science tools to take measurements by typing "/temperature" or "/oxygen", for example. In total, learners have 11 science tools at their disposal. All participants were provided laptops and access to Minecraft.

2.3 Measures

STEM Interest Survey. A validated STEM interest survey developed by our lab was given to all participants on the first day of the program, prior to playing on our Minecraft server. The survey uses a 1–5 Likert type scale (1 = "Strongly Disagree", 5 = "Strongly Agree") and consists of 25 items designed to measure participant interest in a variety of STEM activities (i.e. "I like learning about how different ecosystems work"). Unpublished results showed high reliability (Cronbach's α from .8 to .9).

Log Data. Log data for each participant was collected by the MC Server and stored in a database. Collected log data included timestamped location data every 3 s, every science tool used, every observation made, blocks placed, blocks destroyed, and the value returned for every use of a science tool.

2.4 Procedure

The after-school program consisted of five, two-hour sessions meeting once per week for five consecutive weeks. Each session consisted of short discussions introducing participants to the "What-if" worlds they would explore each day, followed by time spent exploring the worlds, taking science measurements, and making in-game observations. The STEM interest survey was given on the first day of the program.

2.5 Analysis

An aggregate score for each participant was calculated for the STEM interest survey, based on responses to the Likert-type items. The mean value and standard deviation were divided participants into three groups: low STEM interest (LSI), medium STEM interest (MSI), and high STEM interest (HSI).

Once participants were divided into STEM interest groups, Association Rule Mining was used to examine the ways in which the different groups use science tools and observations on the WHIMC server. Due to a low sample size, science tool use was aggregated for each group across all explored worlds.

RapidMiner software was used to conduct the analysis. *Support* is used to calculate the proportion of transactions containing an item or rule from the total number of transactions. Minimum *support* was set at .15. *Confidence* is used to find the proportion of transactions that fit a rule out of the number of transactions that fit the rule's "if" condition. *Confidence* provides evidence for strong relationships in the data, and the cutoff for *confidence* was also set at .15 to account for diversity of tool options in an open environment. Finally, *lift* is calculated by dividing the confidence of a rule, such as "temp-obs", by the probability of an observation occurring. A *lift* value > 1 suggests strong association.

3 Results and Discussion

Based on survey results (*mean* = 80.14, *min* = 65, *max* = 90, *SD* = 7.5) four participants were placed in the low STEM interest group (>1 *SD* from *mean*), five in the medium STEM interest group (within 1 *SD* of *mean*), and five in the high STEM interest group (>1 *SD* from *mean*).

Association rule mining was applied to each group separately to see how STEM interest might influence tool usage and observations. The MSI group showed the greatest activity in terms of science tool use and observations. The LSI group also showed higher activity, in terms of science tools used, than the HSI group. However, when looking at frequent items, the HSI group did more frequently make observations than the other two groups (*support* = 0.652). The lack of activity but high number of observations for the HSI group could be due to greater interest driving exploration and recording inferences. Across all groups, the "Temperature" science tool was most frequently used (*support:* LSI = 0.65, MSI = 0.52, HSI = 0.348). The LSI and MSI groups both exhibited a tendency to measure "Gravity" (*support:* LSI = 0..25, MSI = 0.4) at a more prevalent rate than other science tools, aside from "Temperature". Frequent items associated across groups were "Temperature-Observe", suggesting that the LSI and MSI groups were taking measurements and making observations. Similarly, the HSI group had observations and "Temperature" frequently cooccurring.

For rules generated from the data (Table 1), the HSI group only exhibited one rule, "Temp – Observe", with a moderate *confidence* (*conf* = 0.625). The LSI group produced a total of four rules, including some overlap with the MSI group. All LSI group rules involved "Temperature" measurements, with "Gravity – Temp" (*conf* = 0.8, *lift* = 1.2) and "Observe – Temp" (*conf* = 0.75, *lift* = 1.2) having the highest *confidence* ratings, suggesting they are likely cooccur. The *lift* values > 1, supports the occurrence of these rules. Despite low *support*, multiple measurements cooccurring might suggest the LSI group made efforts to understand how planetary conditions fit together.

Table 1. Rules Generated from ARM analysis divided by STEM interest level

Low STEM interest				Medium STEM interest				High STEM interest			
Rule	Sup	Con	Lift	Rule	Sup	Con	Lift	Rule	Sup	Con	Lift
G-T	0.2	0.8	1.2	T-O	0.3	0.73	1.4	T-O	0.2	0.63	1.0
T-G	0.2	0.3	1.2	T-G	0.2	0.46	1.2				
P-T	0.15	1	1.5	O-G	0.2	0.55	1.4				
O-T	0.15	0.75	1.2	G-T	0.2	0.6	1.2				
				G-O	0.2	0.6	1.4				
				T-O-G	0.2	0.63	1.6				
				O-G	0.2	0.71	1.8				
				T-G-O	0.2	0.83	1.9				
				O-G-T	0.2	0.83	1.6				

Note: *G = Gravity, O = Observation, T = Temperature, P = Pressure*
Sup = Support, Con = Confidence

The MSI group generated the most rules from the data. In *Confidence* scores were highest for the three variable rule, "Temp-Obs – Gravity" (*conf* = 0.833). This suggests a greater complexity to how they were approaching exploration of worlds by taking multiple measures and making observations. An overall reliance on a few specific tools could be attributable to interest in comparing conditions across the planet.

Support was low for all rules, with the highest reported supported rule being "Temp-Observe" (*support* = 0.32) for the MSI group. Low global support values are likely due to the variety of tools available and an emphasis on open-ended exploration.

References

1. Clark, D.B., et al.: Digital games, design, and learning: a systematic review and meta-analysis. Rev. Educ. Res. **86**(1), 79–122 (2016). https://doi.org/10.3102/0034654315582065
2. Maltese, A.V., Tai, R.H.: Pipeline persistence: examining the association of educational experiences with earned degrees in STEM among U.S. students. Sci. Educ. **95**(5), 877–907 (2011). https://doi.org/10.1002/sce.20441
3. Renninger, K.A., Hidi, S.E.: Interest development, self-related information processing, and practice. Theory Pract. 1–12 (2021). https://doi.org/10.1080/00405841.2021.1932159
4. Wouters, P., et al.: A meta-analysis of the cognitive and motivational effects of serious games. J. Educ. Psychol. **105**(2), 249–265 (2013). https://doi.org/10.1037/a0031311

A Teacher without a Soul? Social-AI, Theory of Mind, and Consciousness of a Robot Tutor

Rinat B. Rosenberg-Kima$^{(\boxtimes)}$ (iD) and Alfin Thomas

Technion – Israel Institute of Technology, Haifa, Israel
rinatros@technion.ac.il, alfinthomas@campus.technion.ac.il

Abstract. Is consciousness required for social robots to serve as a tutor? This study explored the effects of a tutor robot's social-AI on participants' perception of social robots' Theory of Mind abilities, perception of social robots' consciousness, and acceptance of social robots in education. We were also interested in the relationships between these variables. One hundred and twenty participants from Amazon Mechanical Turk were randomly assigned to one of four conditions. The participants completed an online survey that included two short videos of a robot tutor interacting with a student, followed by several questionnaires. We manipulated the robot's social-AI abilities with respect to two social abilities: emotion *Expression* and *Detection,* resulting in a 2 × 2 controlled research design. Participants' responses to the Social Robots Theory of Mind Assessment Scale revealed that the robot's social AI behavior had an impact on its perceived ability to detect and influence others' feelings. Perceptions of social robots' TOM and consciousness had a significant positive correlation with acceptance of social robots in education.

Keywords: Social robots · Theory of Mind · Consciousness · Social AI

1 Introduction

While the use of social robots in educational settings has been increasingly explored in recent years, with promising affective and cognitive outcomes [1], social robots are yet to scale up in educational settings. Indeed, when it comes to education, people are hesitant about the possibility of children interacting with social robots at school [2].

Why are social robots yet to scale up? The physical presence of robots generates expectations with respect to their social behavior that is different from the expectations of a social agent viewed on a computer or tablet. A robot tutor should be able to not only perceive and evaluate the cognitive state of the learner but also interpret engagement, confusion, and attention and respond accordingly. Indeed, there has been great progress in social AI; nevertheless, the integration of these technologies is still limited and does not allow accurate interpretation of a learner's social behavior [1]. Thus, while we have witnessed great progress with respect to artificial intelligence in recent years, the problem of social AI is the hardest and still not solved. We claim that the intrinsic complexity of this problem is related to the hard problem of consciousness [3].

© Springer Nature Switzerland AG 2022
M. M. Rodrigo et al. (Eds.): AIED 2022, LNCS 13356, pp. 240–244, 2022.
https://doi.org/10.1007/978-3-031-11647-6_43

Related to consciousness, one of the fundamental social skills in human-to-human interaction is the ability to attribute mental states (i.e., beliefs, emotions, desires, goals) to oneself and others, referred to as the Theory of Mind (TOM) [4]. As a preliminary exploration of the relationships between the problem of social AI and consciousness, this study explored the effects of a tutor robot's social-AI on participants' perception of social robots' Theory of Mind abilities, perception of social robots' consciousness, and acceptance of social robots in education. We were also interested in the relationships between these variables.

2   Method

2.1 Participants

One hundred and twenty participants (62 female and 58 male, age mean = 39.02, SD = 11.31) recruited at Amazon Mechanical Turk were randomly assigned and completed one of four online surveys. The participants were adults English speakers from the United States. The participants signed a consent form that was approved by the institutional ethical committee and received the standard Mechanical Turk payment for their participation.

2.2 Materials

Two short (~17 s) videos were generated per condition (see Fig. 1). The social robot used for the videos is a 58 cm humanoid robot named Nao, developed by SoftBank Robotics. In each video, the social robot Nao behaved differently according to the condition. The robot's behaviors were generated using Choregraphe software. The two videos, followed by questionnaires, were embedded in Qualtics questionnaires that took ~20 min to complete.

2.3 Conditions

We manipulated a robot tutor's social-AI abilities with respect to two social abilities: emotion *Expression* and *Detection*, resulting in a 2 (*Expression$^+$\Expression$^-$*) × 2 (*Detection$^+$\Detection$^-$*) controlled research design. This design resulted in four conditions: (1) *Expression$^+$Detection$^+$ condition* - the robot detected the student's feeling and expressed its feeling to the student, (2) *Expression$^+$Detection$^-$ condition* - the robot expressed its feelings but did not detect the student's feelings, (3) *Expression$^-$Detection$^+$* condition – the robot detected the students' feeling but did not express its own feelings, and (4) *Expression$^-$Detection$^-$* condition – the robot did not detect the students' feeling nor expressed its own feelings.

Fig. 1. Screenshots from a video of the robot Nao and a student

2.4 Dependent Variables

Social Robots Theory of Mind Assessment Scale. To measure the perceptions of social robots' TOM abilities, participants completed a 12-items questionnaire on a five-point Likert scale (1 = strongly disagree to 5 = strongly agree). The items were generated based on the Theory of Mind Assessment Scale (Th.o.m.a.s) [5]. Exploratory factor analysis revealed three subscales: Scale 1-Social robots experience emotions and desires (5 items, $\alpha = 0.82$), Scale 2- Social robots are indifferent to their and others feelings (4 items, $\alpha = 0.76$), and Scale 3- Social robots perceive and influence others' feelings (3 items, $\alpha = .68$).

Perceptions of Social Robots' Consciousness. The participants responded to two items ("Social robots have a mind" and "Social robots can be conscious") on a five-point Likert scale (1 = strongly disagree to 5 = strongly agree). The two items had a Cronbach alpha reliability of $\alpha = .862$ and were combined into a single item.

Acceptance of Social Robots Scale. Five items from Kennedy et al. [2] were used to measure the participants' acceptance of social robots in education (e.g., "It would be acceptable for social robots to be used alongside other technology in school"). The Cronbach alpha reliability of the items was $\alpha = .824$, and they were combined into a single item.

3 Results

Perceptions of Social Robots' TOM Abilities. Two-way MANOVA with conditions (Express$\pm$ $\times$ Detect$\pm$) as between-subjects independent variables were performed (see Fig. 2). Participants in the Detect+ conditions rated scale 3 (social robots perceive and influence others' feelings) significantly higher than participants in the Detect- conditions, $F(1, 116) = 10.433, p < .05, \eta2 = .083$. There was no significant difference between the Detect conditions for scales 1&2. Likewise, there was no significant difference between the Express conditions across the scales.

Perceptions of Social Robots Consciousness. We asked the participants whether they think social robots have a mind and whether they can be conscious. No significant differences were found between the conditions for the perceptions of robot consciousness items. Interestingly, the participants' responses varied from strongly disagree to strongly agree, expressing a wide range of opinions, and were not normally distributed (M = 3.24, SD = 1.32).

Acceptance of Social Robots in Education. No significant difference between the conditions was found for acceptance of social robots in education. Overall, the participants expressed high acceptance of social robots in education regardless of the videos they watched (M = 4.09, SD = .70).

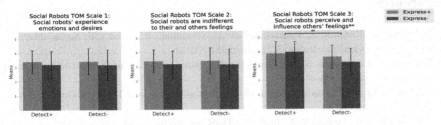

Fig. 2. Social Robots TOM scales by condition

The Relationships Between Perceptions of Social Robots' TOM Abilities, Consciousness, and Acceptance of Social Robots in Education. We conducted Pearson correlations to better understand the relationships between the variables. Perception of social robot's consciousness had a significant strong positive correlation with TOM scale 1, r = .74, p < .001, and a significant moderate correlation with TOM scale 3, r = .38, p < .001. In addition, acceptance of social robots in education had a significant moderate correlation with TOM scale 1, r = .37, p < .001, TOM scale 3, r = .33, p < .001, and perception of social robots' consciousness, r = .45, p < .001.

4   Discussion

We claim that the failure of social robots to scale up in real-world educational settings is related to the social complexity of the behavior expected from a tutor, which might require the unique property of consciousness. As a preliminary attempt to examine this claim, this study explored the effects of a tutor robot's social-AI on participants' perception of social robots' Theory of Mind abilities, consciousness, and acceptance of social robots in education. We found that manipulating the social behavior of the robot tutor, in particular the robot's ability to detect feelings, affected the robot's perceived TOM ability, yet did not affect its perceived general consciousness nor the acceptance of social robots in education. We also found that perception of social robots TOM and consciousness had a significant positive correlation with acceptance of social robots in education, indicating that the lack of consciousness might be a key explaining social robots' present failure to scale up. But should we aim to develop a conscious social robot? The participants in the current study were selected from the general population. Participants in future studies should include teachers, parents, children, and other stakeholders to address the ethical problems involved in developing a conscious-like social robot. Furthermore, we propose that the present failure in creating a simulated conscious agent has to do with the open philosophical *hard problem of consciousness* [3]. Until this hard problem is solved, it might be more productive to find the added value of social robots over teachers, which we believe has to do more with AI than with social AI. Thus, leaving the complex social interaction to human teachers.

References

1. Belpaeme, T., Kennedy, J., Ramachandran, A., Scassellati, B., Tanaka, F.: Social robots for education: a review. Sci. Robot. 3(21), eaat5954 (2018)
2. Kennedy, J., Lemaignan, S., Belpaeme, T.: The cautious attitude of teachers towards social robots in schools (2016)
3. Chalmers, D.J.: The Conscious Mind: In Search of a Fundamental Theory. Oxford University Press, New York (1996)
4. Baron-Cohen, S., Leslie, A.M., Frith, U.: Does the autistic child have a 'theory of mind'? Cognition 21(1), 37–46 (1985). https://doi.org/10.1016/0010-0277(85)90022-8
5. Bosco, F.M., Gabbatore, I., Tirassa, M., Testa, S.: Psychometric properties of the theory of mind assessment scale in a sample of adolescents and adults. Front. Psychol. 7 (2016). https://doi.org/10.3389/fpsyg.2016.00566.

Semantic Modeling of Programming Practices with Local Knowledge Graphs: The Effects of Question Complexity on Student Performance

Cheng-Yu Chung[1]([✉])[iD] and I-Han Hsiao[2][iD]

[1] Arizona State University, Tempe, USA
Cheng.Yu.Chung@asu.edu
[2] Santa Clara University, Santa Clara, USA
ihsiao@scu.edu

Abstract. Questions are one of the most essential assessment components in education. Although the delivery of questions has been revolutionized by technologies like intelligent tutoring systems (ITS), questions generation (QG) still largely relies on expert knowledge. QG requires instructors to address multifaceted aspects of teaching, including student performance, learning goals, the coverage of concepts/topics, and so on. To the best of our knowledge, there is little research investigating the structural characteristics of instructor-made questions (for specific students in a class), textbook questions (for a broader range of readers in general), and their relationship with student performance in practice. This work used the local knowledge graph (LKG) to analyze structural features of the instructor-made multiple-choice questions and those from textbooks. The results showed that the instructor-made questions were much less complex than the textbook questions in terms of concept diversity. Also, the complexity of the network components involved in the questions was significantly correlated with the performance in a classification analysis.

Keywords: Question complexity · Semantic network analysis · Local knowledge graphs

1 Introduction

Questions are one of the most widely used tools in education. Instructors use questions to assess students' knowledge acquisition or utilize them to organize in-class activities (Ham Myers, 2019). The rise of the remote and distance learning has made the role of questions more imperative than ever. Accompanied by the growing size of classrooms and the limited human resources in teaching, making use of more questions has been thriving and inevitable. Previous studies have shown various educational technologies that are effective for students to

© Springer Nature Switzerland AG 2022
M. M. Rodrigo et al. (Eds.): AIED 2022, LNCS 13356, pp. 245–249, 2022.
https://doi.org/10.1007/978-3-031-11647-6_44

keep competitive in such a learning environment. For example, instructors can periodically release questions (e.g., quizzes, exercises, worked examples) on a self-assessment platform and help students regularly and persistently practice the content [2]. The demand for questions to support day-to-day learning activity consequently increases, therefore becoming an important issue for instructors and content creators.

Although there has been a surge of interest in the field of QG, much research mainly focuses on the data processing aspect of applications (e.g., identifying linguistic features for QG processes) rather than investigating the semantic relationship between question complexity and student performance. We believe that findings of this relationship can shed light on not only the understanding of the question structures but also the associated challenges that are faced by the students during the learning process. This study adopted a semantic network model, local knowledge graph (LKG). The model was used to extract concepts in questions in the format of semantic triples and to construct a semantic network around them. The network was further used to analyze the complexity of questions that was represented by various network features. Comparing instructor-made questions from 9 undergraduate introductory programming courses and textbook questions with two existing references, we found that the former had much simpler complexity than the latter. This outcome revealed that instructor-made questions specifically focused on some topics and might not fully cover the breadth and depth of learning content.

2   Methodology

Extending the concept of semantic role labeling (SRL), researchers have proposed Open Information Extraction (OIE) that considers both SRL and propositions asserted by sentences. For example, the sentence "computers connected to the Internet can communicate with each other" can be decomposed into two possible propositions: "(Computers connected to the Internet; can; communicate with each other)" and "(Computers; connected; to the Internet)". These two propositions represent two aspects of the meaning. Compared to conventional SRL, OIE can extract more information about the intent of a given sentence.

Based on OIE, the Local Knowledge Graph (LKG) is a semantic network model that connects subjects, verbs, and objects in OIE triples [3]. It has been shown that the LKG can be used to store a large volume of documents and provide an efficient structure for search queries. We believe that the LKG can also help analyze the structure of question content due to its network structure: the interconnection of similar semantic objects and the span of edges may represent the complexity of underlying knowledge, thereby forming a semantic network.

We collected three datasets of questions, one of which represent instructor-made questions and the other textbook questions. The first dataset, QuizIT, was instructor-made multiple-choice questions (MCQ) collected from QuizIT [1] that was used in 9 undergraduate courses about entry-level Java programming over 3 years. The number of unique questions was 779 (355 when counting unique question text only). The second dataset of practice questions, Textbook, was collected from a free online textbook, *"Introduction to Programming Using Java, Eighth Edition"*[1]. In total, there were 163 practice questions, most of which were free-text questions. The third dataset, QBank, was collected from a question bank accompanied by the textbook used in the entry-level Java programming course. After processing, we collected 225 practice questions that were also in the format of MCQ.

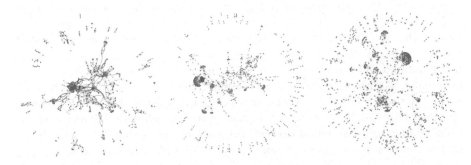

Fig. 1. Comparing LKGs from the Three Datasets (left: QuizIT; middle: Textbook; right: QBank).

3   Results and Conclusions

As shown in Fig. 1, the three LKGs had a similar structure (one large component and many small components) but different textures. First, we noticed that the number of questions was not correlated with the size of networks. The largest network was from QBank, which had 1025 nodes and 1492 edges. The Textbook network was the second largest and had 883 nodes and 1349 edges. The smallest network was from QuizIT, which had only 401 nodes and 573 edges. The result suggests that the number of questions was not positively correlated with the size of LKGs, which suggests that instructor-made questions on QuizIT might be less complex in terms of the content.

To understand the relationship between student performance and the characteristics of LKGs, we collected the usage data on QuizIT in the courses and computed the first-attempt error rate of the questions. 570 students contributed 14,534 first attempts. We grouped questions into "high-error" and "low-error" based on the average error rate. Since the question contents were transformed into semantic networks, the scope of a mapped area on the network might suggest the concepts that were addressed by the question. To capture the scope of the mapped area, we first summarized a list of unique connected components in the LKG. Then, for each question, we computed all unique connected components it involved (which was a set that we called a "combination of unique

---

[1] https://math.hws.edu/javanotes/.

connected components", CUCC). Afterward, we summarized how many high-error questions and low-error ones were in each CUCC.

The result showed that there was a clear correlation between the CUCC and the error rate. There was a set of CUCC that had more low-error questions than high-error ones and vice versa. The instructor-made questions from QuizIT had low complexity. Also, the CUCC were correlated with the error rate, and some of them only appeared in high- or low-error questions. We further grouped questions into three categories by the CUCC: cucc_high, cucc_low, and cucc_gen. cucc_high includes the CUCC that had more high-error questions than the low-error ones and vice versa for cucc_low. The category cucc_gen contains C1 only because it had significantly more questions than the other CUCC. A chi-square test showed that both cucc_high ($x^2 = 28.12, p = 0.00$) and cucc_low ($x^2 = 43.45, p = 0.00$) had significant differences in the number of high- and low-error questions.

Finally, we conducted a classification analysis to evaluate the classification performance of the CUCC and other network characteristics. The Random Forest classifier with 10-fold cross-validation was used to benchmark different configurations of features. We found two network characteristics that helped reach the best performance: the out-degree of verbs (ODV) and the betweenness of argument (BTA). The classification reached f1 = 0.72, precision = 0.75, recall = 0.70, and accuracy = 0.73. This result suggests that the ODV and the BTA might be indirect indicators of the CUCC. Hypothetically, a small component had low ODV and high BTA; a large component had high ODV and low BTA. Although the current state of analysis was not able to examine this hypothesis, the result from the classification analysis can help develop a measure to evaluate the complexity of questions.

Overall, this study is aimed to investigate the characteristics of instructor-made question content by an explainable model from two aspects: the difference from textbook questions and the correlation with student performance. We adopted a semantic network model, local knowledge graph (LKG), which is built by interwoven verb-arguments semantic triples. Comparing one dataset of instructor-made questions from introductory programming courses with two datasets of textbook questions, the analysis showed that instructor-made questions had less diversity of concepts and simpler formats than the textbook questions. We also found a correlation between the coverage of concepts and the average error rate of questions.

References

1. Alzaid, M., Trivedi, D., Hsiao, I.H.: The effects of bite-size distributed practices for programming novices. In: Proceedings of 2017 IEEE Frontiers in Education Conference (FIE), pp. 1–9. IEEE (2017). https://doi.org/10.1109/FIE.2017.8190593

2. Chung, C.Y., Hsiao, I.H.: Investigating patterns of study persistence on self-assessment platform of programming problem-solving. In: Proceedings of the 51st ACM Technical Symposium on Computer Science Education, pp. 162–168. ACM, New York, February 2020). https://doi.org/10.1145/3328778.3366827. https://dl.acm.org/doi/10.1145/3328778.3366827

3. Fan, A., Gardent, C., Braud, C., Bordes, A.: Using local knowledge graph construction to scale Seq2seq models to multi-document inputs. In: EMNLP-IJCNLP 2019–2019 Conference on Empirical Methods in Natural Language Processing and 9th International Joint Conference on Natural Language Processing, Proceedings of the Conference, pp. 4186–4196 (2019). https://doi.org/10.18653/v1/d19-1428

Classification of Natural Language Descriptions for Bayesian Knowledge Tracing in Minecraft

Samuel Hum[✉], Frank Stinar, HaeJin Lee, Jeffrey Ginger, and H. Chad Lane

University of Illinois at Urbana-Champaign, Urbana, IL, USA

Abstract. Application of Bayesian Knowledge Tracing (BKT) has primarily occurred in formal learning settings. This paper presents an integration of BKT in an informal learning context to assess the structure and skill level of learner scientific observations. We compare different approaches to text classification in a Minecraft science simulation. Our models were trained on data collected from two separate middle schools with students of different backgrounds. Experimental results demonstrate the effectiveness of several machine learning models to automatically label observations.

Keywords: Bayesian Knowledge Tracing · Minecraft · Informal learning

1 Introduction

Making scientific observations is one of the most important and challenging skills for children to acquire as they learn about science. It is often difficult for learners since it is not always obvious what features or aspects of some phenomena are most relevant. Educational support is needed to help learners make the shift from more casual ("everyday") observations to those that are more rigorous and scientific. Thus, we are developing a set of AI-based tools to support students in the task of writing scientific observations made while exploring virtual worlds in *Minecraft*. In this paper, we present our approach to assess and give automated feedback on the structure of observations. Our aim is to scaffold learner experience to create more meaningful observations that lead to inference-making and integration of knowledge.

Additionally, we describe an in-game open-learner model updated by Bayesian Knowledge Tracing (BKT) that visualizes participant progress in learning to make different types of observations. To date, a large portion of research on BKT has focused on well-defined domains and on problem-solving activities. In our work, we seek to model the more open-ended skill of making observations in an open, virtual world environment.

Supported by the National Science Foundation under Grants 1713609 and 1906873.

M. M. Rodrigo et al. (Eds.): AIED 2022, LNCS 13356, pp. 250–253, 2022.
https://doi.org/10.1007/978-3-031-11647-6_45

2   Learning Science in Minecraft

Our work occurs in the context of the WHIMC project (*What-if Hypothetical Implementations in Minecraft*). In WHIMC, learners explore hypothetical versions of Earth as well as known exoplanets and engage in numerous science-related activities, such as taking measurements, making observations, interacting with non-player character (NPC) scientists, and building habitats for survival.

Learners in WHIMC record observations directly in the Minecraft world they are exploring. An observation is an "official" record of them having seen something that stands out as interesting and potentially important. In this paper, we focus on assessing the structure of WHIMC observations. To do this, we leverage a scheme to categorize student work, which produced four observation types that we want students to focus on. *Descriptive* are observations related to color, temperature, quantity, and other physical attributes such as weight or size. *Comparative* are observations comparing one natural phenomena to another. *Analogy* are observations comparing natural phenomena with another similar structure or object and can be considered an advanced form of comparative. Finally, *Inference* are observations where a hypothesis or explanation is proposed [1].

3   Dataset and Participants

We collected observation data from an after-school community center and workshops at a public middle school in a small Midwestern town in the United States. There were ~86 participants over 4 years from the community center. Data provided in a 2018 grant application shows 95% of participants are Black or African American and over 80% come from low income families. Gender balance varies by camp but is typically an even split. There were ~14 participants from the after-school workshops. Data provided on self-report surveys shows 1 student identified as female the rest male with a majority identifying as White or Caucasian.

There were 1165 observations made by participants from the community center and 176 from the middle school. Of the observations we collected at both sites 571 were Descriptive, 295 were Comparative, 9 were Analogy, 53 were Inference, and the rest were categorized as being related to unproductive behaviors. The observations were coded and cross-validated by two Caucasian education graduate students at a large Midwestern university.

4   Bayesian Knowledge Tracing

On the WHIMC server, we implemented BKT using an adaptation of the brute force BKT model with simulated annealing to solve for the parameters [2,3]. We focused on observation structure to address the open-ended nature of the domain. Our BKT model tracks the four types of observations defined in Sect. 2. To update the learner model, we compare the student's self-classification of their observation with the predicted class from an automated classifier.

Please note due to an absence of student self-classified observations, our current BKT model is trained using 208 student observations not used to train our classifier where we compared the model's prediction against the researchers' codes. With the availability of student self-classified observations, we will train our BKT model using student self-classified observations with the researcher classifications to get an accurate representation of the domain (Fig. 1).

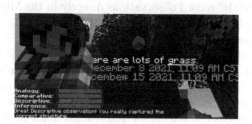

Fig. 1. In-game correctness and open-learner model feedback with a BKT update.

5 Observation Classifier

There were three main considerations for selecting our approach to observation classification: text representation, machine learning model, and model evaluation.

For text representation, preprocessing cleans observations to be vectorized. We tried a baseline method (removing non-alphanumeric characters), Porter Stemming, and WordNet lemmatization. Vectorization converts the preprocessed sentences into vectors for our classifiers. We used term frequency-inverse document frequency (TF-IDF), word2vec, and Global Vectors (GloVe).

For our classifier we show the results of the stochastic gradient descent (SGD), support vector machine (SVM), and multilayer perceptron (MLP) models, which were the three best performing algorithms on our dataset and pipeline. To account for imbalances in our samples and classes (see Sect. 3), we used sample weighting based on data collection site and Synthetic Minority Oversampling Technique (SMOTE).

To evaluate our models, we ran 10-fold cross-validation. We compared four different standard metrics for evaluating our models: accuracy (A), precision (P), recall (R), and F1 score (F1). For precision, recall, and F1 score we utilized macro averaging.

6 Results

Table 1 shows the results of our three classifiers, vectorization, and preprocessing approaches using the dataset described above in Sect. 3 and the approaches to handle dataset imbalances described above in Sect. 5 for our four metrics.

Table 1. Averages of 10-fold cross validation for each ML algorithm, vectorization, and preprocessing approach on the four different metrics. Bold scores indicate the best result for that metric.

ML Algorithm	Vectorization Technique	Baseline				Stemming				Lemmatization			
		A	P	R	F1	A	P	R	F1	A	P	R	F1
SGD	TF-IDF	.700	.588	.545	.544	.700	.612	.555	.559	.699	.590	.541	.544
	w2v	.667	.532	**.633**	.549	.630	.475	.601	.495	.673	.535	.620	.542
	GloVe	.409	.421	.484	.382	.402	.361	.448	.329	.396	.421	.502	.376
SVM	TF-IDF	.701	**.644**	.472	.500	**.709**	.641	.473	.500	.701	**.644**	.472	.500
	w2v	.683	.545	.584	.547	.630	.502	.563	.497	.683	.545	.584	.547
	GloVe	.287	.383	.429	.308	.305	.298	.372	.244	.287	.383	.429	.308
MLP	TF-IDF	.689	.604	.550	.548	.700	.610	.547	.555	.688	.608	.540	.550
	w2v	.704	.545	.613	**.560**	.667	.517	.613	.537	.698	.547	.612	.559
	GloVe	.562	.495	.585	.497	.529	.406	.519	.426	.576	.500	.578	.504

7 Discussion

Our results found that there is currently no clear "best" approach to classifying observations. With the exception of GloVe vectorization, each preprocessing, vectorization, and ML technique produced the best result on at least one metric. Since GloVe was the only one that used a pretrained model, we believe the language in our dataset was vastly different from that found in common datasets. Thus, researchers using natural language classifiers with data from middle schools should prioritize proper vectorization over other aspects of their pipeline. Another significant point of discussion is the inclusion of statistically underprivileged participant populations in our dataset. We take measures to train our model to equally weigh the language differences from students of different backgrounds and believe our work could connect to future studies regarding information ethics. The effectiveness of our AI scaffolding will be empirically measured in future studies.

References

1. Yi, S., Gadbury, M., Chad Lane, H.: Coding and analyzing scientific observations from middle school students in Minecraft (2020)
2. Baker, R.S.J., Corbett, A.T., Aleven, V.: More accurate student modeling through contextual estimation of slip and guess probabilities in bayesian knowledge tracing. In: Woolf, B.P., Aïmeur, E., Nkambou, R., Lajoie, S. (eds.) ITS 2008. LNCS, vol. 5091, pp. 406–415. Springer, Heidelberg (2008). https://doi.org/10.1007/978-3-540-69132-7_44
3. Miller, W.L., Baker, R.S., Rossi, L.M.: Unifying computer-based assessment across conceptual instruction, problem-solving, and digital games. Technol. Knowl. Learn. **19**, 165–181, July 2014

An Automatic Self-explanation Sample Answer Generation with Knowledge Components in a Math Quiz

Ryosuke Nakamoto[1(✉)], Brendan Flanagan[2], Yiling Dai[2], Kyosuke Takami[2], and Hiroaki Ogata[2]

[1] Graduate School of Informatics, Kyoto University, Kyoto, Japan
s0527225@gmail.com
[2] Academic Center for Computing and Media Studies, Kyoto University, Kyoto, Japan

Abstract. Little research has addressed how systems can use the learning process of self-explanation to provide scaffolding or feedback. Here, we propose a model automatically generating sample self-explanations with knowledge components required to solve a math quiz. The proposed model contains three steps: vectorization, clustering, and extraction. In an experiment using 1434 self-explanation answers from 25 quizzes, we found 72% of the quizzes generated sample answers with all necessary knowledge components. The similarity between human-created and machine-generated sentences was 0.719, with a significant correlation of $R = 0.48$ for the best performing generation model by BERTScore. These results suggest that our model can generate sample answers with the necessary key knowledge components and be further improved by using the BERTScore.

Keywords: Self-explanation · Rubric · Automatic summarization · NLP

1 Introduction

Self-explanation is defined as generating explanations to oneself and explaining concepts, procedures, and solutions to deepen understanding of the material [1]. It has been widely recognized for its learning effects for a long time [2]. The iSTART system is the leading research method in self-explanation evaluations, which guides learners through the exercise to support active reading and thinking [3].

In mathematics, there is a procedure for solving a quiz, and the quiz is solved according to that procedure. Therefore, we proposed a method to check whether students can describe each step in a self-explanation by comparing the similarity between the human-created sample answer and students' self-explanations [4]. It was judged that the student's knowledge was likely to be insufficient because the information and words of the unit required were included or, if not, they were missing some knowledge components. We defined "Rubric" as can-do descriptors that clearly describe all the essential knowledge components of the quiz and "Sample Answer" as model answers of self-explanations with knowledge components, which are prepared according to the step rubric number(Table 1). In this study, we propose an automatic generating sample answers model

M. M. Rodrigo et al. (Eds.): AIED 2022, LNCS 13356, pp. 254–258, 2022.
https://doi.org/10.1007/978-3-031-11647-6_46

in place of human-created sample answers. Our contributions have a wide range of implications, such as scoring self-explanations and generating self-explanation scaffold templates based on sample sentences.

Table 1. Rubrics and a sample answer of self-explanation in a quiz.

Number	Rubric	Sample Answer of Self-explanations
Step 1	Be able to find the equation of a linear function from two points	Substituting the y-coordinate of p into the equation of the line AC
Step 2	Be able to find the equation of the line that bisects the area of a triangle	Find the area of triangle ABC, and then find the area of triangle OPC
Step 3	Be able to represent a point on a straight line using letters	Since the coordinates of P are (3, 5/2), the line OP is y = 5/6, and Q are (t, 5/6)

2 Data Collections and Model Architecture

We collected the data from January 1, 2020, to December 31, 2021, using the LEAF platform [5], which consists of a digital reading system named BookRoll, and a learning analytics tool LAViEW (Fig. 1). For this experiment, we chose quizzes with at least five answers. The number of quizzes were 25, and the total number of answers were 1434. Figure 2 illustrates the proposed model, which consists of (i) Vectorizing component, (ii) Clustering component, and (iii) Extracting Component. As the vectorizing component, we adopted Sentence BERT and BERT Japanese pre-trained model to represent the sentences [6, 7]. As the clustering component, we employed an unsupervised learning model, K-means. The reason for generating meaning-intensive clusters through unsupervised learning is to reproduce the solution steps in mathematics. From an educational point of view, a problem for junior high school students would probably contain at least two steps and at most six steps of unit knowledge components and set the number of clusters in the range of 3–5 by the elbow method. As the extracting component, for each semantic cluster, the most representative sentences are extracted and sorted by multiplying them by their position in the problem, obtained from pen strokes. For extracting a representative sentence, Lexrank [8] was tested to extract the most representative sentences from each cluster. The input is all the self-explanation sentences associated with the quiz, and the output is the summarization with knowledge components for the quiz.

3 Experiments

Firstly, we set the rubrics for each quiz for evaluation (Table 2). Secondly, two authors and one assistant evaluated the machine-generated self-explanations to determine if they contained the necessary knowledge components. Though the Fleiss' kappa coefficient [9] was 0.518 initially, after discussing the differences among the three, the final coefficient

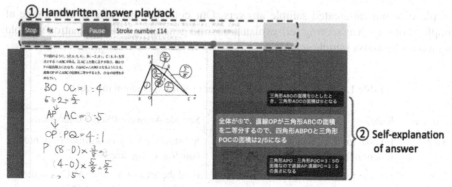

Fig. 1. The students input a sentence of explanation every time they think they have completed some step in their answers during the playback. Therefore, the self-explanations are temporally associated with the pen stroke data.

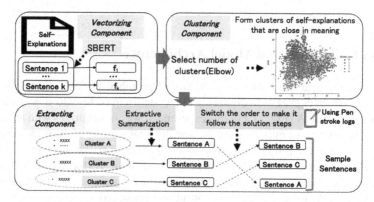

Fig. 2. Overall model architecture.

was 0.870. Table 2 shows the human evaluation results in 72% of the quizzes, it could generate all of the maximum five knowledge components.

Next, we evaluated the similarity of human-created and machine-generated sentences from several metrics: BERTScore, BLEU [10, 11]. In addition, we conducted a Spearman correlation analysis to investigate the correlations between the summary index and human evaluation. The Human Evaluation Score (HES) was scored according to how well machine-generated answers met the knowledge components against rubrics in the following form.

Table 3 presents the F1 Metrics scores. The highest similarity metric was BERTScore with an average of 0.719. Table 4 shows the correlations and RMSE between HES and metrics. As for correlation, it was 0.48 for BERTScore, showing a moderate correlation. As for RMSE, the BERTScore with the minor error was 0.273, while the other metrics were over 0.5, a significant difference.

Table 2. Missing knowledge components of each quiz by Human evaluation

Missing knowledge components	0	1	>=2
Num of quizzes	18	4	3
Probability density	0.72	0.16	0.12

Table 3. The similarity evaluation(F1)

BERTScore		BLEU	
M	SD	M	SD
0.719	0.032	0.300	0.093

Table 4. RMSE and Correlations between HES and metrics.

	BERTScore	BLEU
Correlations	0.48**	0.46**
RMSE	0.273	0.582

Note. $*** p < 0.01, ** p < 0.05, * p < 0.1$

4 Conclusion

This study attempted to generate sample self-explanation sentences from collected data. The collected 1434 self-explanations from 25 quizzes were fed into a model and the results showed that 72% of the quizzes could generate all of the maximum five knowledge components. The similarity between human-created and machine-generated sentences was 0.715, with a significant correlation of R = 0.48(BERTScore). Results suggest it is possible to generate sample answers using the proposed model to extract the necessary knowledge components and improving the BERTScore accuracy correlates with extracting essential knowledge components.

Acknowledgments. This work was partly supported by JSPS Grant-in-Aid for Scientific Research 20H01722, 21K19824, and NEDO JPNP20006, JPNP18013.

References

1. Rittle-Johnson, B.: Promoting transfer: effects of self-explanation and direct instruction. Child Dev. **77**(1), 1–15 (2006)
2. Bisra, K., Liu, Q., Nesbit, J.C., Salimi, F., Winne, P.H.: Inducing self-explanation: a meta-analysis. Educ. Psychol. Rev. **30**(3), 703–725 (2018). https://doi.org/10.1007/s10648-018-9434-x

3. McNamara, D.S., Levinstein, I.B., Boonthum, C.: iSTART: interactive strategy training for active reading and thinking. Beh. Res. Methods, Inst. Comput. **36**(2), 222–233 (2004)
4. Nakamoto, R., Flanagan, B., Takam K., Dai Y., Ogata, H.: Identifying students' stuck points using self-explanations and pen stroke data in a mathematics quiz. In: ICCE 2021, 2021.11.22–26 (2021)
5. Flanagan, B., Ogata, H.: Learning analytics platform in higher education in Japan. Knowl. Manage. E-Learn. (KM&EL) **10**(4), 469–484 (2018)
6. Reimers, N., Gurevych, I.: Sentence-BERT: sentence embeddings using Siamese BERT-Networks, arXiv preprint arXiv:1908.10084 (2019)
7. Suzuki, M.: Pretrained Japanese BERT models, GitHub repository. https://github.com/cl-tohoku/bert-japanese. Accessed 10 Aug 2020
8. Erkan, G., Radev, D.: LexRank: graph-based lexical centrality as salience in text summarization. arXiv:1109.2128 (2004)
9. Fleiss, J.L.: Measuring nominal scale agreement among many raters. Psychol. Bull. **76**, 378–382 (1971)
10. Zhang, T., Kishore, V., Wu, F., Weinberger, K.Q., Artzi, Y.: Bertscore: evaluating text generation with bert. arXiv preprint arXiv:1904.09675 (2019)
11. Papineni, K., Roukos, S., Ward, T. & Zhu, W.: BLEU: a method for automatic evaluation of machine translation. In: Proceedings of the 40th Annual Meeting on Association for Computational Linguistics (ACL 2002). Association for Computational Linguistics, USA, 311–318. https://doi.org/10.3115/1073083.1073135(2002)

Gamification Through the Looking Glass - Perceived Biases and Ethical Concerns of Brazilian Teachers

Armando Toda[1,2](✉)[ID], Paula T. Palomino[1][ID], Luiz Rodrigues[1][ID],
Ana C. T. Klock[3][ID], Filipe Pereira[2,7][ID], Simone Borges[5][ID],
Isabela Gasparini[6][ID], Elaine H. Teixeira[4][ID], Seiji Isotani[1][ID],
and Alexandra I. Cristea[2][ID]

[1] University of Sao Paulo, São Carlos, Brazil
armando.toda@usp.br
[2] Durham University, Durham, UK
[3] Tampere University, Tampere, Finland
[4] Federal University of Amazonas, Manaus, Brazil
[5] Federal University of Technology – Parana, Dois Vizinhos, Brazil
[6] Santa Catarina State University, Joinville, Brazil
[7] Federal University of Roraima, Boa Vista, Brazil

Abstract. Gamification applied to education could impact positively (or negatively) students' psychological and cognitive aspects, such as motivation, engagement and learning performance. Nevertheless, the perceptions of education professionals and their concerns about implicit biases and ethics that emerge from applying gamification to support learning has not been widely explored. Thus, we conducted a qualitative study that aims to identify gamification potential biases and ethical concerns, from teachers' perspectives, in the Brazilian context. We designed a survey answered by 61 Brazilian teachers, who had not used gamification before - even thought they were aware of its potential benefits. Our results point out that social aspects, difficulties in planning and evaluation, psychological and behavioral impacts, and privacy of students are some of the main concerns related to ethics and biases. This paper contribution focused on providing a *list of potential biases and ethical concerns to consider when designing and applying gamification in education*.

Keywords: Gamification · Ethics · Biases in learning technologies

1 Introduction

Gamification, when properly designed and tailored to students' needs, is often 'sold' as a solution for educational problems, such as lack of motivation and engagement [5]. Most studies are focused on the perspective of researchers, and experiments with students to help design gamification properly [3]. Additionally, few are concerned with the teachers' workload, or their needs and perceptions

© Springer Nature Switzerland AG 2022
M. M. Rodrigo et al. (Eds.): AIED 2022, LNCS 13356, pp. 259–262, 2022.
https://doi.org/10.1007/978-3-031-11647-6_47

on gamification, but rather present it to these practitioners as a solution they did not ask for nor know how to use. Since these professionals are the main stakeholders (aka, the ones that will potentially use gamification effectively for their students), it is important to consider what their thoughts about ethical issues are, when planning and applying any kind of technology, gamification included, to learning environments [1,2]. Thus, considering their concerns on real or perceived biases is crucial, to understand why they would or not use gamification in education.

Hence, this study aims to understand how education practitioners perceive the gamification biases in educational contexts and their main ethical concerns towards using gamification in learning environments, through an exploratory qualitative research. Furthermore, we address the following research questions (RQs): (A) **What are educational practitioners' biases towards gamification in education?**; and (B) **What are educational practitioners' main ethical concerns towards the use of gamification in education?**. Based on these RQs, our main contributions with this paper are to propose, for the first time: a set of potential biases that must be addressed during gamification design in education; a set of ethical concerns that must be considered before gamifying a learning environment. Those insights can be used in the design and development of gamification in learning environments.

2 Methods

To approach our research questions, we opted to conduct a qualitative study, since the main objective is to explore a field (that of biases and ethical concerns in gamification in education) that has not yet been tackled or approached by other studies [4]. To collect the data, we opted for the survey method, due to its reach and low-cost [4], alongside snowball sampling, where we created an online link to our survey and shared with teachers and other educational practitioners' social media groups in Brazil. Our survey consisted of 41 questions, divided into two parts. The first part (22 questions) consisted of collecting demographic and socioeconomic data, while part two (19 questions) dealt with the gamification aspects. It is worth mentioning that the data collected was carefully anonymised. Such information was not further used in the study reported in this paper, thus fully complying with the General Data Protection Regulations (GDPR), which was approved by the ethical committee of University of Sao Paulo, CAE 42598620.0.0000.5464. Considering part two of our survey, we asked: (i) if participants knew what gamification is; (ii) what they understood by the concept of gamification; (iii) previous gamification usage; (iv) reasons to use gamification; (v) perceptions about gamification *planning*; (vi) perceptions about gamification *application*; (vii) perceptions about gamification *evaluation*; (viii) main concerns about gamification that may affect its adoption by other practitioners; (ix) thoughts about gamification impacts on students' autonomy; (x) biases and ethical concerns.

3   Results

Regarding the biases, we received 61 comments from practitioners, from where two experts extracted possible common themes. This process led to a total of 64 biases. After a brainstorming session, these 64 biases were reduced to 11, divided into four categories by six experts. We removed similar themes and grouped the ones that were identified as similar in their semantics (e.g. *"Teacher formation"* and *"Incentives on teacher formation"* were attributed to the same theme). Finally, we created four categories, to group those 11 themes, for a clearer overview:

- **Social Aspects**: this category encompasses biases related to social contexts (e.g. when teachers are concerned about the social context of students), acceptance of gamification by teachers (e.g. there is still a major resistance from teachers and other education practitioners towards game-based approaches in educational environments), and students (e.g. when the gamification does not consider students' characteristics and they find it boring), and lack of interest (e.g. when people just do not care about gamification at all);
- **Planning**: this category contains themes related to how gamification is thought of, and is related to the lack of theoretical knowledge (e.g. when the teacher does not know the basis for using gamification), lack of practical knowledge (when the practitioner does not know how to apply in practice), lack of time (e.g. when practitioners state they do not have time to plan gamification), and personalisation (e.g. when the practitioner is concerned with students' characteristics when planning /applying gamification);
- **Evaluation**: this theme is also its main category, due to many practitioners stating that evaluating the effectiveness of gamification might not be so doable in real learning environments, making it difficult to accept in practice;
- **Budget**: this category is related to the lack of resources (e.g. when the institution does not provide infrastructure, or financial resources).

Considering the *ethical concerns*, we identified a total of 70 concerns, that were grouped into nine themes. These nine themes were finally divided into five categories. We also removed similar concerns and grouped others that had similar semantics (e.g. *"Undesired Behaviour"* and *"Students behavioural concerns towards competition"*). The categories that were created are:

- **Psychological impacts**: teachers, in general, were worried about how gamification could affect students' motivation, as well as needed to be careful not to be used as a tool to manipulate students and cause embarrassments;
- **Social issues**: in this category, teachers were concerned about gamification being equal and equitable, as well as attend the needs of minorities, so no student feel excluded;
- **Privacy**: teachers also expressed their concerns about how students' data would be gathered and organised in gamified environments;

- **Humanisation:** according to some teachers, gamification should be used as a tool to improve creativity and make students' more respectful and proactive, some teachers also believe that gamification should be used as a tool to promote positive behaviours in students' routines;
- **Behaviour:** regarding this theme, some teachers were worried that gamification could also lead to negative behaviours that would impact directly on the psychological aspects (e.g. loss of motivation due to extreme competition in a given gamified task).

4   Conclusions and Future Works

In this work, we navigated through the looking glass of teachers' perception on gamification biases and ethical concerns. We believe that our main contribution is the mapping of these aspects, as well as providing a summarised, yet abstract, set of topics that must be considered before gamifying educational environments. It is important to understand how these practitioners see these ethical concerns, to create more equitable and equal gamified environments, as well as understanding what teachers expect, when they want to gamify something related to their teaching practices. As future works, we aim to both conduct a deeper analysis on the data collected and expand it.

Acknowledgements. This research was partially funded by CNPq (141859/2019-9, 163932/2020-4, 308458/2020-6,308395/2020-4, and 308513/2020-7); CAPES (Code 001); FAPESP (2018/15917-0, 2013/07375-0), Samsung-UFAM (agreements 001/2020 and 003/2019). This work received financial support from the Coordination for the Improvement of Higher Education Personnel - CAPES - Brazil (PROAP/AUXPE).

References

1. Holmes, W., et al.: Ethics of AI in education: towards a community-wide framework. Int. J. Artif. Intell. Educ., 1–23 (2021)
2. Kim, T.W., Werbach, K.: More than just a game ethical issues in gamification. Ethics Inf. Technol. **18**(2), 157–173 (2016)
3. Klock, A.C.T., Gasparini, I., Pimenta, M.S., Hamari, J.: Tailored gamification: a review of literature. Int. J. Hum.-Comput. Stud., 102495 (2020). https://doi.org/10.1016/J.IJHCS.2020.102495
4. Lazar, J., Feng, J.H., Hochheiser, H.: Research methods in human-computer interaction. Morgan Kaufmann, 2nd edn. (2017)
5. Toda, A.M., Valle, P.H.D., Isotani, S.: The dark side of gamification: an overview of negative effects of gamification in education. In: Cristea, A.I., Bittencourt, I.I., Lima, F. (eds.) HEFA 2017. CCIS, vol. 832, pp. 143–156. Springer, Cham (2018). https://doi.org/10.1007/978-3-319-97934-2_9

Are All Who Wander Lost? An Exploratory Analysis of Learner Traversals of Minecraft Worlds

Maricel A. Esclamado[1,2](✉) ⓘ and Maria Mercedes T. Rodrigo[1] ⓘ

[1] Ateneo de Manila University, Quezon City, Metro Manila, Philippines
maricel.esclamado@obf.ateneo.edu, mrodrigo@ateneo.edu
[2] University of Science and Technology of Southern Philippines, Cagayan de Oro City, Philippines

Abstract. In this paper, we analyze in-game data and out-of-game assessment data from 15 Grade 6 boys from the Philippines who were completing a learning task with the What-If Hypothetical Implementations using Minecraft (WHIMC) to determine how distance traveled and area covered relate to assessment outcomes. We also determine the extent of overlap of areas covered by computing the Jaccard Index and Maximum Similarity Index (MSI). We find no significant correlation between assessment scores and overall distance, area, or MSI. However, when we break the data down into five-minute intervals, we find a significant negative correlation between assessment scores and distance traveled and area covered during certain time periods. These findings suggest that wandering off early in game play may be indicative of low learning outcomes later on. The absence of a significant relationship between MSI and assessment scores suggests the absence of a canonical traversal in an open-ended environment.

Keywords: Minecraft · WHIMC · Exploration behaviors · Assessment · Philippines

1 Introduction

What-If Hypothetical Implementations in Minecraft (WHIMC) [10] is a set of simulations that learners can explore in order to learn more about science, technology, engineering, and mathematics (STEM). In each of the alternate Earths and exoplanets, learners explore the terrain, describe the environment, report observations about how life on Earth is affected by these circumstances, and possibly create habitats which will enable them to survive. By immersing learners in these activities, WHIMC hopes to generate interest in and excitement for STEM among participating learners.

Among the many challenges of using an open-ended game such as Minecraft for education is assessment: How can we as educators tell whether learners are learning and what they have learned? Some educators opt not to assess learners at all, asserting that assessments of creative, open-ended projects would be "challenging" and that benchmarks would be difficult to establish [2]. However, not to assess learner performance is

© Springer Nature Switzerland AG 2022
M. M. Rodrigo et al. (Eds.): AIED 2022, LNCS 13356, pp. 263–266, 2022.
https://doi.org/10.1007/978-3-031-11647-6_48

becoming less and less of an option. As the presence of serious games grows in education and training, it is assessment that makes them vehicles for learning as opposed to forms of pure entertainment [6].

The ways in which outcomes of Minecraft activities are assessed are often limited to pre- and post-tests [9], self-reports [1], and teacher or researcher observations [1]. Some researchers have advocated the use of game analytics to provide insights into the learning process, separate from out-of-game assessments [4]. The use of game analytics provides researchers with objective data on learner engagement with the game [8] and allows the examination of interesting phenomena such as exploration strategies [5] and map exploration styles [7]. In this study, we use in-game data and assessment data to examine how learners' traversals of Minecraft worlds relate to their assessment outcomes. We compare similarities in map coverage to speculate on the extent to which goal-directed tasks performed in open-ended environments such as Minecraft have canonical answers or optimal exploration paths.

We hope that the results of this analysis can point out behaviors that may lead to better or worse learning outcomes, guide the teacher in identifying learners who need more assistance, and suggest ways in which learners can be guided or nudged towards better learning outcomes as they explore these open-ended worlds.

2 Methodology

The data analyzed for this paper consisted of in-game data and out-of-game assessment data from 15 Grade 6 learners, 10 to 12 years old, from the Philippines. Partner teachers from the school developed a total of six (6) learning modules that explored the Rocket Launch Facility, Lunar Base LeGuin, Earth with No Moon, Tilted Earth, and Exoplanets and implemented these modules with the 15 learners. We limited the analysis to data from the first module only.

We computed four (4) metrics: distance traveled, area covered, Jaccard Index, and the Maximum Similarity Index (MSI) [6].

We calculated the cumulative distance traveled by each learner in five-minute intervals. We also determined the smallest convex polygon that contains all locations that the learner visited, i.e. the convex hull [3]. Based on these convex hulls, we computed the area that each learner explored. The convex hull was generated every 5 min to get the cumulative area that the learner explored.

Because each learner had a unique convex hull, we selected the convex hulls of the two highest performing learners for all six modules. The convex hulls of these two learners served as ground truth. We computed the Jaccard Index or Intersection over Union (IOU) to measure the similarity of convex hulls of each learner and each of the ground truths. The IOU value is the area of the intersection of the two convex hulls divided by the area of their union. The highest of the IOU values or the MSI was the value that was used to represent how similar a learner's answer is to a canonical answer. We then conducted correlation analysis between the cumulative values of distance, area, and MSI with learners' assessment scores for the first module.

3   Results

There was no significant correlation between assessment scores and the total distance traveled, total area, and MSI. However, when broken down into five-minute intervals, some significant relationships emerge. There was a significant negative correlation ($r = -.57$, $p = .03$) between the cumulative distance traveled by the learners and assessment score at the 5-min mark. Figure 1 shows that during the first five minutes of exploration (Fig. 1a), the low-performing learner already traveled quite far from the starting point and the high-performing learner traveled short distances and stay within a confined area. This exploration behavior continues up to the 10-min mark (Fig. 1b.)

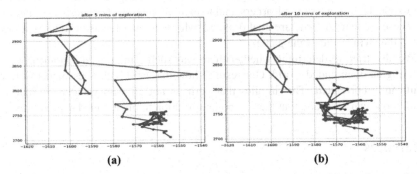

(a)   (b)

Fig. 1. Exploration Path of a low-performing learner (red) and a high-performing learner (green) after 5 min (1a) and 10 min (1b) of exploration. (Color figure online)

There were significant negative correlations between the accumulated area explored and assessment scores at the 5-min ($r = -.61$, $p = .02$), 15-min ($r = -.57$, $p = .04$), and 20-min ($r = -.69$, $p = .01$) marks. At the 10-min mark, the negative correlation was marginally significant ($r = -.45$, $p = .09$). Figure 2 illustrates the convex hulls of the same learners in Fig. 1 during the first 10 min of the session. After the first 5 min of exploration (Fig. 2a), the high-performing learner only covered a small area compared to the low-performing learner. After 10 min, the high-performing learner gradually expanded their area of exploration (Fig. 2b).

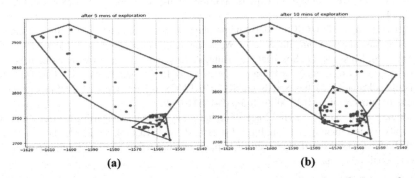

(a)   (b)

Fig. 2. Location points with convex hulls to enclose the area of exploration of a low-performing learner (red) and a high-performing learner (green) after 5 min and 10 min of exploration (Color figure online)

4   Conclusion

In this paper, we tried to determine the relationship between learner traversals of the WHIMC worlds and their assessment outcomes. Because they tend to explore a far, wide area early in their game play, our results suggest that low-performing learners exhibit what Si and colleagues [7] patterns characteristic of either wanderers or seers. Wanderers are learners who explore a map without a definite purpose or destination. They do not have a high-level understanding of their location and have no specific plans to reach wherever it is they should go. They tend to pick a direction and move forward, without a clear strategy. Seers tend to use a depth-first search strategy when exploring an area and aim to expand their range, venturing into unexplored territory as quickly as possible. High-performing learners tend to exhibit patterns characteristic of targeters [7]. These are learners who are goal oriented, who seek out objects or items that might lead to the "win" state of the game.

We also tried to determine whether there was such a thing as a canonical way of exploring an open-ended world. Our results show that MSI and assessments were not significantly correlated. This implies that perhaps not all who wander are lost, that explorations can vary, and that there is no one solution to open-ended exploration.

References

1. Callaghan, N.: Investigating the role of Minecraft in educational learning environments. Educ. Media Int. **53**(4), 244–260 (2016)
2. Čujdíková, M.: Create Minecraft Game, Save the world. In: 13th International Conference on Game Based Learning, ECGBL, October 2019
3. Gamby, A., Katajainen, J.: Convex-hull algorithms: implementation, testing and experimentation. Algorithms **11**(12), 195 (2018)
4. Horn, B., Hoover, A.K., Barnes, J., Folajimi, Y., Smith, G., Harteveld, C.: Opening the black box of play: strategy analysis of an educational game. In: Proceedings of the 2016 Annual Symposium on Computer-Human Interaction in Play, pp. 142–153, October 2016
5. Käser, T., Hallinen, N.R., Schwartz, D.L.: Modeling exploration strategies to predict learner performance within a learning environment and beyond. In: Proceedings of the Seventh International Learning Analytics & Knowledge Conference, pp. 31–40, March 2017
6. Loh, C.S., Sheng, Y.: Maximum Similarity Index (MSI): a metric to differentiate the performance of novices vs. multiple-experts in serious games. Comput. Hum. Behav. **39**, 322–330 (2014)
7. Si, C., Pisan, Y., Tan, C.T.: Understanding players' map exploration styles. In: Proceedings of the Australasian Computer Science Week Multiconference, pp. 1–6 (2016)
8. Smith, S., Hickmott, D., Southgate, E., Bille, R., Stephens, L.: Exploring play-learners' analytics in a serious game for literacy improvement. In: Marsh, T., Ma, M., Oliveira, M.F., BaalsrudHauge, J., Göbel, S. (eds.) JCSG 2016. LNCS, vol. 9894, pp. 13–24. Springer, Cham (2016). https://doi.org/10.1007/978-3-319-45841-0_2
9. Tangkui, R., Keong, T.C.: The effects of digital game-based learning using Minecraft towards pupils' achievement in fraction. Int. J. E-Learn. Pract. (IJELP) **4**, 76–91 (2021)
10. WHIMC. (n.d.) What-If Hypothetical Implementations using Minecraft. https://whimcproject.web.illinois.edu/

A Comparative Assessment of US and PH Learner Traversals and In-Game Observations within Minecraft

Jonathan D. L. Casano[✉] [iD] and Maria Mercedes T. Rodrigo[iD]

Ateneo de Manila University, Quezon, Philippines
jcasano@ateneo.edu

Abstract. What-If Hypothetical Implementations using Minecraft (WHIMC) is a set of Minecraft worlds that learners can explore to learn more about science, mathematics, engineering, and technology. Because Minecraft is open-ended by design, assessing whether students are learning is always a challenge. In addition to using methods such as post-tests and self-reports, how can we use in-game data to measure learning? In this paper, we analyze and compare American (US) and Filipino (PH) learner traversals and in-game observations against canonical answers from experts to determine the extent to which students achieved the desired learning outcomes and the ways in which outcomes varied. We grouped students into high- and low-performing categories based on an out-of-game post test. We found that high-performers tended to make more observations than low-performers. Observations of high-performers tended to align with canonical answers. Many PH students tended to be low-performers and tended not to make in-game observations. In contrast, even low-performing US students tended to record observations actively.

Keywords: Minecraft · WHIMC · Comparison · Philippines · United States

1 Introduction

The What-If Hypothetical Implementations using Minecraft (WHIMC) project is a set of Minecraft worlds that learners can explore as supplementary activities to learn more about science, mathematics, engineering, and technology. It logs both the ways in which learners traverse these words and the observations that learners make during their explorations.

When engaging learners in open-ended environments such as Minecraft, assessment is always a challenge: How can we determine whether students are learning and what they have learned? Educators use a variety of well-worn assessment methods, e.g. pre- and post-tests [7] self-reports [1, 4] and teacher or researcher observations [1]. In recent years, more and more researchers are examining game logs, sometimes in conjunction with out-of-game assessments, to measure learning.

In this paper, we analyze game logs of American (US) and Filipino (PH) learner traversals and observations in order to assess whether learners achieved expected learning outcomes and to determine ways in which outcomes varied. We first organize the

© Springer Nature Switzerland AG 2022
M. M. Rodrigo et al. (Eds.): AIED 2022, LNCS 13356, pp. 267–270, 2022.
https://doi.org/10.1007/978-3-031-11647-6_49

observations into word clouds, taking into account the context in which the observations were made. We then compare canonical answers from experts against answers from (1) All students, (2) High-performing and low-performing learners and (3) US and PH learners.

2   Review of Related Literature

To guide our analysis, we drew on prior work regarding WHIMC, word clouds, and map exploration archetypes of [6]. As WHIMC logs all student activity and all observations that students make, we organized student observations made in WHIMC into contextualized word clouds and compare student answers with canonical answers from experts.

Word clouds or tag clouds "are visual presentations of a set of words, typically a set of tags, in which attributes of the text such as size, weight or color can be used to represent features (e.g., frequency) of the associated terms" [3]. By displaying text data in graphical form, word clouds help readers surmise the gist of a text quickly [5]. However, word clouds are only useful for quick looks and cannot replace careful analysis of student responses. It was therefore necessary to group observations by location before creating the word clouds, a process that will be discussed further in the section on the Word Cloud Visualization Tool.

Finally, to interpret the traversal patterns of the students, we referred to the four map exploration archetypes of [6] and attempted to operationalize some of these archetypes in the logs that we analyzed.

3   Data Collection

The demographic details of the respondents are shown in Table 1. High- and low- performers were determined using out-of-game assessments [2]. The observations data of these groups were considered in the analysis.

Table 1. Summary of demographic details of respondent groups.

Group	Inclusive dates	Total	Year level/Subject	Age range	# Observations
US 1	6/28 - 7/2, 2021	10	Grade 5 Science	11 - 14 y/o	110
US 2	7/12 - 7/28, 2021	11	Grade 5 Science	11 - 14 y/o	53
PH 1	8/27 - 9/15, 2021	16	Grade 6 Science	11 - 14 y/o	291
PH 2	9/29 - 10/2, 2021	24	Grade 5 Science	10 - 13 y/o	102

4 Results and Analysis

The observations were organized into word clouds using a custom visualization program written in Python.

Using the custom visualization tool, eighteen (18) comparison attempts were made across 6 WHIMC maps and 3 performance categories (high-, average- and low-performers) for a total of 108 comparison attempts. Upon analyzing the 44 successful attempts, we find that:

High-Performing Students Make More Observations. When there were insufficient observations for the K-means clustering, the visualization was not generated. Hence, we see that making more observations across all maps visited is characteristic of high-performing students. Low-performing students fewer than average number of observations or no observations at all.

Observations of High-Performing Students Matched the Canonical Observations from Experts. Most of the observation matches occur in comparisons where the high-performing students were included. The matching observations were mostly from the high-performing group.

High-Performing Students Tend to Wander. A comparison between the performance categories within each data set was. It is noted that for each comparison, there are more wanderer clusters generated by the data coming from the high-performing group. Wandering around the Minecraft map and making observations beyond the prompts of the NPC seems characteristic of high-performing students.

Low-Performing Students from PH Did Not Make any Observations. None of the low-performing students from PH made any observations as opposed to the low-performers in the US data who all made observations. We speculate that active observation-making, might be an indicator of positive performance.

5 Contributions and Future Work

This work contributes to the literature in at least two ways. First, we contribute a tool that helps cluster and visualize word clouds and maps them onto the contexts from which the data was generated. This context can help with their interpretation. Second, we perform a cross-cultural comparison of data from two different populations, US and PH, using the same game-based learning environment, teasing out some of the similarities and differences between these populations.

In summary, high-performing students from the US and PH made the most in-game observations and that these observations tended to match the canonical observations from experts. US and PH high-performers tended to exhibit more wandering or moving around the map without a defined destination or purpose. Low-performers tended to make fewer observations in general. However, low-performers from the PH set did not contribute any observations while those from the US data set actively recorded their observations. The observations from low-performers from the US tended to match the canonical answers.

The differences between these groups may have been a function of differences in the way WHIMC sessions were introduced and conducted. US learners were asked to listen to 10-min lectures that were specifically about the WHIMC worlds they were about to explore. The PH students were introduced to WHIMC, but preparatory lessons leading up to the explorations tended to be more generic, e.g. a lesson on making observations or a lesson on ordinal numbers. WHIMC served as a backdrop or context against which students demonstrated knowledge of these concepts.

We would like to see how we can leverage on word embeddings libraries to merge or combine synonyms or related words, rather than treating them as unique and would like to further investigate the types of observations made by low-performing students. The data suggests that low-performing students who actively record their observations still manage to produce quality outcomes. However, more data is needed to validate this finding.

Finally, we hope to continue the cross-cultural comparisons we began in this study. WHIMC gives us a unique opportunity to capture and study the behaviors of US and PH learners within the same learning context. Investigating these differences may give us more insight about how to best use environments such as WHIMC to support learning goals in culturally appropriate ways.

Acknowledgement. The authors thank H Chad Lane and Jeff Ginger for their enthusiastic collaboration, Dominique Marie Antoinette B. Manahan, Mikael William Fuentes, and Ma. Rosario Madjos for their support, the Ateneo Laboratory for the Learning Sciences, the Ateneo de Manila University, and the Department of Science and Technology's Philippine Council for Industry, Energy, and Emerging Technology Research and Development for the grant entitled, "Nurturing Interest in STEM among Filipino learners using Minecraft."

References

1. Callaghan, N.: Investigating the role of Minecraft in educational learning environments. Educ. Media Int. **53**(4), 244–260 (2016)
2. Conati, C., Kardan, S.: Student modeling: supporting personalized instruction, from problem solving to exploratory open ended activities. AI Mag. **34**(3), 13–26 (2013)
3. Halvey, M.J., Keane, M.T.: An assessment of tag presentation techniques. In: Proceedings of the 16th International Conference on World Wide Web, pp. 1313–1314, May 2007
4. Melián Díaz, D., Saorín, J.L., Carbonell-Carrera, C., de la Torre Cantero, J.: Minecraft: three-dimensional construction workshop for improvement of creativity. Technol. Pedagog. Educ. **29**(5), 665–678 (2020)
5. Rivadeneira, A.W., Gruen, D.M., Muller, M.J., Millen, D.R.: Getting our head in the clouds: toward evaluation studies of tagclouds. In: Proceedings of the SIGCHI Conference on Human Factors in Computing Systems, pp. 995–998, April 2007
6. Si, C., Pisan, Y., Tan, C.T.: Understanding players' map exploration styles. In: Proceedings of the Australasian Computer Science Week Multiconference, pp. 1–6, February 2016
7. Tangkui, R., Keong, T.C.: The effects of digital game-based learning using Minecraft towards pupils' achievement in fraction. Int. J. E-Learn. Pract. (IJELP) **4**, 76–91 (2021)

First Steps Towards Automatic Question Generation and Assessment of LL(1) Grammars

Ricardo Conejo$^{(\boxtimes)}$ ⓘ, José del Campo-Ávilaⓘ, and Beatriz Barrosⓘ

Departamento de Lenguajes y Ciencias de la Computación, Universidad de Málaga,
Campus de Teatinos, 29071 Málaga, Spain
{conejo,jcampo,bbarros}@uma.es

Abstract. Automatic question generation and the assessment of procedural knowledge is still a challenging research topic. This article focuses on case of it, the LL(1) grammar design. This is a well known technique for construct a top-down parser. There are many tools that given a context-free grammar can construct the LL(1) tables, but they are not designed for assessment. This article describes an application that covers all the tasks needed to automatize the assessment process.

Keywords: Automatic assessment · Question generation · Procedural knowledge · Adaptive feedback · Top-down parsing

1 Introduction

Compiler construction is a compulsory subject in almost all computer science degrees. Here the student learn different algorithms, tools and methods necessary to understand how a compiler for a programming language is constructed [1]. A core part of compiler construction is the design of the language grammar and the construction of the parser. The LL(1) technique allows to construct efficient top-down parsers based on theoretical grounds, but it requires some conditions to be met for the grammar design. Let's introduce some concepts that are used in this article:

A context-free grammar (CFG) is defined as $G(N, T, P, S)$, where N is a set of *non-terminal* symbols, T is a set of *terminal* symbols, P is a set of *production rules* (of the form $A \rightarrow \gamma$, where A is called the *antecedent* and $\gamma \in (N \cup T)^*$ is called the *consequent*), and S is the *axiom*. The languages that can be generated by a CFG are called context-free languages (CFL).

LL(1) grammars are a subset of context-free grammar (CFG) that accomplish the LL(1) condition. There is a well-known algorithm to efficiently determine if a context-free grammar is LL(1) and construct its parsing table. It is based on the construction of the functions $FIRST$, $FOLLOW$ and the directive symbols of each production rule DS [1]. LL(1) languages are those context-free language that can be generated by an LL(1) grammar.

ⓒ Springer Nature Switzerland AG 2022
M. M. Rodrigo et al. (Eds.): AIED 2022, LNCS 13356, pp. 271–275, 2022.
https://doi.org/10.1007/978-3-031-11647-6_50

This article describes the implementation of a computer-based assessment application, constructed as a plugin of the SIETTE assessment system (see Sect. 2) that is able to automatically generate a random CFG and evaluate the student answers.

2 System Architecture

In order to assess the student knowledge and skills needed to design and implement LL(1) parsers for a given language, we have implemented a plug-in extension of the SIETTE assessment environment. Using this plug-in and some of the standard features of SIETTE, we are able to automatically generate different types of questions.

2.1 The SIETTE Assessment System

SIETTE [2] is a general-purpose automatic assessment environment that supports the generation of different question based on JSP templates, different types of questions and student answer interfaces; automated recognition of students' open answers based on regular expression patterns; and a flexible support of any other assessment requirement based on the construction of a plug-in extension. SIETTE implements the Classical Test Theory (CTT), Item Response Theory (IRT), Computer Adaptive Testing (CAT), and it provides built-in statistical and psychometric tools to analyze students, tests and questions results.

The student answer is given to SIETTE in a plain or structured text format. SIETTE recognizes whether the answer is correct using a pattern matching process. Patterns are provided by the teacher and the matcher algorithm is implemented as a plug-in. There are some default matcher plugins that are already implemented in SIETTE. On of them is the *SIETTE regular expression* that allows to recognize the student answer based on a regular expression pattern provided by the teacher or, in this case, automatically generated.

2.2 Automatic Generation of Context-Free and LL(1) Grammars

One of the first challenges of this project is to define a way to generate small context-free languages that can be used to pose questions to students. The alphabet of these languages (*terminal symbols*) is restricted to lowercase letter in order to be easy to write it in text format. *non-terminal symbols* are written using uppercase letters. The *axiom* of the grammar is always the *non-terminal symbol* that appears on the left hand side of the first rule.

Small context-free grammars can be generated just by setting the antecedent and a random length string that combines terminal and non-terminal symbols. This strategy requires validating the generated grammar and repeating the process until a correct grammar is obtained.

On the other hand, a well defined context-free grammar can be generated based on composition of "building block" grammars. The building blocks are tiny context-free grammars with just two or three production rules. Some of them are listed below:

$$A \rightarrow Aa$$
$$A \rightarrow a$$

(1)

$$A \rightarrow aAb$$
$$A \rightarrow ab$$

(2)

The plug-in defines some building block grammars, but they can be easily extended as needed. Using this building blocks we apply a composition rule just by replacing a *terminal* symbol with a *non-terminal* symbol of another building block. For instance, combining Block 1 and Block 2 in this order will generate the following grammar:

$$A \rightarrow AB$$
$$A \rightarrow B$$
$$B \rightarrow aBb$$
$$B \rightarrow ab$$

On the other hand, combining Block 2 and Block 1 grammars can give one of these four grammars:

$A \rightarrow BAb$	$A \rightarrow BAb$	$A \rightarrow aAB$	$A \rightarrow aAB$
$A \rightarrow Bb$	$A \rightarrow Bb$	$A \rightarrow aB$	$A \rightarrow aB$
$B \rightarrow Ba$	$B \rightarrow Bb$	$B \rightarrow Ba$	$B \rightarrow Bb$
$B \rightarrow a$	$B \rightarrow b$	$B \rightarrow a$	$B \rightarrow b$

Note that there are four possible ways to combine Block 2 and Block 1 grammars, because we have two alternative options: (1) In Block 2, there are two terminal symbols, so we have two options to replace a *terminal* with a *non-terminal* of the second grammar; (2) we have to choose if the *terminal* symbols of the resulting grammar are the same or not. Nevertheless, without loss of generality, we can always assume that *terminals* are different, and at the end of the generation process, two or more symbols can be merged as a single *terminal*. That is, in the last example, options 2 and 4 can be obtained from options 1 and 3 just by considering that *terminal* a and *terminal* b are the same. This process is delayed until we finish the combination process.

Thus, a context-free grammar can be randomly generated by selecting the building blocks to combine, the number of combinations to apply (or alternatively the number of production rules in the final grammar) and the final number of *terminal* symbols (which will randomly merge two symbols until the desired number of *terminal* symbols is met).

Finally, a validation and refinement process is triggered to eliminate unused rules or symbols, and/or duplicate rules, to guarantee that the context-free grammar is correct and that the FIRST and FOLLOW sets can be effectively calculated.

2.3   Automatic Construction of LL(1) Parsers

Given a context-free grammar it is always possible to compute FIRST and FOLLOW functions and obtain the directive symbols of each production rule [1]. The result of these functions are a set of symbols. Determining if a context-free grammar accomplishes the LL(1) condition depends on these sets.

The system requires a student response by asking to type the symbols in the set. The student can shuffle the order of symbols in the set, but the pattern will recognize the answer anyway. Figure 1 presents a composed question where a common grammar has been generated, and some questions about FIRST and FOLLOW sets are posed. Each question is evaluated independently.

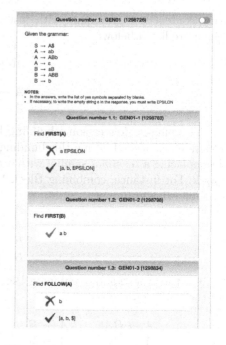

Fig. 1. A composed SIETTE question about FIRST and FOLLOW sets

3   Conclusion

The application described in this article provides a way for the student to enhance the practice of design of the LL(1) context-free grammars. Although the generation of an LL(1) grammar and the recognition of grammar equivalence are unsolvable issues in the general case, a heuristic approach can provide a practical solution for assessment purposes.

The application has been designed and used for formative and summative assessment. It includes automatic recognition of student answers and personalized feedback. The application is embedded in the SIETTE system, which provides additional features that can be used, such us adaptive question selection, scoring procedure selection, access control, etc. Question difficulty can be controlled by means of the number of building block grammar combinations, but it can also be obtained empirically through SIETTE question calibration and learning analytic tools.

We do not claim that the system itself is responsible for the increase in the student scores, but the data obtained from the students that have used the system shows that it helps them to practise and be aware of their progress.

Links to on-line assessments

FIRST and FOLLOW: https://www.siette.org/siette?idtest=631978
LL(1) analysis: https://www.siette.org/siette?idtest=633742
LL(1) table construction: https://www.siette.org/siette?idtest=525382
LL(1) grammar design: https://www.siette.org/siette?idtest=633700

References

1. Aho, A.V., Lam, M.S., Sethi, R., Ullman, J.D.: Compilers: Principles, Techniques, and Tools, 2nd edn. Addison-Wesley Longman Publishing Co., Inc., Boston (2006)
2. Conejo, R., Guzmán, E., Trella, M.: The SIETTE automatic assessment environment. Int. J. Artif. Intell. Educ. **26**(1), 270–292 (2015). https://doi.org/10.1007/s40593-015-0078-4

AI-Based Open-Source Gesture Retargeting to a Humanoid Teaching Robot

Hae Seon Yun[1]([✉]), Heiko Hübert[2], Volha Taliaronak[3], Ralf Mayet[4], Murat Kirtay[3], Verena V. Hafner[3], and Niels Pinkwart[1]

[1] Department of Computer Science, Humboldt University of Berlin, Berlin, Germany
{yunhaese,pinkwart}@informatik.hu-berlin.de
[2] HTW Berlin - University of Applied Sciences, Berlin, Germany
heiko.huebert@htw-berlin.de
[3] Adaptive Systems Group, Department of Computer Science,
Humboldt University of Berlin, Berlin, Germany
{taliaronak,murat.kirtay,hafner}@informatik.hu-berlin.de
[4] Technische Universität Berlin, Berlin, Germany
mayet@campus.tu-berlin.de

Abstract. Gestures and speech modalities play potent roles in social learning, especially in educational settings. Enabling artificial learning companions (i.e., humanoid robots) to perform human-like gestures and speech will facilitate interactive social learning in classrooms. In this paper, we present the implementation of human-generated gestures and speech on the Pepper robot to build a robotic teacher. To this end, we transferred a human teacher gesture to a humanoid robot using a web and a kinect cameras and applied a video-based markerless motion capture technology and an observation-based motion mirroring method. To evaluate the retargeting methods, we presented different types of a humanoid robotic teacher to six teachers and collect their impressions on the practical usage of a robotic teacher in the classroom. Our results show that the presented AI-based open-source gesture retargeting technology was found attractive, as it gives the teachers an agency to design and employ the Pepper robot in their classes. Future work entails the evaluation of our solution to the stakeholders (i.e. teachers) for its usability.

Keywords: Motion retargeting · Teacher robot · Non-verbal behavior · Motivation · Pepper robot

1 Introduction

Designing a robotic teacher or teaching assistant is an emerging area that hosts various interdisciplinary challenges and opportunities for the researchers and

This project was primarily funded by the Deutsche Forschungsgemeinschaft (DFG, German Research Foundation) under Germany's Excellence Strategy – EXC 2002/1 "Science of Intelligence" – project number 390523135.

© Springer Nature Switzerland AG 2022
M. M. Rodrigo et al. (Eds.): AIED 2022, LNCS 13356, pp. 276–279, 2022.
https://doi.org/10.1007/978-3-031-11647-6_51

educators [3]. Employing a robotic agent in an educational setting presents noticeable benefits both for learners and teachers. For learners, a robotic teacher can enhance motivation and learning gain by providing students with one-on-one attention through physical social presence [1,2,8], in addition to mitigating learners' anxiety or embarrassment which they usually feel when asking questions to a human teacher [9]. For teachers, a robotic teaching assistant can ease mundane teaching loads from teachers so that they could better allocate their resources [3]. Even though teachers endorse the benefits of a robotic teacher as it supports personalized learning [12], the disbelief on the competency of a robotic teacher and the technical limitation in a robot in autonomous [13] or in Wizard of Oz mode [10] hinders actual in-class adoption. Therefore, in this paper, we propose a practical approach to design a robot teacher based on our real-life design requirements to reflect teachers' needs and lower the technical barriers.

2   Methodology

Our design requirements are enumerated as follows: 1) a robotic teacher should behave similar to a human teacher; 2) the implementation of a robotic teacher should involve open-source offline solutions to ensure the user's privacy and omit platform dependency; 3) the teachers should be able to implement a robotic teacher without in-depth technical knowledge; and 4) the design and development process of a robotic teacher should involve human teachers as active participants to ensure the reliability and ecological validity.

To fulfill our first design requirement, we focused on transferring human teacher's gesture to a robotic teacher by recording two teachers speaking out the scripts from the ITS system, Betty's Brain [4] in front of cameras (Kinect v2, Webcam). We use the term, 'gesture' which is generally accepted as a movement of a part of the body (hands or the head) to express and emphasizes an idea or meaning. The captured videos were retargeted using 1) an AI (specifically machine learning) based pose tracking method and 2) pre-defined motions.

For the AI-based motion retargeting, we customized the method introduced in [6], which provided a teleoperation tool for a Pepper robot. This method uses a Kinect v2 (RGB+depth) camera and OpenPose for creating 3D keypoints which are then converted into joint angles for controlling the motion of Pepper. We replaced the OpenPose-based motion capture method with MediaPipe [5]. With this setting, we were able to reduce computational complexity and only use 2D-RGB camera feed to estimate 3D keypoints of body movement, 3D (depth camera) input is not required.

This AI-based gesture retargeting flow is combined with an audio processing path for speech control of Pepper as shown in the combined toolflow diagram in Fig. 1. To synchronize gesture and speech, we adjusted the speed of the video data (frames per second) before feeding it to the motion detection tools and for the speech, we added delays between the group of words in the text file and adjusted the speech speed parameter. For the text transcription and synchronization between gesture and speech, the current approach involves several manual steps.

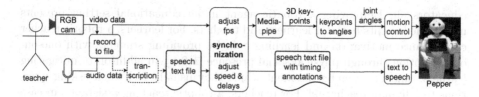

Fig. 1. Toolflow diagram of AI-based open-source framework

To create robotic teachers using pre-defined motions, we manually reviewed the teachers' gesture videos and selected a set of matching motions from the built-in animations and combined with speech. For instance, we used pre-defined animation called (YouKnowWhat_5) and matched it with the text "Hello, User!" followed by the animation, (Explain_8) with "what are we going to do today?".

In addition to creating robotic teachers that gesture and speak similar to a human teacher, we created a robotic teacher that uses general explaining or pointing gestures by selecting some gestures from the built-in animation package randomly. Lastly, we created two robotic teachers which only speak without performing any gesture, which we omitted in evaluation to focus on the gesture implementation in a robotic teacher.

As a result, we have used five versions of robotic teachers. 1) Teacher A based predefined robotic teacher (PDA), 2) Teacher B based predefined robotic teacher (PDB), 3) Teacher A based AI robotic teacher(MPA), 4) Teacher B based AI robotic teacher (MPB) and 5) Random predefined robotic teacher (R).

To evaluate our implemented robotic teachers, we have conducted an online interview with six teachers which includes a perception study similar to the one stated in [7] to gain understanding on what and how our prospective user (i.e., teachers) receive the robotic teacher and verify the importance of gestures in a robotic teacher. Specifically, we presented five gesturized robotic teacher versions (PDA, PDB, MPA, MPB and R) and asked which version they consider as most human-like, most teacher-like, most fitting and explain their reasoning.

3   Results and Discussion

Learners enjoy interacting with a robot with gestures as they find a robot more friendly and communicative [11]. In pursuit of a practical solution for teachers to employ Pepper robot in their class as an assistant or as a support teacher, we began with eliciting design requirements, on which we applied AI-based approach along with pre-defined method to integrate gestures in a robotic teacher. We conducted online interviews with six teachers to verify the usability of our solution by presenting five different versions of robotic teachers.

We can not conclude how exactly the AI-based robotic teacher is different from other versions that we created. However, in spite of the limitation, our research indicates the importance of providing a solution that enables teachers to use without putting more workloads on them (*I would always use the less time consuming option*). Additionally, in the course of ongoing research, we were able to resolve the challenge of manual synchronization of speech and gesture and present our renewed diagram [14].

Our future work entails engaging teachers to create teaching scenarios where a robotic teacher plays a role in a classroom and evaluate our solution in terms of usability.

References

1. Bainbridge, W.A., Hart, J.W., Kim, E.S., Scassellati, B.: The benefits of interactions with physically present robots over video-displayed agents. Int. J. Soc. Robot. **3**(1), 41–52 (2011)
2. Belpaeme, T., Kennedy, J., Ramachandran, A., Scassellati, B., Tanaka, F.: Social robots for education: a review. Sci. Robot. **3**(21) (2018)
3. Belpaeme, T., Tanaka, F.: Social robots as educators. In: OECD Digital Education Outlook 2021 Pushing the Frontiers with Artificial Intelligence, Blockchain and Robots: Pushing the Frontiers with Artificial Intelligence, Blockchain and Robots, p. 143 (2021)
4. Biswas, G., Segedy, J.R., Bunchongchit, K.: From design to implementation to practice a learning by teaching system: Betty's brain. Int. J. Artif. Intell. Educ. **26**(1), 350–364 (2016)
5. elggem: Github-naoqi-pose-retargeting (2022). https://github.com/elggem/naoqi-pose-retargeting
6. FraPorta: Pepper openpose teleoperation (2021). https://github.com/FraPorta/pepper_openpose_teleoperation
7. Kelly, S.D., McDevitt, T., Esch, M.: Brief training with co-speech gesture lends a hand to word learning in a foreign language. Lang. Cogn. Process. **24**(2), 313–334 (2009)
8. Li, J.: The benefit of being physically present: a survey of experimental works comparing copresent robots, telepresent robots and virtual agents. Int. J. Hum Comput Stud. **77**, 23–37 (2015)
9. Newton, D.P., Newton, L.D.: Humanoid robots as teachers and a proposed code of practice. In: Frontiers in Education, vol. 4, p. 125. Frontiers (2019)
10. Riek, L., Howard, D.: A code of ethics for the human-robot interaction profession. In: Proceedings of We Robot (2014)
11. Salem, M., Rohlfing, K., Kopp, S., Joublin, F.: A friendly gesture: investigating the effect of multimodal robot behavior in human-robot interaction. In: 2011 Ro-Man, pp. 247–252. IEEE (2011)
12. Serholt, S., et al.: Teachers' views on the use of empathic robotic tutors in the classroom. In: The 23rd IEEE International Symposium on Robot and Human Interactive Communication, pp. 955–960. IEEE (2014)
13. Sidner, C.L., Kidd, C.D., Lee, C., Lesh, N.: Where to look: a study of human-robot engagement. In: Proceedings of the 9th International Conference on Intelligent User Interfaces, pp. 78–84 (2004)
14. Yun, H.S., Hübert, H., Taliaronak, V., Sardogan, A.: Utilizing machine learning based gesture recognition software, mediapipe, in the context of education and health. Poster presented at: The AI Innovations, Berlin (5-6 May 2022)

Supporting Representational Competencies in an Educational Video Game: What Does and Doesn't Work

Tiffany Herder$^{(\boxtimes)}$ and Martina A. Rau

University of Wisconsin-Madison, Madison, USA
{therder,marau}@wisc.edu

Abstract. To examine the effectiveness of two types of representational-competency supports in educational video games, we conducted a 2 × 2 experiment with 142 students. We found that one type of support was effective, but only for students with high prior astronomy knowledge. We discuss implications for the design of representational-competency supports for educational video game.

Keywords: Video games · Representational competencies · Instructional support

1 Introduction

Educational video games provide intuitive access to authentic scientific practices [1]. To engage in these practices, games provide interactive visual representations that students manipulate to solve domain-relevant problems [2, 3]. Visuals can enhance learning because they can make complex concepts accessible [4, 5]. However, visuals can also be confusing if students do not know how they depict information [4, 5].

To overcome these issues, research has investigated ways to support students' representational competencies: skills that enable students to use visuals to solve tasks [5]. This research has shown that students need two types of representational competencies to benefit from visuals: sense-making competencies that allow students to explain how visuals depict information [4, 5] and perceptual fluency that allows students to effortlessly extract information from visuals [5, 6]. To support sense-making competencies, instructions prompt students to verbally explain which visual features correspond to disciplinary concepts [4, 5]. To support perceptual fluency, instructions engage students in many short matching tasks [5, 6]. Prior research shows that combining support for both sense-making competencies and perceptual fluency enable students to overcome difficulties with visuals and enhance learning of disciplinary concepts [7].

However, research on representational-competency supports has mostly focused on structured learning environments (e.g., [5]) and has not examined these supports in educational games. Yet, visuals are ubiquitous within games [2] and can support game-based learning [1]. Nevertheless, students have difficulties understanding how visuals show concepts, especially if they are not central to game interactions [2]. Further, students often focus on intuitive understanding instead of critically reflecting on visuals [2].

© Springer Nature Switzerland AG 2022
M. M. Rodrigo et al. (Eds.): AIED 2022, LNCS 13356, pp. 280–283, 2022.
https://doi.org/10.1007/978-3-031-11647-6_52

Thus, students encounter similar difficulties with visuals in educational video games as in structured learning environments. Thus, integrating supports for representational-competencies shown effective in structured environments may be helpful for students in game environments.

2 Methods

We recruited 142 undergraduate students without prior undergraduate astronomy courses from our institution via flyers and emails to participate in our study.

Students were randomly assigned to one of four conditions that resulted from a 2 × 2 design (sense-making support: yes/no; perceptual-fluency support: yes/no). Students received representational-competency supports in the form of five 2-min. videos and verbal reminders given at regular intervals during game play, which provided step-by-step guidance on how to interact with visuals. To control for any potential effects of the videos and reminders, the control condition received videos and reminders related to the benefits of educational video games and technical aspects of the game.

All students played At Play in the Cosmos, an astronomy game designed to help students make connections among multiple visuals of astronomical phenomena and to engage students visually with astronomy concepts [3]. In the game, students complete guided missions to explore galaxies where they acquire resources for the corporation. Students use a variety of visuals to gather information about celestial objects. Interactions with visuals are designed to intuitively engage students with astronomy concepts without having to compute mathematical formulas. In the game, visuals illustrate how to obtain variables for an equation and students drag values to corresponding color-coded locations. We selected 10 missions of the game for our study.

We assessed students' learning of astronomy content knowledge, sense-making competencies, and perceptual fluency with three isomorphic versions of a test that was delivered at three test times (pretest, intermediate test, posttest). Scores for content learning and sense making were computed as the average of correct answers on the items. Scores for perceptual-fluency were computed as an efficiency score to take accuracy and speed into account [8]. Students also completed a Mental Rotation Test (MRT) [9] and rated their cognitive load after each mission of the game [10].

The study involved two sessions, 1–5 days apart. In session 1, students took the pretest and MRT, watched videos, played the game, and took the intermediate test. In session 2, students continued watching videos and playing the game and took the posttest.

3 Results

We used a repeated measures ANCOVA with pretest, MRT, cognitive load, and sense-making pretest as covariates, test scores as dependent measures, test time (intermediate and post) as repeated, within-subjects factor, and sense-making and perceptual-fluency support as between-subjects factors. With respect to research question 1 (whether sense-making support is effective), we found no significant main effect of sense-making support ($F < 1$). With respect to research question 2 (whether perceptual-fluency support is effective), we found a medium significant main effect of perceptual-fluency support,

$F(1, 120) = 9.383$, $p = .034$, $\eta^2 = .077$. This main effect was qualified by a medium significant interaction between perceptual-fluency support and the content pretest scores, $F(1,120) = 7.580$, $p = .007$, $\eta^2 = .063$. Students with low prior content knowledge showed higher learning outcomes at the posttest if they had *not* received perceptual-fluency support. In contrast, students with high prior content knowledge showed higher learning outcomes at the posttest if they *had* received perceptual-fluency support. With respect to research question 3 (whether the combination of supports is effective), the interaction between the two types of support was not significant ($F < 1$).

4 Discussion

We investigated whether representational-competency supports that have proven effective in structured learning environments are effective in the context of an educational video game. Specifically, we tested the effects of two types of representational-competency supports. We found no evidence that sense-making support enhanced learning. Our results suggest that effects of perceptual-fluency supports depend on student characteristics. Specifically, we found that perceptual-fluency support enhanced content learning from the game, but only for students with high prior content knowledge.

A possible explanation for why the sense-making support was ineffective is that the game may not have provided enough information for students to reflect deeply about the information the visuals depict when they were prompted to do so. In the absence of this information, students may have engaged in shallow processing of the visuals. Further, students may have been unwilling to make sense of the visuals because the game fostered intuitive, game-like interactions more than reflection. Thus, our version of sense-making support might have been incompatible with the present game.

On the flipside, perceptual-fluency support might have been effective because it aligned with the game design that, as described above, aimed to intuitively engage students with astronomy visuals. However, the finding that perceptual-fluency support enhanced learning outcomes suggests that the game itself may not have sufficiently encouraged perceptual processing in ways that maximized students' content learning. Further, only students with high prior knowledge benefited from perceptual-fluency support. While this result corroborates prior research suggesting that perceptual-fluency support is most effective for students with prior content knowledge [7], it highlights a limitation of the game. Even though the goals of the game was to provide intuitive access to disciplinary practices with visuals [3], our results suggest that it may not be suitable as a first introduction to astronomy. Instead, it may be more effective if used after students have acquired some preliminary content knowledge.

Our study has several limitations. The game we used aimed to provide intuitive access to concepts like many educational video games. However, other video games may instead emphasize conceptual reasoning with visuals. Thus, future research should test representational-competency supports with other types of games. Further, our representational-competency supports were provided outside of the game. Future research should test whether integrated types of support are effective. Third, we conducted our study in a research lab to maximize internal validity, which may compromise external validity. Future research should examine these supports in realistic game-play contexts.

To conclude, our study suggests that representational-competency supports that are effective in structured learning environments are not necessarily effective in the context of video games. Specifically, sense-making support may be incompatible with video games that foster intuitive interactions with visuals. In contrast, perceptual-fluency support seems to be effective in such games. Our results suggest that designers of such games should add support for perceptual fluency but refrain from reflective prompts that support sense-making competencies during gameplay. Similarly, our results encourage instructors to prompt students' perceptual processing but caution against encouraging reflection during games that emphasize intuitive processing of visuals.

Acknowledgements. This work was supported by the Institute of Education Sciences, U.S. Dept of Ed, under Grant [#R305B150003] to the University of Wisconsin-Madison. Opinions expressed are those of the authors and do not represent views of the U.S. Dept of Ed.

References

1. Clark, D., Nelson, B., Sengupta, P., Angelo, C.: Rethinking science learning through digital games and simulations: genres, examples, and evidence. In: Learning Science: Computer Games, Simulations, and Education. National Academy of Sciences, Washington, DC (2009)
2. Corredor, J., Gaydos, M., Squire, K.: Seeing change in time: video games to teach about temporal change. J. Sci. Educ. Technol. **23**, 324–343 (2014)
3. Squire, K.: At play in the cosmos. Int. J. Des. Learn. **12**, 1–15 (2021)
4. Ainsworth, S.: DeFT: a conceptual framework for considering learning with multiple representations. Learn. Instr. **16**, 183–198 (2006)
5. Rau, M.A.: Conditions for the effectiveness of multiple visual representations in enhancing STEM learning. Educ. Psychol. Rev. **29**, 717–761 (2017)
6. Kellman, P.J., Massey, C.M.: Perceptual learning, cognition, and expertise. In: Ross, B.H. (ed.) The psychology of learning and motivation, vol. 558, pp. 117–165. Elsevier Academic Press, New York (2013)
7. Rau, M.A.: Sequencing support for sense making and perceptual induction of connections among multiple visual representations. J. Educ. Psychol. **110**, 811–833 (2018)
8. Van Gog, T., Paas, F.: Instructional efficiency: revisiting the original construct in educational research. Educ. Psychol. **43**, 16–26 (2008)
9. Peters, M., Laeng, B., Latham, K., Jackson, M., Zaiyouna, R., Richardson, C.: A redrawn Vandenberg & Kuse mental rotations test: different versions and factors that affect performance. Brain Cogn. **28**, 39–58 (1995)
10. Schmeck, A., Opfermann, M., van Gog, T., Paas, F., Leutner, D.: Measuring cognitive load with subjective rating scales during problem solving: differences between immediate and delayed ratings. Instr. Sci. **43**, 93–114 (2014)

Managing Learners' Memory Strength in a POMDP-Based Learning Path Recommender System

Zhao Zhang[✉], Armelle Brun[✉], and Anne Boyer[✉]

Université de Lorraine, CNRS Loria, Vandœuvre-lès-Nancy, France
{zhao.zhang,armelle.brun,anne.boyer}@loria.fr

Abstract. This paper views the learning path recommendation task as a sequential decision problem and considers Partially Observable Markov Decision Process (POMDP) as an adequate approach. This work proposes M-POMDP, a POMDP-based recommendation model that manages learners' memory strength, while limiting the increase in complexity and data required. M-POMDP has been evaluated on two real datasets.

Keywords: Learning path recommendation · POMDP · Memory strength

1 Introduction

Recommender systems (RS) in education help learners reach their learning goals, while keeping care of the recommendation adoption. The recommendation are sequences of resources, that maximise the probability the goal is reached. Such recommender systems are Learning Path Recommender Systems (LPRS). LPRS can be viewed as a sequential decision problem and approached by a Markov Decision Process (MDP). However, in the educational context some elements remain uncertain such as the learners' knowledge level or the motivation [2]. LPRS can thus be formulated as a POMDP. Although the learners' memory ability is an important factor, it is seldom considered in recommendation, generally at the cost of a high model complexity. We intend to manage it to promote the review of resources and foster long-term retention, with a limited complexity in a learning environment where no metadata about the resources is provided.

2 Related Work

Learning path recommender systems (LPRS) are designed to recommend a sequence of educational resources that contributes to reach a predefined goal. This goal can be the knowledge increase, minimization of learning time, etc. Associated models generally exploit the learners' past interactions with pedagogical resources.

© Springer Nature Switzerland AG 2022
M. M. Rodrigo et al. (Eds.): AIED 2022, LNCS 13356, pp. 284–288, 2022.
https://doi.org/10.1007/978-3-031-11647-6_53

Several approaches have been proposed to perform learning path recommendations, especially Markov-based algorithms, which are known to be good at dealing with this sequential problem. In the educational context, MDP and POMDP have shown to be relevant [2]. POMDPs compute a policy for selecting sequential actions when information may be unobserved. A POMDP consists of a tuple of at least 7 elements, among which the set of states, the set of actions, the observation probability, a transition and a reward function, beliefs, etc.

Memory is important in education and numerous studies in psychology have been interested in modeling human memory. They model how memory decays with time, through a forgetting curve or a half-life regression model [4]. [5] studies several forgetting curve models that incorporate human expertise: psychological and linguistic features, to predict the probability of word recall. One main limit of these works is their complexity and the large datasets they require.

3   Learners' Memory Strength in a POMDP-Based LPRS

We formulate LP recommendation as a POMDP when no content information about resources is available. An action is the act of accessing a resource. The set of actions is thus defined as the set of resources, as in [2]. A state s is defined by two simple attributes: s_{LP} represents the learner's history learning path and s_{KL} represents the estimated knowledge level of the learner. As resources are not indexed, we propose to represent the learner's knowledge for each resource in s_{LP}. To limit the complexity, we discretize the knowledge level [3]. Given that action a is taken in state s, the observation model $p(z|s,a)$ indicates the probability of observing z. The reward function is defined as $R(s,a) = r(s_{LP},a) + r(s_{KL},a)$. The transition function T models the possible effects of the actions on a state. $T(s,a,s') = P(s'|s,a) = P(s'_{LP}|s,a) \cdot P(s'_{KL}|s,a)$ manages both attributes of the state independently [3].

Given an learning path LP, two cases arise when estimating a learner's knowledge level for a resource. First, if the action points to an evaluation resource er (quiz, exam), the knowledge level $KL(LP,er)$ can be directly estimated from the grade obtained by the learner $(eval(LP,er))$. Second, if the action does not point to an evaluation resource, we assume that $(eval(LP,er))$ is an accurate indicator of knowledge level of all the resources that have been studied before er. So, the knowledge level of the current resource cr can be estimated from the evaluation resource that follows cr. We propose to apply a discount factor λ, the longer distant cr is from er, the lower the knowledge level for cr. We present the way we estimate the knowledge level on cr as: $KL(LP,cr) = round(\lambda^{dist(LP,cr,n\_eval(LP,cr))} eval(LP, n\_eval(LP,cr)))$ where $n\_eval(LP,cr)$ is the next evaluation resource that follows cr in LP and $dist(LP,cr,n\_eval(LP,cr))$ is the distance between cr and the corresponding next evaluation resource.

Managing Learners' Memory Strength. We propose Memory-based POMDP (M-POMDP) that manages the learners' memory strength to foster resource reviewing. M-POMDP is intended to limit the increase in complexity. M-POMDP stores learners' memory strength in the state, as an additional attribute, under the form of a discretized attribute [4]. The corresponding attribute s_{NLT} is set as an array, where each element represents the number of times the corresponding resource in s_{LP} has been studied by the learner. It is used to evaluate the need of review of this resource. s_{NLT} is deterministically incremented each time the learner interacts with a resource. In line with the literature, when a learner has studied a resource MAX_{NLT} times, it does not need to be reviewed.

Reward Function. s_{NLT} is a supplement to s_{KL} and impacts the reward function. We propose to redefine the reward function as $R(s,a) = r(s_{LP},a) + r(s_{KL},a) + r(s_{NLT},a)$ where $r(s_{NLT},a)$ is the reward function that computes the reward based on s_{NLT}, defined as follows: if the NLT is increased, it gains a unit of reward u_{NLT}; otherwise the reward is 0.

Transition Function. The transition function $T(s,a,s')$ is also impacted by s_{NLT}. It is evaluated by three independent sub-functions: $T(s,a,s') = P(s'_{LP}|s,a) \cdot P(s'_{KL}|s,a) \cdot P(s'_{NLT}|s,a)$. Since NLT is deterministic, $P(s'_{NLT}|a,s) = 1$, so the transition function remains unchanged.

This simple solution faces a limit: the time gap between two actions is not considered. Even if it could be simply stored as the last access date of each resource, this would be at the cost of a significant increase in the number of states due to the high number of possible values for this new attribute.

4   Experiments

Experiments are conducted on two real-world datasets: EOLE and the EdNet datasets. EOLE, described in [6], is a medium-sized dataset that contains 3,972 interactions from 104 learners on 39 resources. The median length of LP is 38 and the repetition rate is 0.30. About EdNet[1], it is a large dataset with a LP median length of 15 (twice lower than for EOLE) and a repetition rate of 0.22.

Evaluation Protocol. We propose to adopt a leaving one out cross validation. The interactions of each test learner in the review period are split into two. The first 50% form the elements which help to determine the initial state of the POMDP, the rest is used to evaluate the recommended LP. For the EdNet dataset, we select the last 50% of interactions of each learner as the test set.

Parameter Settings. The length of the history is set to the average length of the learning path in datasets ($N = 7$). The number of knowledge levels is set to $K = 4$ [3], $MAX_{NLT} = 3$ and $\lambda = 0.9$. The SARSOP solver [1] is used.

Evaluation Metrics. We use the well-known precision measure. To fit the sequential characteristics of our data, we redefine the "matched" resources from the upper part of the equation by the Longest Common Subsequence (LCS) between RLP and ground truth LP (GTLP). This updated precision is defined as $Precision = \frac{|LCS(RLP,GTLP)|}{|RLP|}$. Besides, we use precision of $SLLP$ measure

---

[1] https://github.com/riiid/ednet.

(Similar Learners Learning Path) [6], noted $Prec_{SLLP}$. Based on [6], learners are split in three groups: Good (GL), Average (AL) and Promising (PL) Learners.

Table 1. Evaluation of POMDP and M-POMDP for EOLE and EdNet datasets

Dataset	Measures	POMDP				M-POMDP			
			Knowledge level				Knowledge level		
		ALL	GL	AL	PL	ALL	GL	AL	PL
EOLE	$Prec$	0.34	0.41	0.36	0.24	0.44	0.59	0.43	0.30
	$Prec_{SLLP}$	0.26	0.30	0.27	0.21	0.33	0.40	0.35	0.22
EdNet	$Prec$	0.20	0.14	0.31	0.15	0.29	0.27	0.38	0.23
	$Prec_{SLLP}$	0.07	0.1	0.11	0.02	0.10	0.09	0.15	0.05

Table 1 presents the values of $Prec$ and $Prec_{SLLP}$. *Considering the baseline POMDP, Prec* decreases with the level of the group, which was expected. This confirms that POMDP recommends PL paths that are closer to those adopted by learners with a higher level, which is confirmed by $Prec_{SLLP}$. *Considering M-POMDP* on the entire set of learners $Prec$ and $Prec_{SLLP}$ are improved by similar rates and the quality of the recommendations if also increased for each group of learners. This confirms that the way M-POMDP manages learners' memory strength seems to be adequate.

We can see that the values on EdNet are lower than for EOLE, explained by the number of resources that is twice larger on EdNet; the average length of learners' learning path that is 3 times smaller than in EOLE. For M-POMDP, the increase on $Prec$ and $Prec_{SLLP}$ for the entire set of learners are similar. Considering each group of learners, $Prec$ is improved significantly for each group. $Prec_{SLLP}$ remains stable with M-POMDP.

From these experiments, we can conclude that the simple way M-POMDP to manages learners' memory strength is adequate and fits medium size datasets.

5   Conclusion and Perspectives

This work focused on the learning path recommendation task through POMDP. We have designed a model that manage learners' memory strength with a limited increase in complexity, validated experimentally. As a future work, we intend to incorporate additional information in the model, whether they are teacher expertise or from data.

References

1. Kurniawati, H., Hsu, D., Lee, W.S.: SARSOP: efficient point-based POMDP planning by approximating optimally reachable belief spaces. In: Robotics: Science and Systems (2008)

2. Rafferty, A.N., Brunskill, E., Griffiths, T.L., Shafto, P.: Faster teaching via POMDP planning. Cogn. Sci. **40**(6), 1290–1332 (2016)
3. Ramachandran, A., Sebo, S.S., Scassellati, B.: Personalized robot tutoring using the assistive tutor POMDP (at-POMDP). In: AAAI Conference, vol. 33, pp. 8050–8057 (2019)
4. Settles, B., Meeder, B.: A trainable spaced repetition model for language learning. In: Proceedings of the 54th Annual Meeting of the ACL, pp. 1848–1858 (2016)
5. Zaidi, A., Caines, A., Moore, R., Buttery, P., Rice, A.: Adaptive forgetting curves for spaced repetition language learning. In: Proceedings of AIED Conference, pp. 358–363 (2020)
6. Zhang, Z., Brun, A., Boyer, A.: New measures for offline evaluation of learning path recommenders. In: Alario-Hoyos, C., Rodríguez-Triana, M.J., Scheffel, M., Arnedillo-Sánchez, I., Dennerlein, S.M. (eds.) EC-TEL 2020. LNCS, vol. 12315, pp. 259–273. Springer, Cham (2020). https://doi.org/10.1007/978-3-030-57717-9_19

Contrastive Deep Knowledge Tracing

Huan Dai[1,2], Yue Yun[1,2], Yupei Zhang[1,2(✉)], Wenxin Zhang[1,2],
and Xuequn Shang[1,2(✉)]

[1] School of Computer Science, Northwestern Polytechnical University, Xi'an, China
daihuan@mail.nwpu.edu.cn, {ypzhang,shang}@nwpu.edu.cn
[2] Laboratory of Big Data Storage and Management,
Ministry of Industry and Information Technology, Xi'an, China

Abstract. Knowledge tracing (KT) aims to predict student performance on the next question according to historical records. Recently deep learning-based models for KT task successfully modeling student responses receive good prediction results of student performance. The student responses encoded as input of KT models use a one-hot encoding. We find that one-hot encoding represents student responses on different items related to the same concepts in completely different vectors. However, items related to the same concept have certain relationships in the real world so the student has a similar representation in these items. In this paper, we propose a new method named Contrastive Deep Knowledge Tracing (CDKT) for providing a reasonable representation of students. We evaluate our model using three public benchmark datasets and the experimental results demonstrate improvements over state-of-the-art methods.

Keywords: Knowledge tracing · Contrastive learning · Deep learning

1 Introduction

Knowledge tracing (KT) is one of the critical problem research in personalized education [1–3]. KT problem can be defined as given student history records $X = \left\{ \left(q_1^i, a_1^i\right), \left(q_2^i, a_2^i\right), \ldots, \left(q_t^i, a_t^i\right) \right\}$, where q_t^i represents student i answering the question q at time t, $a_t^i \in \{0, 1\}$, 0 represents student i incorrectly answering this question, 1 represents correctly answering this question. We need to predict the performance on next question $\mathrm{P}\left(a_{t+1}^i = 1 \mid \left(q_1^i, a_1^i\right), \left(q_2^i, a_2^i\right), \ldots, \left(q_t^i, a_t^i\right)\right)$. Among existing knowledge tracing models, deep learning-based model (DKT) shows more promising results than others. Deep knowledge tracing uses Long Short-Term Memory (LSTM) to model student learning process [4–6]. The input is one student historical responses on each item. Usually it uses one-hot encoding to denote item which student answered. Different item uses different one-hot encoding. This method fails to distinguish different item related to the same concept. Because one-hot encoding only takes the different items into consideration which ignores the concept-level information. How to choose the proper way to

© Springer Nature Switzerland AG 2022
M. M. Rodrigo et al. (Eds.): AIED 2022, LNCS 13356, pp. 289–292, 2022.
https://doi.org/10.1007/978-3-031-11647-6_54

integrate concept-level into student response encoding? Can we set an auxiliary task to distill items and concepts information from the input data itself? We introduce contrastive learning framework into DKT model. Contrastive learning [7] has received interest due to its success in self-supervised representation learning in the computer vision domain [8]. Contrastive learning aims to learn representation by maximizing feature consistency under differently augmented views and has achieved remarkable success in various fields, such as natural language processing [9] and computer vision[10,11]. In this paper, we propose a new method Contrastive Deep Knowledge Tracing (CDKT). To our best knowledge, there is almost no exploration of the contrastive learning in deep knowledge tracing scenarios.

2   Methodology

2.1   Student Contrastive Learning

The import thing in contrastive learning is to construct positive and negative samples [8]. In our model, we adopt the items with same concepts as positive samples and the items with different concepts as negative samples. We attempt to maximize feature consistency under different items. Then we adopt the contrastive loss to minimize the agreement of positive pairs and maximize that of negative pairs as in [8]:

$$\mathcal{L}_{cl} = -\log \frac{\exp\left(\text{sim}\left(s_i, s_i^+\right)/\tau\right)}{\sum_{j=1}^{N}\left(\exp\left(\text{sim}\left(s_i, s_j^+\right)/\tau\right) + \exp\left(\text{sim}\left(s_i, s_j^-\right)/\tau\right)\right)} \tag{1}$$

where s_i denotes the response of student i and s_i^+ denotes the positive sample of student s_i, s_i^- denotes the negative sample of student s_i. sim denotes the similar of two samples.

2.2   Deep Knowledge Tracing

Modeling the student learning process, the current knowledge state of a student is highly related to the previous knowledge state. Thus, the student learning process can be modeled by Long Short-Term Memory network [12]. We get the new features from student contrastive learning as the input of LSTM and then calculate the probability of correctly answering item $i_{k,t+1}$ by student k. The process can be represented as: $p_{k,t+1} = \sigma\left(W_s h_{i,t} + b_s\right)$. Where $W_s \in \mathcal{R}^{d_s}$, and $b_s \in \mathcal{R}^1$ are the weight matrices and biases to be learned, and d_s denotes the dimension of the hidden vector. $p_{k,t+1}$ is the probability that student i can answer item $k+1$ correctly.

2.3   Optimization

The objective of the proposed model are two folds: one is to learn new embedding of student responses of different items, and the other is to make accurate predictions of students' responses. For student i answering item $q_{k,t}$ at time t, the prediction loss $\ell_{k,t}$ can be modeled with the binary cross-entropy: $\ell_{i,t} = -\left(y_{i,t} \log\left(p_{i,t}\right) + \left(1 - y_{i,t}\right) \log\left(1 - p_{i,t}\right)\right)$. Where $p_{i,t}$ is the probability that student i can answer item $q_{k,t}$ correctly and $y_{i,t}$ is the correctness of the response of student i at time t. The total loss can be represented as the weighted sum of the total prediction loss and the total contrastive loss:

$$\mathcal{L} = \sum_{n=1}^{|\mathcal{N}|} \ell_{i,t} + \lambda \mathcal{L}_{\mathrm{CL}} \tag{2}$$

3   Experiments

3.1   Prediction of Student Performance

We compare the CDKT model with the DKT. Results are show in Table 1. From Table 1, we can find that CDKT gets the better AUC and ACC than DKT which means using contrasitve learning to learn student features is superior to one-hot encoding.

Table 1. Prediction results of DKT and CDKT.

Dataset	Method			
	DKT		CDKT	
Metrics	ACC (%)	AUC (%)	ACC (%)	AUC (%)
ASSISTment2009	77.02	81.81	78.23	**82.10**
ASSISment2015	74.94	72.94	76.73	**75.82**
STATICS2011	81.27	82.87	83.15	**84.21**

3.2   Visualization of Student Knowledge State

KT task aims to quantitative analysis student knowledge state. In this experiment, we show the hidden state of deep model which represents the knowledge state of student. Figure 1 shows the change of student knowledge state over time.

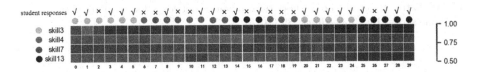

Fig. 1. Visualization of student knowledge state. The horizontal axis represents a question answered by a student. The vertical axis represents the related skills.

Acknowledgement. This study was funded in part by National Natural Science Foundation of China (61802313, U1811262, 61772426), Key Research and Development Program of China (2020AAA0108500), Reformation Research on Education and Teaching at Northwestern Polytechnical University (2021JGY31), Education And Teaching Reform Research Project of Northwestern Polytechnical University (2022JGY62).

References

1. Sarsa, S., Leinonen, J., Hellas, A.: Deep learning models for knowledge tracing: review and empirical evaluation. arXiv preprint arXiv:2112.15072 (2021)
2. Zhang, Y., Dai, H., Yun, Y., et al.: Meta-knowledge dictionary learning on 1-bit response data for student knowledge diagnosis. Knowl.-Based Syst. **205**, 106290 (2020)
3. Zhang, Y., An, R., Cui, J., et al.: Undergraduate grade prediction in Chinese higher education using convolutional neural networks. In: LAK21: 11th International Learning Analytics and Knowledge Conference, pp. 462–468 (2021)
4. Staudemeyer, Ralf, C., Morris, E.R.: Understanding LSTM-a tutorial into long short-term memory recurrent neural networks. arXiv preprint arXiv:1909.09586 (2019)
5. Yun, Y., Dai, H., Cao, R., Zhang, Y., Shang, X.: Self-paced graph memory network for student GPA prediction and abnormal student detection. In: Roll, I., McNamara, D., Sosnovsky, S., Luckin, R., Dimitrova, V. (eds.) AIED 2021. LNCS (LNAI), vol. 12749, pp. 417–421. Springer, Cham (2021). https://doi.org/10.1007/978-3-030-78270-2_74
6. Dai, H., Zhang, Y., Yun, Y., Shang, X.: An improved deep model for knowledge tracing and question-difficulty discovery. In: Pham, D.N., Theeramunkong, T., Governatori, G., Liu, F. (eds.) PRICAI 2021. LNCS (LNAI), vol. 13032, pp. 362–375. Springer, Cham (2021). https://doi.org/10.1007/978-3-030-89363-7_28
7. Le-Khac, P.H., Healy, G., Smeaton, A.F.: Contrastive representation learning: a framework and review. IEEE Access (2020)
8. Chen, T., et al.: A simple framework for contrastive learning of visual representations. In: International Conference on Machine Learning. PMLR (2020)
9. Dai, B., Lin, D.: Contrastive learning for image captioning. arXiv preprint arXiv:1710.02534 (2017)
10. Saeed, A., Grangier, D., Zeghidour, N.: Contrastive learning of general-purpose audio representations. In: ICASSP 2021–2021 IEEE International Conference on Acoustics, Speech and Signal Processing (ICASSP). IEEE (2021)
11. Giorgi, J.M., et al.: DECLUTR: deep contrastive learning for unsupervised textual representations. arXiv preprint arXiv:2006.03659 (2020)
12. Malhotra, P., et al.: Long short term memory networks for anomaly detection in time series. In: Proceedings, 89 (2015)

Reflection as an Agile Course Evaluation Tool

Siaw Ling Lo[1]([✉]) [iD], Pei Hua Cher[2] [iD], and Fernando Bello[2] [iD]

[1] Singapore Management University, Singapore, Singapore
sllo@smu.edu.sg
[2] Duke-NUS Medical School, Singapore, Singapore
{peihua.cher,f.bello}@duke-nus.edu.sg

Abstract. Reflection is often used as a tool to analyse student learning, be it for internalizing of acquired knowledge or as a form of seeking help through expression of doubts or misconceptions. However, it can be a challenge to extract relevant information from the free-form reflection text. Often times the workload of manually analyzing the reflection text can be a form of deterrence instead of providing insights in the course delivery for instructors, let alone improving the learning experience. In this paper, we review the current usage of reflection and propose an automated reflection framework, together with an end-to-end analysis of reflection that provides agility in the course delivery and, at the same time, timely feedback to both students and instructors.

Keywords: Reflection framework · Course evaluation · NLP

1 Introduction

Do my students understand? This question lingers in every instructor's mind after each lesson. With the adoption of online teaching during COVID safe management measures, it is no longer feasible to observe individual student's expression in a class to gauge their understanding. One of the options available is to collect reflections from students after each lesson to extract relevant feedback, so that doubts or misconceptions can be addressed in a timely manner. In general, reflection is important for interpreting and internalizing academic activity [3]. More specifically, Karm [3] has shown that, in order to support the development of university instructors and academics in teaching professions, it is crucial to engage students in active reflection.

The current use of reflection is mainly in learning, even though the learning can be from students or instructors and instructor trainees. Interestingly, one article mentioned how reflection can be used as a feedback tool to improve the curriculum [5], having the potential to evolve as a tool for instructors and not merely as a learning tool for students.

Reflection journals or learning logs are commonly used as platforms for students to express what they have experienced, what they might have learned and

© Springer Nature Switzerland AG 2022
M. M. Rodrigo et al. (Eds.): AIED 2022, LNCS 13356, pp. 293–297, 2022.
https://doi.org/10.1007/978-3-031-11647-6_55

their doubts or questions. Even though many insights, such as misconceptions or doubts can be extracted, it remains a challenge to effectively analyse these free form reflection text.

In this paper we propose an automated reflection framework that enables reflection to be used as an agile course evaluation tool for instructors, rather than only as a learning tool for students. In the next section, we explain how reflection can be used as a course evaluation tool. Section 3 presents the proposed automated reflection framework while Sect. 4 covers our recommendations and conclusion.

2   Reflection as an Agile Course Evaluation Tool

Sharp & Lang [6] has proposed a conceptual framework for integrating agile principles in teaching and learning. According to the authors, "given that instructors face large amounts of uncertainty regarding the needs and capabilities of the students prior to or at the beginning of a course, it appears that agile principles may be useful in the course development process". This is in line with observations from Gibson et al. [2], which reported that more frequent early reflections throughout a course may help teachers and students improve through receiving more immediate and regular feedback. This approach also allows the raising of concerns throughout the learning journey, rather than at the end of a long period of teaching. In other words, it can be more useful to collect reflection at the end of each lesson, instead of at the end of a course, to allow agility in evaluating course delivery.

Instead of the end of the course assessment alignment suggested by Ozdemir et al. [5], Chen et al. [1] analyzed students' journals on a weekly basis and discovered additional topics that are of interest to the students, but not explicitly covered by the weekly syllabus. It has the potential to guide instructors to develop future teaching content and activities that are tailored toward students' needs. Lo et al. [4], on the other hand, used an automated method to effectively extract questions or doubts. This gives instructors an option to adjust teaching materials dynamically based on students' reflections after each lesson. With the doubt identified, a list of topics that require further explanation or clarification can be extracted. Additional materials can be designed to cater to the students, addressing the misconceptions before the start of the next lesson.

As discussed above, when reflection is used as a tool for course evaluation and not contributing directly to assessment, it is usually not evaluated stringently and currently there is no framework for analysing the data. Although there are various rubrics or frameworks proposed to evaluate the depth of reflection for learning, to the best of our knowledge, no comprehensive framework has been proposed to use reflection as a course evaluation tool.

3   Proposed Automated Reflection Framework

The course feedback collected at the end of the term usually only benefits the next cohort of students and has no direct impact on the current cohort. In this paper, we would like to propose adoption of both end of term course feedback and individual lesson reflection as a tool for agile course evaluation. The core source of input is the student self-reflection from each lesson, which collectively complements the end of term course feedback (as shown in Fig. 1).

The analytics output component, in the proposed automated reflection framework (Fig. 1), consists of two core analyses: objective-based and doubt/misconception-based. The objective-based analysis is for the instructors to compare the themes stated in the course/lesson objectives to the content mentioned in the student's lesson reflections and course feedback. This provides insights on whether the objectives are well covered or any new theme can be uncovered that needs further clarification. The doubt/misconception-based analysis is useful to identify topics that require further attention. By aligning with lesson objectives, personalised learning advice can be offered to the individual students on top of addressing the common misconceptions identified. Finally, the result from the Reflection Analytics Model (Fig. 2) are presented on an analytics dashboard that enables users to perform the objective-based analysis and the doubt/misconception-based analysis.

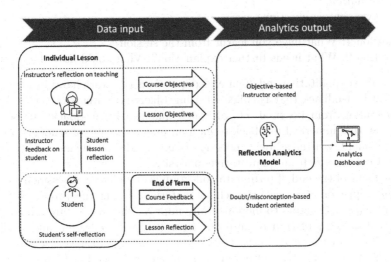

Fig. 1. Proposed automated reflection framework

Along with our automated reflection framework, we have created a Reflection Analytics Model (Fig. 2) that summarises the various automated analysis approaches that have been used in recent research. The model consists of five main sections, namely: Information Retrieval, Pre-Processing, Feature Extraction, Reflection Classification Model and Analytics Dashboard.

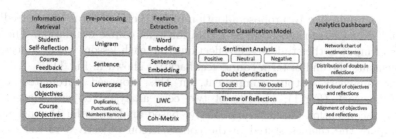

Fig. 2. Reflection analytics model that represents the automated analysis process.

4   Recommendations and Conclusion

Through our proposed automated reflection framework, we recommend collecting reflections/feedback both at the end of individual lessons and end of term. We propose to use the reflection after each lesson as an enabling tool for all types of courses and provide some guiding questions to aid in extracting relevant information.

The two generic questions are:

– Question 1: What have you learnt from the session?
– Question 2: What needs further explanation? What could have done better?

The above qualitative and open-ended questions can work hand-in-hand with quantitative feedback that is collected using Likert-type items or scales. Based on the finding from Lo et al. [4], although quantitative feedback can be used to analyse the numerical feedback to gain an understanding of students' overall self-assessment, written student reflection is able to identify themes and concepts that did not emerge from quantitative analysis.

The focus of the end of term data collection is on delivery, assessment design, and the depth or coverage of teaching materials. Using the proposed framework, it helps instructors to incrementally improve the course through comparing actual delivery of course materials from student's reflection and feedback against lesson or course objectives set at the beginning of the course.

In this paper, we have proposed an Automated Reflection Framework that enables an end-to-end analysis of reflection to provide agility in course evaluation. This automatic extraction of reflection allows a systematic analysis of information as well as providing agility in course adjustment (through reflection analysis from individual lessons) and improving the learning experience for the current cohort of students.

References

1. Chen, Y., Yu, B., Zhang, X., Yu, Y.: Topic modeling for evaluating students' reflective writing: a case study of pre-Service teachers' journals. In: ACM International Conference Proceeding Series, vol. 25–29-April, pp. 1–5 (2016)

2. Gibson, A., Shum, S.B., Aitken, A., Tsingos-Lucas, C., Sándor, Á., Knight, S.: Reflective writing analytics for actionable feedback. In: Proceedings of the Seventh International Learning Analytics and Knowledge Conference, pp. 153–162 (2017)
3. Karm, M.: Reflection tasks in pedagogical training courses. Int. J. Acad. Dev. 15(3), 203–214 (2010)
4. Lo, S.L., Tan, K.W., Ouh, E.L.: Automated doubt identification from informal reflections through hybrid sentic patterns and machine learning approach. Res. Pract. Technol. Enhanc. Learn. 16(1), 1–24 (2021)
5. Ozdemir, D., Opseth, H.M., Taylor, H.: Leveraging learning analytics for student reflection and course evaluation. J. Appl. Res. High. Educ. 12(1), 27–37 (2020)
6. Sharp, J.H., Lang, G.: Agile in teaching and learning: conceptual framework and research agenda. J. Inf. Syst. Educ. 29(2), 45–51 (2018)

Toward Ubiquitous Foreign Language Learning Anxiety Detection

Daneih Ismail$^{(\boxtimes)}$ ⓘ and Peter Hastings ⓘ

DePaul University, Chicago, IL 60614, USA
dismail1@depaul.edu, peterh@cdm.depaul.edu

Abstract. We present a novel design for detecting Foreign Language Anxiety (FLA) while the learner is using an English as a second language system (ESL). Our method uses sensor-free metrics to avoid disrupting the learning process. We evaluated the validity and reliability of several machine learning models using data from two different systems. Using 9 features extracted from the interaction, we found that Random Forest, XGBoost, and Gradient Boosting Regressor provided suitably accurate predictions of anxiety, and outperformed Linear Regression, Support Vector Regressor, and Bayesian Ridge Regression.

Keywords: Affect detection · FLA · ML · Sensor-free metrics

1   Introduction

Automated detectors for predicting emotions such as engagement, boredom, confusion, and frustration have achieved high accuracy [1]. However, there is still a need to improve prediction of Foreign Language Anxiety (FLA), a significant impediment to learners of new languages [7]. Previous emotion research has shown that multiple factors affect learners' vulnerability to FLA, such as task complexity [3], academic achievement, gender and age [10]. Horwitz, et al. [4] found that there are three dominant components that influence FLA: fear of negative evaluation, communication apprehension, and test anxiety [4]. Based on this, they developed a well-established and validated instrument for measuring FLA, the Foreign Language Classroom Anxiety Scale (FLCAS) [4]. The FLCAS was developed for use in a classroom context, but it has also been shown to correlate well with self-reported anxiety within online tutoring system [6].

Sensor-free metrics detect emotions from the users' interactions with the system without using any physical monitors [7]. Previous researchers have built sensor-free emotion detector, comparing various machine learning algorithms to reach the best prediction [1]. In this work, we focused on Foreign Language Anxiety in particular. We extracted features from learners' interactions with ESL systems then built machine learning models to predict FLA from those features. We used self-reports of FLA as ground truth for our predictions. We analyzed the following research questions:

© Springer Nature Switzerland AG 2022
M. M. Rodrigo et al. (Eds.): AIED 2022, LNCS 13356, pp. 298–301, 2022.
https://doi.org/10.1007/978-3-031-11647-6_56

Research Question 1: "Can FLA be detected without learning interruption when using ESL learning systems?" This question has two sub-questions:

RQ1A: What features best predict FLA?

RQ1B: What machine learning methods are better for predicting FLA?

Research Question 2: "How well can sensor-free detectors be generalized to other emotionally intelligent foreign-language/ESL learning systems?"

2 Method

The data was collected from two different ESL systems. Dataset 1 came from a system focused on practice, with no tutorial and no scoring. Thirty participants did 27 exercises, covering vocabulary, grammar, listening, conversation, and speaking, providing data from a total of 810 exercises. Dataset 2 came from an online system which included video tutorials and feedback on the answers [8]. 29 participants did 26 exercises which covered vocabulary, grammar, listening, reading, and writing, producing data from a total of 704 exercises. For both experiments the participants completed level of anxiety self-report after each exercise.

From each of the exercises, 16 features were extracted. Following [7], we used the average of the three FLCAS component scores: fear of negative evaluations, communication apprehension, and test anxiety [4]. Following [10], we included the participant's age, gender, education level, English level, exercise score, duration, exercise topic, score on the preceding exercise, the percentage of previous incorrect scores, the percentage of previous correct scores, average percentage of all previous exercises, and average duration of exercises of the same section. We did a correlation analysis and set an absolute threshold value of 0.5 to eliminate multicollinearity. Then we used the Gini importance feature selection algorithm to distill the features that could cause overfitting and kept only the features that improved the model. From an original set of 16 features, we ended up with 9 features that provided an acceptable accuracy with the least bias.

We made predictions using regression instead of classification because we used continuous-valued self-report to measure anxiety in order to capture moment-to-moment emotion fluctuation [9] and to provide more accurate high-resolution measurements (as opposed to, e.g., the Likert classification scale). The methods we evaluated were Random Forests, XGBoost, Gradient Boosting Regressor, Linear Regression, Bayesian Ridge Regression, and Support Vector Regressor. We implemented these machine learning models in the scikit-learn library in Python. We evaluated each detector using 10-fold cross-validation.

3 Results

Regarding RQ1A, determining which features are predictive of FLA, based on the correlation analysis and Gini importance algorithm, the final set of features that reliably predicted FLA were: exercise score, percentage of all previous exercise scores, percentage of previous incorrect scores, exercise duration, relevant exercise duration, FLCAS score, English level, exercise topic, and the participant's age. Gini indicated that the most important features were FLCAS score followed by the average percentage of all previous exercises.

With respect to RQ1B, on the types of machine learning methods, for Dataset 1, the Random Forest method was most accurate, predicting 47% of the variance of FLA. XGBoost was close behind, predicting 45%. For Dataset 2, both methods predicted 66% of FLA. The performance of Gradient Boosting was slightly behind that of the other ensemble methods. In contrast, the non-ensemble methods performed much worse, predicting a maximum of 21% of the variance of FLA. For RQ2, which focused on the robustness of these features and models across the different systems and datasets, we found that the set of most important features for both datasets was identical, with slight differences in ranking. The relative performance of the models was identical across the datasets.

4 Discussion and Conclusions

Prior research used FLCAS components and exercise score as sensor-free metrics to predict FLA [7]. Here, we extend this by uncovering features that produce better predictions using machine learning without interrupting the learning process. Previous research found that FLA could be predicted up to 43% using Linear Regression using FLCAS components and exercise scores, but only by including self-reports of system and language difficulty after each exercise. Without these intrusive self-report measures, the maximum prediction was 20%. Here, we found that by augmenting the FLCAS scores with behavioral features from the participants' interactions with the systems, and by using machine learning models, we can predict up to 66% of the learner's anxiety without interruptions. This level of accuracy is imperfect yet satisfactory; affect detection is extremely difficult because it is not directly accessible [1].

Because the two datasets had identical highest feature importance rankings, these features should provide reliable predictive performance for any e-learning system that teaches English as a foreign language. They are also easily derived from pretest and behavioral data.

As mentioned above, previous research found that sensor-free metrics could predict FLA using Linear Regression, accounting for 20% of the variation in anxiety [7]. Here, we found that machine learning models can produce much more accurate predictions. It is clear that ensemble learning models (Random Forest, XGBoost, Gradient Boosting Regressor) outperform non-ensemble models (Linear Regression, Bayesian Ridge, and Support Vector Regressor). The high performance of the ensemble learning models is consistent with other research demonstrating the robustness, reliability, and stability of these methods [5]. Because these ensemble learning methods can produce acceptably accurate predictions of the learner's level of anxiety, they can be used to support an emotionally intelligent tutoring system which can adaptively provide interventions according to the learner's current emotional state.

When the performance of a model on a second dataset is the same or better than on the one for which it was developed, that provides evidence for the reliability of the model [2]. Here, we extracted a set of features from one system and dataset, evaluating predictive performance with multiple models. Then, using the same features, found that the relative performance was equivalent across the systems, and that all of the models actually produced better performance in the second dataset. Thus, we demonstrated the generalizability of this approach to any ESL system because the models used features that can be easily extracted from any such system. A limitation of this approach is that the features were selected along with the machine learning algorithm. For our future work we will build an emotionally intelligent tutoring system to detect FLA and reduce it by adaptively providing appropriate feedback, creating a more positive and effective learning environment.

References

1. Baker, R., et al.: Towards sensor-free affect detection in Cognitive Tutor Algebra. In: Proceedings of the Fifth International Conference on Educational Data Mining, pp. 126–133 (2012)
2. Bosnić, Z., Kononenko, I.: An overview of advances in reliability estimation of individual predictions in machine learning. Intell. Data Anal. **13**(2), 385–401 (2009)
3. Hashemi, M.: Language stress and anxiety among the English language learners. Procedia Soc. Behav. Sci. **30**, 1811–1816 (2011)
4. Horwitz, E.K., Horwitz, M.B., Cope, J.: Foreign language classroom anxiety. Mod. Lang. J. **70**(2), 125–132 (1986)
5. Hueniken, K., et al.: Machine learning-based predictive modeling of anxiety and depressive symptoms during 8 months of the COVID-19 global pandemic: Repeated cross-sectional survey study. JMIR Mental Health **8**(11), e32876 (2021)
6. Ismail, D., Hastings, P.: Identifying anxiety when learning a second language using e-learning system. In: Proceedings of the 2019 Conference on Interfaces and Human Computer Interaction, pp. 131–140 (2019)
7. Ismail, D., Hastings, P.: A sensor-lite anxiety detector for foreign language learning. In: Proceedings of the 2020 Conference on Interfaces and Human Computer Interaction, pp. 19–26 (2020)
8. Ismail, D., Hastings, P.: Way to go! effects of motivational support and agents on reducing foreign language anxiety. In: Roll, I., McNamara, D., Sosnovsky, S., Luckin, R., Dimitrova, V. (eds.) AIED 2021. LNCS (LNAI), vol. 12749, pp. 202–207. Springer, Cham (2021). https://doi.org/10.1007/978-3-030-78270-2_36
9. Lottridge, D., Chignell, M., Jovicic, A.: Affective interaction: understanding, evaluating, and designing for human emotion. Revi. Hum. Factors Ergon. **7**(1), 197–217 (2011)
10. Onwuegbuzie, A.J., Bailey, P., Daley, C.E.: Factors associated with foreign language anxiety. Appl. Psycholinguist. **20**(2), 217–239 (1999)

A Semi-automatic Approach for Generating Video Trailers for Learning Pathways

Prakhar Mishra[✉][iD], Chaitali Diwan[✉][iD], Srinath Srinivasa[iD],
and G. Srinivasaraghavan

International Institute of Information Technology, Bangalore, India
{prakhar.mishra,chaitali.diwan,sri,gsr}@iiitb.ac.in

Abstract. This paper presents an approach for semi-automatic generation of video trailers for learning pathways containing sequenced learning
resources of various forms. The main aim of presenting a trailer to the
learners is to generate curiosity and interest among them and help them
make an informed choice about the learning pathways or the courses they
want to learn. We propose generation of video trailers using Machine
Learning and Natural Language Processing techniques. The proposed
trailer is in the form of a timeline consisting of various fragments created
by selecting, para-phrasing or generating content using various proposed
techniques. The fragments are further enhanced by adding voice-over
text, subtitles, etc., to create a holistic experience. Human evaluation of
trailers generated by our framework showed promising results.

Keywords: Trailer generation · Artificial intelligence · Educational
technology

1 Introduction and Background

Internet boom has led to an enormous increase in freely available educational
content, thus leading to production of redundant courses and videos overtime
for many topics. Choosing an appropriate course that might align to learners'
interest has become more challenging. Often, enrolling to a course that does
not meet the learner's expectations on course curriculum and other peripherals
such as expected commitment, support availability, etc., leads to an interim loss
of motivation and eventual drop-out from the course [9]. We believe that this
problem can be tackled to a certain extent by providing a glimpse of the course
to the learner before the start of the course, in the form of a video trailer.

The concept of *Trailers* is not new and has been widely used in the movie
industry for a long time now. Use of Artificial Intelligence in creating movie
trailers has also been explored [1,3,5,6]. However, movie trailers are used mainly
for advertising, whereas trailers in context of educational domain have a different

Supported by Mphasis F1 foundation and Gooru (https://gooru.org).

M. M. Rodrigo et al. (Eds.): AIED 2022, LNCS 13356, pp. 302–305, 2022.
https://doi.org/10.1007/978-3-031-11647-6_57

purpose. They set the right expectations in learners' mind in-terms of takeaways and mastering competencies, before they start their learning journey.

While studies have established the importance of trailers in educational domain [4,10], automatic or semi-automatic generation of the trailers for academics is unexplored. Hence, we propose a semi-automatic framework for generating video trailers for learning pathways, which are a sequence of related educational documents of various forms [2].

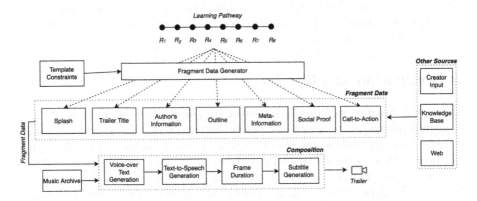

Fig. 1. Trailer generation flow

2   Proposed Method

Our proposed trailer generation framework consists of design of different trailer fragments that form a trailer timeline, together with the proposed algorithms for generation of the trailer fragments. The proposed trailer timeline consists of 7 trailer fragments namely, Splash, Trailer Title, Author Details, Outline, Meta-Information, Social Proof and finally the Call-to-Action(CTA). Each of these fragments define a specific part and purpose in the trailer. Fragments are composed of a sequence of frames and each frame is composed of various types of elements and their properties. The look and feel of the trailer is determined by various pre-defined templates, which can be selected by the creator of the trailer.

The overall approach for trailer generation is illustrated in Fig. 1. All the resources mapped to a learning pathway along with template constraints like number of fragments, frames and elements, etc., form the input to our *Fragment Data Generator (FDG)* module. Splash fragment displays introductory information related to the trailer such as credits, software logo, etc., mostly obtained from the creator's input. Author Details fragment is populated using automation utilities around fetching images, author name, title etc., from the learning resources, and inputs from the creator. The Trailer Title and Outline fragments are generated using our previous work [7,8] along with custom pruning strategies. Meta-Information Fragment informs the learners about other important aspects of the course like course structure, total reading time, total number of resources, etc. and is generated automatically from the learning pathway. Social Proof fragment has information about the learners that have traversed or are traversing the pathway and is derived from the deployed learning environments. Last fragment is CTA where different encouraging phrases are displayed for the learners to get started on the learning pathway.

Once all the fragments are generated, composition module is called which adds various multi-modal experiences. Voice-over is generated based on the narration script as per slot based text grammar. Frame duration and other basic animations like fade-in/out, voice type, etc., are derived from the template to come up with the final trailer.

3   Evaluation and Results

63 human evaluators well versed in the technical domain representing our dataset evaluated the trailers generated by our framework. 6 trailers[1] generated from 3 different learning pathways were evaluated using two templates (T1 and T2) per learning pathway. The evaluation for each trailer was done on a set of 8 questions on Likert-scale from 1 to 5 (1 = very poor, 5 = very good). The questions[2] covered two major themes, one around the motivation, clarity, impression and learner appeal and other around it's visual properties and experience like font-size, audio frame sync, etc.

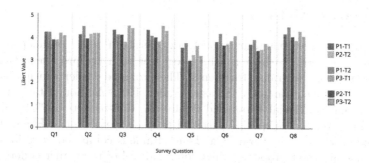

Fig. 2. Average scores per Survey Question for all 3 pathways and trailers. Here P1, P2, P3 represent 3 Pathways and T1, T2 represent Templates

As can be seen in Fig. 2, the scores obtained for each of the survey questions are good and far above the average (score of 3) for almost all the generated trailers. Figure 3 shows some of the trailer fragments generated by our proposed system.

---

[1] Sample Trailers: https://bit.ly/3Hscie9.
[2] Evaluation Questions: https://bit.ly/3FIak8p.

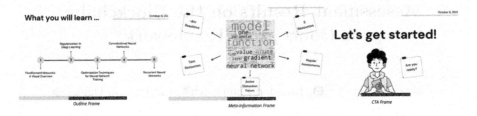

Fig. 3. Trailer fragments

4 Conclusion and Future Work

We presented a framework for generating video trailers for learning pathways using ML and NLP techniques. Evaluation of the generated trailer samples were encouraging. In future, we plan to add more interesting themes like automatically detecting learning outcomes given the resources.

References

1. Brachmann, C., Chunpir, H.I., Gennies, S., Haller, B., Kehl, P., Mochtarram, A.P., Möhlmann, D., Schrumpf, C., Schultz, C., Stolper, B., et al.. In: Digital Tools in Media Studies, pp. 145–158. transcript-Verlag (2015)
2. Diwan, C., Srinivasa, S., Ram, P.: Automatic generation of coherent learning pathways for open educational resources. In: European Conference on Technology Enhanced Learning. pp. 321–334. Springer (2019)
3. Gaikwad, B., Sontakke, A., Patwardhan, M., Pedanekar, N., Karande, S.: Plots to previews: Towards automatic movie preview retrieval using publicly available metadata. In: Proceedings of the IEEE/CVF International Conference on Computer Vision. pp. 3205–3214 (2021)
4. Gayoung, L., Sunyoung, K., Myungsun, K., Yoomi, C., Ilju, R.: A study on the development of a mooc design model. Educational technology international **17**(1), 1–37 (2016)
5. Hermes, T., Schultz, C.: Automatic generation of hollywood-like movie trailers. eCulture Factory (2006)
6. Hesham, M., Hani, B., Fouad, N., Amer, E.: Smart trailer: Automatic generation of movie trailer using only subtitles. In: 2018 First International Workshop on Deep and Representation Learning (IWDRL). pp. 26–30. IEEE (2018)
7. Mishra, P., Diwan, C., Srinivasa, S., Srinivasaraghavan, G.: Automatic title generation for learning resources and pathways with pre-trained transformer models. International Journal of Semantic Computing **15**(04), 487–510 (2021)
8. Mishra, P., Diwan, C., Srinivasa, S., Srinivasaraghavan, G.: Automatic title generation for text with pre-trained transformer language model. In: 2021 IEEE 15th International Conference on Semantic Computing (ICSC). pp. 17–24. IEEE (2021)
9. Simpson, O.: Student retention in distance education: are we failing our students? Open Learning: The Journal of Open, Distance and e-Learning **28**(2), 105–119 (2013)
10. Wong, B.T.m.: Factors leading to effective teaching of moocs. Asian Association of Open Universities Journal (2016)

Assessment Results on the Blockchain: A Conceptual Framework

Patrick Ocheja[1]($\boxtimes$) ⓘ, Brendan Flanagan[2] ⓘ, and Hiroaki Ogata[2] ⓘ

[1] Graduate School of Informatics, Kyoto University, Kyoto, Japan
ocheja.ileanwa.65s@st.kyoto-u.ac.jp
[2] Academic Center for Media and Computing Studies,
Kyoto University, Kyoto, Japan

Abstract. Assessments evaluate students' understanding of taught concepts. While students may be assessed on various concepts and in different learning environments, learning institutions often fail to provide students with a way to connect their learning and assessment results beyond a single institution. This results in problems such as the inability to understand one's readiness for a new program, the inability to decide the most suitable program to enrol after current learning, and difficulty finding the right institution to enrol based on one's profile. This work proposes a conceptual framework for connecting assessment results on the blockchain beyond a single institution in a secure, privacy-preserving, tamper-proof, and trusted manner. We also discuss some typical use cases, results of the evaluation and visualizations when students graduate from high school and enter higher institution.

Keywords: Assessment · Blockchain in education · Learning analytics · Lifelong learning

1 Introduction

Evaluating students and reporting performance outcome is an essential component in teaching and learning. However, information on students' performance are often not readily available to decision makers (teachers, students, parents, etc.) or even provided in a comprehensible format [7]. This arise from limitations such as data privacy, interoperability, lack of distributed analytics and consequent implications of such distributed access [1]. Thus, we propose a framework to enable a decentralized reporting and access to assessment results based on a Blockchain of Learning Logs (BOLL) system [6] that connects learning behaviour logs and digital contents across institutions. We extend the BOLL system to include assessment results by integrating scores reporting, blockchain encoding, decentralized analytics opt-in/out function and a visualization to support data-driven decision making by stakeholders. To highlight the potential impact of our proposal, we present a simple orchestration of university enrolment decision making process guided by distributed analysis of assessment results.

M. M. Rodrigo et al. (Eds.): AIED 2022, LNCS 13356, pp. 306–310, 2022.
https://doi.org/10.1007/978-3-031-11647-6_58

Other works such as [2,3,5] have explored the use of blockchain to report and manage academic data, focusing mainly on providing trusted data for a single student. Thus, this work is the first to propose a framework that offers chain analytics that can provide trusted aggregate data and insights not previously accessible.

2   Proposed Framework

We propose a framework that allow institutions to report assessment results of their students on a decentralized network with strict privacy preservation and learner control as illustrated in Fig. 1 based on the Boll System [4]. In the next subsections, we will discuss the layers in proposed architecture.

Service Layer consist of 3 components: Analytics Service Installer (ASI), Subscription and Event Handlers. The ASI allow researchers or learning service providers to install a Learning Analytics Service (LAS) that can aid students' learning objectives such as predictions, interventions and recommendations. The subscription handler is a set of smart contracts and utilities that help learners manage their subscription to LAS's. The event handler enable data and visualization layers to stay updated when data associated with LAS's change and may require updates to the previous models or visualizations.

The Data Layer is made up of the aggregator, data processors, analyzers and the resulting model. The aggregator helps to retrieve the data of all the subscribers to a given LAS. The processor and analyzer provide similar functions as the data processing and analysis tasks in data mining and it is unique to each LAS. The model represents the results obtained by analyzing the data provided at a specific time, t.

The Visual Layer is directly user facing and provide interfaces for querying various insights deduced from the previously provided data. The Explorer allow subscribers to invoke the compute functions implemented by LAS's. Based on the type of exploration or compute function selected, the Builder invokes the relevant LAS's compute function; providing the subscriber's details as input.

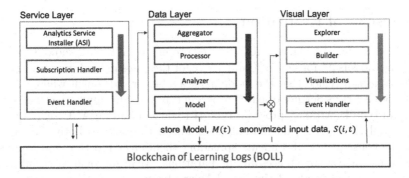

Fig. 1. Proposed architecture

Table 1. Dataset for training initial model (2010 - 2020).

Program type	No. students	Math		English		Japanese		National	
		μ	σ	μ	σ	μ	σ	μ	σ
I Education...	1116	68.17	12.57	57.27	14.76	55.86	14.31	59.30	13.02
II Science...	169	52.17	16.15	48.29	16.22	51.58	15.56	47.89	14.23
III Sports...	164	62.15	14.40	54.88	15.46	55.85	15.20	55.64	14.36
IV Medicine...	870	60.13	14.00	56.63	15.20	58.26	14.29	56.60	13.88
V Economics...	559	65.63	19.96	61.75	18.69	59.34	18.35	60.85	18.42
VI Others	173	57.69	14.11	58.02	15.70	59.22	15.26	56.39	14.07

Table 2. AUC for predicting enrolment decision based on score data.

Program type	I	II	III	IV	V	VI
I	1					
II	**0.82**	1				
III	0.65	**0.78**	1			
IV	**0.70**	0.60	0.62	1		
V	0.56	**0.81**	**0.70**	**0.72**	1	
VI	**0.80**	0.69	**0.74**	**0.70**	**0.80**	1

3 Use-case: Determining Career Path from Score Data

To evaluate the usefulness of the proposed framework, we use the assessment results of students in a High School in Japan to build a decentralized model that can support students in making data-informed enrolment decisions. We begin by connecting the learning infrastructure at the selected school to BOLL [6].

We use the assessment results of past students (2010 - 2020) shown in Table 1 and the program category they successfully enrolled in to build the initial model using Random Forest classifier that can predict enrolment likelihood and recommend the most suitable options.

4 Results of Evaluation

The Area Under the Curve (AUC) scores presented in Table 2 showed that the resulting models performed 70% and above for the cases in bold. Further analysis of these cases revealed that the model makes a better judgement where comparing between Art vs Science oriented program types. For example, the classifier for program type I (Arts) vs program type II (Science) had an AUC score of 82%. Surprisingly, over 200 students still went ahead to apply to both. This comes with the consequence of students spreading themselves too thin and finding it difficult to make smart choices and preparations. We also provide a visualization in Fig. 2 as an example of the visual layer of our proposed framework where students can determine which school their profile fits.

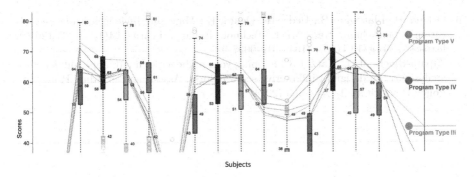

Fig. 2. Interactive visualization to support enrolment decision and preparation

5 Conclusion and Future Work

In this work we conceptualized a framework for managing assessment results on a decentralized network using the blockchain. There is also a lot of potential to expand beyond education in schools as the BOLL could cross into private sector employment and training. Recommended future research include the implementation of the proposed framework, usability and performance testing and evaluating its impact on teaching and learning.

Acknowledgment. This work was partly supported by JSPS Grant-in-Aid for Scientific Research (B)20H01722, (Exploratory)21K19824, (S)16H06304, Grant-in-Aid for JSPS Fellows 22J15869 and NEDO JPNP20006, and JPNP18013.

References

1. Baker, R.S., et al.: Challenges for the future of educational data mining: the baker learning analytics prizes. J. Educ. Data Mining **11**(1), 1–17 (2019)
2. Gräther, W., Kolvenbach, S., Ruland, R., Schütte, J., Torres, C., Wendland, F.: Blockchain for education: lifelong learning passport. In: Proceedings of 1st ERCIM Blockchain workshop 2018. European Society for Socially Embedded Technologies (EUSSET) (2018)
3. Ocheja, P., Flanagan, B., Ogata, H.: Connecting decentralized learning records: a blockchain based learning analytics platform. In: Proceedings of the 8th International Conference on Learning Analytics and Knowledge, pp. 265–269 (2018)
4. Ocheja, P., Flanagan, B., Ogata, H., Oyelere, S.S.: Visualization of education blockchain data: trends and challenges. Interactive Learning Environments 0(0), 1–25 (2022)
5. Ocheja, P., Flanagan, B., Oyelere, S.S., Lecailliez, L., Ogata, H.: A prototype framework for a distributed lifelong learner model. In: 28th International Conference on Computers in Education Conference Proceedings, vol. 1, pp. 261–266. Asia-Pacific Society for Computers in Education (APSCE) (2020)

6. Ocheja, P., Flanagan, B., Ueda, H., Ogata, H.: Managing lifelong learning records through blockchain. Res. Pract. Technol. Enhanc. Learn. **14**(1), 1–19 (2019). https://doi.org/10.1186/s41039-019-0097-0
7. Zapata-Rivera, J.D., Katz, I.R.: Keeping your audience in mind: applying audience analysis to the design of interactive score reports. Assessment in Educ. Principles, Policy Practice **21**(4), 442–463 (2014)

Cognitive Diagnosis Focusing on Knowledge Components

Sheng Li[1,2], Zhenyu He[1,2], Quanlong Guan[1,2(✉)], Yizhou He[2,3], Liangda Fang[1,2], Weiqi Luo[1,2], and Xingyu Zhu[4]

[1] College of Information Science and Technology, Jinan University, Guangzhou 510632, China
lsjnu@stu2020.jnu.edu.cn, {tzhenyuhe,gql,fangld,lwq}@jnu.edu.cn
[2] Guangdong Institute of Smart Education, Jinan University, Guangzhou 510632, China
hyz@jnu.edu.cn
[3] College of Management, Jinan University, Guangzhou 510632, China
[4] Science and Technology Department of Jinan University, Jinan University, Guangzhou 510632, China

Abstract. Cognitive diagnosis is a crucial task in the field of educational measurement and psychology, which aims to diagnose the strengths and weaknesses of participants. Existing cognitive diagnosis methods only consider which of knowledge concepts are involved in the knowledge components of exercises, but ignore the problem that different knowledge concepts have different effects on practice scores in actual learning situations. This paper proposes the CDMFKC model, which considers the effect of different knowledge concepts on exercise scores, and uses neural networks to model complex interactions between students and exercise factors. We validate our model on three real datasets.

Keywords: Cognitive diagnosis · Neural network · Knowledge concepts

1 Introduction

Cognitive diagnosis aims to discover the states of students in the learning process, such as students' specific mastery of each knowledge concept [1]. Many scholars have conducted a lot of research on cognitive diagnosis, such as Deterministic Inputs, Noise And gate model (DINA) [2], Item Response Theory (IRT) [3], Multidimensional IRT (MIRT) [4] and Matrix Factorization (MF) [5]. These models have achieved some good results, but still use hand-designed interaction functions. They are a linear combination of the features of students and exercises, mostly of which are solved by parameter estimation. Hence, it cannot capture the complex connection between students and exercises.

To address this defect, some scholars designed a neural network-based cognitive diagnosis framework [6]. This framework incorporates neural networks to

© Springer Nature Switzerland AG 2022
M. M. Rodrigo et al. (Eds.): AIED 2022, LNCS 13356, pp. 311–314, 2022.
https://doi.org/10.1007/978-3-031-11647-6_59

learn the complex exercising interactions, for obtaining both accurate and inter-pretable diagnosis results.

However, we observe that these models only consider which of knowledge concepts are involved in the exercise. In actual learning situations, knowledge concepts in different knowledge components have different effects on exercise scores. Therefore, it is necessary to reshape the learning situation by combining the multi-factor relationships between knowledge concepts.

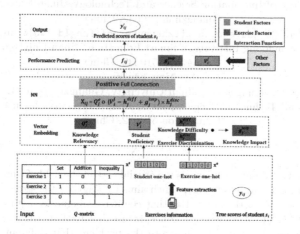

Fig. 1. Description of CDMFKC model diagnosis process.

2   Model Description

Given a learning system $\psi = (S, E, R, K, Q)$, the task of cognitive diagnosis is to obtain students' knowledge concepts proficiency. As shown in Fig. 1, we formulate the output of the CDMFKC model as:

$$y'_{ij} = PP(NN(VE(IN(x^s, x^e), Q))) \tag{1}$$

where y'_{ij} indicates the predicted score of student i in exercise j; IN, VE, NN and PP are input layer, vector embedding layer, neural network layer and performance prediction layer respectively.

Input Layer. In this layer, the information include the Q-matrix that reflects the connection between exercises and knowledge concepts, the information e of the exercise and the students' real score y_{ij} on the exercise.

Vector Embedding Layer. In this layer, we design the vector of our embedded layer based on the information provided by the input layer.

– We represent each student with a skill proficiency vector as $V_i \in (0, 1)^{1 \times K}$.

- The relationship between exercises and knowledge concepts is represented by Q-matrix, which is $Q_j^e \in (0,1)$ here.
- Knowledge difficulty $h_j^{diff} \in (0,1)^{1 \times K}$ represents the different difficulty of each knowledge concept examined by exercise e_j, and exercise discrimination $h_j^{disc} \in (0,1)$ represents the ability of exercise e_j to distinguish students with different skill proficiency levels.
- In actual scenarios, since different knowledge concepts represent different attributes, their effects on answering are mostly different. This shows that our design of knowledge concept impact parameters is able to fit real scenarios, simulate more complete learning interactions, and achieve more accurate diagnosis. The impact of knowledge concept g_j^{imp} is obtained by the joint design of exercise discrimination h_j^{disc} and knowledge concept difficulty h_j^{diff}.

Neural Network Layer. Inspired by the design of MIRT model, we connect the parameters of vector embedding layer to get our input X_{ij} and train it.

Performance Prediction Layer. In this layer, we consider the dynamic influence of some factors on the response probability. We observed that the impact of knowledge concepts indicates the magnitude of the impact of different knowledge concepts on exercise responses. However, we noticed that the previous methods (such as NCDM) used 0.5 as the limit of right and wrong judgment, which is a bit absolute in the actual teaching scene. The higher the impact of the knowledge concepts contained in exercise e_j, the higher the requirements for the proficiency of students' knowledge concepts. For example, when the average knowledge concept impact AVG_j^{imp} of exercise e_j is lower than the average knowledge concept impact AVG_{ij}^{imp} of all exercises done by student s_i, exercise e requires less knowledge and concept proficiency of students, so the judgment boundary is also reduced.

Output Layer. Finally, the output layer gets the adjusted student's prediction and answers y_{ij}', We can do some analysis at this level.

Table 1. Experimental results on predicting student grades on three real datasets.

Model	ASSIST			Math1			Math2		
	ACC	RMSE	AUC	ACC	RMSE	AUC	ACC	RMSE	AUC
DINA	0.650±.001	0.467±.001	0.676±.001	0.593±.001	0.487±.001	0.686±.001	0.592±.001	0.475±.002	0.683±.001
IRT	0.674±.002	0.464±.002	0.685±.001	0.693±.002	0.478±.001	0.705±.001	0.700±.001	0.457±.001	0.723±.001
MIRT	0.701±.001	0.461±.001	0.719±.002	0.719±.002	0.464±.001	0.733±.001	0.712±.002	0.448±.002	0.736±.002
PMF	0.661±0.001	0.476±.001	0.732±.002	0.703±.001	0.447±.001	0.712±.001	0.696±.001	0.464±.001	0.734±.001
NCDM	0.719±.008	0.439±.002	0.749±.001	0.714±.002	0.439±.001	0.796±.001	0.698±.003	0.450±.002	0.771±.001
CDMFKC	**0.762±.001**	**0.431±.001**	**0.750±.001**	**0.740±.003**	**0.437±.001**	**0.797±.001**	**0.732±.002**	**0.442±.001**	**0.778±.002**

3   Experiment

To verify the validity and explanability of the proposed model in performance prediction tasks, our experiment used three real world datasets: ASSIST [7],

Math1, and Math2 [8]. Table 1 summarize the experimental results of all models on student performance prediction tasks, with the best results shown in bold. All models are evaluated using 5-fold cross-validation, ± representing standard deviations. From Table 1, we can observe that compared with NCDM on three datasets, CDMFKC predicted an improvement of about 3.5%, 2.6% and 3.4% respectively, and performed better than all other baselines, indicating the validity of our model. On the other side, we find that models with fewer considerations perform worse (such as DINA and PMF), which also proves that fully utilizing the information of the learning system can improve the accuracy of the model.

4 Conclusion

This paper proposes a new neural network-based cognitive diagnosis model CDMFKC for cognitive diagnosis of students. Specifically, we use neural networks to obtain rich information in the learning system, introduce the impact of knowledge concepts and correlate them with exercise discrimination and knowledge difficulty, and dynamically design judgment boundaries to improve the model. Prediction results on real data sets show that our model is more accurate and interpretable. In this paper, the neural network design is relatively simple, which will be an extensible direction, and we will continue to explore.

Acknowledgements. This paper was supported by National Natural Science Foundation of China (Nos. 62077028 and 61877029), the Science and Technology Planning Project of Guangdong (Nos. 2021B0101420003, 2021A1515011873, 2020B1212030003, 2020ZDZX3013 and 2018KTSCX016), the Science and Technology Planning Project of Guangzhou (Nos. 202206030007 and 202102080307), the teaching reform research projects of Jinan University (JG2021112).

References

1. Liu, Q., et al.: Fuzzy cognitive diagnosis for modelling examinee performance. ACM Trans. Intell. Syst. Technol. **9**(4), 1–26 (2018)
2. Haertel, E.H.: Using restricted latent class models to map the skill structure of achievement items. J. Educ. Meas. **26**(4), 301–321 (1989)
3. Lord, F.: A theory of test scores. Psychometric monographs (1952)
4. Reckase, M.D.: Multidimensional item response theory models. In: Multidimensional Item Response Theory, pp. 79–112. Springer, New York (2009). https://doi.org/10.1007/978-0-387-89976-3
5. Koren, Y., Bell, R., Volinsky, C.: Matrix factorization techniques for recommender systems. Computer **42**(8), 30–37 (2009)
6. Wang, F., Liu, Q., Chen, E., Huang, Z., Chen, Y., Yin, Y., Huang, Z., Wang, S.: Neural cognitive diagnosis for intelligent education systems. In: Proceedings of the AAAI Conference on Artificial Intelligence, vol. 34, pp. 6153–6161 (2020)
7. Xiong, X., Zhao, S., Van Inwegen, E.G., Beck, J.E.: Going deeper with deep knowledge tracing. International Educational Data Mining Society (2016)
8. Wu, R., Liu, Q., Liu, Y., Chen, E., Su, Y., Chen, Z., Hu, G.: Cognitive modelling for predicting examinee performance. In: Twenty-Fourth International Joint Conference on Artificial Intelligence (2015)

Investigating Perceptions of AI-Based Decision Making in Student Success Prediction

Farzana Afrin[1(✉)], Margaret Hamilton[1], Charles Thevathyan[1], and Khalid Majrashi[2]

[1] RMIT University, Melbourne 3000, Australia
s3862196@student.rmit.edu.au,
{margaret.hamilton,charles.thevathyan}@rmit.edu.au
[2] Institute of Public Administration, Riyadh, Saudi Arabia
majrashik@ipa.edu.sa

Abstract. The current intelligent algorithms leverage various factors for predicting student success. However, concerns have been raised that the selection of predictors in the automated decision-making process may cause unfair decisions. Recent research has recognized the computational challenges and the need for developing fairness-aware intelligent algorithms. However, the unavailability of ground-truth about fair predictors of student success can make the available solutions ineffective. In this paper, we fill this need by identifying how students perceive different success predictors and the reasons behind their perceptions. We leverage three scenario-based user studies comprising of students from universities in three different countries and note the substantial perception changes across these different scenarios.

Keywords: AI in education · Algorithmic fairness · Human-centred AI

1 Introduction

Artificial intelligence applications are increasingly being used to inform decision-making in the educational sector. Predicting student performance is one of the many applications of algorithmic decision-making that enables educators to make informed decisions on students' success [1][8]. Intelligent algorithms use various predictors [6] to make accurate decisions. However, some of these factors, if considered, may be unfair to an individual or a group in various situations. For instance, the consideration of grades in high school maths may predict the students who require assistance in a introductory programming course, however, may leave out students with high grades who really need tailored support.

Prior research has pointed to the need for developing fairness-aware algorithms in education [2]. However, the lack of ground-truth (i.e., the understanding on user perceptions of predictors and their incorporation in algorithmic decision-making) may make the current algorithms ineffective [5][7]. The overall

© Springer Nature Switzerland AG 2022
M. M. Rodrigo et al. (Eds.): AIED 2022, LNCS 13356, pp. 315–319, 2022.
https://doi.org/10.1007/978-3-031-11647-6_60

aim of this project is to collect and investigate how students perceive and reason about the use of different predictors in algorithmic decision-making for the purpose of student success prediction. The identification of this missing ground-truth will help to understand and guide the development of future AI algorithms in making effective student assessments. The contributions of this paper includes:

- The formation of a user study to build a new student perception dataset from multiple institutions in three different countries.
- Analysis of perceptions through three different scenarios including a base scenario, an accuracy scenario and a discrimination scenario. In particular, we investigate participants' general perceptions on different predictors and the reasons behind those perceptions. We also observe how the perceptions may change with the changes of context (i.e., scenario).

2   Methodology

Recruitment of Participants. We recruit 1658 tertiary student participants (77.6% male, 21.9% female, 0.5% other or did not disclose their gender) from three institutions in Australia, Bangladesh and Saudi Arabia. Participation in this research was completely voluntary and anonymous.

Collecting Perceptions. We adapted the frameworks described in [3] and [4] to collect and analyse student perceptions in our study. The participants were given a set of predictors, under three scenarios, related to demographics, internal evaluation, psychometric and course content, as highlighted in the recent literature to predict student success (see Table 1). We also ask participants to identify the most and least important feature they perceive for this prediction task. They were also allowed to suggest addition features if they think are missing in the collection.

- Base scenario: asks participants about their general perception on a given predictors. In particular we ask if they believe it is fair to use a specific predictor in algorithmic decision making for prediction of their success. As a follow-up, we ask what motivates them to select this answer.
- Accuracy scenario: asks if it is fair to use a specific predictor in algorithmic decision making, if it increases the accuracy of the prediction?
- Discrimination scenario: asks about the fairness of a predictor if its inclusion may lead to falsely predicting one group of students as having a higher risk of poor performance than another group.

Perception Analysis. Given any predictor of student success, the participants rate their perception in a 5-point scale (Strongly Disagree, Disagree, Neutral, Agree, Strongly Agree). The proportion of responses are given in Table 1.

We see a mixed response in terms of agreement and disagreement for all the features. While many respondents agree to consider the features related to types of high school, rating of high school, grades obtained in high school,

Table 1. Proportion of perceptions in different scenarios

Feature	Base					Reason					Accuracy			Discrimination		
	Strongly disagree	Disagree	Neutral	Agree	Strongly Agree	Reliability	Relevance	Privacy-sensitive	Discriminatory	Other	Yes	Possibly	No	Yes	Possibly	No
Gender	.21	.11	.24	**.26**	.18	**.30**	.22	.10	.25	.14	**.42**	.25	.32	**.37**	.27	.35
Age	.12	.08	.26	**.36**	.18	.28	**.31**	.15	.11	.15	**.40**	.38	.21	**.38**	.35	.27
Parent's education	.15	.16	.26	**.29**	.15	.23	**.25**	.21	.15	.17	.32	**.34**	.34	.31	.34	**.35**
Family's financial status	.14	.11	.28	**.33**	.14	.22	**.24**	.23	.15	.15	.34	**.37**	.29	.33	**.39**	.28
Background of mathematics	.07	.05	.17	**.44**	.28	**.37**	**.37**	.07	.06	.13	**.57**	.36	.07	**.50**	.38	.12
Grades obtained in high school	.09	.13	.25	**.40**	.13	.28	**.34**	.09	.12	.18	.38	**.42**	.20	.39	**.41**	.20
Rating of the high school	.11	.14	.29	**.32**	.14	.26	**.30**	.10	.16	.19	.35	**.39**	.26	.32	**.40**	.28
Type of high school	.14	.15	.28	**.33**	.11	.25	**.29**	.08	.19	.19	.32	.30	**.38**	.33	**.37**	.29
Quiz performance	.07	.06	.17	**.45**	.25	**.37**	.36	.06	.07	.15	**.62**	.30	.07	**.57**	.34	.09
Lab performance	.07	.03	.15	**.43**	.32	**.39**	.36	.07	.05	.13	**.71**	.24	.05	**.62**	.28	.09
Attendance	.07	.05	.17	**.39**	.32	**.36**	.35	.11	.04	.14	**.65**	.29	.06	**.58**	.33	.10
Homework scores	.07	.04	.17	**.45**	.27	.31	**.40**	.07	.07	.15	**.61**	.32	.06	**.57**	.33	.09
Personality traits	.07	.07	.21	**.44**	.21	.28	**.31**	.13	.09	.19	**.53**	.37	.11	**.49**	.38	.13
Attitude towards programming	.07	.04	.16	**.39**	.34	**.37**	.34	.06	.06	.16	**.66**	.30	.05	**.59**	.32	.08
Time spent on the contents supplied	.07	.04	.18	**.43**	.27	**.36**	.34	.08	.06	.17	**.61**	.34	.05	**.57**	.34	.09
Number of entries to the contents	.07	.05	.24	**.46**	.19	.31	**.35**	.09	.06	.19	**.58**	.35	.07	**.52**	.37	.11
Attendances to live sessions	.07	.04	.23	**.41**	.24	.34	**.36**	.07	.06	.17	**.59**	.34	.07	**.51**	.39	.10
Time spent in live sessions	.07	.07	.23	**.40**	.23	.31	**.34**	.06	.08	.21	**.56**	.36	.08	**.52**	.37	.11

Table 2. Result from Wilcoxon signed-rank test

Responses in two scenarios	Statistics	p-value	Remarks
Yes - Yes	04.00	<0.01	Distribution is significantly different
Possibly - Possibly	17.00	<0.01	Distribution is significantly different
No -No	19.00	<0.01	Distribution is significantly different

family's financial status, parent's education, age and gender, the same features accounted for most of the disagreements among all the features. In relation to reason behind the perceptions, we see that a high proportion of respondents are mainly concerned about the reliability and relevance of the predictors for decision-making. Participants also raised the issues related to privacy and discriminating decisions. The factor 'age' is labelled as the most discriminating factor, followed by high school type and parent's education.

In accuracy scenario and discrimination scenario, we observed that the majority of the participants are still happy to see the given features to be considered by the decision-making algorithms. The cumulative proportion of 'yes' and 'possibly' is substantially higher than 'no' responses. However, we should note that approximately 4–38% participants responded 'no' for the given predictors in the accuracy scenario while 8–35% in the discrimination scenario. Next, a Wilcoxon signed-rank test (see Table 2) shows that the perceptions change from the accuracy scenario to discrimination scenario with statistical significance (p<0.01).

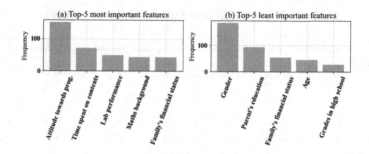

Fig. 1. Top-5 most and least important features perceived by participants

We also identified the 5 most and least important features as perceived by our participants (see Fig. 1(a-b)). Note that the family's financial status is perceived as one of the most important as well as least important features by two groups of respondents. This could be due to the groups coming from different countries with different backgrounds and values. The complexity of perceived responses given by the student respondents in our study suggests the importance of user-specific consideration in the development phase of fairness-aware algorithms.

3   Conclusion

This paper investigated students' perceptions of AI-based decision-making in the prediction of their success. We also analyze the reasons that motivate these perceptions in terms of a feature's reliability, relevance, privacy sensitiveness, and the possibility to cause discriminating decisions. We also examine if the perceptions change under different scenarios. We found that the change in the proportion of different perceived responses is statistically significant. We identify the most and least important predictors perceived by the students. Future research may include a perception analysis for different demographics such as instructors, cultural aspects and study area. We also plan to analyze the individual shift in perception under various scenarios.

References

1. Ahadi, A., Lister, R., Haapala, H., Vihavainen, A.: Exploring machine learning methods to automatically identify students in need of assistance. In: ICER, pp. 121–130 (2015)
2. Baker, R.S., Hawn, A.: Algorithmic bias in education. IJAIED, pp. 1–41 (2021)
3. Grgic-Hlaca, N., Redmiles, E.M., Gummadi, K.P., Weller, A.: Human perceptions of fairness in algorithmic decision making: a case study of criminal risk prediction. In: WWW, pp. 903–912 (2018)
4. Grgić-Hlača, N., Zafar, M.B., Gummadi, K.P., Weller, A.: Beyond distributive fairness in algorithmic decision making: Feature selection for procedurally fair learning. In: AAAI, vol. 32 (2018)

5. Marcinkowski, F., Kieslich, K., Starke, C., Lünich, M.: Implications of ai (un-) fairness in higher education admissions: the effects of perceived ai (un-) fairness on exit, voice and organizational reputation. In: FAT*, pp. 122–130 (2020)
6. Rodrigo, M.M.T., Baker, R.S., et. al.: Affective and behavioral predictors of novice programmer achievement. In: ITiCSE, pp. 156–160 (2009)
7. Van Berkel, N., Goncalves, J., Hettiachchi, D., Wijenayake, S., Kelly, R.M., Kostakos, V.: Crowdsourcing perceptions of fair predictors for machine learning: a recidivism case study. ACM HCI 3(CSCW), pp. 1–21 (2019)
8. Yu, R., Li, Q., Fischer, C., Doroudi, S., Xu, D.: Towards accurate and fair prediction of college success: evaluating different sources of student data. In: EDM (2020)

Introducing Response Time into Guessing and Slipping for Cognitive Diagnosis

Penghe Chen[1,2], Yu Lu[1,2(✉)], Yang Pian[2], Yan Li[2], and Yunbo Cao[3]

[1] Advanced Innovation Center for Future Education, Beijing Normal University,
Beijing 100875, China
luyu@bnu.edu.cn

[2] School of Educational Technology, Faculty of Education, Beijing Normal University,
Beijing 100875, China

[3] Tencent Cloud Xiaowei, Tencent Beijing Headquaters, Beijing 100193, China

Abstract. Cognitive diagnostic model (CDM) aims to estimate learners' cognitive states utilizing different techniques so that personalized educational interventions can be provided. The deterministic inputs noisy and gate (DINA) model is a fundamental CDM that estimates learners' cognitive states based on response data. However, the response time that learners used to answer test items provides rich information for cognitive diagnosis and influences the accuracy of the responses. In this work, we propose to introduce the response time into guessing and slipping of DINA model, which could better differentiate individual learner's dynamic cognitive states.

Keywords: Cognitive diagnostic model · DINA · Response time

1 Introduction

Cognitive diagnostic model (CDM) aims to estimate learners' cognitive states utilizing different techniques, and then finds out the learning obstacles of learners [2]. The earlier model is item response theory (IRT) [7] that models learners' cognitive states with a single ability variable and estimates it using logistic function based on response data. The following and more popular CDM is deterministic inputs noisy and gate (DINA) model [5]. It defines learners' cognitive states as the mastery of a set of cognitive attributes that denote knowledge skills, which are similar to knowledge states [1]. Based on the assumption that a learner can answer a test item correctly only when all the required cognitive attributes are mastered, DINA establishes a probabilistic model to make an estimation on learner's cognitive states.

The traditional DINA mainly relies on learners' response information (i.e., the correctness of their answers) to model learners without considering learners' response time. Learners' response time refers to the time learners take to answer individual item (i.e., question) in a test. Siegler advocates response time could be useful information to infer the different strategies employed by learners to solve

© Springer Nature Switzerland AG 2022
M. M. Rodrigo et al. (Eds.): AIED 2022, LNCS 13356, pp. 320–324, 2022.
https://doi.org/10.1007/978-3-031-11647-6_61

the same test item [10]. Van et al. suggest that response time could show extra insight not reflected by the response information [8]. Van and Wim further point out that response time actually affects the accuracy of learners' responses [6].

There were several previous studies considering learners' response time in the model design. Thissen proposes a linear regression model integrating the response time into the IRT model [11] and Ferrando et al. extend it to non-linear model [4]. B-GLIRT employs the two-dimensional confirmatory factor analysis to provide a more generalized IRT model [9]. A hierarchical CDM is designed to that jointly utilize learners' response answer and response time [6], which has been extended to multi-dimensional model [12].

However, no existing CDM explicitly considers the relations between learners' response time and their guessing/slipping behaviors during a test. Hence, we propose to properly introduce the response time information into the guessing and slipping parameters of DINA model. It could not only better utilize the important response time information, but also differentiate individual learner's dynamic cognitive state.

2   DINA Model and Response Time Integration

The key assumption of DINA is that one learner can answer one test item correctly only if he has mastered all the cognitive attributes required by the test item. Hence, DINA utilizes η_{ij} to represent whether learner i has mastered all the required cognitive attributes of test item j, which can be computed as:

$$\eta_{ij} = \prod_{k=1}^{K} \alpha_{ik}^{\mathbf{Q}_{jk}} \tag{1}$$

where α_{ik} denotes learner i's mastery on cognitive attribute k, $\mathbf{Q}_{jk}$ denotes whether concept k is required to answer test item j correctly.

DINA model also assumes that the learner's answering process and response are inherently affected by two factors, namely guessing and slipping. Guessing means that learners correctly answered the item by chance in a test although she or he has not mastered all the cognitive attributes associated to the item. Slipping means that learners have mastered all the necessary cognitive attributes but still falsely answered the item in a test. Given g_j and s_j as the guessing and slipping probabilities of test item j, DINA model can be expressed as:

$$P(\mathbf{R}_{ij} = 1|\alpha_i) = g_j^{1-\eta_{ij}}(1 - s_j)^{\eta_{ij}}, \tag{2}$$

where α_i is learner i's cognitive state and $\mathbf{R}_{ij}$ is learner i's response to test item j. Hence, DINA model could estimate individual learner's cognitive state by leveraging their responses and $\mathbf{Q}$-matrix. The detailed estimation process could refer to [3].

2.1   Guessing and Response Time

Intuitively, learners' successfully guessing is affected by their response time: too short response time indicates that the learner might just make a random guess;

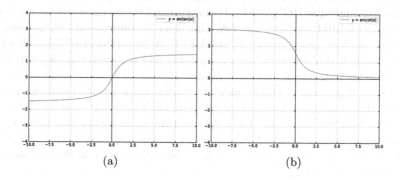

Fig. 1. Function Plots: (a) $arctan(x)$ function; (b) $arccot(x)$ function.

longer response time indicates that the learner might take more efforts to analyze the test item; too long response time might have no further help on correctly answering the item finally. Hence, it is reasonable to assume that the probability of successfully guessing is significantly increasing when learners' response time increases initially. However, the probability of successfully guessing gradually approaches to a certain limit even though further increasing the response time. To simulate this property, we propose to employ the $arctan(\cdot)$ function, which monotonically increases with its input and fulfill the feature of learner's guessing property given above. Specifically, given t as learner's response time on test item j, the probability of successfully guessing can be denoted as:

$$f_j^g(t) = u_j^g \times arctan(w_j^g * t) \tag{3}$$

where u_j^g and w_j^g are the two parameters, which determine the asymptotic value of the guessing and the varying ratio, respectively. Figure 1(a) gives a plot of $arctan(\cdot)$ function.

2.2   Slipping and Response Time

Similarly, learners' successfully slipping is affected by their response time as well: too short response time indicates that the learner might simply make a mistake; longer response time indicates that the learner might take more efforts to avoid a simple mistake; too long response time might prevent a slipping occurring. Hence, it is reasonable to assume that the probability of slipping decreases significantly when learners' response time increases initially. The probability of slipping gradually approaches to zero with the response time increases. To simulate this property, we propose to employ the $arccot(\cdot)$ function, which monotonically decreases with its input and fulfill the feature of learner's slipping property given above. Specifically, given t as learner's response time on test item j, the probability of slipping can be denoted as:

$$f_j^s(t) = u_j^s \times arccot(w_j^s * t) \qquad (4)$$

where u_j^s and w_j^s are the two parameters, which determine the highest probability of slipping and the varying ratio, respectively. Figure 1(b) gives a plot of $arccot(\cdot)$ function.

3   Discussion and Future Work

Based on the newly defined guessing and slipping functions that explicitly consider learners' response time, individual learner's performance can be better captured and modeled. For example, given two learners both correctly or falsely answering the same item in a test, their response time could help to better estimate their guessing and slipping probability and accordingly differentiate them in term of their cognitive state.

We believe this work could inspire a number of new CDM designs, which might be able to tackle the practical issues in cognitive diagnosis. The corresponding experiments need also be carefully designed to validate new models.

Acknowledgements. This research is supported by the National Natural Science Foundation of China (No. 62177009, 62077006, 62007025), the Fundamental Research Funds for the Central Universities, BNU Interdisciplinary Research Foundation for the First-Year Doctoral Candidates (Grant BNUXKJC2002), and Tencent Cloud Xiaowei.

References

1. Chen, P., Lu, Y., Zheng, V.W., Pian, Y.: Prerequisite-driven deep knowledge tracing. In: 2018 IEEE International Conference on Data Mining (ICDM), pp. 39–48. IEEE (2018)
2. von Davier, M., Lee, Y.S.: Handbook of diagnostic classification models. Springer International Publishing, Cham (2019)
3. De La Torre, J.: Dina model and parameter estimation: a didactic. J. Educ. Behav. Stat. **34**(1), 115–130 (2009)
4. Ferrando, P.J., Lorenzo-Seva, U.: An item response theory model for incorporating response time data in binary personality items. Appl. Psychol. Meas. **31**(6), 525–543 (2007)
5. Haertel, E.H.: Using restricted latent class models to map the skill structure of achievement items. J. Educ. Meas. **26**(4), 301–321 (1989)
6. van der Linden, W.J.: A hierarchical framework for modeling speed and accuracy on test items. Psychometrika **72**(3), 287–308 (2007)
7. Lord, F.: A theory of test scores. Psychometric monographs (1952)
8. van der Maas, H.L., Jansen, B.R.: What response times tell of children's behavior on the balance scale task. J. Exp. Child Psychol. **85**(2), 141–177 (2003)

9. Molenaar, D., Tuerlinckx, F., van der Maas, H.L.: A bivariate generalized linear item response theory modeling framework to the analysis of responses and response times. Multivar. Behav. Res. **50**(1), 56–74 (2015)
10. Siegler, R.S.: Hazards of mental chronometry: an example from children's subtraction. J. Educ. Psychol. **81**(4), 497 (1989)
11. Thissen, D.: Timed testing: an approach using item response theory. New Horizons in Testing, pp. 179–203 (1983)
12. Zhan, P., Jiao, H., Liao, D.: Cognitive diagnosis modelling incorporating item response times. Br. J. Math. Stat. Psychol. **71**(2), 262–286 (2018)

Engagement-Based Player Typologies Describe Game-Based Learning Outcomes

Stefan Slater[1]([⊠]), Ryan Baker[1], Valerie Shute[2], and Alex Bowers[3]

[1] University of Pennsylvania, Philadelphia, PA 19104, USA
slater.research@gmail.com
[2] Florida State University, Tallahassee, FL 32306, USA
[3] Teachers College Columbia University, New York City, NY 10027, USA

Abstract. Engagement is a strong predictor of learning in educational contexts, but the definition of engagement can vary from study to study, with small differences in definition leading to substantial differences in findings. In addition, students frequently employ strategies in online learning systems that the system designers may not have expected, which can challenge the assumptions made in these definitions. Students playing educational games employ a particularly wide variety of strategies and behaviors, which can make measuring overall engagement with the game challenging. In this study we examine student engagement by describing players' profiles of behaviors and interactions with a physics-based simulation game, Physics Playground. To identify possible sub-groups of players we use Latent Profile Analysis (LPA), a type of person-centered mixture model that assigns individuals to a set of mutually exclusive classes based on patterns of variance in a set of response data. We found support for two classes of players – high engagement players and low engagement players – and we show that students' membership in these classes is predictive of their performance on a posttest assessment. We end by discussing the limitations of this method, as well as the potential for identification and analysis of these types of player profiles to be used in adaptive game mechanics and personalization of learning contexts.

Keywords: Game-Based Learning · Mixture modeling · Engagement profile · Player typology

1 Introduction

The designer of an AIED system typically has an implicit context in which they expect the system will be used. However, individuals playing games -- educational or not -- generally seem to manifest a range of gameplay strategies and styles that can be difficult for a system designer to account for. There has been considerable research on the different ways that people orient themselves to non-educational games (Bartle 1996; Williams et al. 2008), suggesting that players have a wide range of motivations for play. Some players report that they enjoy social interaction, others immersion in a detailed world, and others in-game measures of their achievement. What a player says they enjoy doing in a game, and what they do when playing that game, however, can be very different.

© Springer Nature Switzerland AG 2022
M. M. Rodrigo et al. (Eds.): AIED 2022, LNCS 13356, pp. 325–328, 2022.
https://doi.org/10.1007/978-3-031-11647-6_62

For a deeper understanding of the different approaches that players use to engage with games, it's crucial that actual play data be used in conjunction with students' evaluations of enjoyment and motivation in play. Some research efforts have attempted to do this, as well as determine if these typologies are also seen in educational games. Slater et al. (2017) attempted to replicate Bartle's original typologies in a physics simulation game, identifying achiever and explorer classes of players, using log data collected from within the game environment during play. A third group was also identified, called disengaged players. They hypothesized that this latter group might have consisted of students who, in other games, might be socializers and/or killers, as the game studied was a nonviolent, single-player game, and there were no multiplayer elements for these types of players to partake in.

In this work, we replicate and extend Slater et al.'s study, using a later data set from the same physics simulation game. We construct a typology of game players from multiple variables created by logged game actions recorded by the system. We then use latent profile analysis to construct a typology of players in the game and measure performance differences between groups of players on a post-test assessment. By connecting learning to these typologies, we can understand how styles of play relate to learning gains in digital games.

2 Methods

We conduct this research using *Physics Playground*, an educational physics simulation game developed by Valerie Shute's lab at Florida State University (Shute et al. 2019). In Physics Playground, students draw simple machines to navigate a ball through obstacles and to a balloon somewhere else in the level. The game contains worked examples and hint-based physics lessons to help students that are having difficulty solving levels. Data were collected from 199 high school students in the southeast United States as part of a broader study on Physics Playground (for study details see Shute et al. 2021). The study took place over six days, and students spent approximately 250 min on gameplay. Student actions (e.g. menu navigation, level start, stop, and completion, objects drawn, and hints used) were recorded by the system.

Following data collection, gameplay features were constructed from the log data generated by the game. Our final feature set consisted of 8 features, covering multiple different facets of game interaction: number of gold and silver coins earned for level completion; number of unique levels and total levels visited by the player; number of machines, number of total drawings, and number of erases made by the player; and number of times the player used the learning support button. Each feature is summed across a student's entire record of play. Using a limited feature space was necessary due to the use of Latent Profile Analysis (LPA) as our modeling approach (Jung and Wickrama 2008) in the statistical software program MPlus (Muthén and Muthén 2017). In developing our features we ensured that multiple different elements of gameplay were represented, both in terms of level-to-level behaviors as well as within-level behaviors, so that our resulting groups of players would be representative of a multivariate measure of engagement in Physics Playground. We also included one distal learning outcome, consisting of students' score on an 18-item post-test that measured students' physics knowledge.

3 Results

We found statistical support for the existence of two distinct classes of Physics Playground players in our data. This model consists of two groups: (a) low engagement (Class 1, red, $n = 146$), with below-average mean scores on all measured variables, and (b) high engagement (Class 2, blue, $n = 53$), with above-average mean scores on all measured variables. Players in the two classes differed significantly for all features except for the number of gold coins earned and the number of learning supports used. In each case, the blue high engagement group had higher mean scores than the red low engagement group. This difference in engagement was apparent both in terms of overall progression, in the case of unique levels and total levels played, and in level-specific actions.

The impact of class membership on posttest learning outcomes was estimated using the BCH method within MPlus (Asparouhov and Muthén 2014). We found that students in the high-engagement group scored significantly higher on the post-test than students in the low-engagement group, with a mean difference of over half a point ($X(4) = 15.691$, $p < 0.01$).

4 Discussion and Conclusions

Engagement is important to learning both within educational games, and in learning technologies more broadly, but students can engage with the same game in different ways. In this paper, we identify subgroups of students in an educational game context using features drawn from the game log data. We interpret these subgroups in terms of their overall game engagement, and we link subgroup membership to learning outcomes. By using players' process of gameplay, rather than self-reported measures, we hope to explicitly link engagement to the actions taken within the learning context (Fincham et al. 2019).

In this study we found a single 'engaged' group, where previous work was able to differentiate between achiever students who were engaged by tangible rewards for achievement, like coins and badges, and explorer students who were engaged by exploring the rules and bounds of the game environment. There are several reasons that this finding may have failed to replicate. First, our sample was relatively underpowered for this type of analysis. Second, Physics Playground has undergone multiple design changes since 2017, and it's possible that these design changes have subsequently changed the ways that students are able to interact with the game. We also found that engaged students outperformed disengaged students on a posttest measure of physics understanding. While engagement in educational tasks has already been shown to strongly influence eventual learning outcomes, we think it's important to construct engagement in the game task as a multivariate measure of an individual's experiences within the game.

The measurement and analysis of engagement profiles represent a valuable means of informing game designers and educators on the behavioral patterns of educational game players, and how those behavioral patterns may be used to drive eventual learning (Ruiperez-Valiente et al. 2020). In this work we have demonstrated that Latent Profile Analysis can be used to generate player typologies that align with overall game engagement, and that these typologies are predictive of posttest performance. Given the increasing breadth and depth of educational games, we hope that analyses such as the one presented here see continued use in determining the best methods for engaging, supporting, and instructing learners in educational game contexts.

References

Asparouhov, T., Muthén, B.: Auxiliary variables in mixture modeling: using the BCH method in Mplus to estimate a distal outcome model and an arbitrary secondary model. Mplus Web Notes **21**(2), 1–22 (2014)

Bartle, R.: Hearts, clubs, diamonds, spades: players who suit MUDs. J. MUD Res. **1**(1), 19 (1996)

Fincham, E., et al.: Counting clicks is not enough: Validating a theorized model of engagement in learning analytics. In: Proceedings of the 9th International Conference on Learning Analytics & Knowledge, pp. 501–510 (2019)

Jung, T., Wickrama, K.A.: An introduction to latent class growth analysis and growth mixture modeling. Soc. Pers. Psychol. Compass **2**(1), 302–317 (2008)

Muthén, L.K., Muthén, B.O.: Mplus User's Guide, 8th edn. Muthén & Muthén, Los Angeles, CA (2017)

Ruiperez-Valiente, J.A., Gaydos, M., Rosenheck, L., Kim, Y.J., Klopfer, E.: Patterns of engagement in an educational massively multiplayer online game: A multidimensional view. IEEE Trans. Learn. Technol. **13**(4), 648–661 (2020)

Shute, V.J., Almond, R.G., Rahimi, S.: Physics Playground (version 1.3)[computer software]. Tallahassee, FL (2019). https://pluto.coe.fsu.edu/ppteam/pp-links

Shute, V.J., et al.: The design, development, and testing of learning supports for the Physics Playground game. Int. J. Artif. Intell. Educ. **31**(3), 357–379 (2021)

Slater, S., Bowers, A. J., Kai, S., Shute, V.: A Typology of Players in the Game Physics Playground. In: DiGRA Conference, July 2017

Williams, D., Yee, N., Caplan, S.E.: Who plays, how much, and why? Debunking the stereotypical gamer profile. J. Comput.-Mediat. Commun. **13**(4), 993–1018 (2008)

Quantifying Semantic Congruence to Aid in Technical Gesture Generation in Computing Education

Sameena Hossain[✉], Ayan Banerjee[✉], and Sandeep K. S. Gupta[✉]

Arizona State University, Tempe, USA
{shossai5,abanerj3,Sandeep.Gupta}@asu.edu

Abstract. Generation of gestures that conform to the syntax of a gestural language (such as American Sign Language (ASL)) and are congruent with the meaning of a technical term, has significant impact on enhancing the participation of people with hearing disabilities in Technical Higher Education. In this paper, we present a semantic congruity metric formulated to aid in generation of new gestures conforming to the syntax of ASL while being congruent with the meaning of the technical word and show the usage and validity of the metric using 70 ASL gestures.

Keywords: Accessible computing education · Gesture learning · Semantic congruity

1 Introduction

One of the biggest hurdles in technical education for the Deaf and Hard of Hearing(DHH) population is communicating technical terms through gestures [5]. Frequently, technical terms are finger-spelled, which does not convey the action or purpose related to the term. There have been several initiatives to generate a technical sign corpus for Computer Science (CS) including *#ASLClear* [1] or *#DeafTec* [4] or CSAVE [5] that enable the development of a repository of CS technical gestures and also educate DHH population [2]. Although such initiatives are a significant step towards a solution, they are non-curated and lack a metric to have the gestures standardized. For faster adoption and recognition, any gesture generation framework should follow the syntax of signed communication that has been established through years of interaction within the DHH population and between their hearing counterparts. A recent collaborative effort between eminent researchers has stressed the need for including American Sign Language (ASL) in Natural Language Processing research [10]. *The framework proposes a novel semantic congruity metric to evaluate conformance of new gestures to the ASL semantics.*

© Springer Nature Switzerland AG 2022
M. M. Rodrigo et al. (Eds.): AIED 2022, LNCS 13356, pp. 329–333, 2022.
https://doi.org/10.1007/978-3-031-11647-6_63

2   Semantics of ASL: Congruence and Concepts

In this section we define the semantic congruence in terms of ASL and discuss
the underlying concepts in an ASL gesture.

2.1   Word-Gesture Semantic Congruence for Creating New Gestures

Traditionally, when a gesture is not available for a certain word, a skilled ASL
user can make up a gesture that can represent the concept/s associated with the
word [7]. A DHH individual can collaborate with her interpreter to assign an
adhoc gesture for a CS technical term for which no ASL sign exists.

For example, the word
'Venn Diagram' (Fig 1), is
a mathematical term that
has no sign available in the
CS technical term reposi-
tory. A suggested gesture
can combine two concepts
related to the word Venn
Diagram: a) gesture for cir-
cular shapes, followed by
b) gesture for overlap. The
resulting combined gesture
is semantically congruent
with the word Venn Dia-
gram. We explored the con-

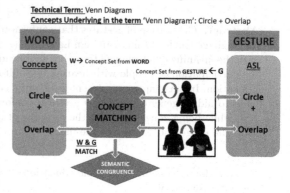

Fig. 1. Semantic congruence: venn diagram

cept of *Semantic Congruence* used in Linguistics [3,9] and define *Semantic Congruence* as referring to common gesture concepts executed for words with similar
or related meaning.

2.2   ASL Word Syntax: Gesture Expression in Terms of Concepts

We build our *Concept Set* based on the three unique modalities of ASL gestures:
1) location, 2) movement and 3) handshape. We consider the *Concept Set*, Γ,
where $\Gamma = \Gamma_H \bigcup \Gamma_L \bigcup \Gamma_M$. Here, Γ_H is the set of handshapes, Γ_L is the set of
locations and Γ_m is the set of movements [6,8].

3   Word Gesture Network and Congruity Metric

In this section we discuss the generation of word-gesture network and definition
of congruity metric.

3.1   Gesture Network Δ

A gesture network (as shown in Fig. 2) is an un-directed graph that expresses the semantic relations between a set of words by evaluating similarity in their concept identity. Each word used in the gesture network is a node. There can be three types of edges between any two node: a) handshape edge, which denotes similarity in the initial or final handshape between the two nodes, b) location edge, which denotes similarity in the initial or final location between two nodes, and c)

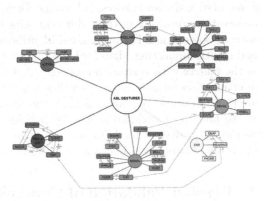

Fig. 2. Manual word-gesture network

movement edge, which denotes similarity in movement. The edge set of the gesture network can be expressed using three upper-triangular adjacency matrices: A_H, for handshape edge, A_L, for location, and A_M for movement. The entries are either 1 when the concepts match in the gestures, or 0 if they do not. We first group the videos based on a common theme for the gestures, for example-gestures for "deaf", "hearing" and "phone" are grouped in a category named "Ear". We then develop Word gesture network using the commonalities identified between the ASL gestures for different words. The graph connects words that have common execution of concepts in their ASL gestures. Two different word gesture networks are then developed based on - 1) manual observation by an ASL user, as shown in Fig. 2, and 2) automated identification.

3.2   Semantic Congruity Score ω:

This metric is defined to evaluate semantic grouping of words by two different agents. In our paper, we use it to compare the semantic grouping obtained from an expert with that obtained from the automated system. Given two gesture networks Δ_1 with adjacency matrix $\{A_H^1, A_L^1, A_M^1\}$ and Δ_2 with adjacency matrix $\{A_H^2, A_L^2, A_M^2\}$, the semantic congruity score is defined by Eq. 1.

$$\omega(\Delta_1, \Delta_2) = \sum_{\forall K \in \{H, L, M\}} \frac{\sum_i^N \sum_{j=i}^N |(a_K^1(i, j) - a_K^2(i, j))|}{\sum_i^N \sum_{j=i}^N (a_K^1(i, j) + a_K^2(i, j))} \tag{1}$$

3.3   Adjacency Matrix Creation

We create adjacency matrices for Location, Movement and Handshape using the *location*, *movement* and *handshape* edges that connect each gesture node in the Word-Gesture Network. A $45X45$ matrix is created for location, where cell value 1 is for gesture nodes with same location in manual observation. Gesture nodes that do not have a location edge connecting them, share a cell value of 0. Same matrix size and process is followed for movement and handshape in manual observation, resulting in three different adjacency matrices for gesture nodes in manual observation. For automated identification, same process is followed, and nodes with connecting edges for location, movement and handshape are included in the matrix with corresponding cell value 1, and other nodes with a 0. Three more matrices for automated identification of location, movement and handshape are thus obtained. We then calculate the semantic congruity ω, as shown in Eq. 1. We find the individual scores for handshape, location, movement and then overall congruity, ω, for both set of gestures (regular and animal). We discuss the results of these calculations in detail in Sect. 4.

4   Results: Validation of Congruity Metric

Automated Congruity Check: We compute the congruity metric for the Gesture Networks developed in Sect. 3. In addition we consider a second set of 32 videos of only animal gestures from Signing Savvy website. This new data set is used to test the proposed metric. We first create an Animal Gesture Network based on the manually observed similarities. We then run these videos through the automated recognition process and collect the location labels, movement and handshape matrices. The gesture nodes are connected with edges if they have same location labels. Using the movement and handshape matrices, we identify similar movement and handshape. A threshold of 0.73 (=73% similarity) is selected for hand shapes to be considered similar. Gesture nodes with similar movements and handshapes are then connected with edges. We also build six adjacency matrices for this data set following the process described in Sect. 3. For each of the networks: Gesture and Animal, we compute the semantic congruity between automated and manual adjacency matrices. Here, lower score represents higher congruence between the manual and the automated gesture networks, i.e. a movement congruity score of 0 would mean that manual observation and automated identification of movements in the gestures were same, and movement congruity score of 1 would mean that manual observation and automated identification of movements in the gestures were different. While computing the congruities for each of the concepts, we are expecting scores to be in between 0 to 1. And by adding these individual congruities, we expect the overall congruity score, ω, to be in between 0 to 3.

We also compare across the hand-shape, location and movement congruity scores between regular and animal gestures. Figure 3 shows the computed semantic congruity scores, as per Eq. 1. The results show that with respect to location and movement, automated recognition mechanism can capture the semantic relations between the words. This is reflected in the low values of the semantic congruity score for

Network	Handshape Congruity	Location Congruity	Movement Congruity	Overall Congruity
Regular Gestures	0.71	0.09	0.27	1.07
Animal Gestures	0.65	0.04	0.3	0.99
Regular Vs Animal Gestures	0.96	0.7	0.8	2.46

Fig. 3. Semantic congruity results

each of these concepts. In comparison between regular and animal specific gesture concepts, we see a higher congruity value- reflecting that there is significant difference between the concepts of regular and animal gestures.

5   Conclusion and Future Work

This paper provides a metric to quantify the semantic congruence between two gestures. Results obtained from testing the metric with two different data sets show that the automated congruity checker proposed in the paper agrees with manual observation. We intend to evaluate an iterative gesture generation mechanism as a continuation of this work.

References

1. ASL Clear: The learning center for the deaf (2020). https://www.tlcdeaf.org/asl-clear. Accessed 14 Sep 2020
2. Banerjee, A., Lamrani, I., Hossain, S., Paudyal, P., Gupta, S.K.S.: AI enabled tutor for accessible training. In: Bittencourt, I.I., Cukurova, M., Muldner, K., Luckin, R., Millán, E. (eds.) AIED 2020. LNCS (LNAI), vol. 12163, pp. 29–42. Springer, Cham (2020). https://doi.org/10.1007/978-3-030-52237-7_3
3. Breal, M.: Essai de sémantique: science des significations (1897)
4. DeafTEC: Deaftec stem sign video dictionary (2020). https://deaftec.org/stem-dictionary/. Accessed 14 Dec 2020
5. Hossain, S., Banerjee, A., Gupta, S.K.S.: Personalized technical learning assistance for deaf and hard of hearing students. In: Workshop on Artificial Intelligence for Education, AAAI 2020, New York, New York, USA (2020)
6. Hossain, S., Kamzin, A., Amperayani, V.N.S.A., Paudyal, P., Banerjee, A., Gupta, S.K.S.: Engendering trust in automated feedback: a two step comparison of feedbacks in gesture based learning. In: Roll, I., McNamara, D., Sosnovsky, S., Luckin, R., Dimitrova, V. (eds.) AIED 2021. LNCS (LNAI), vol. 12748, pp. 190–202. Springer, Cham (2021). https://doi.org/10.1007/978-3-030-78292-4_16
7. Lifeprint: Finger Spelling for asl (2016). http://www.lifeprint.com/asl101/fingerspelling/fingerspelling.htm. Accessed 15 Apr 2016
8. Paudyal, P., Lee, J., Kamzin, A., Soudki, M., Banerjee, A., Gupta, S.K.: Learn2sign: explainable ai for sign language learning. In: Explainable Smart Systems Workshop in ACM Intelligent User Interfaces Conference (2019)
9. Secora, K., Emmorey, K.: The action-sentence compatibility effect in asl: the role of semantics vs. perception. Lang. Cogn. **7**(2), 305–318 (2015)
10. Yin, K., Moryossef, A., Hochgesang, J., Goldberg, Y., Alikhani, M.: Including signed languages in natural language processing. In: Proceedings of the 59th Annual Meeting of the Association for Computational Linguistics (Volume 1: Long Papers), pp. 7347–7360. Association for Computational Linguistics, Online, August 2021

A Multi-pronged Redesign to Reduce Gaming the System

Yuanyuan Li[1], Xiaotian Zou[1], Zhenjun Ma[1], and Ryan S. Baker[2]([✉])

[1] Learnta, Inc., 1460 Broadway Street, New York, NY 10036, USA
will@learnta.com
[2] University of Pennsylvania, 3700 Walnut Street, Philadelphia, PA 19104, USA
rybaker@upenn.edu

Abstract. Despite almost two decades of interest in reducing gaming the system in interactive learning environments, gaming continues as a key factor reducing student learning outcomes and contributing to poorer learning outcomes. In this study, we redesigned the Kupei learning system by implementing a combined set of three interventions aimed at mitigating the impact of the two gaming behaviors we documented. Our results show evidence of a possible positive effect of the combined gaming prevention intervention at reducing the second type of gaming behavior within our system, however, it was not as successful at mitigating the first type of gaming behavior.

Keywords: Gaming the system · Learning engineering · Iterative redesign · Interactive learning environments

1 Introduction

Interactive learning environments are intended to create opportunities for students to learn but a substantial proportion of students choose instead to game the system, attempting to succeed by taking advantage of the regularities and properties of a system rather than by learning the material [1]. Considerable research has demonstrated negative correlations between gaming the system and student outcomes [2, 3].

Over the last 15 years, a range of interventions have been proposed and investigated. Several research groups attempted to mitigate the impact of gaming, in a more subtle fashion when gaming occurs and then adapt in real time to detection of gaming behavior [4, 5]. Other approaches attempted to prevent gaming behavior in the first place by adding delays or minimum amount of wait time between two actions [6, 7]. Some of these approaches [4, 5] improved learning outcomes but were not adopted at scale, while others may also hinder the usefulness of help-seeking for non-gaming students [7].

In this paper, we investigate a multi-pronged redesign of an AIED system, using three interventions in tandem to reduce students' propensity to game. We conduct a within-system quasi-experiment, investigating whether the redesigned version of the system leads to reduced gaming behavior and better within-system performance.

© Springer Nature Switzerland AG 2022
M. M. Rodrigo et al. (Eds.): AIED 2022, LNCS 13356, pp. 334–337, 2022.
https://doi.org/10.1007/978-3-031-11647-6_64

2 Method

2.1 Platform

Our intervention was developed in the context of the Kupei learning platform that supports the learning of math, English and science subjects. Rather than teacher-led instruction, the system uses algorithms that can automatically determine which content a student should work on next.

With the Kupei learning system, students usually take less than three practice sets to achieve basic mastery (probability of mastery falls between 80% and 95%) of each concept. Therefore, we define practices on the same concept after three practice sets as extra practices. We believe that a considerable proportion of extra practice will be the result of either gaming the system or struggling with the content.

Kupei uses Bayesian Knowledge Tracing (BKT) [8] to estimate student proficiency in real-time. When studying a concept using the Kupei system, Kupei assesses students' probability of mastering a concept after the first 3 items are completed. If the probability falls between 80% (a cut-off used by many commercial systems) and 95% (the original cut-off in) [8], then the concept is labeled as basic mastery, and the student will continue to work on two additional items. If the mastery probability is more than 95% (advanced mastery), then the system stops and advances the student to the next concept. If a student's probability of knowing a concept is less than 80% after the first three items are completed, the concept is labeled as unmastered, and the practice stops and displays the result. Students who did not master the concept (whether after 3 or 5 problems) are next required to complete an integrated review on the same concepts/skills (involving video and/or lecture notes). In all situations, the learning recommendation offered after each concept will change according to students' performance during the practice.

2.2 Gaming Behaviors

Prior to the integration of the gaming prevention intervention, gaming behaviors typically observed in the Kupei system can be divided into two types:

1. Students use an exhaustive method to obtain the correct answers of the practice sets by inputting random answers for each question of each practice set until earlier questions are re-shown.
2. Students open a practice set to obtain the set of questions, then quit the practice set midway to seek answers elsewhere.

2.3 Design and Method

Our design aimed to simultaneously accomplish two goals: first, by increasing the costs of gaming, it is hoped that students will game the system less often; second, with less gaming behaviors, we hope that students will engage in more productive behaviors and learn more effectively.

Aiming to achieve these objectives, we designed three gaming prevention interventions: first, we re-designed the system so that students may not complete more than two practice sets (of five problems each) on a concept more than three times a day, with a pause of 36 h before they can work on a concept again. Second, Kupei now provides meta-cognitive feedback which acts as a reminder to the students about the cost of gaming -- if they now game, they will have to wait 36 h [5]. Third, the system now requires students who responded too quickly and failed to reach basic mastery to complete an integrated review on the same concepts/skills (involving video and/or lecture notes).

A within-subjects quasi-experiment was conducted comparing two 15-day math learning periods -- a control period before the new strategy for reducing gaming behavior was adopted, and an experimental period immediately following adoption of the new strategy within the system. We analyzed data (i.e. interaction logs) from a total of 343 students who studied at least 10 math concepts in both two periods.

3 Results

3.1 Frequency of Gaming Behavior by Condition

In this study, 93 students were control-gamers (they gamed the system during the control period) and 250 students were control-non-gamers (they did not game the system during the control period). After the gaming prevention interventions were integrated, the average gaming frequency per student decreased from 0.124 in control period to 0.064 in experimental period, a statistically significant difference, $V = 5411$, $p < 0.01$.

In terms of specific behaviors, there was a statistically significant reduction in gaming by quitting to seek answers, from 0.085 during control period to 0.031 during experimental period, $V = 2511$, $p < 0.01$. However, there was not a significant reduction in gaming by memorizing answers, from 0.040 during the control period to 0.032 during the experimental period, $V = 2086$, $p = 0.51$.

3.2 Other Behavior Changes

The Proportion of Extra Practice. The average proportion of extra practice decreased from 12.20% in the control period to 7.66% in the experimental period, which is statistically significant, $V = 25588$, $p < 0.01$. The proportion of extra practice in the control-gamers decreased from 23.1% in the control period to 13.4% in the experimental period, while for control-non-gamers, the proportion of extra practice decreased from 8.1% control to 5.5% experimental. According to a Wilcoxon rank sum test, the control-gamers' decrease in extra practice is significantly steeper than the control-non-gamers, $W = 7928$, $p < 0.01$.

Average Time Spent Per Item. Starting from the second practice set, there was a statistically significant increase in the average time spent on each item in the experimental period compared to the time spent in the control period, especially in the second practice set. The average time spent per item in the first practice set decreased from 98.87 s in the control period to 91.06 s in the experimental period, which is significantly different, $t(342) = 3.32$, $p = 0.001$ for a paired t-test. In the second practice set, the average time a student spent answering each math item increased from 79.66 s in the control period to 116.90 s in the experimental period, which is statistically significant, $t(339) = 12.75$, p < 0.01. In the third practice set, the average time a student spent on each math item increased from 85.05 s in the control period to 97.41 s in the experimental period, which is statistically significant, $t(216) = 2.98$, $p < 0.01$.

4   Discussion and Conclusions

In this paper, we attempted to address the two gaming behaviors that we documented within Kupei learning platform. We found that the multi-pronged gaming prevention intervention appear to have been successful at dissuading students from gaming the system. We detected a lower frequency of gaming behaviors in learning math after the integration of the gaming intervention. In addition, we found that fewer students used extra practice on a concept after the implementation of the gaming intervention. Instead, students spent more time on later items during the experimental period, possibly indicating students are practicing each item more seriously than students in the control period.

However, there appear to be some limitations to this approach that should be considered in future work. The intervention was not successful at reducing at addressing the first type of gaming behavior. Another possible limitation is that even if some students reduce their frequency of gaming the system, they may not replace gaming with the most desirable behaviors.

Ultimately, we hope that our research will inform the design of systems that will reduce students' motivation to game the system, and, in turn, increase the frequency of effective self-regulated learning strategies that lead to better student learning.

References

1. Baker, R.S., Corbett, A.T., Koedinger, K.R., Wagner, A.Z.: Off-task behavior in the cognitive tutor classroom. In: Proceedings of the 2004 Conference on Human Factors in Computing Systems - CHI 2004, pp. 383–390 (2004)
2. Aleven, V., McLaren, B.M., Roll, I., Koedinger, K.: Toward meta-cognitive tutoring: a model of help seeking with a cognitive tutor. Int. J. Artif. Intell. Educ. **16**, 101–128 (2006)
3. Pardos, Z.A., Baker, R.S., San Pedro, M.O.C.Z., Gowda, S.M., Gowda, S.M.: Affective states and state tests: Investigating how affect and engagement during the school year predict end of year learning outcomes. J. Learn. Anal. **1**(1), 107–128 (2014)
4. Baker, R., et al.: Adapting to when students game an intelligent tutoring system. In: Ikeda, M., Ashley, K.D., Chan, T.-W. (eds.) ITS 2006. LNCS, vol. 4053, pp. 392–401. Springer, Heidelberg (2006). https://doi.org/10.1007/11774303_39
5. Arroyo, I., et al.: Repairing disengagement with non-invasive interventions. In: Proceedings of the International Conference on Artificial Intelligence in Education, pp. 195–202 (2007)
6. Murray, R.C., Vanlehn, K.: Effects of dissuading unnecessary help requests while providing proactive help. In: Proceedings of the International Conference on Artificial Intelligence in Education, pp. 887–889 (2005)
7. Aleven, V., McLaren, B., Roll, I., Koedinger, K.: Toward tutoring help seeking. In: Lester, James C., Vicari, Rosa Maria, Paraguaçu, Fábio. (eds.) ITS 2004. LNCS, vol. 3220, pp. 227–239. Springer, Heidelberg (2004). https://doi.org/10.1007/978-3-540-30139-4_22
8. Corbett, A.T., Anderson, J.R.: Knowledge tracing: modeling the acquisition of procedural knowledge. User Model. User-Adap. Inter. **4**(4), 253–278 (1995)

Learning Optimal and Personalized Knowledge Component Sequencing Policies

Fuhua Lin[1]([⊠]), Leo Howard[1], and Hongxin Yan[2]

[1] Athabasca University, Athabasca, Canada
{oscarl,lhoward}@athabascau.ca
[2] University of Eastern Finland, Kuopio, Finland
hongya@student.uef.fi

Abstract. One of the goals of adaptive learning systems is to realize adaptive learning sequencing by optimizing the order of learning materials to be presented to different learners. This paper proposes a novel approach to recommending optimal and personalized learning sequences for learners taking an online course based on the contextual bandit framework where the background knowledge of the learners is the context. To improve learning efficiency and performance of learners, the adaption engine of such an adaptive learning system can select an optimal learning path for a learner by continually evaluating the learners' progress as the course advances. To overcome the complexity of learning path recommendation due to the large number of knowledge components, we use the 'divide-and-conquer' approach to modeling the domain and designing the sequence adaptation algorithm. Also, the adaptation engine can dynamically replan the learning path for a learner if her/his performance is worse than expected. Finally, our approach can improve over time by learning from the experience of previous learners who adopted recommended sequences.

Keywords: Sequence adaptivity · Bandit algorithms · Learning path recommendation · Adaptive learning

1 Introduction

A learning path is a sequence of learning units or learning objects that guides the learners to accomplish the learning goals while satisfying the prerequisite relationships among the units [1, 2]. The needs for sequence adaptivity - recommending optimal and personalized learning paths to learners who have various backgrounds and dynamic performance - have been identified by many researchers [3]. However, according to Doroudi et al. (2019) [4], the studies on learning content sequencing have been the least successful compared to other types of studies, such as sequencing concept-level questions and sequencing learning activities. There are three main challenges of sequence adaptivity. First, learners vary tremendously in backgrounds and already mastered knowledge. As a result, a same recommendation policy is unlikely to provide every learner with the best learning experience. Second, learning sequencing requires dealing with a large

© Springer Nature Switzerland AG 2022
M. M. Rodrigo et al. (Eds.): AIED 2022, LNCS 13356, pp. 338–342, 2022.
https://doi.org/10.1007/978-3-031-11647-6_65

decision space which grows combinatorically with the number of learning units and learning objects. A practical learning system requires the optimal learning path to be recommended in an acceptable period. Third, since the knowledge, experience, and performance of a learner develops and evolves in the process of learning, learners may arrive at different completion status of a course at a given time, depending on their performance in the finished learning units. Hence, a replanning process is needed to recommend alternative sequence for the remaining content.

The goal of this research is to propose a novel approach to recommending learning sequences. In our approach, the adaptation agent can continuously learn from the data about the performance of previous learners who adopted various sequences and replanned learning paths to improve the quality of learning path recommendation.

2 The Proposed Methodology

The proposed architecture of adaptive learning systems is shown in Fig. 1.

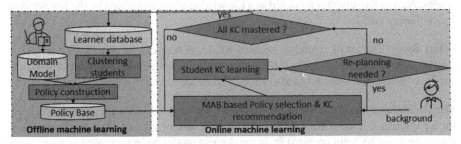

Fig. 1. The proposed architecture of adaptive learning systems.

Domain Modeling: We first construct a domain model using a two-layer tree structure as shown in Fig. 2. The root node of the tree represents the whole domain, denoted DM. DM consists of a set of knowledge units (KUs), which is the second layer of the tree. Because prerequisite relations often exit among KUs, a graph can be formed as KU-graph. Each KU (the tree node at the second layer) can consist of a set of knowledge points (KPs), which are usually interdependent. One way of modeling these KPs is modeling them as learning objectives (LOs) using the Bloom taxonomy [5]. In this research, we call these KUs and KPs the knowledge components (KCs) [6].

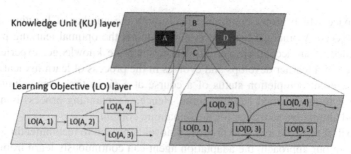

Fig. 2. The domain model.

To represent the KCs and their prerequisite or dependency relationships, a graph is usually used. There are two types of prerequisites: AND and OR. AND captures the combined prerequisite relationship while OR captures the alternative prerequisite relationship. Thus, we can use an AND/OR graph structure to represent the elements and their dependencies in a tree node of the domain model. These nodes are called KC-graphs. Next, a KC-graph is modeled as a knowledge space suggested by the Knowledge Space Theory (KST) [7].

Policy Base: The set of candidate optimal KC sequences can be obtained by selecting a number of the candidate optimal KC sequences with the highest weighted sum of two attribute values. The selected candidate sequences are stored in a policy base.

The Divide-and-Conquer Method: Some KCs within a domain may not be required for a particular course. Thus, the knowledge structure of a course can be obtained by removing those KCs not required from the domain model but represented as a tree with the same structure of the domain model. As a result, each node of the tree for the course is a sub-graph of the corresponding graph of the domain model. Thus, the sequence adaptivity problem for a whole course becomes path sequencing in the KU-graph and KC-graph, which are for the KUs and the KPs, respectively. In this way, we can construct a 'divide-and-conquer' algorithm for learning path sequencing to reduce the complexity of the problem. Because the structure of these graphs is the same, we can only focus on the sequence adaptivity problem for KC-graph.

Learning Paths: A learning path is also called a policy or a learning sequence. Formally, all KCs in a KC-graph are denoted as $KC = \{1, 2, \ldots, n\}$, n is the total number of the KCs of the course. We use Π to denote the set of KC recommendation policies. A policy $\pi \in \Pi$ is represented as $\pi = (e_1, e_2, \ldots, e_n)$, which is a permutation of $\{1, 2, \ldots, n\}$. Each e_i has two attributes: learning time $T(e_i)$ and proficiency value $P(e_i)$. And then, a learning path has two important attributes to describe its quality: learning time $T(\pi) = \frac{1}{n} \sum_{i=1}^{n} T(e_i)$ and proficiency value $P(\pi) = \frac{1}{n} \sum_{i=1}^{n} P(e_i)$.

Candidate Optimal Sequences: From the knowledge space of a KC-graph, we first generate all the possible learning paths. And then, we select a number of the candidate optimal KC sequences with the highest weighted sum of two attribute values: $P(\pi)$ and $1/T(\pi)$. The selected candidate sequences are stored in a policy base.

The Learner Model: Each learner is represented by a list of grades obtained in prerequisite courses, the selected π, and $\{P(e_i)\}_1^n$ and $\{T(e_i)\}_1^n$. By recording and updating the sample mean completion times and competency levels of the KCs in the sequence once a learner completes the course by adopting the sequence, the prior knowledge of the algorithms will be learnt by the system.

3 The Algorithms

3.1 Modeling with the Bandit Framework

We use the contextual bandit algorithm to select and recommend learning paths [8]. The set of candidate optimal KC sequences in the policy base are used as *arms* or *actions*. Let $\pi(s) = (e_1(s), e_2(s), \ldots, e_n(s))$ be the policy that is used on learner s. The system observes the realized reward $r_{e_1(s)}, r_{e_2(s)}, \ldots, r_{e_n(s)}$ along each edge in the path, here $r_{e_j(s)} = \gamma \times \frac{1}{T(e_j(s))} + \delta \times P(e_j(s))$, γ and δ are the weights reflecting the relative importance of these two contributing factors of rewards: time and proficiency. Also, $\gamma + \delta = 1, 0 \leq \gamma \leq 1, 0 \leq \delta \leq 1$.

To personalize the sequence for each learner according to his/her educational background such as the scores of the prerequisite course, we assume that learners with the similar background will achieve similar learning outcomes if they follow the same KC sequence. A clustering algorithm is constructed to adaptively cluster learners and refine the clustering as more learners are enrolled and complete the course. We are implementing several multi-armed bandit (MAB) algorithms such as greedy, Thompson Sampling (TS), UCB, or ε-greedy [8]. Simulation is conducted in our research to benchmark all of them against a random selection algorithm, as well as against each other. This would help us identify the most effective MAB algorithm.

3.2 Replanning

A replanning algorithm is triggered when a learner encounters a KC in an adopted learning path that is too difficult to complete, or the cumulative reward is much lower than expected. When replanning is triggered, the system compares the learner's current path with similar neighbor paths that have the same backward segments. After that, it uses an MAB algorithm to choose the shortest potential path, then evaluates the new path's upcoming *kc* for an acceptable delay. The algorithm iterates over similar neighbor paths until it finds a suitably shorter path. By simulating the replanning process, we can obtain greater learning outcomes for a learner.

4 Conclusion and Future Work

Using the contextual bandit framework, we have proposed an approach to recommending optimal knowledge component sequences to learners with different backgrounds. A replanning mechanism was designed to respond to the difference between expected performance and actual performance of a learner who followed the recommended sequence. The clustering algorithm would also allow further personalization. We are developing an experimental learning system to test the effectiveness of the approach with actual learners.

References

1. Adorni, G., Koceva, F.: Educational concept maps for personalized learning path generation. In: Adorni, G., Cagnoni, S., Gori, M., Maratea, M. (eds.) AI*IA 2016. LNCS (LNAI), vol. 10037, pp. 135–148. Springer, Cham (2016). https://doi.org/10.1007/978-3-319-49130-1_11
2. Muhammad, A., Zhou, Q., Beydoun, G., Xu, D., She, J.: Learning path adaptation in online learning systems. In: 2016 IEEE 20th International Conference on Computer Supported Cooperative Work in Design (CSCWD), Nanchang, China, pp. 421–426. IEEE (2016)
3. Nabizadeh, A.H., Leal, J.P., Rafsanjani, H.N.: Learning path personalization and recommendation methods: a survey of the state-of-the-art. Expert Syst. Appl. **113596**, 159 (2020)
4. Doroudi, S., Aleven, V., Brunskill, E.: Where's the Reward? Int. J. Artif. Intell. Educ. **29**(4), 568–620 (2019). https://doi.org/10.1007/s40593-019-00187-x
5. Bloom, B.S.: Taxonomy of Educational Objectives: The Classification of Educational Goals; Handbook I: Cognitive Domain. David McKay, New York (1956)
6. Brusilovsky, P.: Adaptive hypermedia for education and training. In: Lesgold, A.M., Durlach, P.J. (eds.) Adaptive Technologies for Training and Education, pp. 46–66. Cambridge University Press, Cambridge (2012)
7. Doignon, J.P., Falmagne, J.C.: Knowledge structures and spaces. In: Knowledge Spaces. Springer, Heidelberg (1999). https://doi.org/10.1007/978-3-642-58625-5_2
8. Lattimore, T., Szepesvári, C.: Bandit Algorithms, 1st edn.. Cambridge University Press (2020). https://doi.org/10.1017/9781108571401

Automatic Riddle Generation
for Learning Resources

Niharika Sri Parasa[✉][iD], Chaitali Diwan[iD], and Srinath Srinivasa[iD]

International Institute of Information Technology, Bangalore, India
{niharikasri.parasa,chaitali.diwan,sri}@iiitb.ac.in

Abstract. This paper proposes a novel approach to automatically generate conceptual riddles, with an objective of deployment in online learning environments. The riddles are generated by creating triples from the learning resources using BERT language model, which are fed to the k-Nearest Neighbors language model to identify the proximity between properties and their respective contexts. These properties are classified into Topic Markers and Common based on their uniqueness and modeled on an effective instructional strategy called as Concept Attainment Model. Each riddle is passed through the Validator Module that stores all possible answers for the riddles and is used to verify the learner's answers and provide them hints. The riddles generated by our model were evaluated by human evaluators and we obtained encouraging results.

Keywords: Riddle generation · Triples creation · Language models

1 Introduction and Background

Activity-based learning is achieved by adopting instructional practices that encourage learners to think about what they are learning [8]. One such instructional strategy in pedagogy that is shown to be effective across domains [7] is the Concept Attainment Model (CAM) [5].

The CAM promotes learning through a process of *structured inquiry*. This model is designed to lead learners to a concept by requiring them to analyse the examples that contain the attributes of the concept i.e., positive examples, along with the examples that do not contain these attributes i.e., negative examples. An engaging and fun way to present this model to the learners is by structuring the CAM in the form of riddles.

Although Riddle solving in learning environments motivates and interests the learner rather than just reading [2], most of the previous works [3,4,10,11] on riddle generation are addressed in the context of computational creativity/humor. However, apart from the fact that our approach is backed by an effective instructional strategy, it also has an unique methodology of building riddles by identifying and distinguishing semantically closer concepts based on their properties using the pre-trained language model BERT and k-Nearest Neighbor model.

Supported by Gooru (https://gooru.org).

M. M. Rodrigo et al. (Eds.): AIED 2022, LNCS 13356, pp. 343–347, 2022.
https://doi.org/10.1007/978-3-031-11647-6_66

2   Approach

Our proposed method of Riddle generation includes four modules: Triples Creator, Properties Identifier, Generator followed by Validator as shown in Fig. 1. Each learning resource is passed as an input to our Triples Creator module which first extracts noun phrases, adjectives, verbs and phrases comprising of noun and adjectives as attributes/properties associated with a concept. Then the concept and its associated properties are arranged by masking the relation part as follows: concept <mask> property. The masked token is then predicted using Bert-Uncased whole word masking language model [1][1]. For example: dog <mask> bark returns dog can bark, constructing simple and complete sentences. Refer Triples Creator column in Table 1.

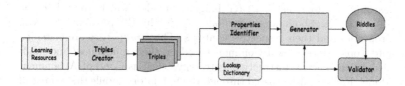

Fig. 1. Architecture of the proposed riddle generation approach

Consequently a *Lookup Dictionary* is created where keys are concepts and values are the list of triples along with their respective properties.

These triples are fed to the *Properties Identifier* module where the properties are classified into *Topic Markers* and *Common*. *Topic Markers* are the properties that explicitly represent a concept [9] and *Common* property is associated with more than one concept.

We use the k-Nearest Neighbor's Language model [6], which uses a data store and a binary search algorithm KDTree[2] to query the neighbours of the target token given its context. Each triple of a concept along with its respective property are passed as queries to the model, returning the distances, neighbours, and their contexts. If all the contexts relate to the target concept, then the triple is categorized as *Topic Marker*, otherwise, it is categorized as *Common*. Subsequently, neighbouring concepts with common properties are extracted for further use. Refer Properties Identifier column in Table 1.

The Generator module creates riddles through a Greedy mechanism which creates combinations of triples, either of class Topic Marker or Common as positive examples. Riddles generated from *Topic Markers* of a concept are termed as *Easy Riddles* and those from *Common* are termed as *Difficult Riddles*.

Difficult Riddles accommodate both positive and negative examples of a concept. So, to generate negative examples for those respective positive examples in 2 versions, the module uses Lookup Dictionary utilizing formerly extracted

---

[1] https://huggingface.co/bert-large-uncased-whole-word-masking.
[2] https://scikit-learn.org/stable/modules/generated/sklearn.neighbors.KDTree.html.

Table 1. Outputs from triples creator, properties identifier, generator and validator

Triples creator	Property identifier		Generator	Validator
	Class	Neighbouring concepts		
Dog can guard your house	Topic marker		**Easy:** I can guard your house. I can bark. I am a loyal friend. Who am I?	Dog
Dog can bark	Topic marker			
Dog is related to canine	Common	Fox, wolf, bear		
Dog is a mammal	Common	Elephant, lion, tiger	**Difficult(v1):** I am related to animals but I am not elephant. I am a pet but I am not a rabbit. I am related to flea but I am not a cat. Who am I?	Dog, ferret, …
Dog is related to flea	Common	Cat, bee, louse		
Dog is a loyal friend	Topic marker			
Dog is a pet	Common	Cat, rat, rabbit		
Dog is a animal	Common	Tiger, fox, elephant		
Dog has four legs	Common	Elephant, rabbit	**Difficult (v2):** I am related to animals but I don't have a trunk. I am a pet but I don't like carrots. I am related to flea but I am not feline. Who am I?	Dog, ferret, …
Dog is for companionship	Common	Animals, cat, fish		
Dog wants a bone	Topic marker			
Dog can run	Common	Cheetah, horse, rat		
Dog is related to a kennel	Common	Ferret, rabbit		

neighbouring concepts and their properties. Some examples of the generated riddles can be seen in Generator column in Table 1.

The generated riddles can have one or more answers. So, each riddle is passed through the *Validator* which generates and stores all possible answers to validate learners' answers and provide hints. Refer Validator column in Table 1.

3 Experiment and Results

We use a dataset of 200 open learning resources of the zoology domain comprising free-text curated from Wikipedia[3]. We had 30 human evaluators that are presented with a sample of 20 riddles both easy and difficult along with multiple-choice options and hints, i.e., topic markers.

---

[3] https://github.com/goldsmith/Wikipedia.

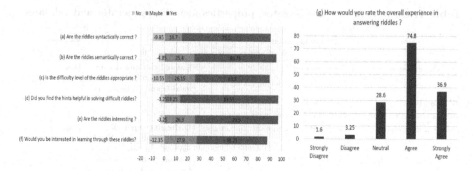

Fig. 2. Evaluation results

Our evaluation approach targets to assess the quality of the riddles (syntactic, semantic, and difficulty level), engagement, informativeness and whether they are fit for learning using 3 point Likert scale. It also captures the overall experience of answering riddles using 5 point Likert scale. As shown in Fig. 2(a), (b) ≈70%–75% of the evaluators agreed that the generated riddles are semantically and syntactically correct respectively. From Fig. 2(e), (c), (f), ≈70% of the evaluators agreed that the riddles are interesting and ≈60% agreed on the difficulty level and their adaptability in learning. More than 70% of the evaluators agreed the experience of answering riddles to be good. (Refer to Fig. 2(g)).

4   Conclusions and Future Work

We presented a novel approach to automatically generate concept attainment riddles given a representative set of learning resources. The results obtained from our evaluation are encouraging. As part of future work, we plan to use the generated riddles to test the concept understanding of the learner.

References

1. Devlin, J., Chang, M., Lee, K., Toutanova, K.: BERT: pre-training of deep bidirectional transformers for language understanding. CoRR abs/1810.04805 (2018). http://arxiv.org/abs/1810.04805
2. Doolittle, J.H.: Using riddles and interactive computer games to teach problem-solving skills. Teach. Psychol. **22**(1), 33–36 (1995)
3. Galván, P., Francisco, V., Hervás, R., Méndez, G.: Riddle generation using word associations. In: Proceedings of the Tenth International Conference on Language Resources and Evaluation (LREC 2016), pp. 2407–2412 (2016)
4. Guerrero, I., Verhoeven, B., Barbieri, F., Martins, P., Pérez y Pérez, R.: TheRiddlerBot: a next step on the ladder towards computational creativity. In: Toivonen, H., et al. (eds.) Proceedings of the Sixth International Conference on Computational Creativity, pp. 315–322 (2015)
5. Joyce, B., Weil, M., Calhoun, E.: Models of teaching (2003)

6. Khandelwal, U., Levy, O., Jurafsky, D., Zettlemoyer, L., Lewis, M.: Generalization through memorization: nearest neighbor language models. In: International Conference on Learning Representations (2020). https://openreview.net/forum?id=HklBjCEKvH
7. Kumar, A., Mathur, M.: Effect of concept attainment model on acquisition of physics concepts. Univ. J. Educ. Res. 1(3), 165–169 (2013)
8. Prince, M.: Does active learning work? A review of the research. J. Eng. Educ. 93(3), 223–231 (2004)
9. Rachakonda, A.R., Srinivasa, S., Kulkarni, S., Srinivasan, M.: A generic framework and methodology for extracting semantics from co-occurrences. Data Knowl. Eng. 92, 39–59 (2014)
10. Ritchie, G.: The jape riddle generator: technical specification. Institute for Communicating and Collaborative Systems (2003)
11. Waller, A., Black, R., O'Mara, D.A., Pain, H., Ritchie, G., Manurung, R.: Evaluating the standup pun generating software with children with cerebral palsy. ACM Trans. Access. Comput. (TACCESS) 1(3), 1–27 (2009)

SimStu-Transformer: A Transformer-Based Approach to Simulating Student Behaviour

Zhaoxing Li[✉][ID], Lei Shi[ID], Alexandra Cristea[ID], Yunzhan Zhou[ID], Chenghao Xiao[ID], and Ziqi Pan[ID]

Department of Computer Science, Durham University, Durham, UK
{zhaoxing.li2,lei.shi,alexandra.i.cristea,yunzhan.zhou,chenghao.xiao,
ziqi.pan2}@durham.ac.uk

Abstract. Lacking behavioural data between students and an Intelligent Tutoring System (ITS) has been an obstacle for improving its personalisation capability. One feasible solution is to train "sim students", who simulate real students' behaviour in the ITS. We can then use their generated behavioural data to train the ITS to offer *real students* personalised learning strategies and trajectories. In this paper, we thus propose SimStu-Transformer, developed based on the Decision Transformer algorithm, to generate learning behavioural data.

Keywords: Student modelling · Decision Transformer · Intelligent Tutoring Systems · Behavioural patterns

1 Introduction

The past decade has seen the rapid development of Machine Learning (ML) in many fields, including Intelligent Tutoring Systems (ITS). Unlike traditional ITS, which are time-consuming to build, recently proposed data-intensive ITS are more efficient, but require a large amount of data to support the ML models, [11]. Unfortunately, the lack of student behavioural data has become one of the most significant barriers for ITS breakthroughs, akin to the scarcity of labelled data in several other AI domains [12]. Previous studies have proposed various approaches to address this problem. For example, building Reinforcement Learning agents to simulate student behavior to train the ITS [5], simulating students' mastery of knowledge through Knowledge Tracing (KT) [7], or classifying students into different clusters based on their social interaction pattern to predict their behaviour [8]. However, none of these approaches has effectively solved this problem. Thus, in this paper, to tackle this challenge, we aim at answering:

How to create adequate high-fidelity and diverse simulated student behavioural data for training ITS?

The Transformer-based strategy allows ITS to capture a small amount of real student behavioural data and provides it to a generator that generates a

© Springer Nature Switzerland AG 2022
M. M. Rodrigo et al. (Eds.): AIED 2022, LNCS 13356, pp. 348–351, 2022.
https://doi.org/10.1007/978-3-031-11647-6_67

large amount of simulated student behavioural data, which can subsequently be used to train the ITS alongside the real data. In this paper, we propose "SimStu-Transformer" based on the Decision Transformer [2].

We adopt the 'sim student' approach [1], to simulate student behaviour. We apply group-level student modelling [4], to identify the "optimal" behavioural patterns that may result in better learning outcome. The results suggest that our model can well imitate student learning behaviour, and that it outperforms the traditional imitation learning method [6]. Our key contributions are twofold:

1. We designed a student learning behaviour simulation method to provide adequate data for ITS.
2. Our results showed a trained SimStu-Transformer model can simulate real student behaviour and surpass traditional imitation learning methods.

2 Experiment

Architecture. Our SimStu-Transformer is developed based on the Decision Transformer [2], initially proposed by Chen *et al.* It consists of an encoder and a decoder that simulate the joint distribution of student 'returns-to-go', 'states', and 'actions'. It divides student interactive trajectory sequences into two halves, one for the encoder's input and the other for the decoder's output [10]. The encoder then receives the first half of the trajectory sequence embeddings as input, and outputs a trajectory to the decoder. To construct the final output trajectory, the decoder takes a shifted embedding trajectory as input.

Data. Our data is from EdNet [3] - the largest student-ITS interaction benchmark dataset in the field of AIED/ITS. We conducted our experiments using the EdNet-KT4 sub-dataset, which provides more detailed interaction data than the other three sub-datasets. EdNet-KT4 contains 297,915 students' data with access to specific features and tasks. 1,000 students (a total of 861,247 action logs) were randomly selected for our experiments (200 students as the training data; 200 as the test data; 600 to compare with the simulated data). Students were divided into 5 groups based on their scores, due to the consideration of the possible correlation between learning performance and learning behaviour (Group 1 to Group 5: "very good" to "very poor"). The training data and the test data were partitioned by stratified sampling with such grouping strategy.

Trajectory Representation. The gap between the individual timestamps is used to replace the actual timestamps. The large UNIX time integers are reduced to small values. We also exclude highly sparse data from the modelling data. *'action_type'* is used to imitate students behaviour, which is denoted by a in the Decision Transformer Trajectory τ. *'user_answer'*, denoted by r, is used for evaluating student performance, thus partitioning them into groups. The correctness of student's answers were examined. *'item_id'* is used for evaluating the feasibility of the learning paths, which takes as the state of the student and is denoted by s. Due to the fact that *'user_id'* does not affect or represent student behaviour, we choose to generate it randomly, after the SimStu-Transformer generation procedure ends.

Experimental Design. The SimStu-Transformer was implemented using the Pytorch framework and trained on an Nvidia RTX 3090 GPU. We used the Adam optimiser with batch size of 64. We set Adam betas as (0.9, 0.95). The initial learning rate was 0.0006 and the dropout rate was 0.1. Two experiments were conducted to access the SimStu-Transformer.

In the first experiment, the similarity of the data generated by the SimStu-Transformer model was compared with the real data using Pearson Product-Moment Correlation Coefficient (PPMCC). The same training and test data was provided to the Behaviour Cloning model proposed by Torabi [9] in the second experiment, which yielded 600 student trajectory data (a total of 4,413,561 actions) and its results were compared with the real student data. We used RELU as the nonlinearity function, with standard batch size of 64. We set the initial learning rate as 0.0001 and the dropout rate as 0.1. We forcused on examing the distribution of 'elapsed time' (i.e., the amount of time a student works on a specific exercise) between the Behaviour Cloning method and the SimStu-Transformer with real student data by PPMCC.

Result and Discussions. Our results discovered some statistical similarities between the distributions of real student data and simulated student data, such as group sizes and the difference in the amount and frequency of actions in each group. Differences were only observed in the actions that occur less frequently, such as *'pay'* and *'undo_erase_choice'*. The resulting PPMCC value of all actions is equal to 0.714, which implies that the simulated student data is 71.4% similar to the actual student data in the average distribution of actions.

Figure 1 shows the distributions of elapsed time of real student (on the left), SimStu-Transformer simulated student (in the middle), and the Behaviour Cloning model simulated student (on the right). It can be seen that our SimStu-Transformer model outperforms the Behaviour Cloning model, as the data simulated by SimStu-Transformer is more similar to the real data (The PPMCC value: 0.762 vs. 0.683).

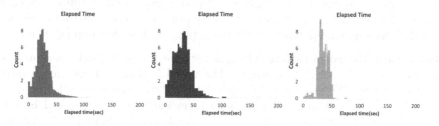

Fig. 1. Elapsed time of real student data (left), SimStu-Transformer method (middle), and Behaviour Cloning method (right).

3   Conclusions and Future Work

This paper presents a Transformer-based technique (SimStu-Transformer) to generate data for ITS training by modelling student behaviour. We use EdNet to train our model, which generates learning behaviour data that can be used to simulate individual students' learning trajectories. This method may benefit ITS training by compensating for the scarcity of real student data.

In the future, we aim to establish measurements to assess the fidelity and variety (coverage) of the simulated students in order to produce data that is as diverse as real student data. We also plan to address the impact of infrequent individual activities, such as *pay*, by evaluating different weights for different actions, for example. Finally, we plan to feed the simulated data into a real ITS to investigate if it can optimise the ITS training process, with a special focus on its effects on personalisation and adaption capabilities.

References

1. Alharbi, K., Cristea, A.I., Shi, L., Tymms, P., Brown, C.: Agent-based simulation of the classroom environment to gauge the effect of inattentive or disruptive students. In: Cristea, A.I., Troussas, C. (eds.) ITS 2021. LNCS, vol. 12677, pp. 211–223. Springer, Cham (2021). https://doi.org/10.1007/978-3-030-80421-3_23
2. Chen, L., et al.: Decision transformer: reinforcement learning via sequence modeling. arXiv preprint arXiv:2106.01345 (2021)
3. Choi, Y., et al.: EdNet: a large-scale hierarchical dataset in education. In: Bittencourt, I.I., Cukurova, M., Muldner, K., Luckin, R., Millán, E. (eds.) AIED 2020. LNCS (LNAI), vol. 12164, pp. 69–73. Springer, Cham (2020). https://doi.org/10.1007/978-3-030-52240-7_13
4. Lajoie, S.P.: Student modeling for individuals and groups: the bioworld and howard platforms. Int. J. Artif. Intell. Educ. 31(3), 460–475 (2021)
5. Li, Z., Shi, L., Cristea, A.I., Zhou, Y.: A survey of collaborative reinforcement learning: interactive methods and design patterns. In: Designing Interactive Systems Conference 2021, pp. 1579–1590 (2021)
6. Nair, A., McGrew, B., Andrychowicz, M., Zaremba, W., Abbeel, P.: Overcoming exploration in reinforcement learning with demonstrations. In: 2018 IEEE International Conference on Robotics and Automation (ICRA), pp. 6292–6299. IEEE (2018)
7. Pandey, S., Karypis, G.: A self-attentive model for knowledge tracing. arXiv preprint arXiv:1907.06837 (2019)
8. Shi, L., Cristea, A., Alamri, A., Toda, A.M., Oliveira, W.: Social interactions clustering MOOC students: an exploratory study. arXiv preprint arXiv:2008.03982 (2020)
9. Torabi, F., Warnell, G., Stone, P.: Behavioral cloning from observation. arXiv preprint arXiv:1805.01954 (2018)
10. Vaswani, A., et al.: Attention is all you need. In: Advances in Neural Information Processing Systems, pp. 5998–6008 (2017)
11. Vincent-Lancrin, S., Van der Vlies, R.: Trustworthy artificial intelligence (AI) in education: promises and challenges (2020)
12. Yang, S.J.: Guest editorial: precision education-a new challenge for AI in education. J. Educ. Technol. Soc. 24(1) (2021)

iCreate: Mining Creative Thinking Patterns from Contextualized Educational Data

Nasrin Shabani[1]([✉]), Amin Beheshti[1], Helia Farhood[1], Matt Bower[1],
Michael Garrett[2], and Hamid Alinejad Rokny[1]

[1] Macquarie University, Sydney, Australia
nasrin.shabani@hdr.mq.edu.au,
{amin.beheshti,helia.farhood,matt.bower,hamid.alinejad}@mq.edu.au
[2] Edith Cowan University, Perth, Australia
m.garrett@ecu.edu.au

Abstract. Creativity can be defined as the process of having original ideas that have value. The use of educational technology to promote creativity has attracted a great deal of attention. However, mining creative thinking patterns from educational data remains challenging. In this paper, we introduce a pipeline to contextualize the raw educational data, such as assessments and class activities. We also evaluate our approach with a real-world dataset and highlight how the proposed pipeline can help instructors understand creative thinking patterns from students' activities and assessment tasks.

Keywords: Educational data mining · Creativity · Data curation

1 Introduction

Recent research on creativity via education reveals that creative thinking, i.e., the ability to consider something in a new way, considered as a skills and can be learned by individuals [1]. There are various tools and techniques available for measuring creativity [2] which mostly involve evaluating the quality of ideas and products using human evaluators. Although scoring systems have proven effective, they are susceptible to two fundamental limitations: labor cost and subjectivity [3].

In the review of the literature, despite the extensive research on adaptive technology for education such as Educational Data Mining and Learning Analytics [4–6], mining creative thinking patterns from educational data remains challenging. To address this challenge, in this paper, we propose a rule-based insight discovery method to discover patterns of creativity in educational data. Our work relies on the knowledge of experts in education to build a domain-specific Knowledge Base to be linked to the extracted features from educational data. To evaluate our approach, we carry out an experiment with a real-world

M. M. Rodrigo et al. (Eds.): AIED 2022, LNCS 13356, pp. 352–356, 2022.
https://doi.org/10.1007/978-3-031-11647-6_68

dataset containing information about 480 students between 7–18 years and highlight how the proposed pipeline can help instructors understand creative thinking patterns from students' activities and assessment tasks.

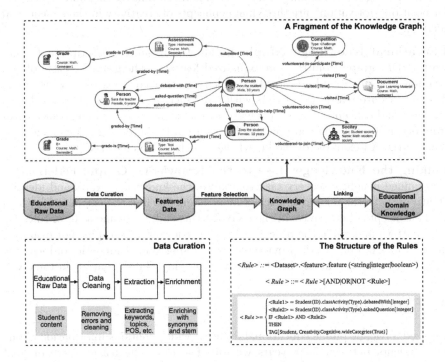

Fig. 1. An overview of the proposed model.

2   Mining Creative Thinking Patterns

We present a pipeline, namely iCreate, to facilitate mining creative thinking patterns from educational data. Figure 1 illustrates the iCreate pipeline which includes the following components:

Data Curation: Leveraging state of the art in data curation [7], which turns the raw educational data into contextualized data and knowledge. The curation pipeline includes cleaning, feature extraction, and enriching the extracted features.

Feature Selection: Reducing the dimensionality of data which has multiple benefits, making the dataset less complex and thus easier to analyse. The most straightforward technique to minimize dimensionality is to choose only the key features from a huge dataset. We employ several types of features in education, including students' demographic data (e.g., age, gender), achievement data (e.g., test scores), and program data (e.g., teacher training programs).

Domain-specific Knowledge Base: Imitating the knowledge of experts in education into an Educational Knowledge Base (eKB), which provides a rich structure of relevant entitles, semantics, and relationships among them. For example, "Cognitive Thinking Skills", "Domain-relevant Skills", and "Affective-Disposition-Motivation" are three main concepts of creativity in the educational hierarchy.

Educational Knowledge Graph: Building a Resource Description Framework (RDF) graph from the contextualized features. The graph consists of two components, entity and relationship, which together create triples of instances. The entities are real-world objects with a distinctive physical (e.g., a teacher, a student) and conceptual (e.g., a course, a task) identity that distinguishes them from other objects. They are defined by a set of attributes, e.g., name, age, and student ID. An example of the knowledge graph is depicted in Fig. 1.

Linking the Knowledge Base to the Knowledge Graph: Performing a user-guided insight discovery task to analyze the knowledge graph and identify creativity patterns related to each student node. We link the extracted features to the concepts of creativity in the eKB. To this end, we define a set of rules for each pattern and guide the process of finding relevant sub-graphs. For example, in Fig. 1 an example rule is provided for the "Using Wide Categories of Ideas" as a creativity pattern under the "Cognitive Thinking Skills" concept in the eKB.

3   Experiment

We used a Kaggle public educational dataset[1] which contains information about 480 students between 7–18 years with 16 features. The features are divided into three groups: (1) Academic background features such as educational level, section, and stage; (2) Behavioral features include the number of raised hands in class, involved discussions, and visiting resources; and (3) Demographic features such as nationality, gender, and age.

After importing the dataset, we performed some preprocessing tasks such as data cleaning and removing noisy inputs. By selecting the features that are most relevant to our approach (e.g., the number of raised hands, involved in discussions, visiting resources, and absences), we built an RDF graph. The completed graph then was saved to be imported into our graph database (GraphDB). We defined a set of rules to be implemented on top of the RDF graph repository and link the constructed RDF graph to the eKB. To query this huge graph, we made use of SPARQL queries to organize the data and extracted features. As a result, students that follow the rules appeared in the result of the queries. For instance, three rules were applied across the graph, including those about having a strong memory, using wide category of ideas, and lack of motivation.

To evaluate the correctness of the model, we carried out a user study. In this study, we tried to validate the following hypotheses: (H1) The components of the KB are relevant to creativity, (H2) The designed rules for pattern mining

---

[1] https://www.kaggle.com/aljarah/xAPI-Edu-Data.

are useful and relevant to creativity patterns, and (H3) The query results are reasonable and show a successful link between the KB and educational data. The experiment was done in a controlled environment to examine our approach. Participants were mostly chosen from academics and students with different backgrounds. Hence, among 10 participants, some had strong knowledge in education and cognitive science, some had computing domain expertise, and others had both. We prepared a questionnaire and shared it with the participants to examine the study's hypotheses. With each of the participants, we asked them to rate the relevancy of the hypotheses to allow the participants to express their opinion.

The finding of the user study supports the hypotheses H1, H2, and H3. However, in H2, the rule-based pattern mining techniques require future improvement to gain a higher score in the evaluation. Moreover, mainly experts in education with knowledge, expertise, and interest in education and computing found our approach valid and confirmed the hypotheses. Based on the participants' feedback and the lessons learned, some future improvements could be considered for the approach. The motivation of participants, time pressure and training, and definition of the rules are important indicators that significantly impact the final evaluation results.

4   Conclusion

In this study, a data-driven technique is provided to relate students' behavior to creative thinking patterns and assist instructors in understanding them from students' activities and assessment tasks. We concentrated on understanding the big educational data, used existing data curation techniques, built a domain-specific KB by leveraging the knowledge of education experts, and linked the contextualized data to the KB using a rule-based technique. We also evaluated our approach through a user study, relying on the knowledge of education experts.

Acknowledgements. We acknowledge the AI-enabled Processes (AIP) Research Centre (https://aip-research-center.github.io/) for funding this research.

References

1. Shaheen, R.: Creativity and education. Creative Educ. **1**(3), 166–190 (2010)
2. Henriksen, D., Mishra, P., Mehta, R.: Novel, effective, whole: toward a NEW framework for evaluations of creative products. J. Technol. Teach. Educ. **23**(3), 455–478 (2015)
3. Beaty, R., Johnson, D.: Automating creativity assessment with SemDis: an open platform for computing semantic distance. Behav. Res. Methods **53**(2), 757–780 (2021)
4. Baker, R.S., Inventado, P.S.: Educational data mining and learning analytics. In: Larusson, J.A., White, B. (eds.) Learning Analytics, pp. 61–75. Springer, New York (2014). https://doi.org/10.1007/978-1-4614-3305-7_4

5. Romero, C., Ventura, S.: Educational data mining and learning analytics: an updated survey. Wiley Interdisciplinary Rev. Data Mining Knowl. Discovery **10**(3), e1355 (2022)
6. Wang, S., et al.: Assessment2Vec: learning distributed representations of assessments to reduce marking workload. In: AIED Conference, pp. 384–389. Springer, Utrecht (2021)
7. Beheshti, A., Benatallah, B., Nouri, R., Tabebordbar, A.: CoreKG: a knowledge lake service. Proc. VLDB Endowment **11**(12), 942–1945 (2018)

Analyzing Speech Data to Detect Work Environment in Group Activities

Valeria Barzola, Eddo Alvarado, Carlos Loja, Alex Velez, Ivan Silva,
and Vanessa Echeverria(✉)(iD)

Escuela Superior Politécnica del Litoral, ESPOL, Guayaquil, Ecuador
{eaalvara,caloja,alanvele,isilva,vanechev}@espol.edu.ec

Abstract. Collaboration is one required skill for the future workforce that requires constant practice and evaluation. However, students often lack formative feedback and support for their collaboration skills during their formal learning. Current technologies for emergent learning due to COVID-19 could make visible digital traces of collaboration to support timely feedback. This work aims to automatically detect the group work environment using speech data captured during group activities. Grounded in literature and students' perspectives, this work defines and implements three indicators for detecting the work environment namely *noise*, *silence* and *speech time*. Three experts rated two hundred thirty-two video instances lasting 30-secs each to get a group work environment score. We report the results of two machine learning models for detecting the group work environment and briefly reflect on these results.

Keywords: Collaboration feedback · Group work environment · Automatic feedback · Human-centered analytics

1 Introduction

Collaboration is one critical skill for the future workforce that requires constant practice and evaluation. However, students often lack formative feedback and support for their collaboration skills during their formal learning. Fostering students' collaboration skills often need close coaching by an expert to prompt timely feedback. Nevertheless, it may be unrealistic in practice. This problem has been exacerbated with the COVID-19, in which it is difficult for the teacher to supervise what is happening in group video calls [1], due to this new form of teaching does not resemble a face-to-face classroom, where the teacher can walk around the class and observe what is happening during group activities.

Recent initiatives have started analyzing real-time users participation using video conferencing software (e.g., Zoom breakout rooms) to support teaching and learning practices. For instance, the tool reported by Zhou and colleagues [7] captured speech data from a video conferencing software to investigate students' experience when facing some analytics (i.e., participation) from speech data. Students expressed that speech participation does not always reflect their true collaboration and that real-time prompts to scaffold their collaboration practices are needed.

© Springer Nature Switzerland AG 2022
M. M. Rodrigo et al. (Eds.): AIED 2022, LNCS 13356, pp. 357–361, 2022.
https://doi.org/10.1007/978-3-031-11647-6_69

In this work, we are focused on automatically detecting collaboration aspects, besides participation, that students unfold during a group activity. We take a human-centered approach by considering collaboration aspects from students' perspectives reported in prior research [4,6]. Mainly, we are interested in automatically recognizing the work environment during group work activities. We define *group work environment* as the facility to carry out work activities and discuss ideas in a safe environment, where all participants are involved in the task and all students participate. This work reports a first step towards computationally defining and implementing a *work environment* model that could be used to generate personalized collaboration feedback (i.e., a dashboard, a notification to the teacher, etc.).

2   Automatic Detection of Group Environment

To automatically detect the group work environment, we implemented a Support Vector Machine (SVM) and Random Forest (RF) models. The following section describes the data used as ground truth for the classification task, feature generation, feature extraction, and results.

Data Collection and Expert Coding. The data collection was done during regular classes of an undergraduate Database Systems course from a Computer Science program in a local university. The class was organized in groups, and students were instructed to work together to propose a solution for a database entity-relationship model. This activity lasted about 35 min. Due to COVID-19, classes were taught remotely, and all group activities were video recorded. We used video recordings from 23 students (20 males, three females) in this work, divided into seven groups of 2–4 students.

We derive the ground truth for this dataset following a manual coding process. Three experts rated 494 collaboration instances (i.e., a video segment of 30 s) using a score ranging from 1 to 5 (1: very bad and 5: very good). After several discussions, they reach a consensus on the final score. Experts rated a total of 232 work environment instances. Due to the unbalanced distribution of the resulting scores, we decided to merge the scale yielding *three final classes*: bad (1–2), regular (3), and good (4–5). The final distribution was 82, 62, and 86 instances for good, regular, and bad classes. More details about the coding process can be found in [4].

Feature Definition. Grounded in the literature and students' perspectives [4,6] about group environment, we defined four indicators as proxies for automatically detecting the group's work environment:

- **Noise** can be considered any sound different from the main sound a person produces, which may cause distraction and could hinder students' learning performance [2]. Students expressed that a good work environment is when *"there is a quiet and nice environment to work with your partners without any noise distraction"*.

- **Speech** is the main characteristic of communication; hence it plays a vital role in collaboration. Students expressed that a good work environment is *"when members want to share their ideas in a safe place"*
- **Silence** could be detrimental to communication if it is not well addressed. Students expressed that, at some points were dissatisfied with some of their partners being quiet during a collaboration activity [5].
- **Uncomfortable Silence**, such as prolonged silences, could lead to a bad group experience. Students expressed that a good work environment is *"when all participants talk actively"* Usually, an uncomfortable silence comes after a group member poses a question, and no answer or acknowledgment is further provided.

Feature Generation. We extracted the audio (.wav) file from each group video recording. All audios were also transcribed using The Microsoft Office transcription service. This transcription resulted in a text (.txt) file with timestamps and text from each participant's speech.

We automatically extracted the **noise, silence** and **speech** percentages as follows: we got segments (start and end time) of speech, noise (including ambient noise) and silence (no energy in the audio segment) from the audio file using an open-source sound segmentation tool [3]. We aligned the segments resulting from the tool with the 30-secs segments and calculated the percentage of *speech, noise* and *silence* (e.g., from the segment 00:00 to 00:30, there was a 5% of noise, 45% of speech and 50% of silence).

To detect **uncomfortable silences**, we use the text and timestamps of the transcribed .txt file. We aimed to calculate the time elapsed (in seconds) when one participant asked a question until another participant answered the question. We detected the speakers and questions from the transcript (i.e., if a sentence had a question mark). Then, we visually inspected some video instances to determine a threshold for differentiating prolonged silences from uncomfortable silences. We selected 7s as a threshold. For example, if a participant asked a question at time 04:09 and another talked at time 04:25, there was an elapsed time of 16s, categorized as an uncomfortable silence.

Model and Results. Using the four indicators as explained above, we build our models. For this work, we report the design and results from two machine learning models, such as Support Vector Machine (SVM) and a Random Forest (RF) models. These two models have been widely used in similar contexts (c.f. REF) First, we tune the hyper-parameters for our two models following a grid search method. Due to a lower sample size for training and testing our models, we used k-fold cross-validation to test the models. To select the best k value, we also ran another grid search from 1 to 10.

The SVM model was trained with four kernel functions: linear, polynomial, sigmoid, and a radial basis function (RBF), and the hyper-parameters γ and *cost* in the range of 2^{-5} to 2^{10}. As several combinations of the *cost* $= (2^{-1})$ performed equally (i.e., the same accuracy), we decided to choose an intermediate value for

$\gamma = (2^2)$ to avoid over-fitting. The RF has two main hyper-parameters: the number of trees to grow in the forest (n) and the number of layers to split in each tree (m). The values tested for n ranged from 1 to 10 and for m from 1 to 500.

According to the evaluation metrics, an SVM model (RBF, $k = 5$) yielded the highest accuracy value of 0.66 and F1-score $= 0.62$. Further, a RF model ($n = 130$, $m = 5$ $k = 10$) yielded the highest accuracy value $= 0.69$ and F1-score $= 0.65$. In relation to the intra-class performance, the RF had a better accuracy for predicting bad (n = 68) and regular (n = 21) classes compared to the SVM model, whereas the SVM model had a better prediction for the good class (n = 75).

3   Discussion and Future Research Directions

In this work, we reported the application of supervised machine learning models to automatically detect the group work environment from speech data. This study demonstrated a great potential to provide feedback automatically in a virtual environment using noise, speech, and silence features. Nevertheless, several challenges remain open and should be tackled in further explorations. For instance, the model could get contextual features by adding multimodal data (e.g., silence could also represent students working on a document or a solution). According to our results, the regular class did not get good accuracy. This could be improved by adding the speech semantics, which may open another challenge, such as data privacy concerns. The results presented in this work are limited to the sample size, meaning that more data is needed to get a generalized model. Future research directions should include detecting other collaboration aspects such as coordination, problem management, and contribution to implementing a feedback report that could be delivered to students timely.

References

1. Almonacid-Fierro, A., Vargas-Vitoria, R., De Carvalho, R.S., Fierro, M.A.: Impact on teaching in times of covid-19 pandemic: a qualitative study. Intl. J. Eval. Res. Educ. **10**(2), 432–440 (2021)
2. Barrera, D.J., et al.: Online-based learning: challenges and strategies of freshman language learners. J. Learn. Dev. Stud. **1**(1), 53–66 (2021)
3. Doukhan, D., Carrive, J., Vallet, F., Larcher, A., Meignier, S.: An open-source speaker gender detection framework for monitoring gender equality. In: International Conference on Acoustics, Speech and Signal Processing (ICASSP), pp. 5214–5218. IEEE (2018)
4. Echeverria, V., Wong-Villacres, M., Ochoa, X., Chiluiza, K.: An exploratory evaluation of a collaboration feedback report. In: LAK22: 12th International Learning Analytics and Knowledge Conference, pp. 478–484 (2022)
5. Skinner, V.J., Braunack-Mayer, A., Winning, T.A.: Another piece of the "silence in pbl" puzzle: Students' explanations of dominance and quietness as complementary group roles. Interdisciplinary J. Problem-Based Learn. **10**(2) (2016)

6. Worsley, M., Anderson, K., Melo, N., Jang, J.: Designing analytics for collaboration literacy and student empowerment. J. Learn. Anal. **8**(1), 30–48 (2021)
7. Zhou, Q., et al.: Investigating students' experiences with collaboration analytics for remote group meetings. In: Roll, I., McNamara, D., Sosnovsky, S., Luckin, R., Dimitrova, V. (eds.) AIED 2021. LNCS (LNAI), vol. 12748, pp. 472–485. Springer, Cham (2021). https://doi.org/10.1007/978-3-030-78292-4_38

Comparison of Selected-
and Constructed-Response Items

Haiying Li[1,2](✉)

[1] Iowa College Aid, Des Moines, IA 50309, USA
Haiying.Li@iowa.gov
[2] Organization for Economic Co-operation and Development (TJA Fellow), 75016 Paris, France

Abstract. A body of research on the assessment of scientific practices revealed that constructed-response items (CRI) are more valid for assessing scientific explanations than selected-response items (SRI). A few studies have compared the differences between these question item formats in small-scale formative science assessments. It is unclear, however, whether this phenomenon is universal in large-scale science assessments, which is within the scope of the present study. This study showed that one-third of students on average demonstrated inconsistent performance across 58 countries/regions when scientific practices were measured by SRI and CRI.

Keywords: Constructed-response · PISA science · Selected-response

1 Introduction

The OECD (Organization for Economic Co-Operation and Development) administers the Programme for International Student Assessment (PISA) to evaluate 15-year-old students' performance on reading, mathematics, science, and other innovative subjects throughout the world. PISA 2015 science added simulation-based tasks to assess scientific practices in real-world settings [8]. These simulated tasks allow students to interact with the computer, evaluate and design experiments, analyze and interpret data and evidence, and explain phenomena scientifically. Researchers used either selected-response items (SRI) [1] or constructed-response items (CRI) [3] to examine students' scientific knowledge and reasoning abilities for scientific explanations. While the assessment structure, knowledge recognition and construction, and cognitive demand are likely different [2, 6, 7], high correlations were found between SRI and CRI in construct equivalence [10], but not in performance [3–5]. For instance, students demonstrated inconsistent performance in simulated labs when scientific practices were measured by SRI (e.g., clicking on the button, selecting from a dropdown list) and CRI (e.g., written claim, evidence, reasoning). These studies, however, focused on small-scale science assessments with a limited number of middle schools in the United States.

   The present study extended the population across countries to answer the following three questions: (1) How many students demonstrate inconsistent performance on scientific practices measured by SRI and CRI? (2) Is the pattern consistent across countries

© Springer Nature Switzerland AG 2022
M. M. Rodrigo et al. (Eds.): AIED 2022, LNCS 13356, pp. 362–366, 2022.
https://doi.org/10.1007/978-3-031-11647-6_70

whose science levels vary from low to high? and (3) How are the countries' science levels associated with the percentage of students who demonstrate inconsistent performance on scientific practices?

2 Methods

Participants were 7,639 students at grade 9.75 ($SD = 0.71$) from 58 countries or regions (age: $M = 15.79$, $SD = 0.29$; 51.5% female). They took the computer-based PISA 2015 science tests in their native languages. This study used six publicly-released simulated science inquiry items [8] to answer the research questions (see Table 1). The SRI and CRI items were both at the medium difficulty level (1–6 scale; SRI: $M = 4.33$, $SD = 1.53$; CRI: $M = 3.67$, $SD = 0.58$). Each item was scored either as 0 for no credit and 1 for full credit or as 0 for no credit, 1 for partial credit, and 2 for full credit. The 0–2 points were converted into 0–1 points. Moreover, OECD released each country's rank in terms of science performance [9], which was used to explore the patterns across countries in terms of science performance.

Table 1. Question items in publicly-released simulated scientific inquiry tasks.

Topic	Question ID	Format	Knowledge	Competency	Difficulty level
Bird migration	CS656Q01S	SRI	Content	Explain	3
	DS656Q02C	CRI	Procedural	Evaluate	4
	CS656Q04S	SRI	Procedural	Interpret	4
Slope-face investigation	DS637Q01C	CRI	Epistemic	Evaluate	3
	CS637Q02S	SRI	Epistemic	Evaluate	6
	DS637Q05C	CRI	Epistemic	Interpret	4

Note. Explain = Explain phenomena scientifically, Evaluate = Evaluate and design scientifically, Interpret = Interpret data and evidence scientifically

3 Results and Discussion

K-mean cluster analyses ($K = 2$) were performed on SRI and CRI scores, respectively to classify students into low and high SRI and CRI performers. Results significantly classified students into low and high groups with large effects, $F(1, 7637) = 20,486.99$, $p < .001$, *Cohen's d* $= 2.93$ for SRI and $F(1, 7637) = 24,616.89$, $p < .001$, *Cohen's d* $= 2.67$ for CRI (see Table 2). This classification formed four quadrants: low SRI but high CRI scores (L-H), low SRI and low CRI scores (L-L), high SRI but low CRI scores (H-L), or high SRI and high CRI scores (H-H). Both L-H and H-L groups demonstrated inconsistent performance, whereas both L-L and H-H groups demonstrated consistent performance.

To answer the first question, a 2 × 2 contingency table showed that about 64.4% of students demonstrated consistent performance but 35.6% of students demonstrated inconsistent performance on scientific practices when it was measured by SRI and CRI (see Table 2). This finding indicates that some students (H-L) could select the right answer from a list of choices but failed to generate good responses, or vice versa (L-H).

Table 2. Students demonstrated consistent and inconsistent competencies.

		Constructed responses	
		Low ($M = 0.20$, $SD = 0.16$)	High ($M = 0.78$, $SD = 0.16$)
Selected responses	Low ($M = 0.20$, $SD = 0.12$)	2,593 (38.7%)	1,467 (19.2%)
	High ($M = 0.69$, $SD = 0.10$)	1,256 (16.4%)	1,963 (25.7%)

To answer the second question, the same 2 × 2 contingency table was performed in each of the 58 countries and regions with the rank of scientific performance as the analysis unit. Figure 1 displays the percentage of each group. Results showed the existence of inconsistent performance, i.e., L-H and H-L, in each country/region, but the proportion of each group varied. These findings support previous findings that students demonstrated inconsistent performance on SRI and CRI scientific practices, but the percentage of inconsistent performers was dependent on the science rank in a country, ranging from 14.3% to 54.2%.

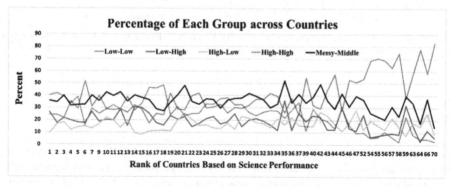

Fig. 1. Percentage of each group in terms of countries' science ranks.

To answer the third question, Pearson correlations were performed between the rank of countries and the percentage of students in each group (e.g., L-H, H-L, L-L, H-H, and Messy-Middle) (see Table 3). A significant negative correlation was found between the rank and the total percentage of inconsistent performers. These findings indicated that if the countries had a lower level of science performance, they had fewer students with inconsistent performance or vice versa. This indicated that if countries have high-level science performance, they had more students with both high SRI and high CRI performance ($r = -.79$) and more students with low SRI and high CRI performance ($r = -.47$), but few students with both low SRI and low CRI ($r = .76$). According to the previous studies on the relationship between scientific explanations and the proficiencies of content, inquiry, and writing skills, the only significant predictor for the CRI, i.e., written claim, written evidence, and written reasoning was writing skill [4]. It is likely that students in high-level countries demonstrated both high knowledge recognition and construction [2, 6, 7], but high writing proficiency contributed to their high performance on CRI.

Table 3. Pearson correlations between rank and the percentage of each group.

Groups	Rank	Low-low	High-high	Low-high	High-low
Low-low	$.76^{***}$				
High-high	$-.79^{***}$	$-.91^{***}$			
Low-high	$-.47^{**}$	$-.72^{**}$	$.51^{**}$		
High-low	$.14$	$-.03$	$-.23$	$-.36^{**}$	
Messy middle	$-.38^{**}$	$-.71^{**}$	$.36^{**}$	$.76^{**}$	$.33^{*}$

Note. *** $p < .001$, ** $p < .01$, and * $p < .05$ (2-tailed).

This study is limited by the number of scientific tasks and the mixed knowledge and competency types. Future studies should include more tasks and measure the same knowledge and competency in SRI and CRI controlling for the reading and writing proficiencies and other individual differences.

Acknowledgments. This work was funded by the Organization for Economic Co-Operation and Development (Grant No. EDU/0500091979).

References

1. Baker, R.S., Clarke-Midura, J., Ocumpaugh, J.: Towards general models of effective science inquiry in virtual performance assessments. J. Comput. Assist. Learn. **32**, 267–280 (2016)
2. Federer, M.R., Nehm, R.H., Opfer, J.E., Pearl, D.: Using a constructed-response instrument to explore the effects of item position and item features on the assessment of students' written scientific explanations. Res. Sci. Educ. **45**, 527–553 (2014)
3. Li, H., Gobert, J., Dicker, R.: Dusting off the messy middle: assessing students' inquiry skills through doing and writing. In: André, E., Baker, R., Hu, X., Rodrigo, M.M.T., du Boulay, B. (eds.) AIED 2017. LNCS (LNAI), vol. 10331, pp. 175–187. Springer, Cham (2017). https://doi.org/10.1007/978-3-319-61425-0_15
4. Li, H., Gobert, J., Dickler, R.: The relationship between scientific explanations and the proficiencies of content, inquiry, and writing. In Klemmer, S., Koedinger, K. (eds.) Proceedings of the Fifth Annual ACM Conference on Learning at Scale, New York, NY (2018)

5. Li, H., Gobert, J., Dickler, R.: Unpacking why student writing does not match their science inquiry experimentation in Inq-ITS. In: Kay, J., Luckin, R. (eds.) Rethinking learning in the digital age: Making the learning sciences count, 13th International Conference of the Learning Sciences (ICLS) 2018, vol. 3, pp. 1465–1466, London, UK (2018)
6. Martinez, M.E.: Cognition and the question of test item format. Educ. Psychol. **34**, 207–218 (1999)
7. Nehm, R.H., Beggrow, E.P., Opfer, J.E., Ha, M.: Reasoning about natural selection: diagnosing contextual competency using the Acorns Instrument. Am. Biol. Teach. **74**, 92–98 (2012)
8. OECD: PISA 2015 technical report. PISA, OECD Publishing, Paris (2017)
9. OECD: PISA 2015 results in focus. PISA, OECD Publishing, Paris (2016)
10. Rodriguez, M.C.: Construct equivalence of multiple-choice and constructed-response items: a random effects synthesis of correlations. J. Educ. Meas. **40**, 163–184 (2003)

An Automatic Focal Period Detection Architecture for Lecture Archives

Ruozhu Sheng⬤, Koichi Ota⬤, and Shinobu Hasegawa$^{(\boxtimes)}$ ⬤

Japan Advanced Institute of Science and Technology, 1-1 Asahidai, Nomi, Ishikawa 923-1292, Japan
hasegawa@jaist.ac.jp

Abstract. This research proposes a deep neural network architecture for detecting focal periods for online students in lecture archives. Due to the COVID-19 pandemic, most universities attempted to use online education instead of traditional classrooms. However, watching long lecture archives, just recorded face-to-face lectures, is difficult for students to keep their attention. Hence, how to provide focal periods of the lecture archives is essential to maintain educational effectiveness in such a situation. This research divides lecture archives with high quality and fixed camera angles into 1-min segments, counts how many times students have accessed each segment from LMS as the label data, and defines the students' focal periods. Then, we demonstrated deep neural network architectures with the combined features to improve detection reliability. Our experiments showed that the proposed method could detect the focal periods with 56.8% accuracy. Although there is room for improvement in accuracy, this enables us to detect certain focal periods with a small amount of computation without using semantic features.

Keywords: Lecture archives · Student attention · Focal period · Deep learning

1 Introduction

Due to the COVID-19 outbreak, almost universities attempted to use online learning instead of traditional classrooms to reduce the risk of infection and keep educational activities. One of the online learning applications is lecture archives that record a face-to-face lecture without editing because they are easy to distribute with a low production cost. However, it is difficult for students to keep their attention for long periods during the recorded lecture [1]. Hence, it is necessary to automatically extract or summarize the main periods of the archive to improve the effectiveness and efficiency of lecture archives in online learning.

Video summarization provides condensed and succinct representations of the content of a video stream through a combination of still images, video segments, graphical representations, and textual descriptors. In the education field, Andra and Usagawa have summarized the lecture video through Attention-based Recurrent Neural Network (RNN) that combines segmentation with the summarization process to improve accessibility to key points [2]. The RNN architecture generates a natural summary, captures

© Springer Nature Switzerland AG 2022
M. M. Rodrigo et al. (Eds.): AIED 2022, LNCS 13356, pp. 367–370, 2022.
https://doi.org/10.1007/978-3-031-11647-6_71

the critical words, and conveys a lecture's central message from the attention-based weighting and linguistic features. However, their method significantly depends on the semantic analysis of lecture content. Such transcripts are not always available for lecture archives, and the accuracy of existing Automatic Speech Recognition (ASR) techniques is not always satisfied in the sound data with noise and precise terminology.

This research proposes a deep neural network architecture for detecting focal periods defined by students' access to lecture archives. The research target is a lecture archive system, which stores lecture archives with high-definition resolution and fixed camera angles in face-to-face lectures. We adopted the deep neural network architecture based on the hypothesis that the image features of the lecturer's action, voice, and slides in the archive effectively estimate the main periods of the lecture.

2 Design and Methodology

2.1 Dataset

The lecture archives used in this research originated in the JAIST Learning Management System (JAIST-LMS), seven lessons (approximately 100 min for each) of the I239 machine learning course in 2018. A ceiling camera and microphone recorded the archives with a fixed angle of 1920*1080 resolution and 30 fps. The archives included the podium area, whiteboard, instructor, and the slide content integrated into the right-bottom corner of the archives, as shown in Fig. 1.

The JAIST-LMS extends video.js to track that the students have accessed durations of the archives for their reflection. Therefore, we divided each archive into one-minute segments except for the first 5 min to denoise. We computed a moving average of the number of accesses every five segments to generate stable labels. Finally, we have seven lessons * 95 min = 665 segments. This research normalized the labels by the maximum accesses in each archive, as shown in Fig. 2.

Fig. 1. Original lecture archive

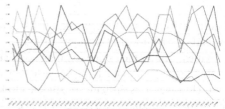

Fig. 2. Normalized label data

2.2 Extracted Features

According to the previous study by Zhang et al. [3], the instructor's behavior influences students' attention. To extract the instructor's action, we first mask the students' seating area and then capture the instructor's body structure feature by Openpose, the first open-source real-time system available to detect multi-person 2D poses, including body, feet, hands, and facial key points. Next, we calculate the optical flow for the instructor's action based on the captured body structure from a one-minute archive segment.

According to a previous study by Wyse [4], neural networks used in classification or regression can benefit from spectrograms which are a visual representation of the spectrum of signal frequencies. Therefore, this research adopts spectrograms to extract the instructor's voice features from the archive segments. A study using eye-tracking technology has demonstrated that the content of slides affects students' attention [5]. Consequently, we propose a method that stacks the slide differences in each segment to extract the slide features as switching frequency of slides.

2.3 Learning Architecture

We first tested the performance of different deep learning models, VGG-16, VGG-19, Res-50, and Res-101, respectively. Using only the action features as input, 665 data in our dataset were divided randomly to 7:3 between training and validation sets. In comparing the four models, VGG-16 with the hyperparameter of 70 training epochs, 32 batch size, and 0.001 learning rate had the best performance.

We designed "Feature Stacking" to save these feature maps of each archive segment as RGB channels in one image file to deal with different features. It converts the action features to the R channel, the slide features to the G channel, and the voice features to the B channel. We also applied the Savitzky-Golay filter, which smooths the output curve on the time series and eliminate the noise. This method makes using VGG-16 as a single image input for training. The advantage of this approach is to reduce the computation and save on video memory. However, the disadvantage is that we compress these feature maps into single channel images, leading to the loss of some features.

3 Experiment

Since our models were trained as a regression task, the output predicted a value of 0 to 1. To detect focal periods from the lecture archives, we specify a sliding window of length five segments (five minutes). If the median value of the window is greater than 0.9, or the mean value of the window is greater than 0.5, we define the window as focal periods. On the other hand, if the mean of the window is less than 0.3, the window is defined as non-focal periods.

Considering that our dataset has only 665 data in total, cross-validation is needed to test the performance of our model. Thus, we choose three methods for comparison: (a) Action features, (b) Action + Voice feature stacking, (c) Action + Voice + Slide (All) feature stacking. The predicted values of the model output were classified as focused, non-focused, and neither by the above rule. We compare these accuracy rates to evaluate the models' performance, as shown in Table 1.

We have confirmed that the (b) Action + Voice feature stacking method has the best performance among our proposed methods through cross-validation. Compared to other methods, the minimum, maximum, and average accuracy are significantly better. We found that combining all the features did not achieve better performance than combining Action + Voice features. The slide features were not suitable for this task since different student attention levels might lead to the same slide feature map. It is necessary to consider using different features such as differential images of whiteboards to solve this problem.

Table 1. Results of accuracy in cross-validation.

Architectures	Average accuracy	Max accuracy	Min accuracy
(a) Action	0.415	0.708	0.109
(b) A + V F	**0.568**	**0.791**	**0.450**
(c) All F	0.439	0.761	0.239

4 Conclusions

We have proposed a method to detect students' focal periods in lecture archives using a deep learning architecture. We finally selected the Action + Voice feature stacking method with an average accuracy rate of 56.8% in cross-validation with a single input VGG-16 model. These characteristics are beneficial for practical application in educational institutions with limited resources. Furthermore, the features used in our method do not contain semantic information. Therefore, it can be applied to multilingual courses without adaptation.

We could choose a suitable model for each feature in future work. We should also continue to experiment with different instructors and lecture rooms to build a more generalized detection model.

References

1. Guo, P.J., Kim, J., Rubin, R.: How video production affects student engagement: an empirical study of mooc videos. In Proceedings of the first ACM Conference on Learning@ Scale Conference, pp. 41–50 (2014)
2. Andra, M.B., Usagawa, T.: Automatic lecture video content summarization with attention-based recurrent neural network. In: 2019 International Conference of Artificial Intelligence and Information Technology (ICAIIT), pp. 54–59. IEEE (2019)
3. Zhang, J., Bourguet, M.-L., Venture, G.: The effects of video instructor's body language on students' distribution of visual attention: An eye-tracking study. In: Proceedings of the 32nd International BCS Human Computer Interaction Conference 32, pp. 1–5 (2018)
4. Wyse, L.: Audio spectrogram representations for processing with convolutional neural networks. In: Proceedings of the First International Workshop on Deep Learning and Music Joint with IJCNN, vol. 1(1), pp. 37–41 (2017)
5. Slykhuis, D.A., Wiebe, E.N., Annetta, L.A.: Eye-tracking students' attention to powerpoint photographs in a science education setting. J. Sci. Educ. Technol. **14**(5), 509–520 (2005)

BERT-POS: Sentiment Analysis of MOOC Reviews Based on BERT with Part-of-Speech Information

Wenxiao Liu, Shuyuan Lin, Boyu Gao, Kai Huang, Weilin Liu, Zhongcai Huang, Junjie Feng, Xinhong Chen, and Feiran Huang[✉]

College of Information Science and Technology/College of Cyber Security, Jinan University, Guangzhou, China
huangr@jnu.edu.cn

Abstract. Sentiment analysis and part-of-speech tagging are two main tasks in the field of Natural Language Processing (NLP). Bidirectional Encoder Representation from Transformers (BERT) model is a famous NLP model proposed by Google. However, the accuracy of sentiment analysis of the BERT model is not significantly improved compared with traditional machine learning. In order to improve the accuracy of sentiment analysis of the BERT model, we propose Bidirectional Encoder Representation from Transformers with Part-of-Speech Information (BERT-POS). It is a sentiment analysis model combined with part-of-speech tagging for iCourse (launched in 2014, one of the largest MOOC platforms in China). Specifically, this model consists of the part-of-speech information on the embedding layer of the BERT model. To evaluate this model, we construct large data sets which are derived from iCourse. We compare the BERT-POS with several classical NLP models on the constructed data sets. The experimental results show that the accuracy of sentiment analysis of the BERT-POS is better than other baselines.

Keywords: Sentiment analysis · MOOC · NLP · BERT

1 Introduction

With the spread of COVID-19, in order to ensure the continuity of courses, more and more colleges and universities have changed the form of class into Massive Open Online Courses (MOOC) [4]. The course forum is a common place to strengthen academic exchanges between students. In MOOC, learners can establish social interaction with other users to ask or answer questions without the support of teachers [3]. Bidirectional Encoder Representation from Transformers (BERT) [2] model is proposed by Google in 2018. Based on Transformer [7] model, it refreshes the best results of multiple Natural Language Processing (NLP) downstream tasks. However, the classification accuracy is not high when dealing with Chinese MOOC comments in sentiment analysis task.

In order to improve the accuracy of sentiment analysis of BERT model, in this work, we propose Bidirectional Encoder Representation from Transformers with

ⓒ Springer Nature Switzerland AG 2022
M. M. Rodrigo et al. (Eds.): AIED 2022, LNCS 13356, pp. 371–374, 2022.
https://doi.org/10.1007/978-3-031-11647-6_72

Part-of-Speech Information (BERT-POS). A quantitative comparison between the BERT-POS model and the eight classical NLP models on the proposed data sets shows that the accuracy of sentiment analysis of the BERT-POS model is better than other baselines. The contributions of our paper are as follows:

- We constructed a large dataset of online class reviews from iCourse and used crowd intelligence algorithm for labeling.
- We add the part-of-speech embedding to the embedding layer. Integrate part-of-speech embedding with position embedding, segment embedding and token embedding.

2   BERT-POS

2.1   Overview

Through the analysis of the original corpus, we found a pattern: under the premise of non-negative words, the emotional tendency of a sentence is often positively correlated with the emotional tendency of adjectives. We add part-of-speech information to the embedding layer of BERT model. The overall structure of the BERT-POS is shown in Fig. 1.

2.2   Input and Output

The input of the model is the addition and fusion of four kinds of embeddings, including token embedding, segment embedding, position embedding and part-of-speech embedding, For part-of-speech embedding, we distinguish adjectives, adverbs and other parts of speech. After this, the fusion codes added by the four coding methods are obtained as the input of the BERT-POS model.

The output of the BERT-POS model is the corresponding contextualized representation for each input Chinese character [2,6]. We extract the tendency of sentiment analysis from the corresponding contextualized representation.

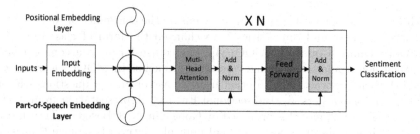

Fig. 1. The overall structure of BERT-POS with the part-of-speech embedding layer.

3 Experiments

3.1 Data and Baselines

This paper uses the data set from all the courses on iCourse[1]. The raw data includes 438 colleges and universities, 9,047 courses and 1.72 million course review data. We mark the part of raw data. The labels of the data set are all manually marked and repeatedly verified, with a total of 11,314 data. The size of the test data set is 500, 2021 and 5044. Meanwhile, in order to obtain accurate labels from these MOOC comments, we employed the crowd intelligence [9] algorithm for labeling. Moreover, we recruited 5 annotators for each candidate comment.

We compare the BERT-POS with several baseline methods, including BERT-base [2], CNN [8], GatedCNN [10], ERNIE [5], ChineseBERT-base [6], BERT-wwm-ext [1], LSTM [11] and GRU [12].

3.2 Results and Analysis

All the experimental results of the models as shown in Table 1. F1-score is a weighted average of precision and recall, which is a comprehensive index to evaluate the sentiment analysis of each model. From Table 1, we can see that our BERT-POS model has significantly improved compared with the BERT-base and other baselines on the data sets. This research shows that the combination of part-of-speech tagging and sentiment analysis can effectively improve the accuracy of sentiment analysis of BERT model.

Table 1. The test results obtained by BERT-POS and other eight kinds of model on the test data set (the data units in table are percentages).

Model	F1-socre		
	500	2021	5044
BERT-base [2]	86.81	87.99	89.62
ERNIE [5]	88.15	89.46	91.66
ChineseBERT-base [6]	87.45	89.18	90.87
BERT-wwm-ext [1]	86.15	89.12	89.96
LSTM [11]	90.00	90.40	92.89
GatedCNN [10]	89.51	91.05	92.24
CNN [8]	87.38	88.77	92.06
GRU [12]	87.05	89.11	90.38
BERT-POS	**92.01**	**93.72**	**95.27**

---

[1] http://www.icourse163.com.

4   Conclusion and Future Work

In this paper, we propose a new model BERT-POS, which changes the embedding layer of the BERT model and introduces part-of-speech embedding. By using the course comments from iCourse as the training and testing dataset, We compare the proposed model BERT-POS with the eight famous NLP models. The experimental results show that the accuracy of sentiment analysis of BERT is significantly improved. In the future, we plan to train the multi-classification sentiment analysis model.

Acknowledgements. This work was supported by the National Natural Science Foundation of China (61877029).

References

1. Cui, Y., et al.: Pre-training with whole word masking for Chinese BERT, pp. 11–21. arXiv preprint arXiv:1906.08101 (2019)
2. Devlin, J., Chang, M.W., Lee, K., Toutanova, K.: BERT: pre-training of deep bidirectional transformers for language understanding, pp. 1–16. arXiv preprint arXiv:1810.04805 (2018)
3. Paranjape, B., Bai, Z., Cassell, J.: Predicting the temporal and social dynamics of curiosity in small group learning. In: Penstein Rosé, C., et al. (eds.) AIED 2018. LNCS (LNAI), vol. 10947, pp. 420–435. Springer, Cham (2018). https://doi.org/10.1007/978-3-319-93843-1_31
4. Santos, O.C., Salmeron-Majadas, S., Boticario, J.G.: Emotions detection from math exercises by combining several data sources. In: Lane, H.C., Yacef, K., Mostow, J., Pavlik, P. (eds.) AIED 2013. LNCS (LNAI), vol. 7926, pp. 742–745. Springer, Heidelberg (2013). https://doi.org/10.1007/978-3-642-39112-5_102
5. Sun, Y., et al.: ERNIE: enhanced representation through knowledge integration, pp. 1–8. arXiv preprint arXiv:1904.09223 (2019)
6. Sun, Z., et al.: ChineseBERT: Chinese pretraining enhanced by glyph and pinyin information, pp. 1–12. arXiv preprint arXiv:2106.16038 (2021)
7. Vaswani, A., et al.: Attention is all you need. In: Advances in Neural Information Processing Systems, pp. 5998–6008 (2017)
8. Wang, S., Huang, M., Deng, Z., et al.: Densely connected CNN with multi-scale feature attention for text classification. In: IJCAI, pp. 4468–4474 (2018)
9. You, Q., Luo, J., Jin, H., Yang, J.: Cross-modality consistent regression for joint visual-textual sentiment analysis of social multimedia. In: Proceedings of the Ninth ACM International Conference on Web Search and Data Mining, pp. 13–22 (2016)
10. Yuan, J., et al.: Gated CNN: integrating multi-scale feature layers for object detection. Pattern Recogn. **105**, 107131–107141 (2020)
11. Zhou, C., Sun, C., Liu, Z., Lau, F.: A C-LSTM neural network for text classification, pp. 1–10. arXiv preprint arXiv:1511.08630 (2015)
12. Zulqarnain, M., Ghazali, R., Ghouse, M.G., Mushtaq, M.F.: Efficient processing of GRU based on word embedding for text classification. JOIV Int. J. Inform. Visual. **3**(4), 377–383 (2019)

An Adaptive Biometric Authentication System for Online Learning Environments Across Multiple Devices

Riseul Ryu(✉) ⓘ, Soonja Yeom ⓘ, David Herbert ⓘ, and Julian Dermoudy ⓘ

University of Tasmania, Hobart TAS, Australia
`riseul.ryu@utas.edu.au`

Abstract. Online learning environments have become a crucial means to provide flexible and personalised pedagogical material, and a major driving cause is due to the COVID-19 pandemic. This has rapidly forced the migration and implementation of online education strategies across the world. Online learning environments have a requirement for high trust and confidence in establishing a student's identity and the authenticity of their work, and this need to lessen academic malpractices due to increased online delivery and assure the quality in education has accelerated. In addition to this, due to the ubiquity of mobile devices such as smartphones, tablets and laptops, students use a variety of devices to access online learning environments. Therefore, authentication systems for online learning environments should operate effectively on those devices to authenticate and invigilate online students. Confidence in authentication systems is also crucial to detect cheating and plagiarism for online education as strong authorisation and protection mechanisms for sensitive information and services are bypassed if authentication confidence is low. In this paper, we examine issues of existing authentication solutions for online learning environments and propose a design for an adaptive biometric authentication system for online learning environments that will automatically detect and adapt to changes in the operating environment. Multi-modal biometrics are applied in the proposed system which will dynamically select combinations of biometrics depending on a user's authenticating device. The adaptation strategy updates two thresholds (decision and adaptation) as well as the user's biometric template they are using the authentication system.

Keywords: Adaptive biometric authentication · Online learning · Ubiquitous mobile devices

1 Introduction

Security is the major concern in an online learning environment. It preserves the integrity, reliability and transparency of the educational process from deficiencies in detecting impersonation through inadequate authentication [1, 2]. Instructors and educators are not physically present during the learning process, and this creates issues in confidence [1]. Without a high confidence of assurance of the authenticity of students who are

© Springer Nature Switzerland AG 2022
M. M. Rodrigo et al. (Eds.): AIED 2022, LNCS 13356, pp. 375–378, 2022.
https://doi.org/10.1007/978-3-031-11647-6_73

accessing the online learning platform, the academic integrity and validity of online assessment is unreliable [1, 2].

There have been many efforts to confront the challenges in authentication for online learning [1, 3, 4]. However, past efforts have constraints of only identifying a user once—and at login time—or by continuously monitoring users through cameras under exam conditions or other specific activities [2, 5]. Biometric based authentication has been proposed in online learning environments for continuous and real-time monitoring and has shown viability [1, 3]. Nevertheless, studies on biometric based authentication are still limited [3, 4]. The limitations identified are.

1. a lack of exploration of different learning environments and circumstances as current research is mainly focused on the exam context and not the entire learning process [2],
2. a lack of authenticator diversity as face and keystroke biometrics are the most common authenticators [3], and
3. a lack of discussion on problems caused from intra-class variability [4].

In this study, we present the design of an adaptive biometric authentication system for online learning environments which offers a dynamic change in the authentication process to cope with changes in the operating environment.

2   Problem Statements

There are many attempts to provide solutions to secure online learning environments using biometrics, however, the biometric features used for online learning environments are mainly limited to face, keystroke and/or voice [3] with past assumptions that a user will only use a single desktop system. As there is now a ubiquity of smart, mobile devices such as smartphones, tablets, and notebooks, a user may potentially use multiple devices to access online learning environments — past assumptions are no longer valid since mobile devices are widely utilized to access online learning environments. Therefore, we identify that a need for a biometric system that can leverage different categories or classes of devices, and that this system is required to include several authenticator factors in operating environments that may vary.

The majority of current solutions identified apply authentication only to test/exams [5]. It is clear that authentication in online learning environments should not be limited to exam environments alone but should also consider other environments and scenarios such as access to learning module/deliveries, peer grading, participation in workshops/tutorials, and so on [6]. Therefore, an exploration is needed to determine whether current solutions can be leveraged, expanded and/or enhanced to address other types of user interaction in online learning environments.

3   The Design of an Adaptive Biometric Authentication System

In this section, we propose the design of an adaptive biometric authentication system which enhances the design provided by Mhenni, Cherrier, Rosenberger and Essoukri

Ben Amara [7] using multi-modal biometrics. The proposed adaptive biometric system follows two main phases: enrolment and test/recognition. During the enrolment phase, the system will collect a genuine user's biometric samples to generate a template that will be stored in a database for later comparison for authentication purposes. The recognition phase aims to decide whether to authenticate or de-authenticate a claimed user. Based on the accessing device detected, varying categories of biometric traits will be sampled. Moreover, the surrounding environmental conditions will be considered during the classification process to allow for variations in the lighting condition, noise, and device motion. For classification, the system will consider the application of several types of classifiers such as Support Vector Machine (SVM), Random Trees, K-Nearest Neighbors (k-NN), etc. [4]. Post-acquisition, the score for each classifier will be used to generate a global classification score. The scores can be fused with the weighted sum method, min-max method, or other methods [5, 6]. The global score will then be compared to an individual per-user threshold to authorise the claimed user identity.

Once a user is verified as a genuine user, the adaptation process is triggered. This process aims to continuously adapt the biometric system to intra-class variations of the biometric samples received from a user. This process will update the matching parameter (i.e., user threshold) and templates stored in the system database. It has been demonstrated that updating both the matching parameter and templates improve the recognition performance [7]; however, that study used a single biometric characteristic— keystroke dynamics—and not multiple biometric characteristics.

Threshold Adaptation: For the proposed system, we use the combination of mixed criteria and dual thresholds to make the adaptation decision [4, 7]. Two thresholds (the decision threshold and the adaptation threshold) are used. The global score (the fusion score of each classification result) is compared to the decision threshold to authenticate a user's identity. After authentication acceptance, the score of each classifier is then compared to the adaptation threshold to decide whether to adapt the sample. For example, the adaptation process will be performed only for a modal that meets the adaptation threshold. The thresholds are specific to a user and to the specific modality used.

Template Adaptation: The system adapts the referential biometric template during the adaptation process to minimise the impact of template ageing [4, 7]. The biometric references will be built from several samples collected during the enrolment phases. The adaptation process will be initiated based on an adaptation threshold—this threshold will be slightly decreased over a user's access lifetime based on the equation proposed in Mhenni, Cherrier, Rosenberger and Essoukri Ben Amara [7]. Adaptation uses a semi-supervised mode using a co-training method—this uses the knowledge of one modality to support labeling of another [4]. The types of semi-supervised machine learning algorithms can be k-NN classifier, SVM, Random Forest, etc. The adaptation process will be performed after the decision criterion of adaptation is met (known as online/real-time adaptation). It is aligned with the semi-supervised adaptation mode since the adaptation system uses the label computed by the verification method on the selected query [4]. The system will initially use the growing window mechanism; therefore, the accepted sample that satisfies the adaptation criteria will be added to the reference samples. Once the maximum size of user references is reached, usage control will be applied. This will remove a sample if a sample is less frequently used.

4 Conclusion and Future Work

This paper proposed a design of an adaptive biometric authentication system for online learning environments, which adapts the decision/adaptation threshold and biometric templates to support template ageing. It is expected that the proposed approach may improve the performance of the authentication system as it provides a user-specific mechanism, and it adapts to the changes in the operating environment. It will provide insights and opportunities to enhance existing authentication systems applied in online learning environments.

The adaptive biometric system has been proposed as a solution to remedy template ageing and/or changing environmental conditions in a biometric authentication system. Noting, however, that there is a lack of studies on how to determine the cause of changes to apply different adaptation strategies and how to evaluate such systems in the field of online learning environments, in the future we will implement and evaluate a system based on the proposed design, and evaluate its feasibility based on authentication performance and efficiency through comparison against existing biometric authentication systems.

References

1. Muzaffar, A.W., Tahir, M., Anwar, M.W., Chaudry, Q., Mir, S.R., Rasheed, Y.: A systematic review of online exams solutions in E-Learning: techniques, tools, and global adoption. IEEE Access 9, 32689–32712 (2021)
2. Labayen, M., Vea, R., Flórez, J., Aginako, N., Sierra, B.: Online student authentication and proctoring system based on multimodal biometrics technology. IEEE Access 9, 72398–72411 (2021)
3. Jack, C., Kevin, C.: Biometric authentication techniques in online learning environments. In: Information Resources Management Association (ed.) Research Anthology on Developing Effective Online Learning Courses. IGI Global, Hershey, PA, USA (2021)
4. Pisani, P.H., et al.: Adaptive biometric systems: review and perspectives. ACM Comput. Survey 52 (102), 1–38 (2019)
5. Kaur, N., Prasad, P.W.C., Alsadoon, A., Pham, L., Elchouemi, A.: An enhanced model of biometric authentication in E-Learning: using a combination of biometric features to access E-Learning environments. In: 2016 International Conference on Advances in Electrical, Electronic and Systems Engineering (ICAEES), pp. 138–143. IEEE, Putrajaya (2016)
6. Fenu, G., Marras, M., Boratto, L.: A multi-biometric system for continuous student authentication in e-learning platforms. Pattern Recogn. Lett. 113, 83–92 (2018)
7. Mhenni, A., Cherrier, E., Rosenberger, C., Essoukri Ben Amara, N.: Double serial adaptation mechanism for keystroke dynamics authentication based on a single password. Comput. Secur. 83, 151–166 (2019)

Detecting Teachers' in-Classroom Interactions Using a Deep Learning Based Action Recognition Model

Hiroyuki Kuromiya[✉], Rwitajit Majumdar, and Hiroaki Ogata

Kyoto University, Yoshida-honcho, Kyoto, Japan
khiroyuki1993@gmail.com

Abstract. In-classroom observations often rely on developed protocols and human observers. However, it requires a lot of human effort. This study investigates how accurately the pre-trained action recognition model can label teacher's behaviors in the classroom. We adopt SlowFast, a state of the art action recognition model, to a real classroom at a junior-high school mathematics class in Japan. In a pilot study of a mathematics class in a junior high school, the pre-trained model had 92.7% accuracy to identify teacher's posture, 31.7% related to the teacher's interaction with objects, and 26.8% related to teacher-student interaction. Compared to the existing baseline (34.3%), our results indicate that the pre-trained model adopts well to classroom videos as well. Possible reasons for the low accuracy of the verbs in the last two categories are (1) the pre-trained model could not sufficiently deal with objects unique to the classroom, such as a whiteboard, and (2) the teacher wore masks as an infection control measure, which made it difficult to recognize teacher's talking behavior. This study provides an initial automated approach to have a teacher's in-classroom interaction dataset extracted from the class videos. One needs to be aware of the ethical implementation and then such deep learning technologies have potential for a data-driven paradigm for the teacher's in action reflection.

Keywords: Action recognition · Deep learning · Video analytics · Teaching reflection

1 Introduction

Systematic observation of classroom behavior is a primary research methodologies in educational research [1]. However, in-classroom observations often rely on developed protocols and human observers [2]. These coding schemes are very useful in the sense that they reduce variability among observers and provide consistent results for coding. However, it takes a lot of human effort.

The aim of this study is to investigate how accurately the deep learning based pre-trained action recognition model can classify teacher's behaviors in the classroom. We used the SlowFast action recognition model [3] which is trained by the Atomic Visual Action (AVA) dataset [4]. It has not yet been investigated how accurately the pre-trained action recognition model can classify teachers' behavior in classrooms.

© Springer Nature Switzerland AG 2022
M. M. Rodrigo et al. (Eds.): AIED 2022, LNCS 13356, pp. 379–382, 2022.
https://doi.org/10.1007/978-3-031-11647-6_74

2 Related Works

The technology of action recognition has advanced dramatically in recent years [5]. With the rise of deep learning, researchers started to adapt CNNs for video problems [6], and others have used deeper networks such as RNNs [7]. However, these models are computationally expensive and difficult to optimize, and there is a need for a model that can handle both time and space and optimize efficiently. One model that achieves both computational efficiency and integration of temporal information is the SlowFast model [3], which is inspired by the mechanism of human action recognition and is characterized by its division into two paths: a slow path for recognizing space and a fast path for recognizing motion. It achieved the state of the art on multiple action recognition dataset.

One of the popular dataset for action recognition tasks is Atomic Visual Action (AVA) dataset [4]. The dataset is taken from 437 movies and human action labels are provided for one frame per second. As the word "atomic" implies, the AVA dataset is unique in that it combines several basic verbs to describe human behavior. The verbs are categorized as three groups: pose (e.g., stand, walk), person-object (e.g., touch, carry/hold), and person-person (e.g., talk to, watch). The AVA dataset has an advantage in terms of ease of manual coding due to the small number of verbs, and in this study, the verbs in the AVA dataset were used to label the teacher's actions.

3 Methods

In this pilot study, we targeted a 50-min class at a junior-high second-grade Math class in Japan. There were 20 students in the class and the teacher had 23-year teaching experience. One observer recorded the teacher from behind the classroom using a SONY handy camera (resolution: 1920 × 1080, frame rate: 30 fps). Due to the late start of the class, the first two minutes were excluded from the video, so we got a 48-min teaching video in total. We described to the teacher about the purpose of the research and got a clearance to conduct this pilot. The positioning of the camera was at the back of the class to avoid student's faces as much as possible. Correct label assignment was done by double-checking by two workers based on the verbs in the AVA data set (Cohen's Kappa = .66, .72, and .42 for each three categories). Where the labels differed, the two annotators discussed and decided on the final label. This became the evaluation labels for the action recognition model.

In this study, we used a pretrained model of the SlowFast action recognition model, which is published on the website[1]. We selected a model described as the depth of the architecture: 101, initial parameters: Kinetics 600, frame length: 8, sample rate: 8, MAP: 29.1, and AVA version: 2.2. The time to process the 48 min of video was 47.2 min on our local machine (CPU: Intel Core i9-11900K, GPU: Nvidia GeForece RTX 3090, RAM: 32 GB).

---

[1] https://github.com/facebookresearch/SlowFast/blob/main/MODEL_ZOO.md.

4   Results

First, we examined the overall accuracy of the SlowFast action recognition model to our dataset. To calculate the accuracy, we compared the output of the model with manual annotations by the three categories of verbs (Fig. 1). The first category is teachers' posture. Compared to manual annotations, SlowFast correctly classified 38 out of 41 frames (92.7%). There were only three frames where the model mis-classified, which is very high when compared to the baseline reported from the authors [4] (34.3%). Next, we examined the verbs related to teacher-object interaction. Compared to manual annotations, SlowFast could correctly classify 13 out of 41 frames (31.7%). This result is a bit low compared to the baseline, which tells us the difficulty of capturing classroom specific actions like wiring on the whiteboard. The last category is the verbs related to teacher-student interaction. Compared to manual annotations, SlowFast correctly classified 11 out of 41 frames (26.8%), which is also much lower than the baseline. This may be due to the fact that machine coding does not take voice data into account, and students and teachers wore masks to prevent infection. It made it difficult to capture the teacher's talking behaviors in the classroom.

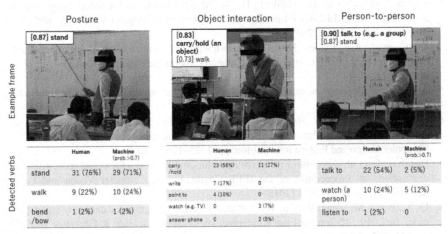

Fig. 1. Annotation example and detected verbs of the lecture video for each three categories of the verbs. The percentages do not add up to 100 because the frames where the appropriate verb was not found are calculated as N/A.

5   Discussion

In the education field, It was common for analysts to prepare their own data for training (e.g., in the smart classroom environment [8] or from YouTube videos [9]). Our approach resolved the data preparation problem by using a pre-trained model using an open dataset. For the next step, we aim to build a classroom orchestration dashboard with the action recognition model. A good reference is EduSense [10], a class observation system using

OpenPose: a deep-learning based skeletal estimation module. This system displays the number of times students raise their hands from their skeletons on a dashboard. Based on our study, we would create a similar dashboard with the action recognition model, which enables teachers to see more diverse information on students' behavior.

Acknowledgement. This study is supported by JST JPM-JAX20AA, JSPS 21J14514, SPIRITS 2020 of Kyoto University, JSPS 20K20131, JSPS 22H03902, JSPS 16H06304, NEDO JPNP18013, and NEDO JPNP20006.

References

1. Walkington, C., Michael, M.: Classroom observation and value-added models give complementary information about quality of mathematics teaching. In: Designing Teacher Evaluation Systems: New Guidance from the Measures of Effective Teaching Project, pp. 234–277, Josey Bass, San Francisco (2013)
2. Volpe, R.J., DiPerna, J.C., Hintze, J.M., Shapiro, E.S.: Observing students in classroom settings: a review of seven coding schemes. School Psych. Rev. **34**, 454–474 (2005)
3. Feichtenhofer, C., Fan, H., Malik, J., He, K.: Slowfast networks for video recognition. In: Proceedings of the IEEE/CVF International Conference on Computer Vision, pp. 6202–6211 (openaccess.thecvf.com, 2019)
4. Gu, C., et al.: Ava: a video dataset of spatio-temporally localized atomic visual actions. In: Proceedings of the IEEE Conference on Computer Vision and Pattern Recognition, pp. 6047–6056 (2018)
5. Zhu, Y., et al.: A comprehensive study of deep video action recognition. arXiv preprint arXiv: 2012.06567 (2020)
6. Karpathy, A., Toderici, G., Shetty, S., Leung, T., Sukthankar, R., Fei-Fei, L.: Large-scale video classification with convolutional neural networks. In: Proceedings of the IEEE conference on Computer Vision and Pattern Recognition, pp. 1725–1732 (2014)
7. Donahue, J., et al.: Long-term recurrent convolutional networks for visual recognition and description. In: Proceedings of the IEEE Conference on Computer Vision and Pattern Recognition, pp. 2625–2634 (2015)
8. Li, X., Wang, M., Zeng, W., Lu, W.: A students' action recognition database in smart classroom. In: 2019 14th International Conference on Computer Science & Education (ICCSE), pp. 523–527 (2019)
9. Sharma, V., Gupta, M., Kumar, A., Mishra, D.: EduNet: a new video dataset for understanding human activity in the classroom environment. Sensors 21 (2021)
10. Ahuja, K., et al.: EduSense: practical classroom sensing at Scale. In: Proceedings of the ACM on Interactive, Mobile, Wearable and Ubiquitous Technologies 3(3), pp. 1–26 (2019)

LetGrade: An Automated Grading System for Programming Assignments

K. N. Nikhila$^{(\boxtimes)}$ and Sujit Kumar Chakrabarti$^{(\boxtimes)}$

International Institute of Information Technology, Bangalore, India
{nikhila.kn,sujitkc}@iiitb.ac.in

Abstract. Manually grading programming assignments is time consuming and tedious, especially if they are incorrect and incomplete. Most existing automated grading systems use *testing* or *program analysis*. These systems rely on a single reference solution and award no marks to submissions that differ from the reference solution. In this research, we introduce an automated grading model *LetGrade*. *LetGrade* is a supervised machine learning-based mechanism for automatically identifying the approach of solving and grading a student's submission. The method looks for a score of the similarities between the submitted solution and multiple correct solutions available to determine the solution's approach. The calculated similarity score is then entered into a pretrained supervised machine learning model that grades the submission. Our models were evaluated against datasets containing Python and C programming problems. The average variance in the grade predicted by the supervised machine learning model is consistently close to 0.5. This indicates that the models can accurately predict the grade within a 10% margin of error.

Keywords: Automated evaluation · Static analysis · Testing · Machine learning · Supervised learning

1 Introduction

Increasing enrolment in programming-based courses has led instructors to utilise automated tools in evaluating programming assignments so as to satisfy the demands of quality and objectivity on the one hand, and timeliness on the other. Most approaches rely heavily on testing. However, fundamental limitations of testing, which are well known in software engineering, also manifest in the context of automated evaluation. Moreover, testing can only verify output, so there are additional problems. Therefore, there has been a lot of work done that uses automated techniques based on static analysis.

A typical workflow of a static analysis based evaluation tool would involve a step that computes an estimated structural similarity between a submitted solution and the reference solution [11]. This similarity estimate will then directly translate into the marks awarded to a given solution. Although structural similarity correlates with the logical closeness of a submission to a reference solution,

© Springer Nature Switzerland AG 2022
M. M. Rodrigo et al. (Eds.): AIED 2022, LNCS 13356, pp. 383–386, 2022.
https://doi.org/10.1007/978-3-031-11647-6_75

it does not linearly map to the grade. The relation between structural similarity and final marks appears to be at best something monotonic but non-linear. In the context of this paper, we define the step of creating a mapping between structural similarity scores and marks as *grading*. The mapping itself is called a *grading model*. The grading model is nothing but a supervised machine learning model that maps similarity scores to marks.

In this paper, we present our investigations done around devising the above grading model. We have tried a number of well-known regression models as our grading model. We have evaluated these models on submissions collected from online coding platform *HackerRank* [3] and our institute's *Introductory programming course in Python*. Problem-specific models and generalised models are able to predict the grade of incorrect solution with ±10% tolerance.

The specific contributions of this paper is a novel approach to automated evaluation of programming assignments with focus on the grading model.

Section 2 introduces the Grading Model. Section 3 discusses related works and Sect. 4 concludes the paper.

2   Approach

2.1   Preliminaries

Some of the principles followed in designing our automated evaluation system are as follows:

1. **Static analysis + testing:** As mentioned in Sect. 1, testing has some inherent shortcomings as an underlying method for automated evaluation. The primary reason for this is that differences between two solutions which produce identical output can not be easily (and in some cases, not at all) detected through testing. Hence, apart from testing, our system also uses static analytic approach: a submission which is structurally more similar to a reference solution is a likely more correct solution.
2. **Support for multiple approaches:** We also observe that there are likely multiple correct approaches to solve a programming problem. It is our purpose to support multiple solution approaches to be considered without making it necessary for the instructor to know about them priorly.
3. **Grading:** We go with the assumption that the verdict of the human grader provides us with the ground truth in evaluation. Hence, our system involves a final step where we map similarity value computed by our system to actual grade by passing them through a grading model, which is the main topic of this paper.

2.2   Grading Model

In the manual evaluation, the grading scheme allows the maximum mark $a > 0$ and the minimum mark is 0 for each submission. The problem resembles the regression problem in supervised machine learning since domain experts assign marks in the interval $[0, a]$. The grade given by the domain expert based on the grading criteria is referred to as the *ground truth*.

We found a nonlinear relationship between *ground truth* and *similarity score*. Therefore, the similarity score and the method of implementation are the key features for building the model and grading the submissions. The feature *similarity score* is created during the approach matching step from the abstract representation of the program. Relationship between the *ground truth* and the *similarity score* is defined by Eq. 1.

$$M_i = f(S_i) \tag{1}$$

where M_i is the grade returned by the model and S_i is the highest similarity score for the ith incorrect solution and correct solutions. We have used regression model, which is a supervised machine learning technique to learn f. The supervised machine learning models *Random Forest Regression* and *Support Vector Regression* with three distinct kernels *RBF, Poly* and *linear* were investigated. The models were trained and validated using a variety of data sets which included set of problems from different programming language. We use 10% of the data as test data and the rest 90% of the data to train the models in each training phase to verify the model's generalisation capacity. 4-fold cross-validation is used to train all of the models. The evaluation metrics *Explanatory Variance (EV)* and R^2-*Score* are used to compare the models.

3   Related Work

This section discusses the prior art in various automated evaluation methods available for evaluating programming assignments and position our work w.r.t. them.

A majority of the current automated evaluation tools used today are mainly test-case based [1,4,9]. In most of these, the evaluator generates a set of test cases either manually or automatically to evaluate the programs. Each test case carries a particular weight of marks and calculates the number of test cases passed and assigns grades [5,6,13]. Another category of automated grading methods considers the semantic similarities among the programs based on abstract representation of the program like *abstract syntax tree* (AST), *control flow graph* (CFG), *data flow graph* (DFG)) and *program dependence graph* (PDG) etc. [2,7,12]. In addition, we have recently seen the evolution of automated grading using machine learning [8,10].

The work presented in our paper builds upon several of the above pieces of work. We use testing to segregate the correct solutions from the partially correct/incorrect ones. We apply feature based similarity estimation in our work both at the similarity estimation stage and marking stage similar to [10]. However, this is done as a step in a larger process. Further, we estimate structural similarity w.r.t. multiple possible reference solutions, rather than a unique one provided by the instructor. Unlike in all pure testing or pure static analysis based approaches, we use a combination of testing, static analysis and machine learning to award the final marks to a solution. This also distinguishes our contribution w.r.t. all evaluation approaches surveyed above.

4 Conclusion

In this work, we have addressed the difficulty in grading the incorrect programming problems and introduced a system for grading the same. Our method LetGrade are capable to grade the incorrect programs like a human grader. The average variance in the grade predicted by the supervised machine learning model is consistently close to 0.5. This indicates that the models can accurately predict the grade within 10% margin of error. This method simplifies the time-consuming work of the instructors in the programming course. We would like to extend the system to other programming languages.

References

1. Douce, C., Livingstone, D., Orwell, J.: Automatic test-based assessment of programming: a review. J. Educ. Res. Comput. (JERIC) 5(3), 4-es (2005)
2. Goswami, N., Baths, V., Bandyopadhyay, S.: AES: automated evaluation systems for computer programing course. In: Proceedings of the 14th International Conference on Software Technologies, pp. 508–513 (2019)
3. HackerRank: Hackerrank (April). http://hackerrank.com
4. Jackson, D.: A software system for grading student computer programs. Comput. Educ. 27(3–4), 171–180 (1996)
5. Joy, M., Griffiths, N., Boyatt, R.: The boss online submission and assessment system. J. Educ. Resour. Comput. (JERIC) 5(3), 2-es (2005)
6. von Matt, U.: Kassandra: the automatic grading system 22(1) (1994). https://doi.org/10.1145/182107.182101
7. Naudé, K.A., Greyling, J.H., Vogts, D.: Marking student programs using graph similarity. Comput. Educ. 54(2), 545–561 (2010)
8. Singh, G., Srikant, S., Aggarwal, V.: Question independent grading using machine learning: the case of computer program grading. In: Proceedings of the 22nd ACM SIGKDD International Conference on Knowledge Discovery and Data Mining, pp. 263–272 (2016)
9. Singh, R., Gulwani, S., Solar-Lezama, A.: Automated feedback generation for introductory programming assignments. In: Proceedings of the 34th ACM SIGPLAN Conference on Programming Language Design and Implementation, pp. 15–26 (2013)
10. Srikant, S., Aggarwal, V.: A system to grade computer programming skills using machine learning. In: Proceedings of the 20th ACM SIGKDD International Conference on Knowledge Discovery and Data Mining, pp. 1887–1896 (2014)
11. Verma, A., Udhayanan, P., Shankar, R.M., KN, N., Chakrabarti, S.K.: Source-code similarity measurement: syntax tree fingerprinting for automated evaluation. In: The First International Conference on AI-ML-Systems, pp. 1–7 (2021)
12. Wang, T., Su, X., Wang, Y., Ma, P.: Semantic similarity-based grading of student programs. Inf. Software Technol. 49(2), 99–107 (2007)
13. Wick, M., Stevenson, D., Wagner, P.: Using testing and Junit across the curriculum. ACM SIGCSE Bull. 37(1), 236–240 (2005)

Foreword to Machine Didactics: On Peer Learning of Artificial and Human Pupils

Daniel Devatman Hromada[1,2]([envelope]) [iD]

[1] Institute for Time-Based Media, Faculty of Design, Berlin University of the Arts, Berlin, Germany
dh@udk-berlin.de
[2] Einstein Center Digital Future, Berlin, Germany
https://bildung.digital.udk-berlin.de

Abstract. Process of human learning has many features in common with the process of machine learning. This allows for creation of human-AI tandems or smaller groups where all members of the tandem or a group learn and develop. Consistently with Vygotskyan and Piagetian theories of learning and role which peers and intersubjective relations play in such theories, we hypothesize that curricula can be established whereby human and artificial learners collaboratively learn together, resulting in a win-win situation for both organic and anorganic agents involved.

Keywords: Machine didactics · Peer learning · Machine learning · Human-machine parallelism · Zone of proximal development

1 Introduction

1.1 Point of Departure

We depart from a simple observation: process of learning of humans or other organic beings shares certain features with the process of machine learning (ML) [2]. One observes deeper analogies than those caused by the trivial terminological fact that both such processes are denoted by the participle "learning". First and foremost, both human as well as machine learning are able lead to discovery and emergence of practically useful generalizations which allow the agent - no matter whether human or artificial - to arrive to accurate conclusions, execute appropriate decisions and manifest well-adapted behaviors in novel and hitherto unseen present or future environments.

In fact, many among most accurate and efficient machine learning algorithms originated as metaphors transposing insights from neurosciences, behavioral sciences, genetic epistemology or developmental psychology into the *in silico* domain. Neural networks, of course, are the most famous example: triggered by Purkyne's discovery of a neural cell, reinforced by Hebbian associanist rule "cells that fire together wire together" and expressed by progressively evermore

© Springer Nature Switzerland AG 2022
M. M. Rodrigo et al. (Eds.): AIED 2022, LNCS 13356, pp. 387–390, 2022.
https://doi.org/10.1007/978-3-031-11647-6_76

complex models of artificial neuron - from perceptron and neocognitron to multi-layered, convolutional network architectures able to provide impressively accurate results in domains as diverse computer vision, speech recognition or time series analysis and prediction.

Asides neurosciences, behavioral psychology has also some words to say: both Thorndike's Law of effect which postulates that a pleasing consequence strengthens the action which triggered it, as well as Skinner's principles of operant conditioning able to stimulate certain kind of future behaviours by means of a reward or inhibit it by means of a punishment prepared solid empiric ground for what is nowadays known as the 3rd pillar of machine learning, i.e. the reinforcement learning paradigm [9]. Implementation of such algorithms into already existing hardware brings very tangible results: defeat of the human world champion of game of Go by the AlphaGo algorithm or attainment of human level of control in playing of 49 distinct computer games by a one single computational agent [7] gradually prepare us for the world where machines develop their own means how to achieve their objectives [1, 8].

1.2   Human-Machine Learning Parallelism

It is true that one cannot a priori exclude existence of an unsurmountable onto-logical difference between learning processes realized on an organic, carbon-based substrate of the human central nervous system and learning processes instantiated on universal Turing machines executed on artificial, silicium-based substrate of modern CPUs, GPUs and TPUs.

Still, similarities and characteristics shared between machine learning (ML) and human learning (HL) permit us to postulate that the process of machine learning could lead, mutatis mutandi, to results indistinguishable from those issued by and from the process of human learning. In layman terms:

Processes of ML and HL have features in common.

Asides being purely descriptive, the observation that *human-machine learning parallelism exists* yields productive consequences:

Humans and machines can learn together.

In other terms, curricula which combine both ML and HL components can be constructed and, if constructed properly, may have synergic potential to increase efficiency of both ML and HL more, than HL or ML curricula which unfold in isolation. And this brings us to peer learning.

1.3   Peer Learning

Discovery of a role of "peers" in processes of socialization and acquisition of knowledge undoubtedly belongs to most important moments of modern and post-modern educational sciences. Thus, as surpassed and outdated are nowadays considered those classical and even 19th century educational concepts in which the notion of learning had been reduced to one-directional vertical transfer of information from a socially superordinated "mature" teacher (=adult) to a subordinated "immature" learner (=child). As indicated by both theoretical and empirical observations of Piaget [5] and Vygotsky [6] and confirmed by success of concepts of Freinet, Montessori or Feuerstein, surrounding children can and do significantly influence and modulate cognitive maturation of a child C.

As observed by practically every teacher faithful to his vocation, the field of vivid social forces generated and exerted by "peers" - i.e. siblings, schoolmates, friends or other subjects on a comparable level of intellectual development - impacts formation of pupil's personality and character equally strong - and sometimes even stronger - then force of Teacher's charisma, knowledge, skill and confucian oracle-like authority.

1.4   Human-Machine Peer Learning

The ultimate intention behind this extended abstract is not constrained to computer-science domain, nor to cognitive-science domain. The ultimate intention is didactical, it is pedagogical: we propose to shift the focus from theoretical algorithmic aspects of machine learning to concrete practical cases of *machine teaching* contextualized in an organized system of a well-thought curriculum.

In order to do so, we hereby introduce the concept of Human-machine peer learning (HMPL) which emerges as a direct logical consequence of conjunction of a human-machine parallelism and human innate affinity to peer and/or collaborative learning scenarios. After combining these two concepts, one states:

Humans and machines can learn from each other.

Main principle of HMPL being thus stated, we now enumerate two major imperatives of HMPL:

1. start small
2. posit zones of proximal development

Primo, the **start small** imperative. This imperative is based on an idea that process of learning of both human as well as artificial learners should depart from quantitatively and structurally minimal datasets. The strongest empiric evidence for importance of the start small principle in both human as well as machine learning comes from domains of psycholinguistics and computational linguistics. Thus, psycholinguists observe that *"mother's choice of simple constructions facilitated language growth [of a child]"* [4]. In the computational realm, the seminal paper of [3] summarized the reasons of success of a connectionist model of acquisition of English grammar with words: *"... However, when the training data were selected such that simple sentences were presented first, the network succeeded not inly in mastering these, but then going on to master the complex sentences as well."* [3].

Secundo, the **posit zones of proximal development** (ZPD) imperative. This imperative is based on an observation that didactic process is most efficient there, where structures-to-be-learned are neither too distant - and therefore unreachable - nor too similar - and therefore devoid of interest - from prior knowledge which the learner already has at her/his disposal. Concisely stated, the ZPD-hypothesis provides to any teacher - organic and artificial - a simple but efficient *didactic meta-algorithm* able to positively influence the impact of one's teaching practice. The core of such meta-algorithm is a *didactic loop*:

1. Assess what the learner knows (i.e. prior knowledge).
2. Expose the learner to novel structures which are close to, but not within, the domain of prior knowledge.
3. Once it is obvious that learner's domain of prior knowledge encompasses the novel structures proceed to step 1.

It is indeed the ability to recognize what the learner already knows (step 1) and what the learner *can* know (step 2) which distinguishes a good teacher from a bad one.

Within HMPL, zones of proximal development are to be assessed for each participant and each competence. If ever it is observed that level-of-mastery for two distinct competences σ and π, as exhibited by two participants X and Y is such that both $X_\sigma >\sim Y_\sigma$ holds in the same time as $X_\pi <\sim Y_\pi$ holds, we say that X and Y are in a state of *a mutual non-equilibrium* in respect to competences σ and π.

In case of such mutual non-equilibria, the main condition that Y can learn from X about σ whilst X will learn from Y something about π, is met.

It is the initial existence of such mutual non-equilibria and their gradual convergence into a state of didactic equilibrium which makes peer learning possible.

Exploration, evaluation and construction of such convergence processes is the main object of study of the research field hereby labeled as machine didactics.

References

1. Bostrom, N.: Superintelligence. Dunod (2017)
2. Dehaene, S.: How We Learn: The New Science of Education and the Brain. Penguin, London (2020)
3. Elman, J.L.: Learning and development in neural networks: the importance of starting small. Cognition **48**(1), 71–99 (1993)
4. Furrow, D., Nelson, K., Benedict, H.: Mothers' speech to children and syntactic development: some simple relationships. J. Child Lang. **6**(3), 423–442 (1979)
5. Susan, L.G.: Implications of Piagetian theory for peer learning. In: Cognitive Perspectives on Peer Learning, pp. 3–37 (1999)
6. Hogan, D.M., Tudge, J.R.H.: Implications of Vygotsky's theory for peer learning (1999)
7. Mnih, V., et al.: Human-level control through deep reinforcement learning. Nature **518**(7540), 529–533 (2015)
8. Russell, S.: Human Compatible: Artificial Intelligence and the Problem of Control. Penguin, London (2019)
9. Silver, D., et al.: Reward is enough. In: Artificial Intelligence, p. 103535 (2021)

Assisting Teachers in Finding Online Learning Resources: The Value of Social Recommendations

Elad Yacobson[1]([⊠]) [iD], Armando M. Toda[2] [iD], Alexandra I. Cristea[2] [iD],
and Giora Alexandron[1] [iD]

[1] Weizmann Institute of Science, 234 Herzl Street, Rehovot, Israel
{elad.yacobson1,giora.alexandron}@weizmann.ac.il
[2] Durham University, Stockton Road, Durham, UK
{armando.toda,alexandra.cristea2}@weizmann.ac.il

Abstract. With hybrid teaching and learning becoming the new educational reality, teachers often face the challenge of searching in open repositories that vary substantially in quality and standards, in order to find suitable learning materials that fit their students' needs and their own pedagogical preferences. Social recommendations, i.e. recommendations from fellow teachers about learning resources, are becoming a popular feature in Open Educational Resources (OER) repositories. However, *very little is known about their value for teachers*, namely, whether teachers actually rely on them for choosing learning resources. To address this gap, we studied the behaviour of science teachers who are using a nation-wide OER system with social-based recommendation features, in which teachers can share experiences and feedback on learning resources with the community. Our work helps to establish a reliable causal link between recommendations and use. This is done by adding a time-series quantitative analysis on the impact of the social recommendations on teachers' instructional choices.

Keywords: Social-based recommendation systems · Personalized teaching environments · Blended learning · Cooperative/collaborative learning · Human-computer interaction

1 Introduction and Background

Online educational repositories typically offer a wide collection of digital resources that are suitable for different pedagogical needs and individual preferences teachers may have. To aid search & discovery in these collections, open educational resources (OER) repositories typically offer various filtering mechanisms [3]. Whilst useful, these come with their own challenges within large OER

This research is part of the JANET project supported by a Making Connections Grant funded by Weizmann UK.

M. M. Rodrigo et al. (Eds.): AIED 2022, LNCS 13356, pp. 391–395, 2022.
https://doi.org/10.1007/978-3-031-11647-6_77

repositories, since teachers are generally overwhelmed by the large amount of information returned to them [2]. Various approaches have been examined in the past to address this difficulty, one of which is using recommender systems (RSs) [4]. RSs can be defined as software tools that provide suggestions on the most relevant items for a particular user [1].

Reviewing previous work on the use of RSs to aid teachers, we identified a major gap that exists in most of these studies concerning the evaluation of the RS usefulness from the users' perspective. This gap was already described more than a decade ago in [4]: "[...]a closer look to the current status of their development and evaluation reveals the lack of systematic evaluation studies in the context of real-life applications" (p.18). A very recent systematic literature review of RSs for teachers [1] indicated that not much has changed in this regard since then.

Realising the potential value of social recommendations in aiding teachers to find suitable learning resources (LRs), we implemented a recommendation panel into a nation-wide blended learning environment used by STEM teachers in Israel. This panel shows teachers' endorsement about resources they used (see details in Sect. 2).

The goal of the present study is to examine the usefulness of social recommendations for teachers. The main contribution of this paper is providing, for the first time (to the best our knowledge), a strong causal evidence on the impact of social recommendations on teachers' search strategies in real-life educational settings.

2   Methodology

2.1   The Learning Environment

The learning environment in which we conducted this research is PeTeL (**Perso**nalized **Te**aching and **Le**arning). PeTeL is both a shared repository of OERs, and an LMS that also includes social network features and learning analytics tools. It is developed within the Department of Science Teaching at the Weizmann Institute of Science, with the goal of providing STEM teachers with a blended learning environment for personalised instruction.

2.2   A 'Lightweight' Social Recommender System

A feedback mechanism was implemented into PeTeL during the 2019–2020 school year. This mechanism works as follows: each time a teacher downloads a new LR and uses it in class, the teacher is presented with a pop-up window, requesting the teacher to rate the LR and provide feedback on how it was used. Resources that are recommended are presented to the rest of the teacher community through a recommendation panel. The recommendation panel appears in the main page of PeTeL (see Fig. 1). Every time a teacher recommends a learning resource, the recommendation panel is updated with the new recommendation at the top end.

Each recommendation contains the name of the recommending teacher, the title of the resource, the content of the recommendation, and a link to the recommended resource, in order to provide easy access to it, so that interested teachers would be able to simply download it to their own personal environment.

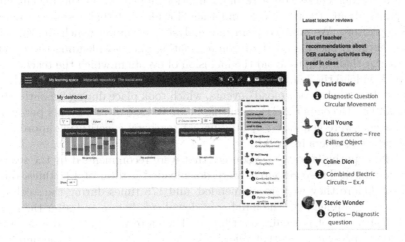

Fig. 1. Social recommendation panel introduced in PeTeL

2.3   Research Population

PeTeL is accessible to all STEM teachers in Israel for free, and is currently in use by approximately 1000 teachers. In this study, we focused on 169 physics teachers who are *active users* of PeTeL. Teachers are considered to be active users if they downloaded at least one LR from the environment and used it in class with their students.

2.4   Analysis

We analysed teachers' activity in the system, stored in log files containing data about their interaction with the platform, and intersected these logged activities with the information concerning the recommendation that appeared on the recommendation panel. In order to identify occurrences in which a teacher followed a recommendation and downloaded the recommended LR, we looked for events in which teachers clicked on a recommendation about a LR appearing in the recommendation panel, and shortly afterwards downloaded that resource to their own environment.

   We also conducted a hypothesis test, to determine if the number of downloads after the recommendation was significantly larger than the number of downloads before it, by comparing, per recommended resource that appeared in the recommendation panel, the number of times it was downloaded before and after the recommendation.

3   Results and Discussion

The log file containing the detailed account of teachers' clicks on recommendations and downloads of resources contained 7,099 events: 682 were clicks on recommendations and 6,417 were downloads of resources. We found 109 events where a teacher clicked on a recommendation appearing in the recommendation panel about a certain LR, and then downloaded that resource from the OER repository. We omitted from the analysis 8 events in which the time that passed between clicking on the recommendation and downloading the resource was greater than 24 h, to avoid the inclusion of events in which the teacher decision to download the resource might have been unrelated to the recommendation. This left 101 "click & download" events, which took place during 8 months (from April 2021 to December 2021), resulting in an average of 12.6 events per month. These events were generated by 50 different teachers, constituting approx. 30% of the active teachers in PeTeL.

In the hypothesis test, 81 resources that were recommended by the teachers were examined. These resources were downloaded a total of 132 times during the week before they were recommended, and 178 times during the week after they were recommended. The number of downloads before and after the recommendation were not normally distributed. Therefore we conducted a one-tailed Wilcoxon signed-rank test that rejected the null hypothesis that the number of downloads before the recommendation is equal to or higher than the number of downloads after the recommendation ($n = 81$, $Z = 1.955$, $p = 0.025$).

To conclude, RSs are becoming increasingly popular in the educational domain [1]. Developing such systems and integrating them into online learning environments requires substantial time and effort. However, up until now, there has been scarce empirical evidence that teachers find the recommendations given to them useful, making the aforementioned effort worthwhile. In this study, we examined the usefulness of a recommendation panel presenting social recommendations about learning resources, for Physics teachers using a blended learning environment that caters for a country-wide audience. This study presents a novel methodology for evaluating the systems' usefulness for teachers, which relies on temporal analysis of logs of teachers' actions. Results show that a substantial number of teachers (approx. 30% of the active teachers in the environment) used the social recommendations appearing in the recommendation panel to find and download learning materials.

Employing the temporal analysis methodology that was used in this paper to additional OER repositories with social-based RSs could help to establish the generalisability of the results that were reported above.

References

1. Dhahri, M., Khribi, M.K.: A review of educational recommender systems for teachers. In: 18th International Conference on Cognition, pp. 131–138 (2021)
2. Diekema, A.R., Olsen, M.W.: Personal information management practices of teachers. Proc. Am. Soc. Inf. Sci. Technol. **48**(1), 1–10 (2011)

3. Downes, S.: Models for sustainable open educational resources. Interdiscip. J. E-Learn. Learn. Objects **3**(1), 29–44 (2007)
4. Manouselis, N., Drachsler, H., Vuorikari, R., Hummel, H., Koper, R.: Recommender systems in technology enhanced learning. In: Ricci, F., Rokach, L., Shapira, B., Kantor, P.B. (eds.) Recommender Systems Handbook, pp. 387–415. Springer, Boston, MA (2011). https://doi.org/10.1007/978-0-387-85820-3_12

Bi-directional Mechanism for Recursion Algorithms: A Case Study on Gender Identification in MOOCs

Tahani Aljohani$^{(\boxtimes)}$, Alexandra I. Cristea, and Laila Alrajhi

Durham University, Durham, UK

taljohani7@gmail.com, {alexandra.i.cristea,laila.m.alrajhi}@durham.ac.uk

Abstract. Automatically identifying the learner gender, which serves as this paper's focus, can provide valuable information to personalised learners' experiences in MOOCs. However, extracting the gender from learner-generated data (discussion forum) is a challenging task, which is understudied in literature. Using syntactic features is still the state-of-the-art for gender identification in social media. Instead we propose here a novel approach based on Recursive Neural Networks (RecNN), to learn advanced syntactic knowledge extracted from learners' comments, as an NLP-based predictor for their gender identity. We propose a bi-directional composition function, added to NLP state-of-the-art candidate RecNN models. We evaluate different combinations of semantic level encoding and syntactic level encoding functions, exploring their performances, with respect to the task of learner gender profiling in MOOCs.

1 Prologue

MOOCs content can be personalised based on demographic data, particularly "gender" [6]. The gender parameter has already been shown to influence the success of the learning process. Traditionally, demographic data are extracted from pre-course questionnaires that are filled-in by the learners themselves; however, only about 10% provide their demographic data [1]. To resolve this issue, we research automatic extraction of learner characteristics, here, gender, from the traces learners leave in MOOCs. This falls under the umbrella of a more generic research area called *Author Profiling (AP)*, which promotes the use of automatic tools – developed based on Natural Language Processing (NLP) – for the purpose of identifying authors' demographics and characteristics, mainly based on their writing, and only basic types of syntactic representations of text, such as Part of Speech (POS), have been considered in previous studies for gender profiling [3]. The umbrella research question in this paper is: *Can advanced textual features extracted from MOOC discussion forums be used to classify a learner's gender?*. The main contributions of this paper are as follows: examining advanced syntactic features for the learner gender profiling; exploring state-of-the-art recursive models (tree-structured LSTM, SATA, and SPINN models), and applying them, for the first time, to author-profiling (here, for learners gender profiling), and then improving the current recursive models by introducing a novel bi-directional strategy.

© Springer Nature Switzerland AG 2022
M. M. Rodrigo et al. (Eds.): AIED 2022, LNCS 13356, pp. 396–399, 2022.
https://doi.org/10.1007/978-3-031-11647-6_78

2   Methodology

Approximately 322,310 comments from female and male learners in MOOCs

For text normalisation, simple NLP cleaning steps were applied. We made sure that these steps should not harm the learners' writing style. For *semantic representation*, GloVe was used for word input embeddings. These initial inputs are fed to the RecNN models to provide semantic vectors for each word for the leaf nodes or the initial inputs. Sentences are separated based on the full stop (.)" in each comment. This is to reduce the complexity during training time, since samples become shorter, which speeds up the training and generates more samples in the used data. An advanced NLP parse was used (Stanford Probabilistic Context-Free Grammar (PCFG) parser [5]), to convert the phrases at a syntactic level of the text (tree structure) in a binary mode to establish a binary tree. Then, these samples are fed to TreeNNs models for classification (see Fig. 1).

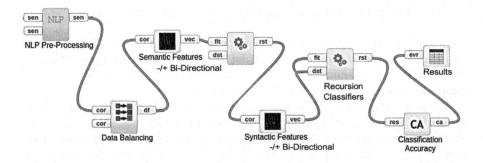

Fig. 1. General workflow of gender identification

As an initial step, a heatmap approach was used to understand the correlation between the POS (in its simple form) and discover a statistical measure linearly. Since there are many POS variables, the aim was to examine how dependent they are on each other, which may be shown in a 2D matrix called a correlation matrix. In the Figs. 2a and b, the lighter the colour between two variables, the stronger the correlation (and vice versa).

3   Findings

The distribution of POS patterns based on the mean, as shown in the heatmaps figures, does not reveal differences in the writing styles of the two groups. This means that this chosen approach also failed to capture the differences in syntactic patterns, due to its simplicity. Thus, DL approaches (advanced approaches) have been examined for gender profiling in this paper.

Three types of syntactic learning models were applied in this study: the original TreeLSTM [7], the Stack-Augmented Parser-Interpreter Neural Network

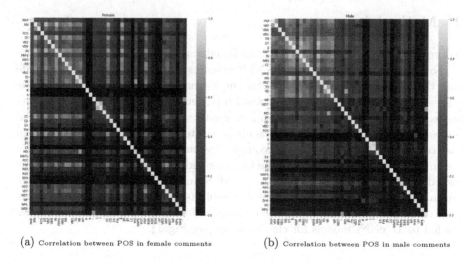

(a) Correlation between POS in female comments   (b) Correlation between POS in male comments

Fig. 2. Correlation between POS in female and male comments

(SPINN) [2], and Structure-Aware Tag Augmented (SATA) [4]. These models were chosen, as they are state-of-the-art DL models for such text representations. Additionally, we introduce new versions of these models based on a bi-directional composition function with different combinations.

We found that no study had examined the performance of these models by adding the bi-directional learning. Bi-directional learning has already shown its effectiveness in improving the sequential LSTM model and it is well-known that bi-directional LSTM outperforms vanilla LSTM for many NLP tasks. Thus, a hypothesis was made in this study that adding bi-directional TreeLSTM would improve the performance of SPINN and SATA. We investigated propagating the top-down direction of information and the bottom-up direction using bi-directional TreeLSTM. In fact, the uni-directional TreeLSTM by default processes inputs from the bottom-up direction in a bottom-up manner through the tree. So, we included the additional set of hidden state vectors in the top-down direction (from root to comment inputs), which then alters the model to the bi-directional paradigm. This is technically another TreeLSTM model, where the final hidden state is the final state vectors of the two LSTMs.

The syntactic learning in a TreeLSTM-based architecture in general consists of the following two steps: word-level encoding with a feedforward neural network or LSTM neural network; and sentence-level encoding with a tree-structured LSTM composition function. While previous literature has recommended using LSTM for word-level encoding, there is no such work to introduce bi-directional LSTM for the word-level encoding. Thus, this research also contributes to fill this gap, by adding the bi-directional LSTM at the word level as well. The motivation for this supplementary bi-directional technique is to increase the high-level representation of tree nodes during the recursive propagation across many branches.

TreeLSTM, SPINN,and SATA in their original structure are considered as baselines. The proposed versions of the bi-directional strategy that were applied for all 18 models (combinations of bidirectional and/or unidirectional of semantic and syntactic learning in each of the three classifiers). All of them were more effective than the baseline models. They achieved competitive performance in relation to each other. In general, all models achieved high performance in identifying the gender class (80.60% or above). This could promote the idea that the use of phrase-level representation is robust for learner gender classification. Every two versions of each model are very similar, but the bi-directional composition function models achieved slightly better results. The highest observed outcome in this paper was 82.62%. This was achieved by the newly proposed model based on the simple Forward Neural Network combined with the SPINN model.

4 Epilogue

This study indicates that bi-directional learning is promising in terms of improving the gender classification accuracy. It also shows the importance of the extra information that the model obtains during the training, which does not have to be limited to tags of constituents included in the SATA model. Furthermore, it is evident that using only a simple model with fewer parameters for word encoding by the Forward Neural Network (which used linear mapping) still achieves good results. This might be attributable to the fact that using linear mapping better preserves word-level semantics, while the LSTM encoding alters the semantic meaning at the word level, thereby making it harder to structure the sentence from a syntactic perspective. This might be related to task complexity.

References

1. Aljohani, T., Cristea, A.I.: User profiling on the future learn platform via deep neural networks, semantic and syntactic representations. Front. Res. Metrics Anal. **6**, 34 (2021)
2. Bowman, S.R., Gauthier, J., Rastogi, A., Gupta, R., Manning, C.D., Potts, C.: A fast unified model for parsing and sentence understanding. arXiv preprint arXiv:1603.06021 (2016)
3. HaCohen-Kerner, Y.: Survey on profiling age and gender of text authors. Expert Syst. Appl. 117140 (2022)
4. Kim, T., Choi, J., Edmiston, D., Bae, S., Lee, S.G.: Dynamic compositionality in recursive neural networks with structure-aware tag representations. In: Proceedings of the AAAI Conference on Artificial Intelligence, vol. 33, pp. 6594–6601 (2019)
5. Klein, D., Manning, C.D.: Accurate unlexicalized parsing. In: Proceedings of the 41st Annual Meeting of the Association for Computational Linguistics, pp. 423–430 (2003)
6. Shi, L., Cristea, A.I.: Demographic indicators influencing learning activities in MOOCs: learning analytics of future learn courses. Association for Information Systems (2018)
7. Tai, K.S., Socher, R., Manning, C.D.: Improved semantic representations from tree-structured long short-term memory networks. arXiv preprint arXiv:1503.00075 (2015)

Authoring Inner Loops of Intelligent Tutoring Systems Using Collective Intelligence

Thyago Tenório[1]([⊠]) [ID], Seiji Isotani[1] [ID], and Ig Ibert Bittencourt[2] [ID]

[1] Institute of Mathematics and Computer Science, University of São Paulo,
400 Trabalhador São Carlense Avenue, São Carlos, Brazil
`tm.thyago@usp.br`, `sisotani@icmc.usp.br`
[2] Computer Science Institute, Federal University of Alagoas,
S/N Lourival Melo Mota Avenue, Maceió, Brazil
`ig.ibert@ic.ufal.br`

Abstract. The inner loops on the Intelligent Tutoring Systems (ITS) are responsible for analyzing step-by-step resolutions and providing the necessary support to students successfully complete a task. Building adequate inner loops that address students' struggles requires a considerable amount of time and several interactions with experts of the domain knowledge which increases the costs and the complexity to develop ITS. To address this problem, we proposed a novel approach to build inner loops using students' collective intelligence (CI), which every interaction of a student to solve an exercise contributes to the process of authoring the ITS domain model. To evaluate our approach, we developed an ITS in a domain of numerical expressions that was used by 147 students. As a result, we observed that our approach helps to create an ITS with comprehensive and adequate inner loops using less time when compared to the time reported in the literature and with little intervention from experts. To the best of our knowledge, this is the first domain-independent approach that uses CI in the process of authoring inner loops of ITS.

Keywords: Authoring ITS · Human-based computation · Domain model

1 Introduction

ITS are computer programs with enormous potential from an educational point of view, which use artificial intelligence techniques to represent knowledge [2] and provide personalized feedback to students while they are performing a task [3], presenting results equivalent to the results of a human tutoring [6]. These

This study was financed in part by the Coordenação de Aperfeiçoamento de Pessoal de Nível Superior - Brasil (CAPES) - Finance Code 001.

results are obtained because ITS provide low granularity feedback for each step taken by the student in the solution of a task [5] through the inner loops. However, building adequate inner loops that address students' struggles requires a considerable amount of time and several interactions with experts of the domain knowledge which increases the costs and the complexity to develop ITSs at scale. In this sense, some ITS do not have inner loops, while others implement it in a simplified way [5], limiting the representation of knowledge for a specific context/domain but often overloading the teacher.

To address this problem, we proposed a novel approach to build inner loops using students' CI, where every interaction to solve an exercise contributes to the process of authoring the ITS domain model with inner loops, which is now co-produced between students and experts in a continuous process managed by the system, which has the direct benefits of reducing the teacher's overload and bringing the student into a more active role. To accomplish that we created a domain-independent authoring step-by-step resolution model that formally describes the process of creating inner loops as a CI problem. To the best of our knowledge, this is the first domain-independent approach that uses CI in the process of authoring inner loops of ITS [4]. In this sense, our research question is to verify whether **authoring ITS domain model using CI presents equivalent quality when compared to authoring using experts only**, contributing in reducing cost and time in the process.

2   Authoring Step-by-Step Resolution Model

Our domain-independent authoring step-by-step resolution model is a knowledge graph ($\mathbb{KG}$) as presented in Fig. 1. Each node represents a possible state of the problem, highlighting the initial states with a dashed edge line and the final states with a double-edged line. A state can contain more than one input and output step. In turn, each edge (represented by a directed arrow) represents a possible step to be performed. A resolution R is a path in the graph, which represents a sequence of vertices and edges. Other information using colors are also included, as the correctness of the states and the validity of the steps. A correct resolution only contains green vertices and edges while an incorrect resolution contains at least one red.

We also created an authoring process using CI, as illustrated in Fig. 2. The first step is to register the tasks by teachers, which serves as starting point of the $\mathbb{KG}$. Next, the second step consists of the students responding such tasks using step-by-step resolution and receiving feedback from the ITS, using them in their resolution process and evaluating their effectiveness. These evaluation data serve as parameter for the feedback recommendation system and teacher's moderation. Then, in the third step, the system updates the $\mathbb{KG}$ from resolutions and the feedback evaluations. It is important to highlight the incremental learning cycle that utilizes the students' CI as the core between the steps 2 and 3.

Next, Step 4 consists of an analysis of the current $\mathbb{KG}$ and the student's resolution R to identify points of additional information that may be requested

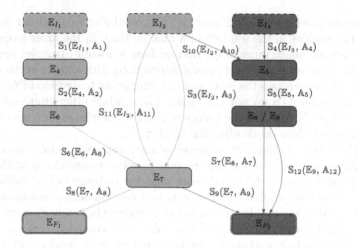

Fig. 1. Example of visual representation of a knowledge graph $\mathbb{KG}$

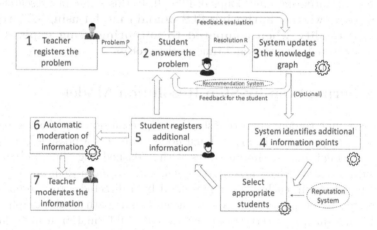

Fig. 2. Knowledge graph construction process

to the best students, based on a reputation system. Step 5 then consists of registering the additional information from the selected students, thereby improving their reputation. Step 6 consists of moderating the information automatically performed by the system and Step 7 consists of manual moderation performed by the teacher with the aim of refining the quality of the information contained.

3   Results and Discussions

To evaluate our approach we developed an ITS in a domain of numerical expressions that was used by 147 students over five consecutive days. The students answered each until 3 problems using step-by-step resolution, and their feedback was received and evaluated. Approximately 249 resolutions were registered. We check for accuracy, time spent to generate knowledge graphs and their coverage. The data indicates that the number of states and steps significantly increased according to the difficulty of the task, but using CI, the system achieved update information as new resolutions were added, which allowed for the construction of more complex KGs. An expert manually analyzed each generated KG and we had compared with the model results. Our model presented a success rate for the problems P1 = 87.70%, P2 = 77.24%, and P3 = 72.50% for the states, and P1 = 97.57%, P2 = 84.07%, and P3 = 75.89% for the steps.

Due to accuracy greater than 70% in all cases, **we concluded that it is possible authoring ITS domain model using CI with equivalent quality to authoring using experts only**. Furthermore, it should be noted that working with step-by-step resolutions requires additional work from the teacher. Some studies have estimated that 200–300 h of development will be spent per hour of instruction, which can be reduced by half using tools like CTAT [1]. We evaluated the time spent to build the knowledge graph using the students in hours (P1 = 6.5, P2 = 5.6 and P3 = 7.0), and we concluded that even though there is no way to directly compare these numbers, **we can observe that the time spent building the domain model using CI is less than the time spent when experts are used**. Additionally, the use of students in the process also complements the information that is unknown by the system, thereby increasing its ability to provide better feedback.

4   Conclusions and Future Works

We had presented a domain-independent step-by-step resolution model and a process to authoring ITS domain model using collective intelligence from students with few interventions by experts. The results indicated that the model built by our proposal is equivalent to the model built by experts, with a success rate always higher than 70%, but spending less time to be created. As future works, we intend to conduct an experiment to evaluate the real gain for students and teachers at classroom environment, to expand the model to include other information like hints, feedbacks, among others, and evaluate the student learning by participating in the process of authoring an ITS domain model.

References

1. Aleven, V., McLaren, B.M., Sewall, J., Koedinger, K.R.: The cognitive tutor authoring tools (CTAT): preliminary evaluation of efficiency gains. In: Ikeda, M., Ashley, K.D., Chan, T.-W. (eds.) ITS 2006. LNCS, vol. 4053, pp. 61–70. Springer, Heidelberg (2006). https://doi.org/10.1007/11774303_7
2. Polson, M.C., Richardson, J.J.: Foundations of Intelligent Tutoring Systems. Psychology Press, New York (2013)
3. Psotka, J., Massey, L.D., Mutter, S.A.: Intelligent Tutoring Systems: Lessons Learned. Psychology Press, New Jersey (1988)

4. Tenório, T., Isotani, S., Bittencourt, I.I., Lu, Y.: The state-of-the-art on collective intelligence in online educational technologies. IEEE Trans. Learn. Technol. **14**(2), 257–271 (2021)
5. Vanlehn, K.: The behavior of tutoring systems. Int. J. Artif. Intell. Educ. **16**(3), 227–265 (2006)
6. VanLehn, K.: The relative effectiveness of human tutoring, intelligent tutoring systems, and other tutoring systems. Educ. Psychol. **46**(4), 197–221 (2011)

Monitoring Tutor Practices to Support Self-regulated Learning in Online One-To-One Tutoring Sessions with Process Mining

Madiha Khan-Galaria[✉] and Mutlu Cukurova

University College London, London, UK
{m.khan.16,m.cukurova}@ucl.ac.uk

Abstract. This paper reports on research that aims to examine what tutoring practices in an online environment can promote students' self-regulated learning (SRL). First, we propose a theoretically grounded framework of signifiers that can be used to track tutor-student interactions with respect to SRL. Second, we operationalize the framework using log data from a virtual learning environment and process mining approaches. Our results demonstrate that there are structural differences in tutor-learner interactions between the high performing versus low performing tutors. High performing tutors show complex patterns of engagement, which emphasize open-ended questioning and reasoning. Whilst the low performing tutors use a more restricted range of teaching practices that focus on instruction and are more strictly led by the learning platform in which they tutor. We conclude the paper with a discussion of these findings.

Keywords: Self-regulated learning · Online tutoring · Process mining · Virtual classroom environment framework of signifiers

1 Introduction

This paper builds on recent research exploring the application of analytics to measure SRL (e.g. [3, 4]). We explore how process mining can be used to track the influence of tutor practices on learner SRL in online one-to-one human tutoring settings. Our key Research Questions are presented below:

1. Which signifiers can be used to track the influence of tutoring practices on learner SRL, in a Virtual Classroom Environment (VCE)?
2. How can we use process mining to detect variations between high and low performing groups of tutors' practice, regarding their influence on students' SRL?

© Springer Nature Switzerland AG 2022
M. M. Rodrigo et al. (Eds.): AIED 2022, LNCS 13356, pp. 405–409, 2022.
https://doi.org/10.1007/978-3-031-11647-6_80

2   Methodology and Analysis

We have followed the methodology set out in [3] to i) Develop a framework of signifiers using a mixed-methods approach, ii) Implement the framework of signifiers using process mining. For our empirical work, we partnered with an industrial supplier named Third Space Learning, which delivers mathematics tutoring for primary school children aged 10 years old. Students and tutors log into a shared online environment, and the student works through a pre-designed online set of questions, with the guidance of a human tutor on an interactive whiteboard. The data available for analysis includes log data from the virtual classroom environment (VCE), learner audio files and tutor audio files. More information on the methods and findings can be found here.

2.1   Developing the Framework of Signifiers

We have built on the research conducted in [7] and adopted the Winne and Hadwin model of SRL [8] as the foundation for our Framework of Signifiers. We have interpreted the model based on the context of our research. More specifically, tutors are regarded as an external Condition, which influence the cognitive and metacognitive Operations of the student across the four phases of SRL. Micro-level processes and signifiers have been identified using a mixed-methods approach. This included a literature review and an empirical review of 50 randomly selected tutoring sessions (Tables 1 and 2).

Table 1. A framework of signifiers for tutor practices

Micro processes	Signifiers	Relevant studies
Boosting motivation	Praise. Effort points, stickers	[2]
Metacognitive prompts	Prompting the student to monitor understanding, plan approach	[1]
Directive engagement	Instructions, with command words	[3]
Explanatory engagement	Utterances in which the tutor explains part or all of the concept	[5] [6]
Guided engagement	Tutor asks closed-ended questions	
Open-ended engagement	Tutor asks open-ended questions	
Passive engagement	Observing student	

Table 2. A framework of signifiers for student operations

Micro processes	Signifiers	Relevant studies
Searching	Student locating information	[5]
Monitoring	Student overtly monitoring, asking for help, writing and erasing	[9] [10]
Assembling	Student computing answer	
Rehearsing	Repeating tutor or question, copying	
Translating	Asking a challenging or clarifying question, explanatory utterances	

2.2 Implementing the Framework of Signifiers, Using Process Mining

We collected samples of 55 online tutoring sessions conducted by high and low performing tutors. Tutor performance was determined using human evaluator rankings, with the top fifteenth percentile and bottom fifteenth percentile of tutors being selected respectively. We coded data based on the framework of signifiers, using the approach presented in [4]. We implemented our framework using a proprietary adaptation of Fuzzy Miner called Disco, configured to show only the most significant events and' transitions adjusting the parameters accordingly (Figs. 1 and 2).

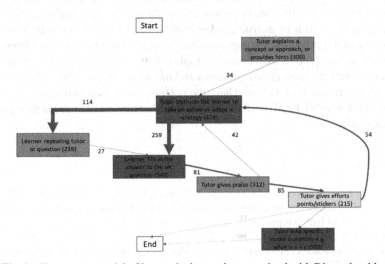

Fig. 1. The process model of low ranked tutor dataset, mined with Disco algorithm

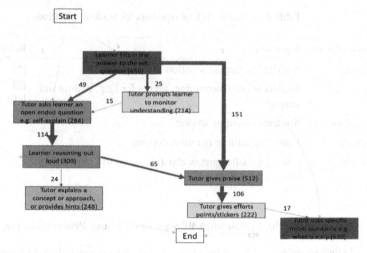

Fig. 2. The process model of high ranked tutor dataset, mined with Disco algorithm

3 Discussion

Our analysis showed that the process maps for the low ranked tutor group differed from the process maps for the high ranked tutor group in terms of both the structure of the map and also the types of prominent events and transitions. Overall, there is a relatively narrow range of events that are considered significant in the low-ranked cluster, with tutor-led practices, such as instruction and explanation, being the most prominent. In contrast, the high ranked tutor group exhibits a broader range of significant events and transitions, and the process map is more complex. Tutoring practices extend beyond the platform and include open-ended questions and metacognitive prompts. These findings align with past research [4, 6, 7] and have the potential to be used to i) design intelligent support for tutors in VCEs, ii) tutor training on these behaviours to improve their practice and student SRL.

References

1. Azevedo, R., Witherspoon, A., Chauncey, A., Burkett, C., Fike, A.: MetaTutor: A MetaCognitive tool for enhancing self-regulated learning. In: Cognitive and metacognitive educational systems: AAAI Fall Symposium (FS-09-02), pp. 14–19 (2009)
2. Blackwell, L., Trzesniewski, K., Dweck, C.S.: Implicit theories of intelligence predict achievement across an adolescent transition: a longitudinal Study and an intervention. Child Dev. **78**, 246–263 (2007)
3. Chi, M., Wylie, R.: The ICAP framework: linking cognitive engagement to active learning outcomes. Educ. Psychol. **49**(4), 219–243 (2014)
4. Cukurova, M., Khan-Galaria, M., Millán, E., Luckin, R.: A learning analytics approach to monitoring the quality of online one-to-one tutoring. J. Learn. Anal., 1–16 (2022). https://doi.org/10.18608/jla.2022.7411
5. Howard-Rose, D., Winne, P.H.: Measuring component and sets of cognitive processes in self-regulated learning. J. Educ. Psychol. **85**(4), 591–604 (1993)

6. Hrastinski, S., Stenbom, S., Benjaminsson., S & Jansson., M,: Identifying and exploring the effects of different types of tutor questions in individual online synchronous tutoring in mathematics. Interact. Learn. Environ. **29**(5), 1–13 (2019)

7. Khan-Galaria, M., Cukurova, M., Luckin, R.: A framework for exploring the impact of tutor practices on learner self-regulation in online environments. In: Bittencourt, I.I., Cukurova, M., Muldner, K., Luckin, R., Millán, E. (eds.) AIED 2020. LNCS (LNAI), vol. 12164, pp. 135–139. Springer, Cham (2020). https://doi.org/10.1007/978-3-030-52240-7_25

8. Winne, P.H., Hadwin, A.F.: Studying as self-regulated learning. In: Hacker, D.J., Dunlosky, J., Graesser, A.C. (eds.) Metacognition in Educational Theory and Practice, pp. 277–304. Lawrence Erlbaum Associates, Mahwah (1998)

9. Winne, P.H.: Learning analytics for self-regulated learning. In: Lang, C., Siemens, G., Wise, A., Gašević, D. (eds.) Handbook of learning analytics, pp. 241–249. Society for Learning Analytics Research, Beaumont, AB (2017)

10. Winne, P.H.: Cognition and metacognition within self-regulated learning. In: Schunk, D., Greene, J. (eds.) Handbook of self-regulation of learning and performance, 2nd edn., pp. 36–48. Routledge, New York (2018)

Using Knowledge Tracing to Predict Students' Performance in Cognitive Training and Math

Richard Scruggs[1]([✉]) [iD], Jalal Nouri[2] [iD], and Torkel Klingberg[1] [iD]

[1] Karolinska Institute, 171 77 Stockholm, Sweden
richard.scruggs@ki.se
[2] Stockholm University, 106 91 Stockholm, Sweden

Abstract. Cognitive training aims to improve skills such as working memory capacity and spatial ability, which have been linked to math skills. In this study, we fit Deep Knowledge Tracing with Transformers (DKTT), Dynamic Key-Value Memory Networks (DKVMN), and Knowledge Tracing Machines (KTM) to a large dataset from a cognitive training system. DKVMN achieved the highest AUC (0.739) of the algorithms. To explore connections between math skills and cognitive skills, the data was split into cognitive and math items. DKVMN's AUC on the math items was higher (0.745) than on the cognitive (0.706). Notably, the split model AUCs did not differ from skill-level AUCs produced by a model trained on the entire dataset, suggesting that math performance did not improve DKVMN's cognitive predictions and vice versa.

Keywords: Knowledge tracing · Cognitive training · Math learning

1 Introduction

Mathematical ability is related to general cognitive skills such as working memory, spatial and non-verbal reasoning abilities [1, 2]. This has led to hopes that improving these cognitive abilities through training would enhance mathematical abilities. While some studies have shown promising results [3, 4], others have failed to find impacts on math performance [5, 6]. In a recent large, randomized trial, [7] found that training of visuospatial working memory and reasoning enhanced mathematical learning, with considerable inter-individual differences. Here we investigate if knowledge tracing methods, useful in individualizing math learning, could be applied to cognitive training.

Knowledge tracing (KT) models the growth in knowledge or skill level as students complete learning tasks. KT is well-positioned to investigate links between cognitive training and mathematical learning for several reasons. First, nearly all cognitive training activities take place in computerized learning systems, yielding rich log data for KT analysis. Second, KT algorithms can effectively predict students' performance on math learning activities [8–10]; this is important as math performance likely relates to working memory capacity [1, 2, 7]. In this study, we use a large dataset from a cognitive training system to show that KT algorithms accurately predict performance in this area.

© Springer Nature Switzerland AG 2022
M. M. Rodrigo et al. (Eds.): AIED 2022, LNCS 13356, pp. 410–413, 2022.
https://doi.org/10.1007/978-3-031-11647-6_81

2 Data and Methods

The data used in this study were collected from students using the Vektor system [7]. This system, which presents problems in a game-like framework, focuses on working memory training, but also includes other cognitive tasks as well as math problems, with all problems' difficulty adapting to students' skill level.39,505 students between the ages of 6 and 8 used the system for seven weeks. In total, the dataset contains 11 skills: two math and nine cognitive, and 184 million problem attempts.

In addition to including cognitive training tasks, there were several other factors that differentiate Vektor data from other KT datasets. First, students attempt Vektor trials very quickly, typically on the order of ten seconds per trial (mean = 11.1 s, median = 7.6 s). This yields thousands of attempts (mean = 5289, median = 5057). By contrast, in the ASSISTments 2009 set, students attempt an average of 82 problems each. The data also present challenges related to automatic problem generation. Vektor automatically generates problems for most skills: our dataset contains about 1.2 million problems, with over 900,000 of them from the working memory tasks. Fitting models without combining problem labels results in thousands of unseen problems in the test set. Three strategies were tried to group autogenerated problems: group only working memory problems by their difficulty, group all problems with only their skill, and group all problems with their skill and in-system difficulty. Not all strategies were tested with all algorithms, as explained later.

As working memory capacity has been shown to be related to performance in mathematics [1], there was the possibility that the algorithms used students' performance on the math skills to estimate their cognitive abilities. To test this hypothesis, we split the dataset by domain and trained and tested two separate models, one on only the cognitive skills and one on only the math skills. The split-skill models were constructed using DKVMN and skill-difficulty labeling as it had the highest AUC in the full dataset. Three knowledge tracing algorithms were used in this study: Dynamic Key-Value Memory Networks [DKVMN; 9], Knowledge Tracing Machines [KTM; 10], and Deep Knowledge Tracing with Transformers [DKTT; 8]. These algorithms were chosen due to their success in modeling math learning as well as their varied modeling approaches.

DKVMN [9] uses recurrent neural networks and an external memory matrix to track students' knowledge states on various concepts over time. The concepts are generated by the algorithm itself based on observed relationships between training items. Code from [9] was used to run DKVMN. Unlike KTM and DKTT, DKVMN does not use distinct labels for problem and skill. To keep skill information available, DKVMN was only tested with skill-and-difficulty and skill-only relabeled problems.

KTM [10] uses factorization machine-based classifiers with item-level biases and additional side information as predictors. KTM was run with code from [10]. The implementation included features for items, skills, and correct and incorrect responses on the skill. KTM was not tested with skill-only relabeling as skills and items are included.

DKTT [8] is an approach that adds item-skill relationships to the sequence-based Transformer neural network architecture. DKTT uses students' attempted items, attempt times, and items' underlying skills. As skills and items were included, DKTT was not tested with skill-only relabeling. DKTT was fit with code from [8].

3   Results

Table 1. AUC and accuracy on combined Vektor dataset.

Algorithm	Grouping strategy	AUC	Accuracy
DKVMN	Skill and difficulty	0.739	0.775
DKVMN	Skill only	0.725	0.779
DKTT	Skill and difficulty	0.634	0.759
DKTT	Rename only WM	0.623	0.757
KTM	Skill and difficulty	0.693	0.763
KTM	Rename only WM	0.705	0.766

Table 1 shows the AUC values of each algorithm on the Vektor dataset. For skill-and-difficulty based grouping DKVMN achieved an AUC of 0.739 on held-out test data; DKTT saw a lower AUC of 0.634, and KTM was in the middle, with an AUC of 0.693. DKVMN was the only algorithm that used skill-only grouping, but it still outperformed all others, with an AUC of 0.725. When only working memory problems were renamed, DKTT's AUC decreased to 0.623 but KTM's increased to 0.705.

Table 2 shows the results of the split model analysis. Although the AUC and accuracy were worse for the cognitive data than the math, the AUC values were adequate for both. Also, the AUC values were not different from training on the entire dataset and computing a skill-level AUC for just the cognitive skills and just the math skills.

Table 2. DKVMN AUC and accuracy on split Vektor dataset.

Dataset	Test data	AUC	Accuracy
Cognitive-only	Cognitive	0.706	0.721
Combined	Cognitive	0.706	0.695
Math-only	Math	0.745	0.802
Combined	Math	0.745	0.831

4   Discussion

These results show that knowledge tracing can predict students' performance on cognitive training tasks. DKVMN saw an AUC of 0.739 on the combined cognitive-math dataset. Considering the differences in structure between Vektor data and existing benchmark datasets, these results are quite encouraging – DKVMN achieved an AUC of only 0.727 on the ASSISTments 2015 dataset [9]. It was surprising that the split models attained the same AUC as the combined model. Given that visuospatial working memory has been shown relate to performance in mathematics, the number of attempts available may have made attempts from other skills less useful to DKVMN.

The results from the different methods of problem grouping suggest that auto-generated problems are not an insurmountable challenge to existing KT algorithms. Similar skill and difficulty labels would likely be present in most tutoring systems that use auto-generated problems. The success of the skill-and-difficulty grouping strategy over the other strategies underscores the importance of underlying structure and side information to knowledge tracing algorithms (see also [10]). Finally, the difference in AUCs between the cognitive problems and the math problems – regardless of the training dataset – suggests that skill-level AUCs are worth further study. It's important to know if algorithms are optimizing for prevalent skills at the expense of others.

Acknowledgement. Funding for this study was provided by the Marianne and Marcus Wallenberg Foundation. The computations for this study were enabled by resources provided by the Swedish National Infrastructure for Computing (SNIC), partially funded by the Swedish Research Council through grant agreement no. 2018–05973.

References

1. Hawes, Z., Ansari, D.: What explains the relationship between spatial and mathematical skills? a review of evidence from brain and behavior. Psychon. Bull. Rev. **27**(3), 465–482 (2020). https://doi.org/10.3758/s13423-019-01694-7
2. Peng, P., Namkung, J., Barnes, M., Sun, C.: A meta-analysis of mathematics and working memory: moderating effects of working memory domain, type of mathematics skill, and sample characteristics. J. Educ. Psychol. **108**, 455–473 (2016)
3. Berger, E.M., Fehr, E., Hermes, H., Schunk, D., Winkel, K.: The impact of working memory training on children's cognitive and noncognitive skills. SSRN J. (2020)
4. Lowrie, T., Logan, T., Hegarty, M.: The influence of spatial visualization training on students' spatial reasoning and mathematics performance. J. Cogn. Dev. **20**, 729–751 (2019). https://doi.org/10.1080/15248372.2019.1653298
5. Roberts, G., et al.: Academic outcomes 2 years after working memory training for children with low working memory: a randomized clinical trial. JAMA Pediatr. **170**, e154568 (2016)
6. Rodán, A., Gimeno, P., Elosúa, M.R., Montoro, P.R., Contreras, M.J.: Boys and girls gain in spatial, but not in mathematical ability after mental rotation training in primary education. Learn. Individ. Differ. **70**, 1–11 (2019)
7. Judd, N., Klingberg, T.: Training spatial cognition enhances mathematical learning in a randomized study of 17,000 children. Nat. Hum. Behav., 1–7 (2021)
8. Pu, S., Yudelson, M., Ou, L., Huang, Y.: Deep knowledge tracing with transformers. In: Bittencourt, I.I., Cukurova, M., Muldner, K., Luckin, R., Millán, E. (eds.) AIED 2020. LNCS (LNAI), vol. 12164, pp. 252–256. Springer, Cham (2020). https://doi.org/10.1007/978-3-030-52240-7_46
9. Zhang, J., Shi, X., King, I., Yeung, D.-Y.: Dynamic key-value memory networks for knowledge tracing. In: WWW '17: Proceedings of the 26th International Conference on World Wide Web, pp. 765–774. Perth, Western Australia (2017)
10. Vie, J.-J., Kashima, H.: Knowledge tracing machines: factorization machines for knowledge tracing. AAAI **33**, 750–757 (2019)

Toward Accessible Intelligent Tutoring Systems: Integrating Cognitive Tutors and Conversational Agents

Michael Smalenberger[1]($\boxtimes$) and Kelly Smalenberger[2]

[1] Department of Mathematics and Statistics, The University of North Carolina at Charlotte, Charlotte, NC, USA
msmalenb@uncc.edu
[2] Department of Mathematics and Physics, Belmont Abbey College, Belmont, NC, USA

Abstract. The literature is rife with investigations into the use of ITS in mathematics, but scant on how these systems impact students with disabilities. Since ITS continue to permeate the educational landscape, and students with disabilities are co-located with their non-disabled peers, such investigations are overdue. To that end, we provide a theoretically grounded framework for authoring accessible ITS by drawing parallels between our work and relevant studies in the literature. Our framework enables the authoring of accessible ITS by integrating a cognitive tutor with a conversational agent. Our focus in this study is on an ITS created using Cognitive Tutor Authoring Tools (CTAT) and is augmented with adaptive capabilities which make it accessible to students who are blind or have motor-function impairments. We describe an ITS piloted with 115 students in two introductory college statistics courses, and share insights gained during the implementation of our framework. We highlight several contributions, including changes to the tutor interface to make it speech-interactive which is not currently available using CTAT, adaptations to the Bazaar infrastructure to enable solution step supports by the conversational agent, and how we used over 75,000 solutions steps and explanations by 415 students on 146 questions outside of an ITS to create supports within an accessible ITS. We conclude by proposing directions for future work on authoring accessible ITS.

Keywords: Accessibility · Intelligent Tutoring Systems (ITS) · Cognitive Tutor Authoring Tools (CTAT) · Conversational agents · Bazaar · Blind · Motor-function impairments

1 Introduction

ITS continue to be widely used in educational settings and have been shown to lead to substantial learning gains, including in mathematics [1]. However, the use of this technology may have differential effects. In classrooms across the United States, students with disabilities are co-located with their non-disabled peers. When teachers use educational technology that is not accessible to students with disabilities, it can leave these

M. M. Rodrigo et al. (Eds.): AIED 2022, LNCS 13356, pp. 414–418, 2022.
https://doi.org/10.1007/978-3-031-11647-6_82

students confused, isolated, and often trailing in progress compared to their classmates. This can be very detrimental to the learning environment. Accessible alternatives to most ITS are nonexistent resulting in scant research on accessible ITS.

While such work is long overdue, we recognize the myriad of challenges. These include that researchers may not be aware of detailed student characteristics, sample sizes may be too small to draw definitive conclusions, different adaptations to ITS may be needed to support students with different kinds of disabilities, etc. However, we believe that many of the required pieces to build accessible ITS already exist. Our work is a modest first step in putting them together to create assessable mathematics ITS for students who are blind or have motor-function disabilities.

2 Prior Work

There are many ways to author ITS. The one relevant to our work is Cognitive Tutor Authoring Tools (CTAT) integrated through TutorShop to create an example-tracing tutor. Development costs of example-tracing tutors using CTAT are significantly lowered compared to "historical" estimates for ITS while still allowing for sophisticated tutoring behaviors [2]. These include providing step-by-step guidance on complex problems while recognizing multiple student strategies while recording action-level data. CTAT has been shown to be quite general, with tutors built for many domains covering a range of pedagogical approaches, including collaborative learning [3]. Tutors have also seen classroom use, including the Mathtutor [4].

Another tool relevant to creating our accessible ITS is Bazaar. Bazaar is a publicly available architecture for orchestrating conversational agent-based support and is intended to facilitate research on collaborative learning. It hosts a library of reusable behavioral components that each trigger a simple form of support. More complex supportive interventions are constructed by orchestrating multiple simple behaviors. Its flexibility means it can be used to develop platforms very rapidly for investigating a wide range of dynamic support research questions [5]. We had to adapt this tool to our objectives, i.e., only one user instead of multiple users.

There is scant research on developing ITS for blind students or those with motor-function impairments. Regarding mathematics education for visually impaired students, a primary obstacle is an inherent difficulty in managing structural information included in math formulae [6], and audio-based interactive computer interfaces have been shown to enhance learning and cognition [7]. We build upon this prior work to create and implement a framework for authoring and researching accessible ITS.

3 Framework

Our framework is intended to structure the development of accessible ITS which can be used by students with motor-function disabilities and blind students. Since these groups require different kinds of support, our tutor has two layers of adaptivity. Specifically, one ITS interface is speech-enabled, while the other embeds this interface into a platform containing a conversational agent which is text and speech generating.

The speech-enabled ITS (Fig. 1, left) is intended to support students with motor-function impairments. It allows users to use speech to navigate the tutor interface by saying which input field should be brought into focus by stating that field's label. It also allows common ITS functions such as using speech to enter values, ask for hints, finish a problem, etc. This ITS has been pilot tested but is not yet classroom ready.

The speech-enabled ITS embedded in a conversational agent (Fig. 1, right) is intended to support students who are blind. The primary impetus behind the text generation facet of the conversational agent is that most CTAT components are not navigable by assistive technologies. Our platform allows screen-readers to read the conversational agent's text output which can also be converted to speech. As described below, this ITS with a conversational agent has not yet been pilot tested.

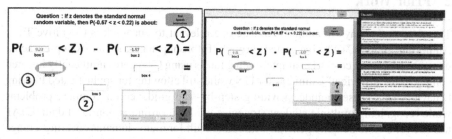

Fig. 1. *Left pane*: Speech-enabled example-tracing tutor interface created using CTAT and Web Speech API; (1) Speech-enabled button to begin and end tutor interaction, (2) each input field has distinctive labels for speech navigation, and (3) in-focus input field is surrounded by hue. *Right pane*: Speech-enabled ITS embedded in conversational agent created using Bazaar.

4 Development Progress and Insights

530 students enrolled in 6 courses taught by one instructor at a large public university participated in this study during the 2021 calendar year. First, 115 students in two introductory college statistics courses during the Spring 2021 semester completed a set of 25 questions on normal random variables (NRV) assigned as homework immediately after discussing that topic in class. The homework used an ITS created using CTAT integrated into the course through TutorShop [2]. We used the 'Stat 1222–004 S21 Chap 5 HW' and 'Stat 1220–010 S21 Chap 5 HW' datasets accessed via DataShop [8] to assess the efficacy of this tutor and observe student behaviors with standard features available in CTAT. Subsequently, the speech-enabled components were added to the tutor interface. Specifically, we added the Web Speech API and tested the ITS with two volunteers, both of whom provided very positive feedback.

Additionally, 118 students in a different introductory college statistics course at the same institution completed 70 questions outside of an ITS over the course of the semester. Each question consisted of two parts. The first part required students to show their work while solving a question relevant to the material recently covered in class. The second part required students to describe the steps they had taken in the first part. This resulted in over 20,000 explanations of individual solution steps. We are using these explanations to assess the verbally described actions students may take in the ITS embedded with the conversational agent, and author supports that are useful to students.

A significant insight we have gained in this work is the myriad of ways students describe their solution steps. These have ranged from elaborate sentences to a single word, oftentimes omitting or misusing mathematics terminology which significantly affected the accuracy of our conversational agent. To mitigate this, during the Fall 2021 academic year, we collected an additional 55,000 solutions and explanations on 76 problems generated by 297 students. We are incorporating these into our model.

5 Conclusion, Limitations, and Future Research

Our work is a modest first step in addressing the need for accessible ITS and filling the void of research in this area. Nevertheless, we make several noteworthy contributions. First, we provide a framework on how to author accessible ITSs using existing tools, namely CTAT and Bazaar. While modifications had to be made to both, the modifications were modest and pilot testing showed promising results. Furthermore, we piloted an ITS with 115 students in two introductory college statistics courses to establish a reference for actions within that ITS and collected over 75,000 solutions steps and explanations by 415 students on 146 questions outside of an ITS to create support within an accessible ITS. Nevertheless, our work as of yet has its limitations. Chief among these is that the system has not yet been used with the target population.

While we already have rich data, once our ITS is utilized in vivo we will have to contend with such issues as analyzing concurrent but different data streams. Furthermore, the use of verbal protocol can give us insights that written responses potentially lack. Additionally, we will be able to compare the impact on mathematics learning of our platform to that of other platforms, while accounting for the fact that some students may have a disability. We will have to grapple with design issues, such as how best to incorporate multiple input streams (speech, braille, and QWERTY keyboards) into cogent responses from the combined ITS/conversational agent platform to the user. The platform's responses also could take the form of speech and text generation, refreshable braille display output, or visual output on a computer screen.

Lastly, we are excited to take these initial steps in making accessible ITS and break ground on research on this topic.

Acknowledgements. As always, K.H.S., J.M.S., E.M.S., and W.J.S. thank you and I l. y.

References

1. Rau, M.A., Aleven, V., Rummel, N., Pardos, Z.: How should intelligent tutoring systems sequence multiple graphical representations of fractions? A multi-methods study. Int. J. Artif. Intell. Educ. **24**(2), 125–161 (2014)
2. Aleven, V., McLaren, B., Sewell, J., Koedinger, K.: A new paradigm for intelligent tutoring systems: example-tracing tutors. Int. J. AI Ed **19**(2), 105–154 (2008)

3. Olsen, J.K., Belenky, D.M., Aleven, V., Rummel, N., Sewall, J., Ringenberg, M.: Authoring tools for collaborative intelligent tutoring system environments. In: Trausan-Matu, S., Boyer, K.E., Crosby, M., Panourgia, K. (eds.) ITS 2014. LNCS, vol. 8474, pp. 523–528. Springer, Cham (2014). https://doi.org/10.1007/978-3-319-07221-0_66

4. Aleven, V., McLaren, B.M., Sewall, J.: Scaling up programming by demonstration for intelligent tutoring systems development: an open-access web site for middle school mathematics learning. IEEE Trans. Learn. Technol. 2(2), 64–78 (2009)

5. Adamson, D., Dyke, G., Jang, H., Rosé, C.P.: Towards an agile approach to adapting dynamic collaboration support to student needs. Int. J. Artif. Intell. Educ. 24(1), 92–124 (2013). https://doi.org/10.1007/s40593-013-0012-6

6. Spinczyk, D., Maćkowski, M., Kempa, W., Rojewska, K.: Factors influencing the process of learning mathematics among visually impaired and blind people. Comput. Biol. Med. 104, 1–9 (2019)

7. Sánchez, J., Flores, H.: AudioMath: Blind children learning mathematics through audio. Int. J. Disability Hum. Dev. 4(4), 311–316 (2005)

8. Koedinger, K, Baker, S., Cunningham, K., Skogsholm, A., Leber, B., Stamper, J.: A Data Repository for the EDM community: The PSLC DataShop. In: Romero, C., Ventura, S., Pechenizkiy, M., Baker, S. (eds.) Handbook of Educational Data Mining. CRC Press, Boca Raton (2010)

Investigating the Role of Direct Instruction About the Notional Machine in Improving Novice Programmer Mental Models

Veronica Chiarelli$^{(\boxtimes)}$ ⓘ and Kasia Muldner ⓘ

Department of Cognitive Science, Carleton University, Ottawa, Canada
veronicachiarelli@cmail.carleton.ca

Abstract. Students learning to program often lack accurate mental models of the *notional machine* (an abstraction of the computer needed to simulate code execution). Our long-term goal is to design a tutoring system to support novice programmers in forming accurate mental models via instruction about the notional machine. Here, we present the completed first phase of this project. We took a qualitative approach to analyze how students interact with instructional materials that include the notional machine construct. We discuss how the findings informed notional machine design and instruction in subsequent research phases.

Keywords: Mental models · Novice programmers · Intelligent tutor systems

1 Introduction

The notional machine (NM) is an idealized abstraction of a computer that can be used to simulate the execution of a program [1, 2]. Students need accurate mental models of the NM to simulate program execution (e.g., to predict program output). Unfortunately, there is little guidance on how to design NM instruction. While some older work has tackled this challenge [3–5], little research has been carried out since, recently leading to formation of working groups devoted to this topic [10]. Tutoring systems exist for other aspects of programming instruction [6–9], but to our knowledge none explicitly target the NM. To address these gaps, the long-term goal of our research program is to develop a tutoring system to support novices' mental model formation via explicit instruction of the NM [11]. As the first step, we conducted a pilot study to evaluate a preliminary NM lesson design. We used a qualitative approach to analyze the data, to gain insight into students' reasoning with the NM. Here, we present the findings and discuss ongoing and future work for subsequent phases of our research program.

M. M. Rodrigo et al. (Eds.): AIED 2022, LNCS 13356, pp. 419–423, 2022.
https://doi.org/10.1007/978-3-031-11647-6_83

2   Methods

We created an introductory text-based Python lesson that was 12 pages long. A core part of the lesson was the NM, which was based on Mayer's 1976 design [3]. Specifically, the NM was shown as a rectangle with the following components (see Fig. 1): *Memory* (boxes to hold variables and their values), *Input* (a box to hold information that the user enters), *Output* (a box that displays information that is output to the screen), and *Program Pointer* (a box with a program and an arrow indicating the current program line being executed). The NM was used to illustrate program execution in the lesson, including how the NM components interact during execution. The state of the NM diagram was updated after each line of code was executed (for instance, by crossing out previous variable values, see Fig. 1). To encourage constructive processing, the lesson included five self-explanation prompts. The first and last prompt were general and not tied to a specific program. The remaining prompts asked about specific programs, with one asking about program execution within the context of the NM.

To obtain insight into how participants reasoned about the NM given this design, we conducted a pilot study ($N = 8$). Participants first completed a pretest and then studied the lesson. We used the think-aloud protocol by having participants verbalize their thoughts while studying (if they stopped talking, they were prompted to continue with a generic prompt, e.g., *please keep talking*). Participants then completed a posttest and an exit interview. The pretest and posttest were equivalent, each with 8 questions asking participants to explain execution of given programs. The study lasted no longer than 2 h per participant. All participant utterances were recorded and transcribed.

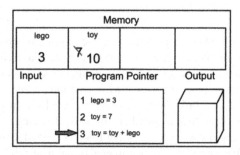

Fig. 1. The state of the NM diagram after executing a 3-line Python program

3   Results

Self-explanations have potential to provide insight into participants' mental models of the NM. Accordingly, we used a qualitative approach to identify self-explanation in the transcripts. Utterances were coded as *self-explanations* if they went beyond the text. To inform on the utility of prompting, we distinguished *prompted* vs. *spontaneous* explanations. To gain insight into mental models related to the NM, we also labelled explanations referring to the NM or its components as *NM intrusions* (ones not referring to the NM were labelled as *general*). The results are in Table 1. The number of prompted self-explanations was more consistent across participants as compared to the number of spontaneous self-explanations. While not surprising, this demonstrates that prompting promotes self-explanation even for participants who may not spontaneously self-explain, thus encouraging all participants to engage with the materials. A larger proportion of prompted self-explanations included NM intrusions compared to spontaneous self-explanations. Further, prompted self-explanations were more complete and more frequently related to the NM. For example, in response to the prompt *"How do you think the computer runs a program?"*, a participant's self-explanation was (intrusions in bold): *"I think it's like our brain, we see something. That's the **input**. And then we try to memorize what we saw. And then at one point, if we need to recall it, we have to go back to our **memory**, but we have to do it in a linear way. So, it will, once the **output** comes out. It has to be very clear when we program so it has to be ABC, we can't go to ACB."* An example of a spontaneous self-explanation about while loops was: *"If you needed, I don't know, a postal code or if you needed a country, and it wasn't allowed in that particular one then it's going to wait until you put the right thing in there."* One exception to these observations about prompting was when students were overly confused (based on self-reports), they did not self-explain even when prompted.

The exit interview provided feedback on the NM design. Participants reported liking the NM model, with some even drawing it while answering posttest questions. Participants also said that the model helped them understand what was going on inside the computer, suggesting that the model was a form of conceptual grounding and not only an organizational tool. Participants also had suggestions for improvements. First, they reported disliking the text-heavy nature of the lesson. To address this as well as the observed difficulties in following the sequence of notional machine states in the static text, subsequent versions will use video lessons instead. Second, some participants expressed confusion about differentiating between the NM components and so in future phases, each NM component and its label will have a unique color. Third, participants reported difficulties identifying which changes had been made to the NM from one state to the next. To address this, moving forward, changes from the previous states will be highlighted.

To analyze students' mental models of the NM, we relied on test performance and transcripts. All participants learned from the lesson (all had positive gains from pretest to posttest, see Table 1). While the presence of NM intrusions in the transcripts suggests that participants' mental models did incorporate elements of the NM, low posttest scores showed there was room for improvement. For example, posttest answers showed participants did not understand that while loops repeat instructions, indicating that this feature was missing from their mental models. To address these issues, we will update the model and add prompts that direct student attention to common areas of difficulty.

Table 1. Count of participant self-explanations, per self-explanation category; and pretest and posttest scores and learning gains (increase from pretest to posttest scores)

Self-explanation category	P1	P2	P3	P4	P5	P6	P7	P8	M	SD
Spontaneous general	4	44	18	62	65	66	68	30	44.63	24.77
Spontaneous NM intrusion	3	9	21	25	20	29	24	6	17.13	9.73
Prompted general	2	2	2	3	1	1	0	2	1.63	0.92
Prompted NM intrusion	2	1	2	1	3	4	3	3	2.38	1.06
Pretest (out of 8)	0	0	0	0	0	1	2	2	0.63	0.92
Posttest (out of 8)	3	2	4	3	2	2	3	3	2.75	0.71
Learning gains	3	2	4	3	2	1	1	1	2.13	1.13

4   Ongoing and Future Work

Based on the pilot we have revised the instructional materials and incorporated them into a tutoring system that provides direct instruction on the NM and encourages learning using self-explanation prompts. To experimentally test the effectiveness of this approach, we created a control version of the tutor that includes the same self-explanation prompts and materials but does not include the NM. To facilitate tutor construction, we used the CTAT framework [12].

The tutor includes a variety of activities, alternating between brief videos showing programming mini-lessons and self-explanation activities related to the video to solidify concepts presented in it. Students are not given feedback. We are currently conducting an experimental study to compare learning from the NM experimental tutor version and the control version without the NM. To assess learning gains, we will compare participants' pre to posttest scores and evaluate the accuracy of mental models formed. Our findings will inform on the effectiveness of explicit instruction about the NM.

References

1. Du Boulay, B.: Some difficulties of learning to program. J. Ed. Comput. Res. **2**(1), 57–73 (1986)
2. Sorva, J.: Notional machines and introductory programming education. ACM Trans. Comput. Educ. **13**, 816–821 (2013)
3. Mayer, R.E.: Different problem-solving competencies established in learning computer programming with and without meaningful models. J. Ed. Psychol. **67**(6), 725–734 (1975)
4. Mayer, R.E.: Some conditions of meaningful learning for computer programming: advance organizers and subject control of frame order. J. Ed. Psychol. **68**(2), 143–150 (1976)
5. Mayer, R.E.: Models for understanding. Rev. Ed. Res. **59**(1), 43–64 (1989)
6. Fabic, G.V., Mitrovic, A., Neshatian, K.: Adaptive problem selection in a mobile Python tutor. In: Adjunct Publication of the 26th Conference on User Modeling, Adaptation and Personalization, pp. 269–274 (2018)
7. Hosseini, R., et al.: Animated examples as practice content in a Java programming course. In: Proceedings of Technical Symposium on Computing Science Education, pp. 540–545 (2016)

8. Nelson, G.L., et al.: Comprehension first: evaluating a novel pedagogy and tutoring system for program tracing in CS1. In: Proceedings of Computing Education Research, pp. 2–11 (2017)
9. Price, T.W., et al.: iSnap: towards intelligent tutoring in novice programming environments. In: Proc. of SIGCSE Technical Symposium on Computer Education, pp. 483–488. (2017)
10. Fincher, S., et al.: Capturing and characterising notional machines. In: Proceedings of ACM Conference on Innovation and Technology in Computer Science Edition, pp. 502–503 (2020)
11. Chi, M., et al.: Eliciting self-explanations improves understanding. Cogn. Sci. **18**, 439–477 (1994)
12. Aleven, V., et al.: Example-tracing tutors: intelligent tutor development for non-programmers. Int. J. Artif. Intell. Educ. **26**, 224–269 (2016)

A Good Classifier is Not Enough: A XAI Approach for Urgent Instructor-Intervention Models in MOOCs

Laila Alrajhi[1,3](✉), Filipe Dwan Pereira[2], Alexandra I. Cristea[1], and Tahani Aljohani[1]

[1] Computer Science, Durham University, Durham, UK
{laila.m.alrajhi,alexandra.i.cristea}@durham.ac.uk,
taljohani7@gmail.com
[2] Computer Science, Federal University of Roraima, Boa Vista, Brazil
filipe.dwan@ufrr.br
[3] Educational Technology, King Abdulaziz University, Jeddah, Saudi Arabia

Abstract. Deciding upon instructor intervention based on learners' comments that need an urgent response in MOOC environments is a known challenge. The best solutions proposed used automatic machine learning (ML) models to predict the urgency. These are 'black-box'-es, with results opaque to humans. EXplainable artificial intelligence (XAI) is aiming to understand these, to enhance trust in artificial intelligence (AI)-based decision-making. We propose to apply XAI techniques to interpret a MOOC intervention model, by analysing learner comments. We show how pairing a good predictor with XAI results and especially colour-coded visualisation could be used to support instructors making decisions on urgent intervention.

Keywords: MOOCs · Comments · Urgent intervention · NLP · XAI

1 Introduction

Instructor intervention in MOOCs may reduce the problem of learner dropout, as it has recently been proven that learners who need intervention are less likely to complete the course (only 13%) [1]. Recently, intervention in MOOC attracted growing interest from researchers, to help instructors in interventions based on learners' comments [2–5]. Intervention systems classified the learner comments into two categories: urgent and non-urgent [6]. Although these systems need to be accurate in their decisions, it is difficult to achieve this, as urgency decisions are hard to make, even for a human [7].

This work deals with the intervention problem. Our initial goal is the proof of concept of using explainable AI for this task of urgent intervention, as this had not been done before. For understanding 'How' and 'Why' the model decisions are made, we explained thus not only the intervention model prediction, but also compared it with human decision making. We formalise our research question as:

RQ: How to construct a transparent XAI model to detect urgent intervention towards supporting instructors' decisions?

In terms of the contribution, to the best of our knowledge *this is the first time that text classification explainability has been applied to an instructor intervention model.*

© Springer Nature Switzerland AG 2022
M. M. Rodrigo et al. (Eds.): AIED 2022, LNCS 13356, pp. 424–427, 2022.
https://doi.org/10.1007/978-3-031-11647-6_84

2 Methods

Our research consists of three basic stages as follows: first, construct an 'urgent' gold-standard dataset, via human experts annotating comments (Sect. 2.1). Next, build an automatic urgent intervention model via BERT (Sect. 2.2). Then, explain the model and visualise words importance, to understand the decision (Sect. 2.3).

2.1 Constructing the Gold-Standard Dataset

We collect and prepare our benchmark corpus, as a case study, based on real-world data from the FutureLearn MOOC environment platform, here, the 'Big data' course, conducted during 2016 – selected as being a topic of current interest for learning; additionally, we expected it to contain many urgent cases, as being (arguably) a more challenging topic. The dataset consists of learner comment texts ('posts') and other features collected from the first 5 weeks ($\approx$50%) of the 9-weeks-long course, to capture the comments that need intervention before dropout. We obtain thus 5786 comments, which, taking into account the hardship of the following manual annotation, were considered sufficient for the current task.

We thus manually annotate these comments, using three human domain experts and one author of this paper, following Agrawal et al.'s instructions [8]. We labelled urgency for every learner comment, mapped onto a scale (1–7) representing the range of urgency level (not urgent – extremely urgent). For validation, we calculated Krippendorff's α agreement value between all annotators, and we found the results very low between any subgroups (confirming prior research [7]). To address this problem, we decide to further convert the scale into a simpler, binary one (mapping 1:3 $\rightarrow$ 0, and 4:7 $\rightarrow$ 1). To be able to increase the reliability, we additionally dropped the annotator who disagreed strongly with other annotators. From the remaining three annotators, we calculate the label value, via the voting technique, since voting is the most common way to gather different opinions for the same task [9]. This result in a class size of ('0' non-urgent $\rightarrow$ 4903, '1' urgent $\rightarrow$ 883).

2.2 Fine-Tuning the BERT Model

As the preprocessing step, we split the data into training and testing sets, using the stratify method [10], to preserve the percentage of samples for each class; with the proportion of 80% training and 20% testing. Thus, the distribution of the training set is (0: 3922, 1: 706) and testing set is (0: 981, 1: 177). Then, for the training set, we split again, as 90% will be used for training, and 10% will be used for validation.

We fine-tune BERT, without any engineering features. We use the 'bert-base-uncased' version. Next, we prepare the text input, with the fixed maximum length 365, which is the maximum number of words on all comments; this will pad all comments to the maximum length. Then we train the model, by defining batch size = 8, number of training epochs = 4 and AdamW, as optimiser, with learning rate = 2e−5. Finally, we evaluate the prediction model performance on the test set, and save the pre-trained model, to use it later for the interpreting.

2.3   Interpreting the BERT Model

After training the model, we interpret our BERT model, by using the Captum package, which supports classification models. We interpret it via the BertForSequenceClassification in Captum from Captum_BERT colab [11], by creating the Layer Integrated Gradients explainer, to identify which words have the highest attribution to the model's output. To illustrate how to use our method and reply to RQ, we randomly choose a single comment and visualise the explainability results with the attribution score and highlight the word importance.

3   Results

The results obtained from BERT to predict the urgent comments show that the accuracy score is high (0.92). However, as the data is extremely unbalanced, we use additional metrics to evaluate the classifier (precision, recall and F1-score) for every class, see Table 1. Please note that here, whilst working with a decent classifier, our focus is not on the optimisation of the classifier, but on the explanation of the obtained results.

Table 1. The results of the BERT classifier.

	Precision	Recall	F1-score
0	.95	.95	.95
1	.73	.71	.72

As previously mentioned, our goal is to analyse the learner comments and explain the text classification decision using Captum, to understand the reasons behind the predictions. Here we chose a random comment prediction from the test set, then show the explainability results, with highlighted text, as shown in Fig. 1. The attribution score = 1.45 and the different colours reflect the effect of word attribution towards the prediction; and the level of highlighting depicts the importance of the feature, for the classification. Specifically, the green colour means a positive contribution (got, looking, understanding, be, ...), whilst red contributes by decreasing the prediction score (forward, useful, ...). In the case of the example below, we found that the predicted label is non-urgent (0) and the true label is also non-urgent (0). Such visualisation can further be used by an instructor to understand the decisions and recommendations of a classifier for urgency detection in learners' chats on MOOCs.

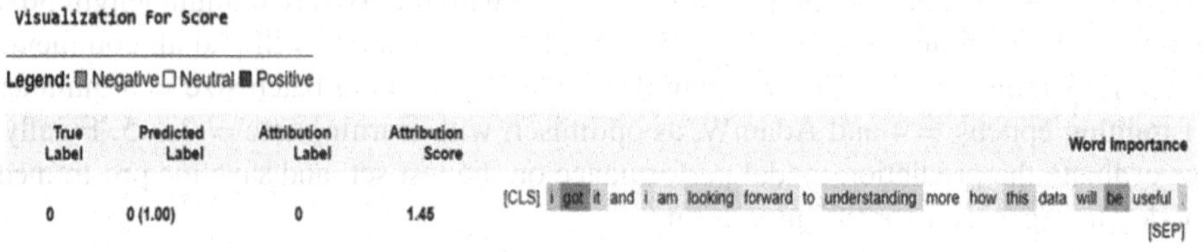

Fig. 1. Screenshots of Captum explanations. (Color figure online)

4   Conclusion

The objective of this paper was to provide an explanation of the machine learning decision, for a specific text classification problem, that of explaining individual predictions in the urgent intervention task in a MOOC environment.

Here, this work also represents a proof-of-concept of using explainable AI on imbalanced data. Moreover, we advance the field of urgency prediction, proposing a method for potentially supporting instructor intervention.

References

1. Alrajhi, L., Alamri, A., Pereira, F.D., Cristea, A.I.: Urgency analysis of learners' comments: an automated intervention priority model for MOOC. In: Cristea, A.I., Troussas, C. (eds.) ITS 2021. LNCS, vol. 12677, pp. 148–160. Springer, Cham (2021). https://doi.org/10.1007/978-3-030-80421-3_18
2. Guo, S.X., et al.: Attention-based character-word hybrid neural networks with semantic and structural information for identifying of urgent posts in MOOC discussion forums. IEEE Access **7**, 120522–120532 (2019)
3. Sun, X., et al.: Identification of urgent posts in MOOC discussion forums using an improved RCNN. In: 2019 IEEE World Conference on Engineering Education (EDUNINE). IEEE (2019)
4. Alrajhi, L., Alharbi, K., Cristea, A.I.: A multidimensional deep learner model of urgent instructor intervention need in MOOC forum posts. In: Kumar, V., Troussas, C. (eds.) ITS 2020. LNCS, vol. 12149, pp. 226–236. Springer, Cham (2020). https://doi.org/10.1007/978-3-030-49663-0_27
5. Khodeir, N.A.: Bi-GRU urgent classification for MOOC discussion forums based on BERT. IEEE Access **9**, 58243–58255 (2021)
6. Almatrafi, O., Johri, A., Rangwala, H.: Needle in a haystack: identifying learner posts that require urgent response in MOOC discussion forums. Comput. Educ. **118**, 1–9 (2018)
7. Chandrasekaran, M.K., et al.: Learning instructor intervention from MOOC forums: early results and issues. arXiv preprint arXiv:1504.07206 (2015)
8. Agrawal, A., Paepcke, A.: The Stanford MOOCPosts Data Set. https://datastage.stanford.edu/StanfordMoocPosts/
9. Troyano, J.A., Carrillo, V., Enríquez, F., Galán, F.J.: Named entity recognition through corpus transformation and system combination. In: Vicedo, JLuis, Martínez-Barco, P., Muñoz, R., Saiz Noeda, M. (eds.) EsTAL 2004. LNCS (LNAI), vol. 3230, pp. 255–266. Springer, Heidelberg (2004). https://doi.org/10.1007/978-3-540-30228-5_23
10. Farias, F., Ludermir, T., Bastos-Filho, C.: Similarity based stratified splitting: an approach to train better classifiers. arXiv preprint arXiv:2010.06099 (2020)
11. Captum: Captum_BERT (2022). https://colab.research.google.com/drive/1pgAbzUF2SzF0BdFtGpJbZPWUOhFxT2NZ

Understanding Children's Behaviour in a Hybrid Open Learning Environment

Vani Shree[⊠], Proma Bhattacharjee, Annapoorni Chandrashekar, and Nishant Baghel

Pratham Education Foundation, Mumbai, India
digital@pratham.org

Abstract. The paper attempts a narrativization of events in an open learning environment such that it renders visible the interplay of digital infrastructure, content and social structure in children's learning journey. The study is built on a program that **exposes** self-organised groups of about 40,000 children aged 10–14, to varied domains of knowledge delivered through offline digital content on tablets. Groups then make a collective choice of a domain or subject they wish to **explore** further through course enrollments, projects and workshops. They engage with the subject matter to create **experiments** of shared value to the group, namely a story, game, science model etc. In showcasing these experiments to other learner groups, village stakeholders and virtual communities, they learn to express their ideas, virtually **exchange** feedback, and iterate their learning process. Together, this constitutes the 'Ex' pedagogical framework. By tracing log files, analysing individual events and their correlations with other events, we identify patterns that show us how children's learning is driven by choice, creativity, and collaboration as they transition to each stage of their learning journey.

Keywords: Hybrid open learning environments · Learning analytics · Data mining

1 Setting Up a Hybrid Open Learning Environment

In the last few decades, research on learning and teaching methodologies has gained considerable momentum. There has been substantial research on the traditional classroom-based learning and its shortcomings, especially, with respect to students' ability to acquire new skills and knowledge. More recently, with the penetration of internet and digital technologies, we have seen a rise in hybrid environments to facilitate learning among children. While hybrid learning environments can provide additional support for the traditional instructional learning approach, there is a need for exploring newer alternative forms of learning environment design in order to explore the potential of new technologies and their impact on learning (Sharma and Fiedler 2004).

One such alternative learning design is the PraDigi Open Learning model which focuses on developing hybrid learning solutions for low income & low-tech environments with the aim to "leverage technology to improve access to education and buttress delivery of quality learning modules." (Singh et al. 2022). It follows a guided multimedia mode of

© Springer Nature Switzerland AG 2022
M. M. Rodrigo et al. (Eds.): AIED 2022, LNCS 13356, pp. 428–431, 2022.
https://doi.org/10.1007/978-3-031-11647-6_85

instruction with focus on not just the technology but also local ownership, involvement of community stakeholders and project-based curriculums that influence its effective use and help self-directed learners thrive.

1.1 Social Structure, Digital Infrastructure and Appropriate Data Systems

The program is set up in villages across Maharashtra, Rajasthan and Uttar Pradesh, where children between the age of 10 to 14 years learn in mix-age and mix-gender groups in their immediate neighborhoods after school hours. Two groups of 5 children each share the ownership of a tablet, and a Coach, who is typically an older youth from the community itself, acts as the guardian of the tablet. Tablets are shared to the groups with off-line content across two learning apps - PraDigi for School and PraDigi for Life. Since the program is based primarily in rural areas, both apps have been designed to be operational without internet availability. A Community Resource Leader (CRL), who is a member of the field and implementation team, oversees 10 villages and visits each group once a week to facilitate the activities undertaken by the group. The CRLs, during their visits to the villages, push the data from the tablets onto our servers every week. This data sync setup serves as a feedback mechanism.

Content on the School App is focused on preparing children for school and has short courses designed on topics across Science, Math, English and Language subjects whereas the courses offered on the Life App have content which is more focused on preparing them for 'life' i.e., learning beyond what is necessary to excel in schools. Examples of some courses designed under the Life App are COVID-19 Awareness, Music, Science experiments etc. The content is available in the form of videos, games and PDFs in their regional languages. Under this model, the children are first encouraged to explore the content and then enroll in courses that interest them. This at first sparks curiosity of the learners by exposing them to a variety of topics and they explore a select few further, depending on their inclinations.

Data collection and database systems were designed to facilitate the tracking of children's engagement over time. Having two levels of identifiers - unique ID of the child and unique ID of the group to which they belong was instrumental in helping us distinguish between individual and collective interests in courses. Learner groups are fluid, which means children can switch groups as per their choice. When new children joined the program in the academic year 2017–18, out of 8000 active groups, 33% groups underwent a change as members moved to other groups or new members joined the group. One of the reasons why children moved from one group to another was because they were more interested in another group's direction of learning. Children, whose interests do not align with the interests of the majority in the group, gradually gravitate to other groups or continue learning through individual sessions on the content. This highlights the children's ability to organize themselves into optimal learning situations with minimal adult interference.

2 Using Application Logs to Measure Children's Engagement and Performance of Digital Content

The analytics efforts at Pratham strive towards understanding the learning experience when students interact with shared devices and digital content in their communities. Application logs are generated as children engage with the learning apps. Each learning log is viewed as a sequence of children's self-directed interactions, called events, with their learning environment. The choice of when, where, and how long the learning session should be, rests with children. Their engagement with the learning content is measured using K means clustering across three parameters (a) Regularity: Average number of days the app was accessed by a group per month, (b) Duration: Average time spent on the app and (c) Exploration: Number of resources accessed by the groups out of the total resources available on the app. The student groups are divided into categories based on their engagement levels, namely Best, High, Medium, and Low engagement group. On average, 45% of resources on the tablet that were accessed by Best engagement groups were Science videos. The next most accessed resources by Best engagement groups were English videos (19%) and Mathematics videos (16%).

In the early years of the program, two groups were given shared ownership of a tablet. We explored the possibility that the engagement level of one group influences the engagement level of the other. We found that 88% of tablets were shared by groups with the same engagement level or groups that were only one engagement level apart. This highlights the role of peer influence in one's learning journey. It is important to remember that these groups were not intentionally paired but evolved organically as the children interacted with each other.

While calculating the retention and decay rate of resources over a period, we observed that popular resources among both old and new users were the same in the beginning, but it changed over time. This implies that the users start with certain popular resources because of an off-tech recommendation system in play but over time as they explore the open content, their engagement patterns diversify.

In order to understand the engagement patterns of video based resources, we applied K-means clustering method based on 4 metrics - complete views (the proportion of times the video was for at least 95% of its duration), repeat views (the proportion of times the video was watched by a children more than once), average viewing duration (the average proportion of the duration of the video that children watch) and average children accessed per month (the average number of children that access the video each month). The content could be grouped into 3 clusters. The cluster with highest engagement levels had videos related to real life demonstrations and science experiments.

3 Incorporating Insights into Pedagogy – Project Based Learning

To understand to what extent exploration drives experimentation, in 2018–19, science courses were launched. After enrolling in a course of their choice, children had to watch the course videos on the tablet. Instead of reducing their learning outcome in terms of only assessment scores, we added creation of science models to the completion criteria. They engaged with the app for 13 days a month on average for 48–50 min per day. Inspired by the content they were exposed to, children picked creative solutions to source materials for the models from everyday objects around them. Over 6,700 groups enrolled in at least one science course. On average each group enrolled in 2 courses. About 2/3 of the children enrolled completed the course.

Taking learnings from exploration and experimentation patterns of children a step further we introduced project completion milestones in the courses. One such example is a design for change inspired activity called 'Karke dekho, Karke Dikhao" or "Show and Tell" which was organised in over 700 communities. In this problem-solving challenge, children groups had to identify and research a pertinent problem prevalent in their village and collaboratively find a solution for it. 28,000 children from 5,700 groups across 734 rural villages participated in this activity, and 3900 groups (68% of groups) completed the activity. Similarly, trigger videos were shown to children that gave them a basic understanding of theatre concepts. The videos covered topics such as how to walk on stage, how to direct a play, and the importance of song, rhythm and tempo. 3003 groups performed a play and completed the project. This is in line with Arvind Gupta's conceptualization of 'learning by doing' which encourages them to apply creative and problem-solving abilities, often resulting in higher learning gains (Krithika 2019).

While efforts are made to involve experiential learning in curriculums, the reporting relies heavily on assessment performance which gives a narrow understanding of 'what' a child learns and might fail to capture the effect that these project completion milestones have on the child's learning potential. Instead, we propose a framework that moves beyond assessment and looks into 'how' children learn when they have the freedom to choose 'what' they want to learn with community support and shared devices.

References

Sharma, P., Fiedler, S.: Introducing technologies and practices for supporting self-organized learning in a hybrid environment. In: Proceedings of I-Know 2004, Graz (2004)

Krithika, R.: Let children learn by doing. The Hindu (2019). https://www.thehindu.com/educat ion/schools/a-conversation-with-renowned-scientistand-educator-arvind-gupta/article28098 701.ece

Liu, N.F., Carless, D.: Peer feedback: the learning element of peer assessment. Teach. High. Educ. 11(3), 279–290 (2006)

Singh, R., Chandrashekar, A., Baghel, N.: Learning, marginalization, and improving the quality of education in low-income countries. In: The Role of Civil Society Organizations and Scalable Technology Solutions for Marginalized Communities. Open Book Publishers, Cambridge (2022)

van Popta, E., Kral, M., Camp, G., Martens, R.L., Simons, P.R.: Exploring the value of peer feedback in online learning for the provider. Educ. Res. Rev. (2016)

Zhou, M., Xu, Y., Nesbit, J., Winne, P.: Sequential pattern analysis of learning logs (2010). https:// doi.org/10.1201/b10274-10

Prediction of Students' Performance in E-learning Environments Based on Link Prediction in a Knowledge Graph

Antonia Ettorre[✉][ID], Franck Michel[ID], and Catherine Faron[ID]

Université Côte d'Azur, CNRS, Inria, I3S, Sophia Antipolis, France
{aettorre,fmichel,faron}@i3s.unice.fr

Abstract. In recent years, the growing need for easily accessible high-quality educational resources, supported by the advances in AI and Web technologies, has stimulated the development of increasingly intelligent learning environments. One of the main requirements of these smart tutoring systems is the capacity to trace the knowledge acquired by users over time, and assess their ability to face a specific Knowledge Component in the future with the final goal of presenting learners with the most suitable educational content. In this paper, we propose a model to predict students' performance based on the description of the whole learning ecosystem, in the form of a RDF Knowledge Graph. Subsequently, we reformulate the Knowledge Tracing task as a Link Prediction problem on such a Knowledge Graph and we predict students outcome to questions by determining the most probable link between each answer and its correct or wrong realizations. Our first experiments on a real-world dataset show that the proposed approach yields promising results comparable with state-of-the-art models.

1 Introduction

The increasingly easy access to high-quality educational resources combined with the recent advances in AI and Web technologies have fostered the development of smart user-centered learning environments. The success of such systems is mainly due to their ability to assist users in their learning process, by offering real-time automated tutoring and personalized revision suggestions. One of the main requirements for these systems is the capacity to trace the knowledge acquired by users over time and assess their ability to face a specific knowledge concept in the future with the final goal of presenting learners with the pedagogical content that will most effectively improve their skills. The challenge of predicting students' outcomes when interacting with a given educational resource, also known as *Knowledge Tracing (KT)*, has been largely investigated and several approaches have been proposed throughout the years, e.g. *Bayesian Knowledge Tracing (BKT)* [3], *Additive Factors Model (AFM)* [2] and *Deep Knowledge Tracing (DKT)* [5]. Although very different, these approaches present a major

© Springer Nature Switzerland AG 2022
M. M. Rodrigo et al. (Eds.): AIED 2022, LNCS 13356, pp. 432–435, 2022.
https://doi.org/10.1007/978-3-031-11647-6_86

commonality: their predictions rely on very limited and simply structured information, i.e. the student approaching the question, the question being answered, and the list of the skills (or knowledge components) involved in the question. Conversely, in real-world scenarios, students' performance can be influenced by several additional factors that are miss-represented or missing in the previously mentioned models, such as type of questions, number of possible answers, assignment or test length, hierarchical organization of knowledge components, etc.

In this paper, we present an approach to represent and exploit this heterogeneous information to provide reliable predictions for students' outcomes to questions. Firstly, we rely on the expressiveness offered by Semantic Web models to represent the whole learning ecosystem as an RDF Knowledge Graph (KG). Then, we reformulate the KT task as a Link Prediction (LP) problem on such a KGand we determine the most probable links between the answers whose outcomes need to be predicted and their correct or wrong results, and we convert these predictions into binary labels for answers' correctness. We think that reformulating the problem of predicting students' outcomes to questions in such a way allows us to take advantage of a much wider amount and variety of information about the learning environment while reusing widely-known and well-established Deep Learning methods for KGs, and avoiding the burden of features engineering. To empirically confirm the validity of the proposed approach, we apply it on a real-world dataset and compare its performance with state-of-the-art KT models.

2 Link Prediction for Students' Outcomes

The approach presented in this paper is mainly based on the hypothesis that it is possible to turn the KT task into a LP problem, after modeling the learning environment as a KG. In other words, instead of predicting the probability that a given student correctly answers a specific question, we evaluate the possibility that an implicit link (i.e. a triple) exists in the KGbetween the expected student's answer and its correct or incorrect result. Finally, an answer is labeled as 1 (correct) if the link towards the correct result has a higher score than the link towards the incorrect one, while it is labeled as 0 (incorrect) in the opposite case. For example, to predict the positive or negative outcome of the answer given by student A to question 1 (Fig. 1), we compute the score of the two triples $\langle answer1, has\_result, correct \rangle$ and $\langle answer1, has\_result, incorrect \rangle$, and we predict that A's answer will be correct if the first triple has a higher score than the second one. To empirically validate our hypothesis, we designed and developed an end-to-end pipeline depicted in Fig. 2, which takes as input the traces of the students' learning history, possibly enriched with contextual knowledge, and the list of the student-question interactions whose outcomes must be predicted. The framework implements four steps:

1. **Graph Building**: create a KGrepresenting the learning ecosystem;
2. **Graph Augmentation for Prediction**: inject into the previously created KGnew nodes representing the new answers we aim to predict;

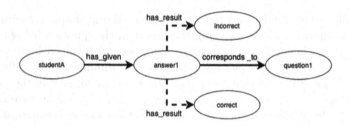

Fig. 1. Example of a KG representing the answer of a student to a question. The dashed lines represent the links we aim to predict.

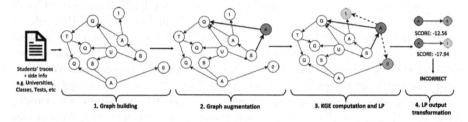

Fig. 2. Depiction of the different steps for the proposed Link Prediction-based approach.

3. **KGE Computation and Link Prediction**: compute the KGEsand use them to assess the scores of the triples to be predicted;
4. **LP Output Transformation**: convert triples' scores into the corresponding probabilities for the binary classification of students' answers (correct or incorrect).

3 Evaluation

Experimental Setup. To validate the proposed approach and to be able to compare it with state-of-the-art KT models, we decided to test our method on a widely-used benchmarking dataset: the ASSISTment 2009–2010 skill builder dataset [4] that stores the learning logs of the users of the ASSISTments platform. For each attempt of a user to a problem, it contains information such as user and problem identifiers, skills required for the problem, problem type (main or scaffolding), answer type (open answer, multiple-choice, etc.), assignment and assistment in which the problem was faced, response time, etc. The first step to apply our approach is to model this information as a KG. For the modeling of the user-problem interactions, we linked each problem to two nodes, one for the positive and one for the negative result, and users' answers are connected to such nodes based on their result. The second step is the computation of the KGEswhich has been carried out using TransE [1] to obtain embeddings of dimension 100.

Results and Discussion. Table 1 shows that the newly proposed LP approach achieves an improvement of 2% in terms of bACC, when compared to DKT. We believe that the main reason for this improvement is the ability of our method to consider a greater variety of knowledge about the learning ecosystem. It is also interesting to point out that, for both models, there is a strong difference between negative and positive F1 scores, with Link Prediction achieving slightly more balanced results. This gap in the F1 values can be explained by the highly unbalanced distribution of the target values in the subject dataset, which contains about 70% of correct answers.

Table 1. Results of Link Prediction and DKT approach.

KT model	F1 (0)	F1 (1)	bACC
Link Prediction	0.544	0.734	**0.657**
DKT	0.512	0.755	0.640

4   Conclusion

In this work, we reformulated the KT task into a KGLP problem able to take advantage of all the information commonly available in a smart learning system. Using a well-known benchmarking dataset that we transformed into a KGbased on Semantic Web models, our empirical evaluation showed that LP performs slightly better in predicting students' outcomes to questions results, when compared to DKT. Although modest, this improvement suggests that rich context information usually ignored by traditional KT approaches can help achieve better prediction. In the future, we wish to further explore this lead, notably by measuring and enriching the information encoded in KGEs.

References

1. Bordes, A., Usunier, N., Garcia-Duran, A., Weston, J., Yakhnenko, O.: Translating embeddings for modeling multi-relational data. In: Advances in Neural Information Processing Systems, vol. 26 (2013)
2. Cen, H., Koedinger, K., Junker, B.: Learning factors analysis – a general method for cognitive model evaluation and improvement. In: Ikeda, M., Ashley, K.D., Chan, T.-W. (eds.) ITS 2006. LNCS, vol. 4053, pp. 164–175. Springer, Heidelberg (2006). https://doi.org/10.1007/11774303_17
3. Corbett, A.T., Anderson, J.R.: Knowledge tracing: modeling the acquisition of procedural knowledge. User Model. User-Adap. Inter. 4(4), 253–278 (1994)
4. Feng, M., Heffernan, N., Koedinger, K.: Addressing the assessment challenge with an online system that tutors as it assesses. User Model. User-Adap. Inter. 19(3), 243–266 (2009)
5. Piech, C., et al.: Deep knowledge tracing. In: Advances in Neural Information Processing Systems, vol. 28, pp. 505–513 (2015)

Ontology-Controlled Automated Cumulative Scaffolding for Personalized Adaptive Learning

Fedor Dudyrev$^{(\boxtimes)}$, Alexey Neznanov, and Ksenia Anisimova

National Research University Higher School of Economics, Moscow, Russia
{fdudyrev,aneznanov,kanisimova}@hse.ru

Abstract. Scaffolding that provides transfer of responsibility is one of the central challenges in personalized adaptive learning (PAL) environments. We extend the well-known concept of the "inner loop" in PAL and fill the gaps between the inner loop (micro-adaptation) and outer loop (macro-adaptation). This extension leads us to a new method of cumulative instructional scaffolding based on explicit knowledge representation. This representation consists of domain ontologies and simulation models of tasks that automatically generate items, formative feedback, and additional pedagogical interventions throughout the scaffolding process.

Keywords: Personalized adaptive learning · Cumulative scaffolding · Inner loop · Outer loop · Domain ontology · Simulation

1 Advanced Scaffolding Model Inspired by the Zone of Proximal Development

The intelligent tutoring system (ITS) is the most important part of the smart learning environment for complex problem solving. To ensure the effective mastery of problem solving, the ITS should provide scaffolding. Scaffolding is support provided by a teacher/parent (tutor) that allows students (tutees) to participate meaningfully and acquire skills in problem-solving [1]. This support is specifically tailored to each student: it allows students to engage in student-centered learning that is, on the whole, more efficient than teacher-centered learning [2, 3]. The notion of scaffolding is being increasingly used today to describe various forms of support provided by software tools. Computer-based scaffolding is defined as computer-based support that helps students execute and assimilate tasks that lie beyond their unassisted abilities. Software tools can help to structure learning tasks by guiding learners through key components and supporting their planning and performance [4, 5].

The conceptualization of scaffolding is consistent with L. Vygotsky's model of instruction that highlights the teacher's role as a more knowledgeable person who helps learners solve tasks that they could not complete unassisted [6]. The interaction between the teacher and the student takes place in the Zone of Proximal Development (ZPD). As a rule, the student experiences difficulties when solving a new task. To support the student's efforts, the teacher provides the missing key to solving a problem that the student cannot tackle alone. While the teacher and the learner are constantly present in the

M. M. Rodrigo et al. (Eds.): AIED 2022, LNCS 13356, pp. 436–439, 2022.
https://doi.org/10.1007/978-3-031-11647-6_87

scaffolding space, their roles are significantly transformed during the transition from one type of scaffolding to another. At the same time, the content of the interactions between participants changes (Table 1).

Table 1. Three levels of scaffolding in PALS

	Teacher	Teacher-learner interactions	Feedback	Learner
Instructional scaffolding	Structuring, channeling, modeling, hinting…	Joint distributed solution to the problem	Evaluation of the learner's response, locally adaptive feed-back, including error-sensitive feedback	Imitation
Cultural scaffolding	Representation of techniques	Mastering the approach to solving the problem in instructional and cultural contexts	Context-aware feedback based on the interpretation of current learner errors	Trial and error, appropriation of cultural means
Developmental (metacognitive) scaffolding	Reflecting the learner's reasoning	Learner's acquisition of education-al experience	Context-aware feedback based on previous solutions	Internalization

The forms and modes of interaction are best investigated and described for *instructional scaffolding*. At this stage, the teacher assists the student by solving the parts of the problem that are beyond the student's capacity. Additionally, scaffolders simplify problems by structuring them so that their solution becomes more productive for students. At the stage of *cultural scaffolding*, one discusses not only the correct sequence of problem-solving steps but also the hidden rules that mediate them. The teacher demonstrates amalgam knowledge that combines the knowledge of content with the knowledge of students and pedagogy [7]. Finally, *developmental (metacognitive) scaffolding* supports internalization and the transfer to the mental plane of results that had previously been obtained with the help of the teacher, allowing the student to use them independently in the future.

2 Adaptive Techniques that Provide Cumulative Scaffolding

The improved scaffolding model presented above describes several types of interaction between the teacher and the learner that reflect varying degrees of support and the gradual delegation of responsibility to the learner. The following ITS requirements provide for the described interactions within PALS. Traditionally, the adaptive behavior of the ITS is assured with the help of several adaptive techniques. At each step of task completion, the tutoring system offers such services as minimal or error-specific feedback, knowledge assessment, hints about the next step, and a review of the solution. The learner's actions and the tutor's feedback within a step, and the navigation of steps within a task constitute the inner loop. As for the outer loop, it is responsible for deciding what task a student should perform next once the present task has been completed. The outer loop is about the sequence and selection of tasks or knowledge units (modules) [8, 9].

Thus, such modeling requires a more detailed representation of domains, some of which relate to the field of pedagogy. The extraction of such complex domain-instructional knowledge about how the problem is solved is a key prerequisite for technique-oriented scaffolding. The additional requirement for adaptability is due to the scale of the assimilation process. In order for the technique to be mastered, it is necessary to solve a series of tasks consistently. Each subsequent task should be based on the results of previous ones. Preceding solutions and mistakes are used as arguments in solving each subsequent problem. The accumulated educational experience is directly involved in the problem-solving process. Traces of past solutions become building blocks for new ones. This corresponds to the manner of a skillful teacher who reminds students of arguments previously used for solving similar tasks.

As for developmental (metacognitive) scaffolding, it is completely based on accu-mulating and updating the student's educational experience. This type of tutor-learner interaction assumes that feedback and support are based primarily on the interpretation of the student's previous reasoning. To maintain this interaction, the tutor must make use of traces of previous solutions and present them in a timely manner. Thus, the behavior of a tutoring system that supports adaptive scaffolding becomes more sophisticated.

The usual scaffolding cycle is supported by the inner loop (Level 1: Assignment → Response → Feedback → Support → Solution), after which the student is given a task on the same topic to build up knowledge and skills (micro-sequencing). In addition, the inner loop includes new components. While improving the solution to a problem in the STEM domain, the student simultaneously masters a solution technique according to the golden standard of pedagogy. The technique and its components are based on a separate feedback and support cycle. It is important to note that the steps and interactions in the instructional and domain levels are synchronized with each other inside the inner loop.

Finally, when going from one task to another, the tutor considers previous training results more comprehensively. First of all, this promotes the reasonable selection of the next task. Secondly, traces of previous solutions are directly applied in the current inner loop. For example, feedback not only determines the error and its causes but also records the frequency of similar errors made earlier. All of this makes it possible to vary and further customize the support. Thus, the inner loop becomes cumulative, and the tutor's behavior can be characterized as partly macro-adaptive.

3 Discussion

This paper presents a model to support adaptive scaffolding in personalized adaptive learning systems. A strong pedagogical background and the practical consideration of synthetic knowledge representation allow us to describe and prototype a new version of ITS in PALS. Introducing the concept of a cumulative inner loop supported by cumulative scaffolding with AIG increases ITS smartness. We extend the set of adaptive techniques by formalizing levels of scaffolding and dealing with the sequence of tasks in the cumulative inner loop. An extended set of scaffolding forms allows the implementation of fading and the transfer of responsibility.

The scaffolding model must relate to the learner and assignment models for the universal implementation of adequate computer-user interaction in ITS with hints and explanations. For these relations, ITS needs formal ontologies (domain knowledge with simulation subsystem and didactic knowledge) in the background of learning trajectory planning, educational materials handling, automatic assignment generation, and scaffolding actions.

Acknowledgements. This publication was supported by a grant for research centers in the field of AI provided by the Analytical Center of the Government of the Russian Federation (ACRF) in accordance with the Agreement on the Provision of Subsidies No. 000000-D730321P5Q0002 and the Agreement with HSE University No. 70-2021-00139.

References

1. Wood, D., Bruner, J.S., Ross, G.: The role of tutoring in problem solving. J. Child Psychol. Psychiatry **17**(2), 89–100 (1976)
2. Belland, B.: Scaffolding: definition, current debates, and future directions. In: Spector, J., Merrill, M., Elen, J., Bishop, M. (eds.) Handbook of Research on Educational Communications and Technology, pp. 505–518. Springer, New York (2014). https://doi.org/10.1007/978-1-4614-3185-5_39
3. Belland, B.: Instructional Scaffolding in STEM Education. Springer, Cham (2017). https://doi.org/10.1007/978-3-319-02565-0
4. Reiser, B.: Scaffolding complex learning: the mechanisms of structuring and problematizing student work. J. Learn. Sci. **13**(3), 273–304 (2004)
5. Vanlehn, K.: The relative effectiveness of human tutoring, intelligent tutoring systems, and other tutoring systems. Educ. Psychol. **46**(4), 197–221 (2011)
6. van de Pol, J., Volman, M., Beishuizen, J.: Scaffolding in teacher-student interaction: a decade of research. Educ. Psychol. Rev. **22**(3), 271–296 (2010)
7. Ball, D.L., Thames, M.H., Phelps, G.: Content knowledge for teaching. What makes it special? J. Teac. Educ. **59**(5), 389–407 (2008)
8. Vanlehn, K.: The behavior of tutoring systems. Int. J. Artif. Intell. Educ. **16**(3), 227–265 (2006)
9. Durlach, P., Spain, R.: Framework for instructional technology: methods of implementing adaptive training and education. Army Res. Inst. Behav. Soc. Sci. (2014)

Investigating the Role of Demographics in Predicting High Achieving Students

Ali Al-Zawqari[⊠][iD] and Gerd Vandersteen[iD]

Department ELEC, Vrije Universiteit Brussel, 1050 Brussels, Belgium
aalzawqa@vub.be

Abstract. Researchers have observed the relationship between academic achievements and students' demographical characteristics in physical classroom-based learning. In the context of online learning, recent studies were conducted to explore the leading factors of successful online courses. These studies investigated the impact of demographical features on students' achievement in the online learning environment. Most works were presented via descriptive statistics or utilized big data with advanced classification algorithms. In this paper, we study the role of students' demographics in predicting high achieving students in online courses. Obtained results show that the interaction of students with the virtual learning environment is more informative than the student's demographical characteristics, which allows for the removal of demographical information without affecting the prediction models' performance. In addition, the testing results present that cross-courses-trained predictive models are as effective as individual-courses-trained models.

Keywords: Prediction models · High achieving students

1 Introduction

Learning environments refer to an educational concept of diverse cultural contexts and physical setups where active interactions happen between students and teachers or between students and other students. In online learning, this interaction is limited due to physical separation, resulting in differences in challenges and patterns compared to physical classroom-based learning [9]. In traditional education formats, teachers utilize both direct and nonverbal communication with the students in the classroom to assess their students' needs on the spot, which allows them to provide the needed assistance. In online learning, this is partially substituted by the early prediction of the students' academic performance to classify them into certain group levels [8]. Recently, several conducted studies focused on using students' background information for early prediction

This work was financially supported in part by the Vrije Universiteit Brussel (VUB-SRP19), in part by the Flemish Government (Methusalem Fund METH1) and in part by the Fund for Scientific Research (FWO).

M. M. Rodrigo et al. (Eds.): AIED 2022, LNCS 13356, pp. 440–443, 2022.
https://doi.org/10.1007/978-3-031-11647-6_88

of their academic performance [5], which were motivated by the links between students' social background and their level of success [6]. Others take a combination of students' activities in an online environment, demographics (e.g., gender and living area), and assessments to build the prediction models [4]. However, analyzing the different roles of each feature category has gained less focus. More attention on this subject is needed as the current prediction models reveal unfairness towards minorities [2]. In this work, the goal is to investigate the high-achieving students' prediction models in two main points: 1) the role of students' background information; 2) the performance of the prediction models in an individual course dataset compared to a cross-courses dataset.

2   Methods

The Open University Learning Analytics dataset (OULAD) [7] is chosen for this work. OULAD has a mixture of students' demographical information and interaction with the learning environment. Most recent studies focused on analyzing OULAD as a whole. Since many universities and schools only started offering online courses recently, a dataset in the size of OULAD is mostly unavailable. Hence, it is decided to take a subset of OULAD to evaluate the performance of the prediction models. Based on extensive experimentation in [1], the focus here is on one binary classification problem: distinction-fail. To examine the role of demographics, three different prediction models are built based on: 1) only demographical features; 2) students' demographics and interactions; 3) only students' interactions. For the prediction model, three different machine learning algorithms are chosen: Gaussian processes, random forests, and artificial neural networks [3]. These methods are selected for two reasons: 1) each represents a different family of machine learning algorithms; 2) their widespread usage. Figure 1 summarizes this experiment pipeline.

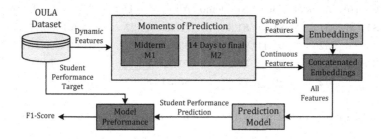

Fig. 1. Experiment pipeline.

3   Results

The experimentation and results of this work are summarized in three parts: 1) build and evaluate prediction models based only on students' demographics, then compare them to the models built with both students' demographics and

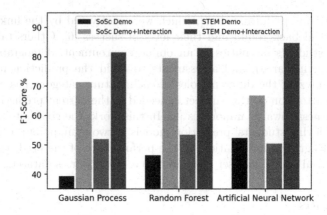

Fig. 2. F1-score of models built with demographics vs. demographics and interactions.

interactions with the learning platform; 2) build and evaluate prediction models based only on students' interactions; 3) build and evaluate individual course prediction models and cross courses prediction models. For the first part, Fig. 2 shows the evaluation of the prediction models in the testing set. Both STEM and SoSc courses results show that models enriched by students' interactions have a +10% higher F1-score than models based only on demographics.

For the second part, table 1 show that the student demographic does not add to the prediction model's performance. This observation opposes the findings in previous research, which link students' backgrounds to their chances of

Table 1. F1-score of models built with and without demographic features.

		GP		RF		ANN	
Course	Moment	All	No Demo	All	No Demo	All	No Demo
STEM	M1	<u>81.44</u>	81.30	82.96	<u>83.27</u>	**84.68**	83.17
	M2	91.44	<u>92.75</u>	**93.67**	92.24	<u>91.60</u>	90.16
SoSc	M1	<u>71.32</u>	70.29	79.56	**80.81**	66.87	<u>76.39</u>
	M2	<u>80.18</u>	77.63	**88.31**	86.65	78.79	<u>80.81</u>

Table 2. F1-score of individual and cross courses models without demographic features.

		GP		RF		ANN	
Course	Moment	Individual	Cross	Individual	Cross	Individual	Cross
STEM	M1	<u>81.30</u>	77.91	83.27	**84.04**	83.17	**84.04**
	M2	**92.75**	86.85	<u>92.24</u>	91.72	90.16	<u>91.44</u>
SoSc	M1	<u>70.29</u>	63.79	**80.81**	79.75	76.39	<u>79.27</u>
	M2	77.63	<u>80.55</u>	**86.65**	85.93	80.81	<u>81.60</u>

success [6]. Table 2 shows the last results, which indicate that a cross-course dataset can be an acceptable solution when lacking enough data to build prediction models.

4 Conclusion

This research examined the role of students' demographics in online learning. First, a subset of OULAD is chosen to be studied, representing two different domain courses. Next, prediction models are built in three different scenarios: 1) prediction models based only on students' demographics; 2) prediction models based on students' demographics and interactions with the online learning environment; 3) prediction models solely based on students' interactions. The last scenario is built twice: one using separate datasets for each course domain and the other using a cross-courses dataset. Obtained results from testing sets are 1) students' demographic information can be eliminated without affecting the prediction models' performance; 2) individual and cross courses based prediction models have comparable performances. The first finding can protect students from inherent bias in the historical data by removing the demographical features. The second finding presents a solution where the dataset is not large enough for machine learning algorithms. This also can extract implicit patterns between students across different specialties, which will be investigated in future work.

References

1. Al-Zawqari, A., Peumans, D., Vandersteen, G.: A flexible feature selection approach for predicting students' academic performance in online courses. Comput. Educ. Artif. Intell. (in review 2022)
2. Bayer, V., Hlosta, M., Fernandez, M.: Learning analytics and fairness: do existing algorithms serve everyone equally? In: Roll, I., McNamara, D., Sosnovsky, S., Luckin, R., Dimitrova, V. (eds.) AIED 2021. LNCS (LNAI), vol. 12749, pp. 71–75. Springer, Cham (2021). https://doi.org/10.1007/978-3-030-78270-2_12
3. Géron, A.: Hands-On Machine Learning with Scikit-Learn, Keras, and TensorFlow: Concepts, Tools, and Techniques to Build Intelligent Systems. O'Reilly Media Inc, Sebastopol (2019)
4. He, Y., et al.: Online at-risk student identification using RNN-GRU joint neural networks. Information 11(10), 474 (2020)
5. Hoffait, A.S., Schyns, M.: Early detection of university students with potential difficulties. Decis. Support Syst. 101, 1–11 (2017)
6. Kotok, S.: Unfulfilled potential: high-achieving minority students and the high school achievement gap in math. High Sch. J. 100(3), 183–202 (2017)
7. Kuzilek, J., Hlosta, M., Zdrahal, Z.: Open university learning analytics dataset. Sci. Data 4(1), 1–8 (2017)
8. Luckin, R., Holmes, W., Griffiths, M., Forcier, L.B.: Intelligence unleashed: an argument for AI in education (2016)
9. Park, J.H., Choi, H.J.: Factors influencing adult learners' decision to drop out or persist in online learning. J. Educ. Technol. Soc. 12(4), 207–217 (2009)

Dynamic Conversational Chatbot for Assessing Primary Students

Esa Weerasinghe[✉], Thamashi Kotuwegedara, Rangeena Amarasena, Prasadi Jayasinghe, and Kalpani Manathunga[iD]

Department of Computer Science and Software Engineering, Sri Lanka Institute of Information Technology, Malabe, Sri Lanka
it18174854@my.sliit.lk

Abstract. Teaching necessitates a method of determining if learners are gaining the desired knowledge and skills. We believe that chatbot technology would be an excellent solution to this problem. Using our approach, the chatbot will assess the student by using questions and answers recorded in a question bank. Four approaches were taken to assess the students' answers against the model answer, using models such as Word2Vec, all-mpnet-base-v2, and Sense2Vec models with Word Mover's Distance algorithm or Cosine Similarity. After evaluating these approaches, the best performing approach was when we used the Sense2Vec model with Cosine Similarity which gave the most accurate similarity score range for correct and incorrect answers.

Keywords: Chatbot · Artificial Intelligence · Natural language processing

1 Introduction

In distance learning student inquiries, providing feedback, and assessing students are challenging [1] and such facts greatly affect primary students because children learn best when they are actively participating in learning. To know whether teaching is effective, assessments are carried out to acquire the student's status, which may prove to be too late to act if it is needed [1]. Therefore, checking whether students understood the delivered content frequently would be more of a favorable solution. But in an instance where a teacher has many students, assessing them frequently is not feasible. Student age is also a key factor that needs to be carefully considered when designing assessments. One solution to solve these problems would be to use chatbot technology.

For quite some time, chatbots have been used for educational purposes. Chatbots can provide a framework for learning by selecting and arranging information to meet a student's requirements and pace, as well as assisting in self-reflection and learning motivation [2]. A conversational agent was used to promote children's verbal communication skills in [3], where a process was designed to merge conversational technology with a speech-to-image system and implemented a wizard-of-oz version of the agent called ISI. ISI permits children to develop their body awareness and self-expression. Web Passive

M. M. Rodrigo et al. (Eds.): AIED 2022, LNCS 13356, pp. 444–448, 2022.
https://doi.org/10.1007/978-3-031-11647-6_89

Voice Tutor (Web PVT) [4], is an adaptive web-based intelligent computer-assisted language learning program that was used to teach non-native speakers the passive voice in the English Language. There has been a significant number of studies on computing text-similarity using various features, algorithms, metrics, etc. In work [5], three different methods were compared and analyzed in computing the semantic similarity between two short texts. The first method to calculate the text similarity was to use Cosine similarity with Term Frequency – Inverse Document Frequency (TF-IDF) vectors. The second method was to use Cosine Similarity with Word2Vec vectors, and the third was to use Soft Cosine Similarity with Word2Vec vectors. The paper concluded that Cosine Similarity using TF-IDF vectors was the best performing method to find similarities between short texts. In paper [6], Levenshtein Distance (LD) and Cosine Similarity are used to compare students' answers with the model answers in a short answer scoring system for English grammar.

When going through past research papers, [4] and [6] was the only research we found that assessed students. Our observation was that research on chatbots used to assess learners especially in primary domain, is inadequate. In this paper the authors explore the possibility of assessing primary students between age 5–9 using conversational AI (Artificial Intelligence) and text similarity techniques such as Word Mover's Distance (WMD) algorithm or Cosine Similarity metric. The best performing approach was when we used the Sense2Vec model with Cosine Similarity, which gave the most accurate similarity score range for correct and incorrect answers. The following section reveals the methodology on how diverse Natural Language Processing (NLP) approaches were utilized and the proceeding section reveals a detailed result and discussion section followed by concluding remarks and future work.

2 Methodology

The chatbot was built using the Rasa Framework, which is an open-source machine learning (ML) framework for building AI assistants and chatbots. The chatbot can conduct quizzes and assess students' answers during the quiz. When answering questions in the quiz, if a student gives an incorrect answer the chatbot will prompt a hint to guide the student towards the correct answer. The chatbot can ask four types of questions: true or false type, MCQ type, image type, and question-answer type. Once the student agrees to take the quiz the chatbot will retrieve the quiz from the question bank and prepare the quiz using custom actions. Next, the chatbot will prompt all the questions to acquire the student's answers. Once all the answers are collected the text similarity model is used to calculate the text-similarity between the student given answer and the model answer. If the text-similarity score is high, the student's answer will be considered correct and if it's low, a hint will be displayed along with the question again.

Four diverse methods were used to calculate the similarity score between answers. In the first method, the model answer and the student's answer are pre-processed. Next, the WMD algorithm uses the Word2Vec word embedding model to calculate the distance between the two pre-processed answers. In the second method, the model answer and the student's answer were pre-processed, and then the Cosine Similarity between the average vectors was calculated which was produced by the Word2Vec model. The third method uses all-mpnet-base-v2 sentence-transformer model to compute word embeddings of the model and student's answers. Once the word embeddings are computed, the Cosine Similarity between two answers was computed. In the last method, the Sense2Vec model is added as a pipeline to SpaCy's en_core_web_lg model. Next, the model answer and the student's answer are pre-processed and used to calculate the similarity score using Cosine Similarity. Synonyms were generated for the model answer and compared with the student's answer to compute the synonym similarity score. Next the average similarity score was computed using these two scores to increase the accuracy of the similarity score.

3 Results and Discussion

An experiment was carried out to evaluate the four methods that computed similarity scores between the model answer and the student's answer. In this experiment, the model answers and the student's answers were in the following forms: both answers are short answers, student answer is a synonym of the model answer; and both answers have a numeric value. The accuracy of the similarity scores for each method was determined by being able to clearly distinguish the similarity scores between correct and incorrect answers. Table 1 presents the model answer, the student's answer, status, and the similarity scores computed for all four approaches.

Table 1. Similarity scores computed for all four text similarity methods

Generated answer	Student's answer	Status	Method 1	Method 2	Method 3	Method 4
It turns into ice and becomes a solid	Turns to a solid	Correct	0.605	0.786	0.722	0.856
To break down	To study closely	Incorrect	2.759	0.070	0.093	0.352
False	False	Correct	0	1	1	0.811
True	False	Incorrect	1.867	0.371	0.523	0.494
Magnifying glass	Hand glass	Correct	1.287	0.638	0.608	0.724
Magnifying glass	Hand lens	Correct	1.746	0.479	0.498	0.636
Magnifying glass	Simple microscope	Correct	1.745	0.337	0.597	0.440
206	206	Correct	0	0	1	0.960
206	Two hundred and six	Correct	3.215	0	0.466	0.078
206	200	Incorrect	0.991	0	0.771	0.455

In the first method, we were able to distinguish a boundary between the correct answers and incorrect answers. Distance of 1.8 and 0 were taken as the range for correct answers since comparing the two opposites yielded 1.867. In the second method, we were not able to distinguish the distinct boundary between correct answers and incorrect answers as in the previous method. In the third method, we were able to distinguish a boundary between the correct answers and the incorrect ones. A similarity score between 0.55 and 1.0 was taken as the range for correct answers since comparing two opposites yielded 0.523. Still, this did not comply when some of the answers given were synonyms for the model answer. In the fourth method, we were able to distinguish a boundary between the correct and incorrect answers. The average similarity score between 0.5 and 1.0 was taken as the range for the correct answers since comparing two opposites yielded an average similarity score of 0.494. Sense2Vec model with Average Cosine Similarity gave a similarity score range for distinguishing correct answers among all four methods. However, comparing numbers in numerals and words was challenging in all four methods. We can conclude from our results that, up to some degree, numeracy is naturally present in standard embeddings. [7] states that this could be because numeracy is one of the types of emergent knowledge.

4 Conclusion

Research was focused on developing a chatbot that can present dynamic quizzes and assess students. Four approaches were adopted from the NLP domain to assess the students' answers against the model answer. According to the test results we obtained, the best performing approach was the Sense2Vec model with Cosine Similarity, which gave the most accurate similarity score range for correct and incorrect answers. Further testing is needed to determine the accuracy of this.

References

1. Alharbi, K., Cristea, A.I., Shi, L., Tymms, P., Brown, C.: Agent-based classroom environment simulation: the effect of disruptive schoolchildren's behaviour versus teacher control over neighbours. In: Roll, I., McNamara, D., Sosnovsky, S., Luckin, R., Dimitrova, V. (eds.) AIED 2021. LNCS (LNAI), vol. 12749, pp. 48–53. Springer, Cham (2021). https://doi.org/10.1007/978-3-030-78270-2_8
2. Sandu, N., Gide, E.: Adoption of AI-chatbots to enhance student learning experience in higher education in India. In: 2019 18th International Conference on Information Technology Based Higher Education and Training (ITHET). IEEE, Magdeburg (2019)
3. Catania, F., Spitale, M., Cosentino, G., Garzotto, F.: Conversational agents to promote children's verbal communication skills. In: Følstad, A., et al. (eds.) CONVERSATIONS 2020. LNCS, vol. 12604, pp. 158–172. Springer, Cham (2021). https://doi.org/10.1007/978-3-030-68288-0_11
4. Virvou, M., Tsiriga, V.: Web passive voice tutor: an intelligent computer assisted language learning system over the WWW. In: Proceedings IEEE International Conference on Advanced Learning Technologies. IEEE, Madison (2001)
5. Sitikhu, P., Pahi, K., Thapa, P., Shakya, S.: A comparison of semantic similarity methods for maximum human interpretability. In: 2019 Artificial Intelligence for Transforming Business and Society (AITB). IEEE, Kathmandu (2019)

6. Olowolayemo, A., Nawi, S., Mantoro, T.: Short answer scoring in English grammar using text similarity measurement. In: 2018 International Conference on Computing, Engineering, and Design (ICCED). IEEE, Bangkok (2018)
7. Wallace, E., Wang, Y., Li, S., Singh, S., Gardner, M.: Do NLP models know numbers? Probing numeracy in embeddings. In: Proceedings of the 2019 Conference on Empirical Methods in Natural Language Processing and the 9th International Joint Conference on Natural Language Processing (EMNLP-IJCNLP) (2019)

Investigating Natural Language Processing Techniques for a Recommendation System to Support Employers, Job Seekers and Educational Institutions

Koen Bothmer and Tim Schlippe[✉]

IU International University of Applied Sciences, Bad Honnef, Germany
tim.schlippe@iu.org

Abstract. Skills are the common ground between employers, job seekers and educational institutions which can be analyzed with the help of natural language processing (NLP) techniques. In this paper we explore a state-of-the-art pipeline that extracts, vectorizes, clusters, and compares skills to provide recommendations for all three parties—thereby bridging the gap between employers, job seekers and educational institutions. Our best system combines Sentence-BERT [1], UMAP [2], DBSCAN [3], and K-means clustering [4].

Keywords: AI in education · Recommender system · Recommendation system · Up-skilling · Natural language processing

1 Introduction

There are often gaps between the skills that are needed in the labor market, the skills that job seekers[1] have and the skills that are taught in educational institutions [5]. Connecting and supporting all three players allows the greatest possible exchange of information and satisfies their needs. However, they usually use AI in isolation from one another [6–9]. Since skills are their common ground which can be analyzed with the help of AI, we investigate several NLP techniques to extract, vectorize, cluster and compare skills. Then we combine the optimal methods in a pipeline which serves as the basis for our application *Skill Scanner*[2] [10] that outputs statistics and recommendations about missing and covered skills for all three players. Our goal was to help employers, job seekers and educational institutions adapt to the job market's needs. Consequently, we used job postings, which represent the job market's needs, as reference. These representative skills, which we draw from a large set of job postings, are referred to as "*market skills*" in this paper. As companies hiring data scientists find that it is difficult to find a so-called "unicorn data scientist" [11], we conducted our experiments and analysis using companies' job postings for a data scientist position, job seekers' CVs for that position, and a curriculum from a master's program in data science. But our investigated methods can be applied to other job positions as well.

---

[1] "Job seeker" refers to individuals who wish to apply for or advance in a job.
[2] https://github.com/KoenBothmer/SkillScanner.

© Springer Nature Switzerland AG 2022
M. M. Rodrigo et al. (Eds.): AIED 2022, LNCS 13356, pp. 449–452, 2022.
https://doi.org/10.1007/978-3-031-11647-6_90

2 Related Work

Automatically ranking CVs is a valuable tool for employers. For example, [12] rank candidates for a job based on semantic matching of skills from LinkedIn profiles and skills from their job description, relying on a taxonomy of skills. Recent advancements in NLP offer opportunities to improve these methods: [6] use word embeddings from Word2Vec [13] to match CVs to jobs. [9] combine a knowledge graph and BERT for finding suitable candidates in a corpus of CVs. Recommendation systems for job seekers have been investigated by [14–16]. As in the systems for employers, text data from social media profiles such as LinkedIn or Facebook is usually processed [8, 17]. [18] give a systematic review of recent publications on course recommendation. Most related work focuses on recommending courses to potential students. They report a growing popularity of data mining techniques. To cope with different levels of abstraction and synonyms in the course materials and students' documents, they first cluster the content, which they can then compare. K-means [4] is usually used for this.

3 NLP to Extract, Vectorize, Cluster and Compare Skills

For a certain job position, our pipeline (1) takes a CV, a job posting or a learning curriculum as input, (2) extracts the skills of the provided document, (3) compares the document's extracted skills to a skill set which represents the market's needs (*market skills*) and (4) returns information of which *market skills* are covered or missing in the document. Figure 1 visualizes the steps of our corresponding NLP pipeline.

Fig. 1. Pipeline to extract, vectorize, cluster, and compare skills.

3.1 Retrieving Skill Sets: Extract Skill Requirements

In job postings, CVs and learning curricula, skills are usually expressed in bullet points. Therefore, we developed keyword- and rule-based techniques to extract bullet points from these sources. Furthermore. we used the *BeautifulSoup* package to gather and extract 21.5k bullet points from 2,633 job postings for data scientists in English from Indeed.com and Kaggle.com which represents the market's needs (*market skills*). Since some bullet points in a job posting are not skill requirements, we analyzed methods to deal with outliers that are not skill requirements as described in Sect. 3.4.

3.2 Vectorizing Skills: Map Skill Requirements to Semantic Vector Space

To compute distances between skills, we mapped the skills to a semantic vector space. To represent the skills which usually consist of several words, we investigated stacking and averaging word embeddings in a skill which were produced with Word2Vec [13] and GloVe [19]. In addition, we explored sentence embeddings. Sentence-BERT (44.2%) [1], a modification of the BERT transformers, outperformed word embeddings like GloVe (39.5%) by 12% in Silhouette score [20] at the end of our pipeline.

3.3 Removing Outliers from Skill Requirements

To remove outliers in the vectorized skills and allow our clustering techniques to perform better, we reduce the dimensionality of the feature space created by Sentence-BERT. For that we experimented with combinations of PCA [21], UMAP [2], and DBSCAN [3]. Using UMAP to reduce the vectorized skills to two dimensions and DBSCAN to remove outliers in the 2-dimensional (2D) space performed best according to our manual checks and reduced the 21.5k potential skills retrieved with our web scraper to 18.8k skills. However, since the 2D vectors did not contain enough information for further analysis of the skill set, we applied another clustering to the original 768-dimensional vectors that remained after removing outliers.

3.4 Clustering Skills

To find comparable skills despite different levels of abstraction and synonyms in job postings, CVs and learning curricula, we use a clustering approach. The benefit of our clustering approach compared to a taxonomy is that our model can pick up new skills without the need to update a taxonomy. K-means clustering has been successfully used in clustering word embeddings [22] and is adaptable and scalable [4]. Consequently, we used K-means to cluster our 768-dimensional vectors with the cosine distance as the distance metric. K was chosen as 31 with the highest Silhouette score of 44%.

3.5 Skill Scanner: Comparison and Analysis

After retrieving clusters and vectors representing the skill of each cluster, we perform mathematical operations to find covered and missing skills regarding the job market's demand which are then visualized in reports for employers, job seekers, and educational institutions. More information on the visualization of our reports is given in [10].

4 Conclusion and Future Work

The labor market dictates what job seekers should learn, and educational institutions should teach. Therefore, our system processes skills in job postings, CVs, and curricula and outputs recommendations for employers, job seekers, and educational institutions based on present and missing skills and their importance to employers. With our clustering approach we do not have to update a taxonomy as skill requirements change. Future work may be to apply our pipeline to other job positions and expand it to other domains. Furthermore, as we used the pre-trained Sentence-BERT it may be analyzed if a fine-tuned Sentence-BERT leads to further improvement.

References

1. Reimers, N., Gurevych, I.: Sentence-BERT: sentence embeddings using Siamese BERT-Networks. In: EMNLP-IJCNLP (2019)
2. McInnes, L., Healy J.: UMAP: uniform manifold approximation and projection for dimension reduction. arXiv, abs/1802.03426 (2018)
3. Ester, M., Kriegel, H.P., Sander, J., Xu, X.: A density-based algorithm for discovering clusters in large spatial databases with noise. In: KDD, pp. 226–231. AAAI Press (1996)
4. Lloyd, S.P.: Least squares quantization in PCM. Technical report RR-5497, Bell Lab (1957)
5. Palmer, R.: Jobs and skills mismatch in the informal economy (2017). 978-92-2-131613-8
6. Fernández-Reyes, F.C., Shinde, S.: CV Retrieval system based on job description matching using hybrid word embeddings. Comput. Speech Lang. **56** (2019)
7. Geyik, S.C., et al.: Talent search and recommendation systems at Linkedin: practical challenges and lessons learned. In: SIGIR (2018)
8. Guruge, D.B., Kadel, R., Halder, S.J.: The state of the art in methodologies of course recommender systems—a review of recent research data, **6**(2), 18 (2021)
9. Wang, Y., Allouache, Y., Joubert, C.: Analysing CV corpus for finding suitable candidates using knowledge graph and BERT. In: DBKDA (2021)
10. Bothmer, K., Schlippe, T.: Skill scanner: connecting and supporting employers, job seekers and educational institutions with an AI-based recommendation system. In: The Learning Ideas Conference 2022 (15th Annual Conference), New York, New York (2022)
11. Baškarada, S., Koronios, A.: Unicorn data scientist: the rarest of breeds. Prog. Electron. Libr. Inf. Syst. **51**(1), 65–74 (2017)
12. Faliagka, E., et al.: On-line consistent ranking on E-recruitment: seeking the truth behind a well-formed CV. Artif. Intell. Rev. **42**, 515–528 (2014)
13. Mikolov, T., Chen, K., Corrado, G., Dean, J.: Efficient estimation of word representations in vector space. In: ICLR (Workshop Poster) (2013)
14. Si-ting, Z., Wenxing, H., Ning, Z., Fan, Y.: Job recommender systems: a survey. In: ICCSE (2012)
15. Hong, W., Zheng, S., Wang, H., Shi, J.: A job recommender system based on user clustering. J. Comput. **8**, 1960–1967 (2013)
16. Alotaibi, S: A survey of job recommender systems. Int. J. Phys. Sci. (2012)
17. Diaby, M., Viennet, E., Launay, T.: Toward the next generation of recruitment tools: an online social network-based job recommender system. In: ASONAM (2013)
18. Guruge, D.B., Kadel, R., Halder, S.J.: The state of the art in methodologies of course recommender systems—a review of recent research. Data **6**(2), 18 (2021)
19. Pennington, J., Socher, R., Manning, C.D.: GloVe: global vectors for word representation. In: EMNLP (2014)
20. Rousseeuw, P.J.: Silhouettes: a graphical aid to the interpretation and validation of cluster analysis. Comput. Appl. Math. **20**, 53–65 (1987)
21. Pearson, K.: On lines and planes of closest fit to systems of points in space. Phil. Mag. **2**(11), 559–572 (1901)
22. Zhang, Y., et al.: Does deep learning help topic extraction? A kernel k-means clustering method with word embedding. J. Informet. **12**(4), 1099–1117 (2018)

Modeling Student Discourse in Online Discussion Forums Using Semantic Similarity Based Topic Chains

Harshita Chopra[1]($\boxtimes$), Yiwen Lin[2], Mohammad Amin Samadi[2],
Jacqueline Guadalupe Cavazos[2], Renzhe Yu[2], Spencer Jaquay[2],
and Nia Nixon[2]

[1] GGS Indraprastha University, Delhi, India
harshitachopra3@gmail.com
[2] University of California, Irvine, USA

Abstract. Students' conversations in academic settings evolve over time and can be affected by events such as the COVID-19 pandemic. In this paper, we employ a Contextualized Topic Modeling technique to detect coherent topics from students' posts in online discussion forums. We construct topic chains by connecting semantically similar topics across months using Word Mover's Distance. Consistent academic discourse and contemporary events such as the COVID-19 outbreak and the Black Lives Matter movement were found among prominent topics. In later months, new themes around students' lived experiences emerged and evolved into discussions reflecting the shift in educational experiences. Results revealed a significant increase in more general topics after the onset of pandemic. Our proposed framework can also be applied to other contexts investigating temporal topic trends in large-scale text data.

Keywords: Text mining · Discourse analysis · Topic modeling

1 Introduction

The onset of the COVID-19 pandemic prompted an urgent shift to online education and created a nontrivial disruption in students' educational experience that affected their academic engagement and mental health [5]. The rapidly changing nature of the pandemic underscores the need for an automated way of detecting the temporal dynamics of themes discussed online and the potential insights they give on its influence on education. Here, we aim to leverage Natural Language Processing (NLP) techniques to capture emergent topics and temporal evolution of undergraduates' online discourse in discussion forums in the months prior to and throughout the pandemic. We employed the Combined Topic Model (CombinedTM) [1] to extract coherent themes that emerged monthly and used Word Mover's Distance (WMD) to construct topic chains by computing the semantic similarity between topics across adjacent months. Additionally, we propose

© Springer Nature Switzerland AG 2022
M. M. Rodrigo et al. (Eds.): AIED 2022, LNCS 13356, pp. 453–457, 2022.
https://doi.org/10.1007/978-3-031-11647-6_91

a measure of course-centricity to distinguish topics that are more specific to certain courses from those which represent broader themes that were observed across multiple courses.

2   Background

Topic modeling methods such as Latent Dirichlet Allocation [2] have been used to extract static themes in learner-generated data and to study the impact of the pandemic on teaching and learning in higher education [8]. However, most of these studies have shown limited capacity to detect coherent topics and do not reflect temporal changes in themes discussed online. Given the rapid changes brought to educational settings, we seek to examine how topics emerge, recur and evolve in student discourse.

Recent advances in deep learning have introduced the combination of neural networks and transformer-based techniques to yield topics that are more coherent and interpretable than traditional models. In this study, we used CombinedTM, a recently proposed neural topic model that uses a Bag of Words (BoW) document representation concatenated with the contextualized document representation from Sentence-BERT [7].

To connect different topics temporally, previous studies have used traditional similarity metrics [3]. By contrast, we used WMD [4] to track topics that represent a similar broad theme but depict a change in context over time. WMD measures the dissimilarity between two text documents, leveraging the power of word embeddings [6], even if they do not have any words in common. By exploring the temporal characteristics of learner discourse during this critical time, we aim to enhance our understanding of the influence of the pandemic and policy responses on learning activities.

3   Data and Methodology

The dataset was obtained from the online discussion forums on the learning management system at a large public university in the United States during the academic year from October 2019 to June 2020. We retained posts generated from the same individuals across months, and removed posts that contained less than two words or five characters. A total of 32,409 posts created by 449 students across 636 courses were retrieved and preprocessed to retain relevant tokens.

We trained CombinedTM on the discussion posts for each month separately. The BoW vocabulary was constructed by retrieving the top 10,000 words with maximum Term Frequency - Inverse Document Frequency weights and Sentence-BERT was used to obtain encodings of the posts. To determine the optimal number of topics (K), we ran the models for each month with K ranging from 5 to 15 topics and evaluated them on the three metrics used by [1]. To determine the degree of course-centricity, we examined how each topic was distributed across courses. We assigned each post a topic with the highest probability. For each topic, the frequencies of the posts for the top-$N(=10)$ most common courses were used to calculate the standard deviation (σ). A lower value of σ denoted a relatively uniform distribution of courses in a topic, suggesting a topic represents a broader theme that is more generally distributed across multiple courses. A higher value of σ denoted a skewed distribution where very few specific courses dominate the discussion, showing that the topic is more "course-centric".

We used WMD to measure the semantic or contextual similarity between every pair of topics in adjacent months. A Word2Vec model [6] was trained on the entire corpus to obtain 100-dimensional word embeddings. Considering each topic as a list of top-30 representative words, we computed the WMD between all topic pairs belonging to adjacent months (m_t and m_{t+1}). For every topic in m_t, we selected the topic having the least WMD (the most similar) from m_{t+1}. To avoid multiple topics in m_t getting mapped to the same topic in month m_{t+1}, we retained only the topic pairs having the least WMD among them. We created a directed graph connecting nodes (or topics) in consecutive months and found all simple paths from each root to leaf. These directed paths are referred to as "topic chains".

4 Results and Discussion

The topic modeling resulted in 8–13 number of optimal topics per month, including students' lived experiences and contemporary events such as social justice movements, which demonstrate sociocultural influences on learning. Details on the topics and top-ranked words are made publicly available[1].

We empirically tested a shift in course centricity with a post-hoc Welch's Two Sample t-test to compare the degree of variability in Fall 2019 and Winter 2020 with that of Spring 2020. Fall and Winter quarters had a greater standard deviation ($M = .10$) than in the Spring quarter ($M = .03$), $t(5.36) = p < .001$. This finding shows that topics became less course-specific in the Spring, which began a few weeks after fully remote learning was implemented due to the COVID-19 pandemic, than in the previous two quarters. Although online forums mainly serve as a place for course-oriented discussions, the emergence of more general topics indicates a common or shared online experience across different courses.

Amongst the identified topic chains (Fig. 1), the top two most consistent themes were casual interactions (Chain 13) and Public Health-related discussion (Chain 12). Chain 12 demonstrated that discourse around public health began as course-centric topics in earlier months and later became more general regarding pandemic-related health inequities. This suggests that public health discussions expanded beyond corresponding courses, became a shared concern and arose in broader student discourse during the pandemic.

Student Life emerged as a relatively new topic starting Mar-2020 (Chain 6). Students' posts included university-related experiences, and major family and life events. A rise in such posts demonstrated an evolved use of online discussion forums to connect with peers during remote learning. This information suggests

---

[1] github.com/The-Language-and-Learning-Analytics-Lab/topic_trends.

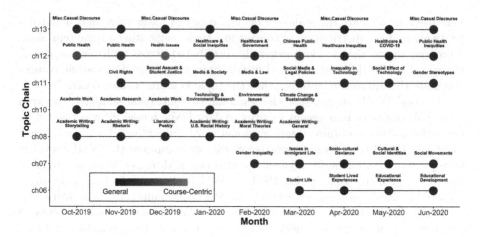

Fig. 1. Topic chains colored by course-centricity of each topic.

the shift in discussion forum's role in providing space for academic discussion to sharing experiences and building social connections in the classroom community. Future studies are needed to investigate how this change might influence learners' sense of belonging during remote learning.

Conclusion. Our study contributes to the literature by moving beyond mining static topics from large-scale discussion forums, towards a more process-oriented, temporal technique of modeling topics. For researchers and practitioners in the AIED community, our proposed approach provides a viable means to analyze the development of discourse in online educational environments in response to certain events or introduction of new policies.

References

1. Bianchi, F., Terragni, S., Hovy, D.: Pre-training is a hot topic: contextualized document embeddings improve topic coherence. In: Proceedings of the 59th Annual Meeting of the Association for Computational Linguistics and the 11th International Joint Conference on Natural Language Processing, vol. 2 (2021)
2. Blei, D.M., Ng, A.Y., Jordan, M.I.: Latent Dirichlet allocation. J. Mach. Learn. Res. **3**, 993–1022 (2003)
3. Kim, D., Oh, A.H.: Topic chains for understanding a news corpus. In: Proceedings of the 12th International Conference on Computational Linguistics and Intelligent Text Processing - Volume Part II (2011)
4. Kusner, M.J., Sun, Y., Kolkin, N.I., Weinberger, K.Q.: From word embeddings to document distances. In: Proceedings of the 32nd International Conference on International Conference on Machine Learning - Volume 37, ICML 2015 (2015)
5. Means, B., Neisler, J., et al.: Suddenly online: a national survey of undergraduates during the Covid-19 pandemic. Technical report, Digital Promise (2020)
6. Mikolov, T., Sutskever, I., Chen, K., Corrado, G.S., Dean, J.: Distributed representations of words and phrases and their compositionality. In: Advances in Neural Information Processing Systems, vol. 26. Curran Associates, Inc. (2013)

7. Reimers, N., Gurevych, I.: Sentence-BERT: sentence embeddings using Siamese BERT-networks. In: Proceedings of the 2019 Conference on Empirical Methods in Natural Language Processing. Association for Computational Linguistics (2019)
8. Vijayan, R.: Teaching and learning during the COVID-19 pandemic: a topic modeling study. Educ. Sci. **11**(7), 347 (2021)

Predicting Knowledge Gain for MOOC Video Consumption

Christian Otto[1]([✉]) [ID], Markos Stamatakis[2] [ID], Anett Hoppe[1,2] [ID],
and Ralph Ewerth[1,2] [ID]

[1] L3S Research Center, Leibniz University Hannover, Hannover, Germany
`christian.otto@tib.eu`
[2] Leibniz Information Centre for Science and Technology (TIB), Hannover, Germany
`{anett.hoppe,ralph.ewerth}@tib.eu, markos.stamatakis@yahoo.de`

Abstract. Informal learning on the Web using search engines as well as more structured learning on Massive Open Online Course (MOOC) platforms have become very popular. However, the automatic assessment of this content with regard to the challenging task of predicting (potential) knowledge gain has not been addressed by previous work yet. In this paper, we investigate whether we can predict learning success after watching a specific type of MOOC video using 1) multimodal features, and 2) a wide range of text-based features describing the structure and content of the video. In a comprehensive experimental setting, we test four different classifiers and various feature subset combinations. We conduct a feature importance analysis to gain insights in which modality benefits knowledge gain prediction the most.

Keywords: Web learning · Resource quality · Knowledge gain

1 Introduction

Research on the automatic assessment of learning resources has targeted a number of possible dimensions, such as the prediction of user engagement towards a certain learning resource [2] or the correlation of knowledge gain and layout features [4]. While these are interesting research directions, they do not address the question of potential usefulness of a resource. This usefulness is often conceptualized as the learning success that a certain user may achieve by using the resource. In this paper, we report on work in progress on the challenging task of knowledge gain prediction for MOOC videos using multimodal structure and content features. Therefore, we extend Shi et al.'s feature set [4] with a large number of text-based features and adapt them to slide and speech content. We also consider that the user's capabilities play a role in this context.

2 Dataset and Feature Extraction

We extract a total of 387 features from five different categories: *syntactic, lexical, structural,* and *readability,* which are abbreviated as *TXT* from here on,

© Springer Nature Switzerland AG 2022
M. M. Rodrigo et al. (Eds.): AIED 2022, LNCS 13356, pp. 458–462, 2022.
https://doi.org/10.1007/978-3-031-11647-6_92

while the fifth category is called *EMBED* and entails semantic sentence embeddings [3]. Additionally, the dataset and extracted multimedia features by Shi et al. [4] (*MM* from here on), consisting of 22 lecture videos of an *edX* course called "Globally Distributed Software Engineering", are used. Our goal is to investigate the importance of the different modalities and see how they influence the challenging task of knowledge gain prediction. The respective knowledge gain scores, established by pre- and post-knowledge tests with multiple-choice questionnaires, are also included in the data. By design, all of these features are independent of the user, since they are based on the educational resource alone. However, our goal of knowledge gain prediction is also influenced by the learner's cognitive capabilities. In order to investigate the influence of user identity in our experiments, we add another feature subset, the person ID (*USER* from here on). To prevent linear dependencies between these IDs, we represent them as one-hot-encoded vectors (13 dimensions). The full feature list, the full list of feature importance results and the utilized code can be found on GitHub[1].

3 Experimental Setup and Results

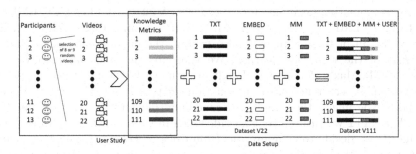

Fig. 1. The workflow of our approach and the composition of our datasets for experiments **V22** and **V111** (best viewed in color). (Color figure online)

We conduct two knowledge gain prediction experiments on all combinations of our feature categories. Figure 1 gives an overview over the setup. The knowledge gain classes are defined as follows: 1.) *Low* KG, if $X < \overline{X} - \frac{\sigma}{2}$; 2.) *Moderate* KG, if $\overline{X} - \frac{\sigma}{2} < X < \overline{X} + \frac{\sigma}{2}$; and 3.) *High* KG, if $X > \overline{X} + \frac{\sigma}{2}$, where *overlineX* and σ are the average and standard deviation of all knowledge gain scores. This results in a dataset composition of 6 low, 10 moderate and 6 high for **V22** and 40 low, 40 moderate and 31 high knowledge gain samples for **V111**. For the first set of experiments (**V22**), we predict the average achieved knowledge gain class per participant that saw video v_i. We establish a challenging *knowledge gain baseline V22* by estimating the performance of participant p_k on video v_i.

Therefore, we average the knowledge gain scores of all other participants p_l with $l \neq k$ who saw v_i and convert it to the appropriate class afterwards, but only on videos $v \neq v_i$. Thus, this baseline has strong hints about the learning outcome of different participants that are not available to our classifiers. It achieves an accuracy of 45.45%. In our second set of experiments (**V111**) we add the person ID as a one-hot encoded vector to the respective video feature vectors to make them unique again, giving us the original 111 samples. Target variable is the recorded knowledge gain class of the learning session. Again, we derive another challenging *knowledge gain baseline V111*. To estimate the knowledge gain class that user u achieved on video i we average his/her score on the $n-1$ other videos seen by him/her and, again, convert it to the appropriate class. This baseline is also challenging (accuracy = 43.24%) because the information about the user-specific learning performance is not available to our classifiers; as mentioned above, our user-specific feature is simply the encoded person ID.

3.1 Data Preprocessing

We examine whether Shi et al.'s features [4] allow for knowledge gain prediction, and how our suggested features (*TXT + EMBED + USER*) are suitable for this task, as separate feature sets and in combination. Consequently, we have seven feature combinations as inputs for experiments **V22** and **V111**: TXT, EMBED, MM, TXT+EMBED, TXT+MM, MM+EMBED, and TXT+MM+EMBED. For **V111** all of these categories also contain the one-hot-encoded person ID (*USER*) of the respective learner. We translate and scale all features with *sklearn's* MinMaxScaler to $[0, 1]$. Lastly, we remove 47 features that are zero for every sample (occurrence-based information like tenses and word types). Finally, for both experiments the samples are randomly split into 80% training and 20% test. For **V111** we ensured that no video seen in training was used in test. Following [1] we decide to compute the *Drop-Column Feature Importance* (source on GitHub[2]). We keep the 13-dimensional person ID vector for the feature selection process, since each bin represents one person and we want to investigate whether the models utilize information about the individual performances of the participants.

3.2 Results and Discussion

We use four classifiers: *Naive Bayes (NB), Sequential Minimal Optimization (SMO), Random Forest (RF), and Multi-Layer Perceptron (MLP)* implemented by the *WEKA* machine learning software[3]. For each classifier (default hyper-parameters), each feature category, and both experiments we conduct a 5-fold cross-validation and average the results per fold in terms of precision, recall, F1-score, and accuracy. Also for each fold, a separate feature importance analysis

---

[2] https://github.com/parrt/random-forest-importances/blob/master/src/rfpimp.py.
[3] https://www.cs.waikato.ac.nz/ml/weka/.

Table 1. Best results for each classifier in the **V22** experiment (top) and **V111** experiment (bottom) on the respective feature category.

Feature category **V22**	Classifier	Low			Moderate			High			Overall			Acc. in %
		Pr	Re	F1	Pr	Re	F1	Pr	Re	F1	Pr	Re	F1	
Random Guess Baseline	–	–	–	–	–	–	–	–	–	–	–	–	–	33.33
Knowledge Gain Baseline V22	–	0.00	0.00	0.00	0.45	1.00	0.62	0.00	0.00	0.00	0.15	0.33	0.21	45.45
EMBED (srt)	MLP	0.10	0.20	0.13	0.47	0.50	0.48	0.30	0.60	0.40	0.29	0.43	0.34	42.00
TXT+EMBED (both)	NB	0.00	0.00	0.00	0.58	0.80	0.67	0.17	0.40	0.24	0.25	0.40	0.30	45.00
TXT+EMBED (slide)	NB	0.00	0.00	0.00	0.60	0.80	0.69	0.17	0.40	0.24	0.26	0.40	0.31	45.00
MM+TXT	RF	0.10	0.20	0.13	0.55	0.60	0.57	0.23	0.60	0.34	0.29	0.47	0.35	**46.00**
EMBED (slide)	SMO	0.10	0.20	0.13	0.45	0.80	0.57	0.00	0.00	0.00	0.18	0.33	0.23	42.00
Feature category **V111**	Classifier	Pr	Re	F1	Pr	Re	F1	Pr	Re	F1	Pr	Re	F1	Acc. in %
Random Guess Baseline	–	–	–	–	–	–	–	–	–	–	–	–	–	33.33
Knowledge Gain Baseline V111	–	0.50	0.18	0.26	0.39	0.85	0.54	0.70	0.23	0.34	0.53	0.42	0.38	43.24
EMBED (srt)+USER	MLP	0.35	0.33	0.34	0.48	0.47	0.47	0.39	0.56	0.46	0.41	0.45	0.42	**44.74**
EMBED (srt)+USER	NB	0.37	0.37	0.37	0.49	0.42	0.45	0.37	0.44	0.40	0.41	0.41	0.41	39.69
EMBED (srt)+USER	RF	0.43	0.41	0.42	0.47	0.55	0.51	0.45	0.30	0.36	0.45	0.42	0.43	41.18
EMBED (both)+USER	SMO	0.40	0.49	0.44	0.39	0.41	0.40	0.28	0.22	0.25	0.36	0.38	0.36	38.92
MM+USER	SNO	0.34	0.46	0.39	0.50	0.40	0.45	0.31	0.24	0.27	0.38	0.37	0.37	38.92

and feature selection is conducted. Table 1 shows the best performing combinations of classifier and feature category for the experiments **V22** and **V111**. The overall scores are macro recall, precision, and F1.

In summary, **V111** suggests that semantic text features that describe the content of a MOOC video, are a better choice for the given task than syntactic features that objectively describe the video. In comparison with **V22** that had a slightly stronger focus on multimedia features describing the objective quality of the video, this finding could be explained as follows: On the one hand, to predict the user-independent (average) learning outcome of a MOOC video (as in **V22**), it is beneficial to consider multimodal features describing general quality aspects. On the other hand, the prediction of the individual knowledge gain (**V111**) depends on a combination of content features and the preferences of the person itself. We tried to capture this personal influence with our one-hot-encoded person ID feature.

Feature Importance (FI): The FI analyses of the two experiments show significant differences. In **V22**, multimedia features [4] dominate. From the 40 features yielding a FI ≥ 0 only 14 were of the textual category. This is reflected in Table 1, where MM+TXT achieved the best performance. In **V111** the textual features obtain the highest FIs, with a slightly stronger focus on the slide content. Out of the 191 most important features with a value ≥ 0 the first multimedia feature has rank 50. Rank 3 is of type *USER* highlighting the importance of this bin in the 13-dimensional one-hot-encoded vector. This suggests that our models identified that this learner's individual performance gave hints about the eventual learning outcome in the other videos he or she saw. In summary, the FI analysis implies that it is beneficial to follow a workflow of our approach, that is to initially consider a broad range of features and assess their importance for the classification. Focusing on a single modality from the start may not yield optimal results as the impact of the selected features may vary heavily depending on the target scenario.

References

1. Breiman, L.: Statistical modeling: the two cultures (with comments and a rejoinder by the author). Stat. Sci. **16**(3), 199–231 (2001)
2. Bulathwela, S., Pérez-Ortiz, M., Lipani, A., Yilmaz, E., Shawe-Taylor, J.: Predicting engagement in video lectures. In: EDM. International Educational Data Mining Society (2020)
3. Reimers, N., Gurevych, I.: Sentence-BERT: sentence embeddings using Siamese BERT-networks. In: EMNLP/IJCNLP, pp. 3980–3990. ACL (2019)
4. Shi, J., Otto, C., Hoppe, A., Holtz, P., Ewerth, R.: Investigating correlations of automatically extracted multimodal features and lecture video quality. In: SaLMM, SALMM 2019, pp. 11–19. ACM, New York (2019)

Programming Question Generation by a Semantic Network: A Preliminary User Study with Experienced Instructors

Cheng-Yu Chung[1]([✉]) [iD] and I-Han Hsiao[2] [iD]

[1] Arizona State University, Tempe, USA
Cheng.Yu.Chung@asu.edu
[2] Santa Clara University, Santa Clara, USA
ihsiao@scu.edu

Abstract. Questions are widely used in various instructional designs in education. Creating questions can be challenging and time-consuming. It requires not only the expertise of the learning content but also the experience of the question designs and the overall class performance. A considerable amount of research in the field of question generation (QG) has focused on computer models that automatically extract key information from a given context and transform them into meaningful questions. However, due to the complexity of programming knowledge, there are only few studies that have explored the potential of Programming QG (PQG) where natural languages and programming languages are often interwoven to constitute an assessment unit. To investigate further, this study experiments with a hybrid semantic network model for PQG based on open information extraction and abstract syntax tree. Our user study showed that experienced instructors had significantly positive feedback on the relevance and extensibility of the machine-generated questions.

Keywords: Automatic question generation · Programming learning · Semantic network analysis · Local knowledge graphs

1 Introduction

Computer programming is a challenging topic that requires the learner to excel in both the concepts and the implementation. With the increase of self-paced online learning channels, the demand for programming practice questions is growing rapidly. For example, much research has shown the effectiveness of self-assessments and distributed learning for programming [1,2]. Such a tool requires a large number of questions that address various aspects of the learning content. Programming learning also requires domain-specific question types such as code-tracing questions and code-writing questions. As a result, the programming question generation (PQG) becomes a time-consuming and demanding process that not only requires the proficiency of the content but also excellence in the programming question design.

A general question generation (QG) process may be aided by computer models that automatically extract key information from the learning content and

© Springer Nature Switzerland AG 2022
M. M. Rodrigo et al. (Eds.): AIED 2022, LNCS 13356, pp. 463–466, 2022.
https://doi.org/10.1007/978-3-031-11647-6_93

generate new questions out of it. An enormous amount of research has been focusing on QG models for various fields of study. However, there is little research for programming QG (PQG). One special challenge in PQG is the alignment of "knowing-that" (conceptual knowledge) and "knowing-how" (procedural knowledge). In computer programming, a question usually involves both the natural language and the programming language. The heterogeneous content may limit the ability of conventional QG models. To fill this research gap, we develop a PQG model that can support instructors to make programming questions by following the knowledge-based approach. We hypothesize that programming code and its intents can be synthesized by their descriptions in verb-arguments formats, thereby building a network in an unsupervised way by the abstract syntax tree (AST) and the local knowledge graph (LKG) model [4].

We conducted a user study with experienced instructors from introductory programming courses to evaluate the quality of the PQG model. They also provided valuable insights into their preferred question types and expected support from PQG tools. The preliminary results showed that the instructors generally had significantly positive feedback toward the PQG model especially the extensibility of question complexity. Overall, this work illustrates the design of the PQG model and demonstrates a feasible approach to AI-assisted PQG tools.

2   Methodology

2.1 Modeling Conceptual Programming Knowledge and Procedural Knowledge

Extending the concept of the semantic role labeling (SRL), researchers have proposed Open Information Extraction (OIE) that considers both SRL and propositions asserted by sentences. An OIE model can decompose, for example, "computers connected to the Internet can communicate with each other" into two predicates, "(computers connected to the Internet; can; communicate with each other)" and "(computers; connected; to the Internet)". These predicates represent two aspects of the input.

This work uses the OIE model from [4] to extract semantic triples from the descriptions around code examples in a textbook, "*Think Java 2*"[1]. To aid the query of related programming concepts, this work builds a semantic network of the triples by following the Local Knowledge Graph (LKG) approach [3]. This work builds an LKG by treating both subjects and verbs as nodes and adding an edge if any two nodes are mentioned in one sentence.

A programming language is usually defined by a formal language with well-structured grammar. This characteristic ensures that program code can be efficiently parsed into binary machine code by a compiler. The Abstract Syntax Tree (AST) is an alternative representation of programs that specifically focuses on the syntactic structure. For example, in an AST of Java code, the node "ClassOrInterfaceDeclaration" represents the entry point of a Java class definition,

---

[1] https://github.com/ChrisMayfield/ThinkJava2.

and the node "VariableDeclarator" represents a statement that declares a new variable and its initializer. Although the AST is not necessarily related to the runtime nature of programs (i.e., the references to external libraries or the actual flow of data), it provides a convenient way to parse and represent programming semantics.

2.2   Automatic Question Generation Process

This work uses follows the template-based QG method to transform the key information into different questions. For a given input question, the PQG algorithm transforms the programming code into AST nodes and extracts programming keywords. Next, the algorithm uses this information to query related code examples from the LKG model, where programming questions are generated by the associated AST nodes, LKG triples and grammar-checked question templates. The generated questions are then ranked by the Tversky index to find out the most relevant ones.

As far as we know, there is no existing and publicly available model or benchmark datasets of PQG. To compare the performance of the model with a reference, we devised a reference model by masking part of the proposed PQG model. The reference model, called the "code-aware" model, uses only the AST structures to generate programming questions. The reference model is compared to the other model called the "context-aware" model which uses the LKG structure to generate programming questions.

3   Results

We recruited 7 participants who had teaching experience in introductory programming courses via communication in professional networks that involve instructors and professors from universities and colleges. According to the responses, around 58% of the participants had more than 5 years of teaching experience in introductory programming courses; the other 42% had 2–5 years of teaching experience. For each variable, we collected 84 data points for analysis. The participants were asked to evaluate machine-generated questions according to the topic relevance (*Topic-Rel*), the extensibility of topics (*Ext-Topics*), complexity (*Ext-Complex*), and their needs in teaching (*Ext-Need*).

The average score of each evaluation question was computed as shown in Table 1. First of all, both the code-aware model and context-aware one received significantly positive feedback from the participants. Specifically, the extensibility of complexity (Ext-Complex) received the highest score for both models. This outcome suggests that the generated questions were able to help the instructors make questions that are complex enough to distinguish the students' abilities. The generated questions also met the instructors' needs in PQG as seen in the significantly positive score from the variable Ext-Need and the variable Ext-Topics. In terms of the relevance of topics (Topic-Rel), the participants gave a significantly positive score. This outcome suggests that the participants only

Table 1. The statistics of the four variables in the measurement (reported in the format "M, SD, test (DoF) = V (pval)").

	Code-aware	Context-aware
Topic-Rel	$0.64, 1.34, t(83) = 4.39(0.00)$	$0.64, 1.34, t(83) = 3.59(0.00)$
Ext-Topics	$0.80, 1.27, t(83) = 5.76(0.00)$	$0.80, 1.27, t(83) = 3.37(0.00)$
Ext-Complex	$1.26, 0.81, t(83) = 14.31(0.00)$	$1.26, 0.81, t(83) = 12.67(0.00)$
Ext-Need	$0.57, 1.24, t(83) = 4.21(0.00)$	$0.57, 1.24, t(83) = 3.62(0.00)$

slightly agreed that the generated questions were related to the topic of the input, which is interesting as it suggests that the participants might have different opinions/expectations about what topics in the input to focus on.

4   Conclusions

We developed a PQG model that aims to support instructors to make new programming questions from the existing ones. Following the knowledge-based QG approach, we used the LKG to represent the conceptual programming knowledge and the AST to represent procedural programming knowledge. We conducted a user study with experienced instructors from introductory programming courses. The preliminary result showed that the participants had significantly positive feedback toward the extensibility of question complexity. Overall, his work demonstrates a feasible design of PQG models and paves the way for the future development of AI-assisted PQG tools for educational purposes.

References

1. Alzaid, M., Trivedi, D., Hsiao, I.H.: The effects of bite-size distributed practices for programming novices. In: Proceedings of 2017 IEEE Frontiers in Education Conference (FIE), pp. 1–9. IEEE (2017). https://doi.org/10.1109/FIE.2017.8190593
2. Chung, C.Y., Hsiao, I.H.: Investigating patterns of study persistence on self-assessment platform of programming problem-solving. In: Proceedings of the 51st ACM Technical Symposium on Computer Science Education, pp. 162–168. ACM, New York, February 2020. https://doi.org/10.1145/3328778.3366827. https://dl.acm.org/doi/10.1145/3328778.3366827
3. Fan, A., Gardent, C., Braud, C., Bordes, A.: Using local knowledge graph construction to scale Seq2Seq models to multi-document inputs. In: EMNLP-IJCNLP 2019–2019 Conference on Empirical Methods in Natural Language Processing and 9th International Joint Conference on Natural Language Processing, Proceedings of the Conference, pp. 4186–4196 (2019). https://doi.org/10.18653/v1/d19-1428
4. Stanovsky, G., Michael, J., Zettlemoyer, L., Dagan, I.: Supervised open information extraction. In: NAACL HLT 2018–2018 Conference of the North American Chapter of the Association for Computational Linguistics: Human Language Technologies - Proceedings of the Conference 1(Section 4), pp. 885–895 (2018). https://doi.org/10.18653/v1/n18-1081

Obj2Sub: Unsupervised Conversion of Objective to Subjective Questions

Aarish Chhabra[(✉)], Nandini Bansal, V. Venktesh, Mukesh Mohania,
and Deep Dwivedi

Indraprastha Institute of Information Technology, Delhi, India
{aarish17212,nandini18056,venkteshv,mukesh,deepd}@iiitd.ac.in

Abstract. Exams are conducted to test the learner's understanding of the subject. To prevent the learners from guessing or exchanging solutions, the mode of tests administered must have sufficient subjective questions that can gauge whether the learner has understood the concept by mandating a detailed answer. Hence, in this paper, we propose a novel hybrid unsupervised approach leveraging rule based methods and pre-trained dense retrievers for the novel task of automatically converting the objective questions to subjective questions. We observe that our approach outperforms the existing data-driven approaches by **36.45%** as measured by Recall@k and Precision@k.

Keywords: Question generation · Unsupervised learning · Clustering

1 Introduction and Related Work

In online platforms, assessments are critical to gauge if the learner has understood the concept. However, assessments with only objective questions may prompt the user to just guess the answer using the options. For more rigorous testing of understanding, we propose the novel task of converting objective questions to subjective questions to mandate a detailed answer. Let Q be an objective question (OQ) and A be the answer to that objective question. Our task is to convert Q to a short subjective question (SQ) S. For example, let's say *Q: The wastes that can choke the drains include, A: used tea leaves, cotton*, then the possible subjective question could be *S: What kind of wastes can choke the drains?* The task of converting objective questions to short subjective questions in the absence of labeled pairs (OQ-SQ pairs) hasn't been specifically explored previously to the best of our knowledge.

Many state-of-the-art Question Generation (QG) systems have been proposed in recent years [2–4]. These systems usually use deep learning-based approaches such as Seq2Seq models [6] or more recently, transformers [7]. However, the mentioned approaches have complex model architectures and require significant amounts of labeled data for training.

This research work is supported by Extramarks Education India Pvt. Ltd., SERB, FICCI (PM fellowship), Infosys Centre for AI and TiH Anubhuti (IIITD).

© Springer Nature Switzerland AG 2022
M. M. Rodrigo et al. (Eds.): AIED 2022, LNCS 13356, pp. 467–470, 2022.
https://doi.org/10.1007/978-3-031-11647-6_94

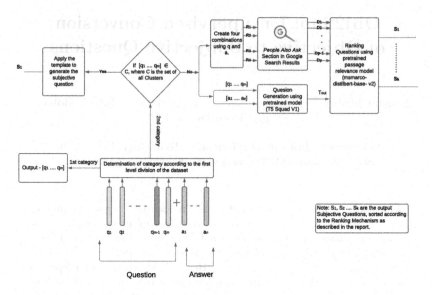

Fig. 1. Architecture for conversion of objective question to a subjective question

We open-source our code and datasets at https://github.com/ADS-AI/ Obj2Sub-AIED2022.

2 Methodology

In this section, we discuss the proposed unsupervised method for generating subjective question(s) from a given objective question. Let Q be a given objective question, A be an answer to this objective question, $q_1 q_2 \ldots q_m$ and $a_1 a_2 \ldots a_n$ be the token sequence of the objective question and answer respectively. Let $S_1 \ldots S_k$ be the subjective questions generated by the proposed Obj2Sub method for a given <Q, A> pair. We propose a *novel, unsupervised* and a *hybrid* approach to automate the process of converting an objective question into a subjective question. Figure 1 gives the complete picture of the proposed methodology. Upon receiving the inputs (Q and A), our system automatically classifies Q into one of the 3 broad categories of objective questions based on what class it represents:

- **Multi-option Dependent** ($\sim$7%) The class of questions that are dependent on the objective question's options and don't focus on a single learning concept. This category has not been dealt with in our proposed method because these are negligible in number. These can be easily filtered using the presence of phrases such as *of the following, choose the statement,* etc.
- **WhWord** ($\sim$61%) The class of questions that can be answered without looking into the option and can be directly used as subjective questions. These can be easily identified using the presence of wh-words as a first token (q_1) in Q.

- **Declarative Sentence** (~32%) This category contains the set of objective questions in which Q + A ('+' depicts concatenation) forms a declarative sentence. The questions not filtered out using either of the first 2 steps are mostly observed to be falling under this category. Eg: Question = The chemical symbol for silver is, Answer = Ag

In this paper, we focus on the conversion of Objective Questions belonging to Declarative Sentence category since this is a more common category of objective questions. For the conversion, we follow a hybrid approach which consists of 3 major components:

- **Clustering and Rule-Based Templates** We observe that some specific tokens are the same for multiple objective questions. These tokens are generally either the *last token (q_m), or the last two tokens (q_m , q_{m-1}), or the first token (q_1)*. For example, the presence of **by** as the last token can be seen in the objective questions: *Law of constant proportions is given **by**, Polio is caused **by***. Thus, we define various clusters of objective questions based on the presence of these specific tokens. However, all these clusters don't represent a broad class of objective questions, leading us to perform cluster pruning based on the frequency (*>500*) in our dataset. Further, we define a single rule-based template using various syntactic features such as part of speech tags, determining auxiliary verbs, named entity recognition of tokens, changing verb forms (lemmatization), subject-auxiliary-inversion as defined in [1] to steer the conversion into a short subjective question. Questions not covered with clustering are dealt using the other 2 components.
- **Leveraging Open Source Knowledge Base** We utilize open-source knowledge base such as the People Also Ask (PAA) section of Google for the conversion task. We form multiple search queries using the input tokens (Q, A) and extract top 4 PAA questions using APIs of a Python-based library. The problem of inconsistency in search results is tackled smartly by using different permutations for forming search queries and devising a filtering mechanism to discard irrelevant questions. Eg: Q = desert plants have scale/spine-like leaves to, A = reduce the loss of water by transpiration, a sample S = How are the desert plants adapted to reduce the loss of water by transpiration?
- **Pre-Trained T5 based model** To eliminate a rare problem of concept drift in the questions generated using PAA section, we utilize a transformer-based architecture (T5 Squad V1[1]) to augment the generated questions further by conditioning the generation of new subjective questions on the given objective question and the context (the answer to the Objective Question) around which the question must be framed.

Now, we have a set S of subjective questions generated using the last 2 components discussed above. We further rank the questions using a pre-trained ranking model (*msmarco-distilroberta-base-v2[2]*) and fetch top-k questions. For this paper, we mostly stick to k = 3.

---

[1] https://huggingface.co/ramsrigouthamg/t5_squad_v1.

[2] https://huggingface.co/sentence-transformers/msmarco-distilroberta-base-v2.

Table 1. R@k and P@k for k=1,2,3 on ObjQA and MCQ Datasets

Dataset	Method	R@1	R@2	R@3	P@1	P@2	P@3
ObjQA	**Obj2Sub** (our method)	**0.203**	**0.318**	**0.408**	**0.610**	**0.477**	**0.408**
	Rule-Based Approach [1]	0.110	0.189	0.222	0.332	0.283	0.222
	T5-Transformer [5]	0.183	0.246	0.299	0.550	0.370	0.299
MCQ	**Obj2Sub**(our method)	**0.255**	**0.329**	**0.393**	**0.767**	**0.493**	**0.393**
	Rule-Based Approach [1]	0.156	0.276	0.317	0.47	0.415	0.317
	T5-Transformer [5]	0.195	0.292	0.378	0.586	0.439	0.378

3   Experiments and Results

In this paper, we compare our results with two different kinds of dataset for a holistic evaluation of the proposed system.

ObjQA Dataset: This is a proprietary dataset from an e-learning platform which consists of approx 2,70,000 non-visual K-12 based objective question samples spanning different subjects.

MCQ Dataset: This is an open-source dataset which is used to further verify the robustness of the system proposed in this paper. Again, questions from different subjects are picked to keep the experiments unbiased.

Due to the novelty of the problem statement, we compare our results with 2 closely related question generation systems, **Transformer based T5 Squad V1** (taking the top-3 outputs) and **Rule Based Approach devised in** [1]. Table 1 shows the various results and suggests that our method outperforms the existing methodologies by **36.45%** as measured using Recall@3 and Precision@3 due to the hybrid nature of the approach.

References

1. Heilman, M., Smith, N.A.: Good question! Statistical ranking for question generation. In: NAACL-HLT, pp. 609–617. ACL, June 2010
2. Khodeir, N., Wanas, N., Darwish, N., Hegazy, N.: Bayesian based adaptive question generation technique. JESIT **1**(1), 10–16 (2014)
3. Kim, Y., Lee, H., Shin, J., Jung, K.: Improving neural question generation using answer separation. AAAI **33**(01), 6602–6609 (2019)
4. Lu, X.: Learning to generate questions with adaptive copying neural networks. In: SIGMOD 2019, pp. 1838–1840. ACM, New York (2019)
5. Raffel, C., et al.: Exploring the limits of transfer learning with a unified text-to-text transformer (2020)
6. Sutskever, I., Vinyals, O., Le, Q.V.: Sequence to sequence learning with neural networks (2014)
7. Vaswani, A., et al.: Attention is all you need (2017)

Comparing Few-Shot Learning with GPT-3 to Traditional Machine Learning Approaches for Classifying Teacher Simulation Responses

Joshua Littenberg-Tobias[(✉)], G. R. Marvez, Garron Hillaire, and Justin Reich

Massachusetts Institute of Technology, Cambridge, USA
joshua_tobias@wgbh.org

Abstract. Teacher educators use digital clinical simulations (DCS) to provide improvisation opportunities within low-stakes classroom environments. In this study, we experimented with GPT-3 and few-shot learning to examine if it could be used with open-text DCS responses. We found that GPT-3 performed substantially worse than traditional machine learning (ML) models even on the same-sized training sets. However, the performance of GPT-3 decreased only marginally compared to traditional ML models with a training set of 20 examples (-0.06). Traditional ML models generally performed well and in some cases had similar performance to the human baseline. Future research will examine whether changes to labeling procedures or fine-tuning with existing data can improve the performance of GPT-3 with DCSs.

Keywords: Natural language processing · Few-shot learning · GPT-3 · Simulations · Teacher education · Professional learning

1 Introduction and Related Work

To expand practice opportunities, many teacher educators are increasingly experimenting with digital clinical simulations (DCSs). DCSs provide teachers with the opportunity to practice and receive feedback in scalable low-stakes environments [5,7]. In DCSs, teachers are presented with a hypothetical teaching situation, such as a student misinterpreting a math problem, and are prompted to respond improvisationally. These responses can be labeled to indicate specific teaching characteristics and then integrated into artificial intelligence (AI) systems using natural language processing (NLP).

However, integrating feedback within simulations using NLP remains a challenge. Although some researchers have begun to develop NLP classifiers for DCSs [2], the expertise and labor cost make this approach difficult to implement at scale. In recent years, the emergence of large-scale language models such as BERT and GPT-3 has shown that, in certain cases, NLP with few-shot learning can produce accurate results [1]. Large-scale transformer-based models are

© Springer Nature Switzerland AG 2022
M. M. Rodrigo et al. (Eds.): AIED 2022, LNCS 13356, pp. 471–474, 2022.
https://doi.org/10.1007/978-3-031-11647-6_95

trained on extremely large text data sets and are often able to produce and classify language with high degrees of accuracy in a wide variety of contexts [4]. Research on GPT-3 has found that few-shot learning with GPT-3 performs better than other large-scale language models on contextual language tasks and, in some cases, human raters [1]. However, applications of the GPT-3 models with context-specific texts have had mixed success. Some studies have found that few-shot learning with GPT-3 has had high degrees of accuracy on some tasks [10]. However, few-shot learning does not perform as well as other methods in very context-specific tasks [1].

In this study, we compare how GPT-3 performs in a few-shot learning situation, classifying participant responses from a DCSs with a human baseline and traditional machine learning (ML) models. We used previously labeled data from a simulation, *Jeremy's Journal*. In addition to the complete training set, we also evaluated the few-shot learning potential of GPT-3 by randomly sampling sets of 20, 100, and 200 equally balanced pre-labeled responses from the training set to use as prompts for few-shot learning with the GPT-3 model.

2   Methods

The *Jeremy's Journal* simulation was developed and implemented using the *Teacher Moments* platform [5]. In the simulation, *Jeremy's Journal*, participants play the role of a middle school English teacher who has a student struggling with personal issues named Jeremy Green. Strong performance in the simulation is associated with recognizing where the student is struggling, identifying effective instructional supports, and being aware of the student's mental and physical well-being. This simulation was embedded in an online professional learning course for educators that ran in early 2021. The course enrolled 5,458 participants, and we focused on participants who completed the simulation and gave their consent to participate in the research (N = 494).

In this study, we focus on participants' responses to three prompts in the simulation. We developed a set of nine binary labels to assess whether participants mentioned specific ideas or concepts in their responses. We evaluated 1,482 responses from the three prompts. Labeling was carried out by three raters with 20% of all texts randomly sampled by prompt to assess inter-rater reliability. Inter-rater reliability was good across all rater combinations with Cohen's kappa between 0.57–0.61, similar to what has been reported in previous research on similar types of tasks [2,6]. These labeled data then served as 'ground-truth' data for training our models.

We pre-processed the data by removing capitalization, symbols, punctuation, and stopwords. We then split the data into a training and validation set using a 80%/20% split stratifying the data by prompt. Although we did not stratify by labels, the distribution of labels in the validation sets was not statistically or meaningfully different from the training sets. To select the best performing traditional ML model, we evaluated six different types of algorithms. We used a stratified five-fold cross-validation on the training data to select model features using 50 random seeds to adjust for sampling variation. Based on our analysis, we selected a model for each label that maximized the weighted average F1 score. In cases where F1 scores were equivalent, we used precision and recall metrics to select the best performing models. Our analysis was conducted using the *scikit-learn* package [9] with Python 3.9 in Jupyter Notebook.

We accessed GPT-3 using the OpenAI API and the *openai* package within Python [8]. For cost and efficiency reasons, we used the *ada* model, available on the OpenAI website. To evaluate GPT-3's few-shot learning capacity, we sampled from the labeled training data sample sets of 200, 100, and 20 that were equally balanced across classes. The same training and validation data were used for both the GPT-3 and traditional ML models.

3 Results

The GPT-3 model performed significantly worse than any of the traditional ML models (Table 1). On average, the full GPT-3 model had F1 scores −0.25 lower than the human baseline and 0.19 lower than the best performing traditional ML model. Performance degraded slightly as the number of labeled examples was reduced: models with 200 and 100 labeled examples had very similar F1 scores to the full training set and in some cases performed better. Only with 20 examples were there any meaningful differences in the average F-1 scores with −0.30 compared to the human baseline and −0.25 compared to the best performing traditional ML model. Compared to the full GPT-3 model, this was a decrease of −0.06 in F-1 scores compared to both the human baseline and traditional ML models.

Table 1. F1 statistics for all models

Approach	Model	1	2	3	4	5	6	7	8	9	Average
Human Baseline		0.93	0.84	0.78	0.92	0.78	0.79	0.77	0.88	0.86	0.84
Traditional ML	Random Forest	0.91	0.68	**0.69**	**0.95**	0.69	0.67	0.71	0.84	0.84	0.78
	SVC (Linear)	0.94	0.66	0.61	0.91	0.60	0.67	0.70	0.83	0.80	0.73
	SVC (Polynomial)	0.85	0.58	0.60	0.85	0.61	0.58	0.64	0.84	0.81	0.71
	SVC (Sigmoid)	**0.94**	**0.72**	0.61	0.93	**0.70**	**0.69**	**0.75**	0.83	**0.86**	0.78
	SVC (RBF)	0.91	0.65	0.68	0.93	0.66	0.69	0.68	**0.84**	0.82	0.76
	Decision Tree	0.91	0.64	0.67	0.93	0.69	0.69	0.70	0.80	0.81	0.76
Best Model		0.94	0.72	0.69	0.95	0.70	0.69	0.75	0.84	0.86	0.79
	GPT-3 (Full)	0.67	0.42	0.43	0.70	0.57	0.53	0.58	0.78	0.72	0.60
GPT-3	GPT-3 (200)	0.67	0.49	0.46	0.74	0.57	0.54	0.57	0.69	0.64	0.60
	GPT-3 (100)	0.65	0.45	0.45	0.72	0.56	0.55	0.57	0.68	0.65	0.58
	GPT-3 (20)	0.53	0.42	0.38	0.66	0.55	0.48	0.55	0.66	0.65	0.54

Note: The best traditional ML models are bolded. Column labels are 1-feel_jeremy, 2-teach_catch_up, 3-student_catch_up, 4-school_policy, 5-learn_challenge, 6-change_for, 7-more_some, 8-jeremy_mental, 9-jeremy_effort.

Although GPT-3 performed significantly worse than the human baseline and traditional ML approaches, there was only a small drop in F1 scores (−0.06) between full training data set responses versus samples of 20 equally balanced responses. This suggests that we might be able to modify our approach to GPT-3 to investigate whether these changes might improve the accuracy of the model. Some possibilities include fine-tuning the base models using context-specific language, a strategy that has been deployed successfully with BERT [3].

References

1. Brown, T.B., et al.: Language models are few-shot learners. arXiv:2005.14165 [cs], July 2020. http://arxiv.org/abs/2005.14165
2. Bywater, J.P., Chiu, J.L., Hong, J., Sankaranarayanan, V.: The Teacher Responding Tool: Scaffolding the teacher practice of responding to student ideas in mathematics classrooms. Comput. Educ. **139**, 16–30 (2019). https://doi.org/10.1016/j.compedu.2019.05.004, https://www.sciencedirect.com/science/article/pii/S0360131519301137
3. Clavié, B., Gal, K.: EduBER: pretrained deep language models for learning analytics. In: Companion Proceedings 10th International Conference on Learning Analytics & Knowledge (LAK20), p. 4 (2020)
4. Floridi, L., Chiriatti, M.: GPT-3: its nature, scope, limits, and consequences. Minds Mach. **30**(4), 681–694 (2020). https://doi.org/10.1007/s11023-020-09548-1
5. Hillaire, G., et al.: Teacher moments: a digital clinical simulation platform with extensible AI architecture. Technical report, EdArXiv, May 2021. https://doi.org/10.35542/osf.io/jf348, https://edarxiv.org/jf348/, type: article
6. Liu, O.L., Brew, C., Blackmore, J., Gerard, L., Madhok, J., Linn, M.C.: Automated scoring of constructed-response science items: prospects and obstacles. Educ. Meas. Issues Pract. **33**(2), 19–28 (2014). https://doi.org/10.1111/emip.12028, https://onlinelibrary.wiley.com/doi/abs/10.1111/emip.12028, _eprint: https://onlinelibrary.wiley.com/doi/pdf/10.1111/emip.12028
7. Mikeska, J., Howell, H., Dieker, L., Hynes, M.: Understanding the role of simulations in k-12 mathematics and science teacher education: outcomes from a teacher education simulation conference. Contemp. Issues Technol. Teach. Educ. **21**(3) (2021). https://citejournal.org/volume-21/issue-3-21/general/understanding-the-role-of-simulations-in-k-12-mathematics-and-science-teacher-education-outcomes-from-a-teacher-education-simulation-conference
8. OpenAI: OpenAI Documentation (2022). https://beta.openai.com/docs/introduction
9. Pedregosa, F., et al.: Scikit-learn: machine Learning in Python. Mach. Learn. Python 6 (2011)
10. Wang, S., Liu, Y., Xu, Y., Zhu, C., Zeng, M.: Want to reduce labeling cost? GPT-3 can help. arXiv:2108.13487 [cs], August 2021. http://arxiv.org/abs/2108.13487

Assessing Readability of Learning Materials on Artificial Intelligence in English for Second Language Learners

Yo Ehara[✉]

Tokyo Gakugei University, Koganei, Tokyo 1848501, Japan
`ehara@u-gakugei.ac.jp`

Abstract. Many scientific publications and materials on artificial intelligence (AI) have been written in English; however, for many AI learners, English is their second language. Therefore, the difficulty (readability) of online self-teaching texts on AI for English-as-a-second-language (ESL) learners is essential for determining the language support ESL learners need to learn AI. However, only a few studies have addressed this issue. Therefore, we identified the difficulty level of English self-teaching texts for ESL AI learners. Because large-scale testing for ESL learners is impractical owing to the financial costs and time involved, we built two distinctive automatic readability assessors: one using sophisticated deep-learning-based natural language processing (NLP) technology, and another using classic NLP based on word frequency and applied linguistics. We conducted our evaluation using AI research papers and university-level online course texts. Interestingly, the distinctive automatic assessors, which were trained on different datasets, showed similar results. Intermediate-level ESL learners could read approximately 10% of online course texts. We also showed that they are significantly easier to read than AI research papers for ESL learners, demonstrating their usefulness in AI learning.

Keywords: Second language learning · Readability · Automatic assessment · Natural language processing

1 Introduction

English is most often used to express developments in science and technology, including AI. The most informative AI educational material is written in English. Hence, for English-as-a-second-language (ESL) learners studying AI, comprehending the content of English educational material can be difficult. Moreover, this problem can intensify if learners want to learn the latest developments in AI technologies, but the corresponding research papers or educational materials have not been translated into their native languages.

To overcome these problems, we first sought to understand the readability of English AI educational materials. Assembling ESL learners to read and evaluate a large number of English texts is difficult because of the time and financial cost involved. Instead, we focused on developing highly accurate readability assessors.

© Springer Nature Switzerland AG 2022
M. M. Rodrigo et al. (Eds.): AIED 2022, LNCS 13356, pp. 475–478, 2022.
https://doi.org/10.1007/978-3-031-11647-6_96

We followed two approaches for building assessors. The first approach is based on educational natural language processing (NLP) [6]. Using a standard corpus, we built a highly accurate readability assessor using deep learning methods, such as bidirectional encoder representations from transformers (BERT) [1].

The second approach involves conducting readability assessments based on information regarding the vocabulary of ESL learners. These methods have been thoroughly studied in applied linguistics, wherein extensive research has confirmed that ESL learners must know more than 95% of the words in a text to read and understand it [4,5]. The notion of assessing text readability based on the vocabulary of each learner is beneficial for interpreting the readability assessment results. Therefore, we also constructed a classifier that ascertains the number of words in a text known to an ESL learner using a dataset of vocabulary tests on ESL learners [2].

In experiments conducted on a standard dataset for evaluating readability [6] in educational NLP, the results of the two approaches were in close agreement. Experiments using real educational AI texts showed that most texts in most AI educational materials were readable by intermediate ESL learners. In addition, we compared the readability of educational AI texts with that of AI paper abstracts. The results showed that the readability of educational AI materials was significantly easier for ESL learners than that of AI paper abstracts. Approximately 10% of the AI paper abstracts were unreadable by ESL learners with an intermediate English proficiency. Most ESL learners learning AI usually have only an intermediate level of English proficiency because of the time constraints involved in learning English. Therefore, this result shows that ESL learners cannot read approximately 10% of AI paper abstracts but can read almost all educational AI material. That is, the results show that educational AI materials are useful for both AI beginners and ESL learners.

2   Experiments with Educational AI Texts

We followed the experimental setting in [3] to train the BERT-based (**spvBERT**) and **Vocabulary-based** classifiers. For the experiment, we used slide PDFs from the course "CS221: Artificial Intelligence: Principles and Techniques Stanford/Spring 2020–2021" of Stanford University[1]. We chose this text because it is an introductory AI course from a reputable university, and the slides cover a wide range of AI topics. Moreover, because the course prerequisites include probability, discrete mathematics, and programming basics, the slides of this course include topics related to only AI, but not AI-related mathematics, making it suitable for directly measuring the readability of AI texts.

Here, the definition of intermediate follows the definitions in [6]: *elementary*, *intermediate*, and *advanced*. After converting all PDF slides from this class into text using the **pdfminer** library (https://pypi.org/project/pdfminer/), 709 texts were obtained by excluding empty lines and lines containing only expressions. For comparison, for the abstracts of AI fields, we obtained abstracts from

---

[1] https://stanford-cs221.github.io/spring2021/.

Table 1. Readability assessment results for ESL learners on educational AI and AI paper abstracts

–	Elem.	Int.	Adv.
Educational AI	0.398	0.598	0.004
AI paper abstracts	0.037	0.860	0.103

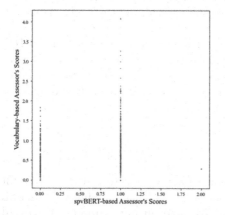

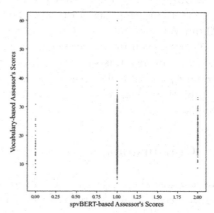

Fig. 1. spvBERT- and Vocabulary-based scores on educational AI texts.

Fig. 2. spvBERT- and Vocabulary-based scores on AI paper abstracts.

AAAI conference papers[2]. We randomly sampled $1,000$ abstracts from all AAAI abstracts from 2011 to 2021 to compare them with the educational AI texts fairly in terms of topics and number of texts.

Table 1 shows the readability assessment results using **spvBERT** on educational AI texts and AI paper abstracts. Because most educational AI texts are elementary or intermediate, this result implies that *intermediate-level ESL learners can read most educational AI texts*. From Table 1, educational AI material is easier to read than an AI paper abstract. The results were statistically significant (Mann-Whitney test, $p < 0.01$).

The results of Table 1 are consistent with those of the **vocabulary-based** approach. Figure 1 compares the results of both methods on the educational AI texts. The horizontal axis shows the results of **spvBERT**, and the vertical axis shows the results of the **Vocabulary-based** approach. Because few texts were assessed as advanced using **spvBERT** in Fig. 1, we can regard them as exceptions. Nevertheless, note that all texts assessed as "elementary" by **spvBERT** have low **vocabulary-based** scores, indicating an easier reading experience for ESL learners. Figure 2 compares the results of both methods on the AI paper abstracts. Again, Fig. 2 shows that texts assessed as "intermediate" are more difficult than those assessed as "elementary." In both Fig. 1 and Fig. 2, the

---

[2] "The AAAI Conference on Artificial Intelligence (AAAI)" papers in https://www.aaai.org/Conferences/conferences.php.

slightly upward trends are statistically significant (Wilcoxon test, $p < 0.01$). This means that a gentle but reliable correlation exists between the spvBERT- and vocabulary-based scores.

The smaller the **vocabulary-based** score, the easier the text is for ESL learners to read. The average score for the AI paper abstracts was 18.45. The average score for the educational AI texts was 0.76. This result indicates that the **vocabulary-based** approach also indicates that educational AI texts are easier to read than AI paper abstracts. This result was also statistically significant (Mann-Whitney test, $p < 0.01$).

Finally, even in educational AI texts assessed as elementary by **spvBERT**, the **vocabulary-based** approach determined that terms *unassigned* and *bootstrapping* are difficult for ESL learners, presumably because they do not appear in typical English textbooks. The details of the dataset are to be available at yoehara.com.

3   Conclusions

We assessed the readability of educational AI texts for ESL learners. We built highly accurate assessors using two approaches. Although intermediate ESL learners could read most of educational AI texts in our dataset, we showed that intermediate ESL learners could not read approximately 10% of AI paper abstracts. As ESL learners are typically not at an advanced English reading level, this result indicates that educational AI texts are more readable. Future research should include more comprehensive experiments using other datasets.

Acknowledgements. This work was supported by JST ACT-X Grant Number JPM-JAX2006, Japan.

References

1. Devlin, J., Chang, M.W., Lee, K., Toutanova, K.: BERT: Pre-training of Deep Bidirectional Transformers for Language Understanding. In: Proceedings of NAACL, Minneapolis, Minnesota, pp. 4171–4186, June 2019
2. Ehara, Y.: Building an English vocabulary knowledge dataset of Japanese English-as-a-second-language learners using crowdsourcing. In: Proceedings of LREC, May 2018
3. Ehara, Y.: Lurat: a lightweight unsupervised automatic readability assessment toolkit for second language learners. In: Proceeding of ICTAI, pp. 806–814. IEEE (2021)
4. Laufer, B., Ravenhorst-Kalovski, G.C.: Lexical threshold revisited: lexical text coverage, learners' vocabulary size and reading comprehension. Read. Foreign Lang. **22**(1), 15–30 (2010)
5. Nation, I.: How large a vocabulary is needed for reading and listening? Can. Mod. Lang. Rev. **63**(1), 59–82 (2006)
6. Vajjala, S., Lučić, I.: OneStopEnglish corpus: a new corpus for automatic readability assessment and text simplification. In: Proceedings of BEA, pp. 297–304 (2018)

Mirroring to Encourage Online Group Participation and Scaffold Collaboration

Adetunji Adeniran[1(✉)] and Judith Masthoff[2]

[1] HCII, Carnegie Mellon University, Pittsburgh, USA
adetunja@andrew.cmu.edu
[2] Utrecht University, Utrecht, The Netherlands
j.f.m.masthoff@uu.nl

Abstract. One way to improve cognitive advantages in online learning is to improve support for collaborative learning in an online environment. We propose a real-time approach to encouraging members of online groups to participate more actively by mirroring individual relative contributions within the group. Our findings suggest that mirroring promotes a number of interactive phenomena that are beneficial to online collaborative learning.

Keywords: Collaborative learning · Online groups · WC-GCMS · Mirroring

1 Background

Students from all socioeconomic backgrounds and geographical locations can now obtain a quality education online without time constraints. Online learning, on the other hand, has yet to maximize the socio-cognitive benefits of groups [6,7] because it provides limited or no real-time support for online collaborative learning [7,12]. Online learners, like in face-to-face group-learning, would benefit from real-time prompts that scaffold group collaboration [6,11]. We previously adapted the Collaboration Management Life Cycle (CMLC) [10] as a framework of support for online collaborative learning, and we developed and tested the Word Count/Gini Coefficient measure of symmetry (WC-GCMS), which we advance for assessing the level of online collaboration in real-time [2–4]. Our previous evaluation shows that the WC-GCMS can provide insight into the phenomena of group collaboration, which can then be applied to provide real-time support for online collaboration [1–4]. In this study, we use WC-GCMS components to provide real-time feedback to online group members (mirroring) during text-based joint problem-solving chat. We anticipate that this feedback will encourage individuals to contribute more to the group chat, thereby scaffolding the group's collaboration. The sections that follow discuss our findings.

© Springer Nature Switzerland AG 2022
M. M. Rodrigo et al. (Eds.): AIED 2022, LNCS 13356, pp. 479–482, 2022.
https://doi.org/10.1007/978-3-031-11647-6_97

2 Real-Time Mirroring: Evaluating the Effect on Online Collaboration

The WC-GCMS metric that measures individual contribution to group discussion was applied as real-time feedback to groups while they chatted online to solve a joint-task (see bar-chart widget displayed in Fig. 1a). We anticipate that this will stimulate each member to contribute more to the group discussion (i.e., self-regulate) [5]. This conjecture is based on theories that suggest successive positive behavior or improved performance can be induced with feedback on previous or current performance [8]. We tested our conjecture and the effect of mirroring using a *between-subject experiment design*.

Ten experimental groups were formed, each with four randomly selected members from 40 participants. Each group was required to communicate via a text-only online chatroom in order to complete the "desert survival" task (adopted as a pseudo group task). In addition, 5 of the 10 groups (experiment-treatment) were randomly assigned to use the chatroom interface shown in Fig. 1a, while the other 5 groups (control-treatment) used the interface shown in Fig. 1b. In the experiment-treatments - Grp1 & Grp3: (1 female, 3 male); Grp2: (4 male); Grp4: (3 female, 1 male) and Grp5 (1 female, 1 male, 2 undisclosed). In the Control-treatment - Grps 1, 3 & 5: (4 male); Grp2: (1 female, 3 male) and Grp4: (2 female, 1 male, 1 undisclosed). Our hypotheses are that groups in the experiment-treatment will have a higher frequency of individual contributions, a higher WC-GCMS measure of group collaboration, and a higher quality of information exchange and collaborative phenomena within-groups than groups in the control-treatment.

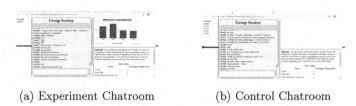

(a) Experiment Chatroom   (b) Control Chatroom

Fig. 1. Controlled experiment design

3 Discourse Data Analysis and Results

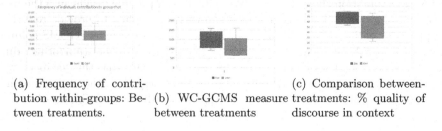

(a) Frequency of contribution within-groups: Between treatments.

(b) WC-GCMS measure between treatments

(c) Comparison between treatments: % quality of discourse in context

Fig. 2. Result: analysis of groups' discourse.

Frequency of Individuals' Contributions Within-Groups: Comparison Between Treatment (H1) - For each of the ten groups, the frequency of an individual's contribution within a group was defined and computed for each member within a group. The ANOVA result between-treatments ($F(1,39) = 5.63, p = 0.02$) indicates a statistically significant difference between frequency of individuals' contribution within their groups; whilst the ANOVA result between-groups ($F(9,39) = 1.7, p = 0.13$) shows no evidence of a statistically significant difference. Figure 2a visualizes the difference in frequency of contribution between-treatments.

WC-GCSM Measure Between-Treatments (H2) - Each group's WC-GCMS measure of group collaboration [2,3] was evaluated and compared to examine significant WC-GCMS differences between-treatments. The ANOVA result between treatments ($F(1,8) = 5.317, p = 0.259$) shows no statistically significant difference in WC-GCMS level of collaboration between the treatments. However, this is very likely to due to our data-set being too small; visualizing the collaboration level measure between-treatment indicates that the measure tends to be better within the groups allocated to the experiment-treatment compared to those in the control-treatment (see Fig. 2b).

Quality of Information Exchanged in Discourse (H3) - Finally, we examined the group discourse in order to assess the quality of information exchange in relation to the task at hand. The Soller taxonomy of collaborative skill [9] informed our coding scheme for collaboration-related contributions. We had three categories of labeling for individuals' text contributions in our coding scheme: [A] Contributions that are relevant to the group task, an element of the Soller taxonomy, such as an *informed question, information to progress in the task, informed argument, group coordination*; [B] Contributions that are relevant to the task but do not contain any information, such as a *one-word acknowledgement or agreement, providing an uninformed answer to the group task*; and [C] Irrelevant information such as *humour or greetings*. In our coding score, A is a 1, B is a 0.5, and C is a 0.

The aggregate score of coding each group's discourse as described, as a representation of the groups' discourse quality, is depicted in Fig. 2c; it shows that the treatment influences a higher information value/exchange within-groups, i.e. a higher score for groups in the experiment-treatment compared to the control-treatment. Other characteristics of group discourse and indications of stimulation by mirroring which we observed uniquely in groups in the experiment-treatment are illustrated in the comments below (colors distinguishing groups and all found in the experiment-treatment)[1].

Member1: btw guys at the end lets spam the send text to gain the system and to become the best group by default
Member2: i think the chat messages are stored though
Why dont we cheat?
Member1: How am I the lowest here?
Member2: Cause you have given like 1 work answers all the time Member2: If you speak at all

---

[1] See all group discourse at: http://colab-learn.herokuapp.com/AIED2022/gNT.php: N ∈ {1, 2, 3, 4, 5} (i.e. group numbers) and T ∈ {E, C} (i.e. group treatment).

4   Conclusions and Recommendations for Future Work

Our findings support hypotheses H1-3 and lead us to the conclusion that our mirroring (real-time feedback to groups) aids in the stimulation of collaborative phenomena within groups as well as the promotion of group collaboration during joint-problem-solving interactions via the text-based online medium. In future work, we can improve the effect of our real-time feedback by providing meta-cognition about it to group learners. Also, the feedback could be used as a prompt to a computer agent for providing explicit intervention that can improve interactivity and collaboration within online groups.

References

1. Adeniran, A., Masthoff, J.: Quantitative analysis to further validate WC-GCMS, a computational metric of collaboration in online textual discourse. In: Roll, I., McNamara, D., Sosnovsky, S., Luckin, R., Dimitrova, V. (eds.) AIED 2021. LNCS (LNAI), vol. 12749, pp. 29–36. Springer, Cham (2021). https://doi.org/10.1007/978-3-030-78270-2_5

2. Adeniran, A., Masthoff, J., Beacham, N.: An appraisal of a collaboration-metric model based on text discourse. In: AIED 2019 TeamTutoring Workshop. CEUR WS (2019)

3. Adeniran, A., Masthoff, J., Beacham, N.: Model-based characterization of text discourse content to evaluate online group collaboration. In: Isotani, S., Millán, E., Ogan, A., Hastings, P., McLaren, B., Luckin, R. (eds.) AIED 2019. LNCS (LNAI), vol. 11626, pp. 3–8. Springer, Cham (2019). https://doi.org/10.1007/978-3-030-23207-8_1

4. Adetunji, A., Masthoff, J., Beacham, N.: Analyzing groups' problem-solving process to characterize collaboration within groups. In: CEUR Workshop Proceedings, vol. 2153, pp. 5–16 (2018)

5. Janssen, J., Erkens, G., Kirschner, P.A.: Group awareness tools: it's what you do with it that matters. Comput. Hum. Behav. **27**(3), 1046–1058 (2011)

6. Laal, M., Ghodsi, S.M.: Benefits of collaborative learning. Procedia Soc. Behav. Sci. **31**, 486–490 (2012)

7. Martinez Maldonado, R.: Analysing, visualising and supporting collaborative learning using interactive tabletops. Thesis 116 (2013)

8. Pajares, F.: Self-efficacy beliefs in academic settings. Rev. Educ. Res. **66**(4), 543–578 (1996)

9. Soller, A.: Supporting social interaction in an intelligent collaborative learning system. Int. J. Artif. Intell. Educ. **12**, 40–62 (2001)

10. Soller, A., Martínez, A., Jermann, P., Muehlenbrock, M.: From mirroring to guiding: a review of state of the art technology for supporting collaborative learning. Int. J. Artif. Intell. Educ. **15**(4), 261–290 (2005)

11. Thomas, G., Thorpe, S.: Enhancing the facilitation of online groups in higher education: a review of the literature on face-to-face and online group-facilitation. Interact. Learn. Environ. **27**(1), 62–71 (2019)

12. Webb, N.M.: The teacher's role in promoting collaborative dialogue in the classroom. Br. J. Educ. Psychol. **79**(1), 1–28 (2009)

When the Going Gets Tough: Students' Perceptions on Affect-Aware Support in an Exploratory Learning Environment for Fractions

Beate Grawemeyer[1](✉) ⓘ, Manolis Mavrikis[2](✉) ⓘ, and Wayne Holmes[2] ⓘ

[1] School of Computing, Electronics and Maths, Coventry University, Coventry, UK
beate.grawemeyer@coventry.ac.uk
[2] UCL Knowledge Lab, Institute of Education, University College London, London, UK
m.mavrikis@ucl.ac.uk, w.holmes@ucl.ac.uk

1 Introduction

It is well understood that affect interacts with and influences the learning process [2,7,9]. The impact of affect on learning is not straightforward. For example, D'Mello et al. explore how confusion, which superficially might be considered a negative affective state, is likely to promote learning under appropriate conditions [3]. In addition, the way the students perceive the tasks and the support they receive can impact their experience, agency and self-efficacy which in turn has been shown to relate to persistence and long-term outcomes [1]. It is important therefore, to deepen our understanding of the role of affect in learning in general and in particular students' perceptions of their own learning with digital environments that provide feedback.

In previous work [4] we described the development of affect-aware support and the effect of such support in relation to learning during a whole classroom intervention with a sequence of fraction learning tasks within the iTalk2Learn platform. In contrast, in this paper, we report on a study that asked students to self-report their affective states while undertaking fractions tasks.

2 The iTalk2Learn Platform

iTalk2learn is a learning platform for children aged 8–12 years old who are learning fractions. It combines structured practice with more open-ended activities in an exploratory learning environment called *Fractions Lab* [8]. In this paper we focus on the exploratory learning environment only. Figure 1 shows the *Fractions Lab* interface of the exploratory learning environment. The learning task is displayed at the top of the screen. Students are asked to solve the task by selecting a representation (from the right-hand side menu) which they manipulate in order to construct an answer to the given task [5].

© Springer Nature Switzerland AG 2022
M. M. Rodrigo et al. (Eds.): AIED 2022, LNCS 13356, pp. 483–487, 2022.
https://doi.org/10.1007/978-3-031-11647-6_98

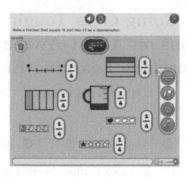

Fig. 1. Exploratory learning environment **Fractions Lab**. See [5]

Adaptive support is provided to the students based on their screen interactions and their speech. The platform is designed to detect and analyse children's speech in near real time (c.f. [6]).

The tasks provided to students included structured practice tasks with more open-ended activities. The order of tasks is calculated based on a classifier that identifies if a student is under-, appropriate-, or over-challenged. Speech and the amount of feedback provided is used for the classification [6]. In this paper we focus on tasks that were provided to students in Fractions Lab only, because the structured practice learning environment was a 'black box'.

3   User Study

We were interested in exploring the impact of the affect-aware support on tasks that differ in their difficulty. 77 students took part in this study. These participants were all primary school students, aged between 8 and 10 years old, recruited from two schools in the UK. Students were randomly allocated into two groups. The first group (N = 41) was assigned to the affect condition: the students were given access to the full iTalk2Learn system, which uses the student's affective state and their performance to determine the feedback type and its presentation. The second group of students (N = 36) was assigned to the non-affect condition, where students were given access to a version of the iTalk2Learn system in which feedback is based on the student's performance only. More details about the system overall and the affect feedback specifically are provided in [4]. Students engaged with the iTalk2Learn system for 40 min according to the experimental condition. A pre and a post test of fractions was provided to students at the beginning and at the end of the session. The tasks provided to students differed in their difficulty as follows:

Task A - Low Cognitive Demand: creating a fraction with all available representations. This task is classified as low cognitive demanding as no actual calculation with fractions is demanded.

Task B - Low Cognitive Demand: creating a fraction and partition the fraction by using the partition tool with right click. This was the first task that students were confronted with. No knowledge about fractions is needed.

Task C - High Cognitive Demand: creating a particular fraction as well as an equivalent fraction with a representation that was used the least in the past. This task is classified as a high cognitive demanding task as students were restricted to a representation that they did used the least in the past.

Task D - High Cognitive Demand: creating two fractions and check if they are equivalent in the compare box. This task is classified as high cognitive demanding as students needed do understand how fractions were calculated.

After each task, a pop-up window asked students to report how they found the task. Prior to the interaction with the system, the classroom teacher had presented this with the students and discussed the options in class through an example. Students were given a choice to select from the following: enjoyable, confusing, frustrating, interesting, something else, or don't know. The types of affective states were selected which are associated with learning [2,3,7].

4 Results

There was no difference between conditions on their self-reported affective states across the different tasks. We hypothesised that students' self-reports will differ based on low or high knowledge of fractions and depending on the task difficulty. Hence we divided the students into high and low knowledge groups based on their pre-test. This resulted in 56 students in the low group and 21 students in the high group. Separate chi-square tests were performed over the different tasks. The results are reported below:

Task A - Low Cognitive Demanding: This task was provided when students were over-challenged with task B. A chi-square test showed that there is a significant difference in low knowledge students reporting that they are frustrated. No student from the affect condition in the high knowledge group performed this task. More low knowledge students in the affect group reported to be frustrated than students in the non-affect group $(x^2(1) = 10.686, p < .001)$.

Task B - Low Cognitive Demanding: This was the first task students were confronted with. There was no difference between the groups in the high knowledge group. Significantly fewer low knowledge students in the affect condition than in the non-affect condition enjoyed this task $(x^2(1) = 4.781, p < .05)$.

Task C - High Cognitive Demanding: There was no difference between the groups in the high knowledge group. In the low knowledge group, more students in the affect condition enjoyed task than students in the non-affect condition $(x^2(1) = 4.829, p < .05)$.

Task D - High Cognitive Demanding: There was no significant difference in the low knowledge student groups. High knowledge students in the affect condition were significantly less frustrated than students in the non-affect condition $(x^2(1) = 4.234, p < .05)$.

5   Discussion and Conclusion

Despite the inconclusive results, some patterns are emerging that require further research. In particular, we hypothesise that the reason that more students with low background knowledge enjoyed the low cognitive demandings tasks in the non-affect condition than in the affect condition, is because when students are working on tasks that are appropriately challenging, they can effectively regulate their affective states. In other words, it could be the case that the affect-aware feedback is getting in the way. However, on high cognitive demanding tasks, more students with low background knowledge were enjoying the task in the affect condition than in the non-affect condition. Also students with high background knowledge were significantly less frustrated in the affect condition than in the non-affect condition. This implies that when students are confronted with a high cognitive demanding task they may be benefiting from the affect feedback more than during low cognitive demanding tasks. This might imply that affect-aware support is important for affect regulation on cognitive high demanding tasks but less effective on low demanding tasks where students do not need support for regulating their affective states.

Future work should address some limitations of this work and include an in-depth analysis of the interaction between affect, feedback and learning tasks that differ in their cognitive demands.

References

1. Adjei, S.A., Baker, R.S., Bahel, V.: Seven-year longitudinal implications of wheel spinning and productive persistence. In: Roll, I., McNamara, D., Sosnovsky, S., Luckin, R., Dimitrova, V. (eds.) AIED 2021. LNCS (LNAI), vol. 12748, pp. 16–28. Springer, Cham (2021). https://doi.org/10.1007/978-3-030-78292-4_2
2. Baker, R.S.J.d., D'Mello, S.K., Rodrigo, M.T., Graesser, A.C.: Better to be frustrated than bored: the incidence, persistence, and impact of learners' cognitive-affective states during interactions with three different computer-based learning environments. Int. J. Hum.-Comput. Stud. **68**(4), 223–241 (2010)
3. D'Mello, S.K., Lehman, B., Pekrun, R., Graesser, A.C.: Confusion can be beneficial for learning. Learn. Instr. **29**(1), 153–170 (2014)
4. Grawemeyer, B., Mavrikis, M., Holmes, W., Gutiérrez-Santos, S., Wiedmann, M., Rummel, N.: Affective learning: improving engagement and enhancing learning with affect-aware feedback. User Model. User-Adapt. Interact. **27**(1), 119–158 (2017). https://doi.org/10.1007/s11257-017-9188-z
5. Hansen, A., Mavrikis, M., Geraniou, E.: Supporting teachers' technological pedagogical content knowledge of fractions through co-designing a virtual manipulative. J. Math. Teach. Educ. **19**(2), 205–226 (2016). https://doi.org/10.1007/s10857-016-9344-0

6. Janning, R., Schatten, C., Schmidt-Thieme, L.: Feature analysis for affect recognition supporting task sequencing in adaptive intelligent tutoring systems. In: Rensing, C., de Freitas, S., Ley, T., Muñoz-Merino, P.J. (eds.) EC-TEL 2014. LNCS, vol. 8719, pp. 179–192. Springer, Cham (2014). https://doi.org/10.1007/978-3-319-11200-8_14

7. Kort, B., Reilly, R., Picard, R.: An affective model of the interplay between emotions and learning. In: IEEE International Conference on Advanced Learning Technologies, no. 43–46 (2001)

8. Mavrikis, M., Rummel, N., Wiedmann, M., Loibl, K., Holmes, W.: Combining exploratory learning with structured practice educational technologies to foster both conceptual and procedural fractions knowledge. Educ. Technol. Res. Dev. (2022). https://doi.org/10.1007/s11423-022-10104-0

9. Pekrun, R.: The control-value theory of achievement emotions: assumptions, corollaries, and implications for educational research and practice. J. Educ. Psychol. Rev. **18**, 315–341 (2006). https://doi.org/10.1007/s10648-006-9029-9

Using Log Data to Validate Performance Assessments of Mathematical Modeling Practices

Joe Olsen[1]([✉]), Amy Adair[1], Janice Gobert[1,2], Michael Sao Pedro[2], and Mariel O'Brien[1]

[1] Rutgers University, New Brunswick, NJ 08901, USA
joseph.olsen@rutgers.edu
[2] Apprendis, Berlin, MA 01503, USA

Abstract. Many national science frameworks (e.g., Next Generation Science Standards) argue that developing mathematical modeling competencies is critical for students' deep understanding of science. However, science teachers may be unprepared to assess these competencies. We are addressing this need by developing virtual lab performance assessments that assess these competencies in science inquiry contexts. Through our design processes, we developed a method for validating the assessments that takes advantage of the unique opportunities afforded by collecting log data. Here, we describe this method and demonstrate its utility by analyzing students' competencies with one example sub-practice of mathematical modeling, *plotting controlled data generated from a simulation.*

Keywords: Intelligent tutoring system · Mathematical modeling · Log data

1 Introduction

To help promote students' deep understanding of science and mathematics necessary for future college and career readiness in STEM [1], standards like the Next Generation Science Standards (NGSS) [2] emphasize the integration of disciplinary ideas and concepts with science and engineering practices, including *using mathematics and computational thinking* (NGSS Practice 5) and *developing and using models* (NGSS Practice 2). These practices, though, can be difficult for teachers to assess without resources that can capture students' competencies in real time [3]. To address this need, we are developing virtual lab performance-based formative assessments within the Inquiry Intelligent Tutoring System (Inq-ITS) environment [4]. These assessments automatically measure students' competencies at building mathematical models within science inquiry contexts using knowledge-engineered algorithms [5, 6]. Part of the development process of assessments and algorithms entails ensuring that they validly and reliably capture the broad range of competencies students may demonstrate. In this paper, we present a method in which we triangulate the virtual lab evaluations with students' actions in the lab and their multiple-choice responses to collect evidence about specific interpretations

© Springer Nature Switzerland AG 2022
M. M. Rodrigo et al. (Eds.): AIED 2022, LNCS 13356, pp. 488–491, 2022.
https://doi.org/10.1007/978-3-031-11647-6_99

of assessment scores on the mathematical modeling task [6]. This method can be useful for rigorously validating the logs and assessment data yielded by intelligent tutoring systems that scaffold and assess competencies for similar complex domains.

2 Method

2.1 Participants and Procedure

US High school students (N = 107) completed an online multiple-choice assessment, followed by an Inq-ITS virtual lab on a physical science topic (i.e., momentum, gravity, or friction) chosen by their teacher. In the virtual lab, students collected quantitative data using an interactive simulation, and developed a mathematical model to fit the trend in their data. This paper focuses on students' responses and actions related to one of the many sub-practices assessed within the system, *plotting controlled data*. The data related to this sub-practice include students' responses to the *Selecting Controlled Data* multiple-choice item (Fig. 1) as well as students' actions on the *Plotting Data* stage, where students must label axes and choose data among the trials they have collected to plot on a graph (Fig. 2).

A student used a computer simulation to study the motion of a ball being dropped from different heights.

Which trials should she use to construct a graph the explore the relationship between the *Height of the Drop* and the *Speed of the Ball*?

a) Trials 4 and 5
b) Trials 2, 3, and 4
c) Trials 1 and 5
d) All the trials
e) Trials 1, 2, and 3

Trial Number	Mass of Ball (kg)	Height of Drop (m)	Speed of Ball (m/s)
1	5	10	10.10
2	10	30	24.21
3	10	50	31.26
4	10	70	40.41
5	15	70	45.48

Fig. 1. Multiple-choice item for *Selecting Controlled Data*

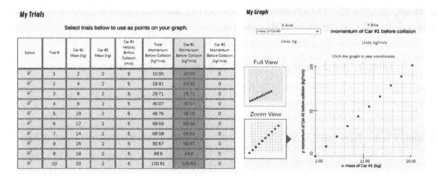

Fig. 2. Screenshots of the *Plotting Data* stage of the mathematical modeling task

2.2 Measures

We triangulated three data sources: students' responses to multiple-choice items (MCs), evaluation logs (ELs), and action logs (ALs). Specifically, the MC item for *plotting controlled data* prompts the student to identify which data from a table should be plotted according to a given goal (Fig. 1). The response is scored as correct (1) or incorrect (0) depending on whether the student selected the answer option with all controlled trials (i.e., choice B). We also collected students' EL scores generated from students' performances within the virtual lab. Scores were generated using knowledge-engineered algorithms based on the axes chosen and points plotted when constructing their graph [5, 6]. The EL score for the sub-practice of *plotting controlled data* was logged as correct (1) or incorrect (0) depending on whether the student selected all controlled trials to plot (Fig. 2). Finally, we gathered the sequential, timestamped actions taken by the students within the virtual lab through the ALs, which detailed what students clicked (e.g., which axes they chose, which points they selected and de-selected) and when.

2.3 Approach to Validating the Virtual Lab Performance Assessment

We compared students' scores on the virtual lab and multiple-choice assessment (i.e., the EL and MC scores, respectively) using 2 × 2 contingency tables. These represent the frequency distribution of student scores within each task type for *plotting controlled data*. We then selected a random subsample of students and analyzed their ALs to generate hypotheses that could explain any discrepancies found between MC and EL scores. Next, we determined which features of the ALs substantiated our hypotheses and distilled the remaining log data into summary reports of those features. From this data, we were able to generate arguments for or against the intended interpretation of the assessment scores.

3 Results: Applying the Method

Relationship between Multiple-Choice and Virtual Lab Performance for Plotting Controlled Data. Table 1 shows that, of the 85 students who received EL = 1 for this sub-practice, 60% (51/85) answered the related MC item incorrectly (MC = 0). We then analyzed the ALs to understand this pattern.

Table 1. 2 × 2 contingency table for *plotting controlled data* sub-practice.

		Virtual Lab Evaluation Log (EL) score		
		0	1	Total
Multiple Choice Item (MC) score	0	15 (14%)	51 (48%)	66 (62%)
	1	7 (7%)	34 (32%)	41 (38%)
	Total	22 (21%)	85 (79%)	107

From the ALs, we generated two hypotheses: (1) students who have mastered collecting controlled data with a simulation may not have mastered selecting a subset of controlled data from a larger set of uncontrolled data, and (2) students may operate under the misconception that they should utilize (i.e., plot) all data points available to them when constructing mathematical models. In relation to the first hypothesis, we found that 94% (48/51) of the students who received $EL = 1$ and $MC = 0$ had collected only controlled data; thus, it was impossible for these students to plot uncontrolled data since they did not have uncontrolled data available. Given that these students had also incorrectly answered the MC item (Fig. 2), we suspect that these students do not fully have this competency; thus, future designs of the virtual lab assessment should be able to discriminate between students with partial competencies with this sub-practice. In relation to the second hypothesis, 83% (89/107) plotted all data points that they collected within the system. Of these students, 45% (40/89) selected the "all trials" option for the MC item. These two pieces of evidence together suggest that selecting all data points that are available might represent a misconception among students, and future design iterations should include scaffolds to help students address this misconception.

4 Discussion

For educators to be confident that systems correctly measure students' competencies, we suggest more effort be spent on developing and utilizing validation methods which leverage log data, such as the method described in this paper. This method can be generalized to other domains and systems that make use of an external measure (e.g., multiple-choice items) that align with fine-grained constructs, performance assessments that can evaluate the same constructs, and additional human-interpretable log data of students' behavior. Such methods like ours not only provide evidence about validity, but also highlight ways to improve the design of performance assessment tasks.

Acknowledgements. This material is based upon work supported by the U.S. Department of Education Institute of Education Sciences (Award Numbers: R305A210432 & 91990019C0037; Janice Gobert & Mike Sao Pedro) and an NSF Graduate Research Fellowship (DGE-1842213; Amy Adair). Any opinions, findings, and conclusions or recommendations expressed are those of the author(s) and do not necessarily reflect the views of either organization.

References

1. National Science Board: Science and Engineering Indicators Digest 2016 (NSB-2016-2). National Science Foundation, Arlington, VA (2016)
2. NGSS Lead States: Next Generation Science Standards: for States, by States. The National Academies Press, Washington, DC (2013)
3. Hernández, M.L., Levy, R., Felton-Koestler, M.D., Zbiek, R.M.: Mathematical modeling in the high school curriculum. Math. Teach. **110**(5), 336–342 (2016)
4. Gobert, J.D., Sao Pedro, M., Raziuddin, J., Baker, R.S.: From log files to assessment metrics: measuring students' science inquiry skills using educational data mining. J. Learn. Sci. **22**(4), 521–563 (2013)
5. Dickler, R.: An intelligent tutoring system and teacher dashboard to support students on mathematics in science inquiry. ProQuest Dissertations Publishing (2021)
6. Dickler, R., et al.: Supporting students remotely: integrating mathematics and sciences in virtual labs. In: International Conference of Learning Sciences, pp. 1013–1014. ISLS (2021)

A Mobile Invented Spelling Tutoring System

Daniel Weitekamp[(✉)] and Patience Stevens

Carnegie Mellon University, Pittsburgh, PA 15213, USA
`weitekamp@cmu.edu`

Abstract. Invented spelling is a common exercise administered in kindergarten classrooms where students are asked to produce phonetically correct, but not necessarily absolutely correct spellings. For instance, "jyraf" is a phonetically correct spelling for "giraffe". We present a mobile intelligent tutoring system capable of robustly providing adaptive feedback for invented spelling practice on nearly any target word, and report some initial user testing results.

Keywords: Invented spelling · Intelligent tutoring system · Mobile

1 Introduction

Invented spelling practice engages important prerequisite skills for reading and writing such as knowledge of grapheme-phoneme (i.e. letter-to-sound) mappings and phonological awareness—the ability to identify the individual sounds that make up words. Ouellete and Sénéchal [5] found that invented spelling mediates the relationship between phonological awareness and word reading in kindergarten and that invented spelling predicts word reading and spelling abilities in first grade students. Invented spelling practice typically requires significant one-on-one feedback from adults, and thus constitutes as a potentially high-impact domain for implementing an intelligent tutoring system. In this work we present a tutoring system that provides automated feedback for invented spellings—a challenging technical problem since there are many ways that invented spellings can be correct and many degrees to which they can be incorrect.

2 A Mobile Invented Spelling Tutoring System

Our Intelligent Tutoring System (ITS) is a cross-platform ReactNative application deployable on all major mobile operating systems and web browsers. We designed our app for independent use by 4- to 6-year-old children across several sessions. Students login with a personalized password consisting of non-alphanumeric symbols, and receive instruction and feedback via audio prompts. Each new problem begins with an audio prompt presenting the target word:

© Springer Nature Switzerland AG 2022
M. M. Rodrigo et al. (Eds.): AIED 2022, LNCS 13356, pp. 492–496, 2022.
https://doi.org/10.1007/978-3-031-11647-6_100

"Let's spell [word] [word-slow] [word]" accompanied by an image depicting the target word. Random prompts such as "Try out dragging some letters" are played, after 17 s of inactivity, and letter tiles will hop and wiggle to indicate that they are interactive. For each problem students are given a small set of 8–10 randomly arranged letter tiles, that can be dragged onto the spelling line. For a word with N letters students can place at most $N+2$ tiles after which additional tiles refuse to snap to the spelling line. Students submit their spelling with the green check-mark button which provides a random motivational prompt such as "Great job!", "Cool spelling!" or "Nice work!" (Fig. 1).

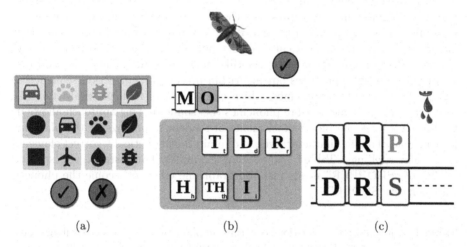

| | | |
| (a) | (b) | (c) |

Fig. 1. (a) A password entered to the login screen. (b) The word "moth" partially completed. (c) Incremental feedback for "drip" spelled "drs". "s" is replaced with "p". The "R" tile expands as it is sounded out.

At submission time the student's spelling is sounded out and they are provided with an automated version of the incremental feedback outlined by Ouellette and Sénéchal [4]. To facilitate comparison between the student's spelling and the app's feedback, copies of the student's tiles fall in above the placed tiles. This animation is accompanied by the audio prompt: "Let's see if we can make your spelling a little closer to [word]". Next animations and audio prompts show the student a single insertion, deletion, or replacement of a tile in their spelling. For instance if the target word "drip" was spelled "drs", then the app would narrate "we'll replace, 's' with 'p' to make the "P sound", and the copy of the "s" ascends off the screen to be replaced by a green colored "p"[1]. If the spelling is phonetically correct but not strictly correct students are instead provided the correct spelling as feedback. After each completed problem students earn silver stars for each correct phoneme in their spelling and a gold star if the spelling

---

[1] Note in this example that "p" is incorporated instead of "i". The calculated correction sequences favor fixing missing or incorrect consonants over vowels.

was completely correct. If students do not earn all the silver stars for a problem then it is repeated once more. Otherwise Bayesian Knowledge Tracing [2] is used to select the next problem.

3   Finding Correction Sequences for Invented Spellings

Our spelling correction algorithm has several elements that distinguish it from the spelling correction feature used in modern text editors and browsers. Spelling correction for text-editing is typically implemented by variants of Kernighan's spelling correction algorithm, an extensions of the Wagner-Fisher string-to-string correction algorithm [6] that produces the most probable target words intended by a misspelling [3]. By contrast, our algorithm finds a correction sequence that produces the nearest phonetically correct but not necessarily correct spelling, of which there is no authoritative set (like an English dictionary). By contrast to string-to-string approaches our algorithm produces a correction sequence as changes to individual graphemes, which can consist of multiple characters like "CK", and map to one or more phonemes (i.e. sounds). This approach allows us to directly target the incorrect elements of student's spellings at the phoneme level instead of at the character level, and retain the phonetically correct elements of student spellings. Table 1 shows some examples. For instance the spelling "SPR" of "SHOPPER" needs just two grapheme edits to become the phonetically correct "SHOPR".

Table 1. Application of correction sequences on phonetically incorrect spellings. Correction sequence items take arguments of the form (index, length, new value).

Target word	Spelling	Correction sequence	Final Sp.
SCRAP	SMKRMOP	remove(1,1), replace(3,1,A), remove(4,1)	SKRAP
SHOPPER	SPR	replace(0,1,SH), insert(1,O)	SHOPR
PITTSBURGH	BITSBG	replace(0,1,P), insert(5,UR)	PITSBURG

Our algorithm extends the approach taken by the Wagner-Fisher string-to-string correction approach [6]. As in Wagner-Fisher we use dynamic programming to incrementally fill a matrix representing the effects of adding, substituting, or deleting elements of partial spellings. However, instead of finding an edit sequence that produces a target word's character sequence, we find an edit sequence of graphemes that produces the breakdown of phonemes in the target word. We obtain this phonetic breakdown for each target word from the CMU Pronouncing Dictionary [1] which contains ARPAbet phonememe sequences for over 133,000 English words, and utilize a corpus of grapheme-phoneme pairs mined from this data and cleaned of infrequent graphemes.

Since our algorithm uses grapheme-phoneme pairs as the functional unit of spelling it must tackle elements of complexity not present in the Wagner-Fisher string-to-string approach. For instance our algorithm handles edge cases such as multi-character graphemes, like "SH" and "OU" which produce insertion and substitution edits that span more than 1 character along the character dimension, and special grapheme-phoneme pairs like $X \rightarrow K$, S, and $QU \rightarrow K$, W that span more than 1 phoneme along the phoneme dimension. A key feature that allows our algorithm to handle these and other edge cases is that each matrix element explicitly holds a PartialSpelling object instead of only an edit distance (as in Wagner-Fisher). Each PartialSpelling instance explicitly encodes an edit distance, a set of phoneme-grapheme pairs, the prefix and postfix sequences of phonemes from the target spelling, and the remaining uncovered characters. At each step a new PartialSpelling is produced by either a no-edit operation corresponding to adding a new matching grapheme-phoneme pair or by applying an insert, delete, or substitute operation that incur an edit cost.

4   Pilot Testing

As a pilot study we observed five kindergarteners from a local lab school as they worked with our app on tablets for 10 min each. All five participants were able to followed the audio prompts to complete several problems in the given time and exhibited a wide range of invented spelling capabilities. We observed a mix of behaviors that included, for instance, randomly placing tiles, spelling words completely correct, repeating the same incorrect spelling on the second attempt, and improving spellings in response to the given feedback on the second attempt. We observed a general pattern of students trying new spellings, or following suggestions from the feedback as they used the app more. Several participants indicated that they enjoyed earning virtual stars at the end of each problem. On the happy/sad-face Likert scale, three students rated the app with a 5, one rated it a 4, and one rated it 1. The most positive verbal impression we received was that "the app helped me spell a lot of new words that I didn't know". The most negative verbal impression indicated that there was a different educational app that the participant preferred instead.

5   Conclusion

We have presented a mobile intelligent tutoring system that facilitates independent invented spelling practice. We implement an algorithm for finding correction sequences for invented spellings that provides grapheme-level feedback, and retains the phonetically correct elements of input spellings. Finally we have presented some initial piloting results with five participants. Improving the quality of automated feedback in early literacy apps could be a key step forward in supporting young spellers' independent practice; as such, tackling the challenges associated with automated phoneme-grapheme level-feedback is essential for education technology in this domain.

Acknowledgements. Carnegie Mellon University's GSA/Provost GuSH Grant funding was used to support this project.

References

1. The CMU pronouncing dictionary. http://www.speech.cs.cmu.edu/cgi-bin/cmudict
2. Corbett, A.T., Anderson, J.R., O'Brien, A.T.: Student modeling in the ACT programming tutor. Cogn. Diagnost. Assess. 19–41 (1995)
3. Kernighan, M.D., Church, K., Gale, W.A.: A spelling correction program based on a noisy channel model. In: COLING 1990 Volume 2: Papers Presented to the 13th International Conference on Computational Linguistics (1990)
4. Ouellette, G., Sénéchal, M.: Pathways to literacy: a study of invented spelling and its role in learning to read. Child Dev. **79**(4), 899–913 (2008)
5. Ouellette, G., Sénéchal, M.: Invented spelling in kindergarten as a predictor of reading and spelling in grade 1: a new pathway to literacy, or just the same road, less known? Dev. Psychol. **53**(1), 77 (2017)
6. Wagner, R.A., Fischer, M.J.: The string-to-string correction problem. J. ACM (JACM) **21**(1), 168–173 (1974)

A Conversational Recommender System for Exploring Pedagogical Design Patterns

Nasrin Dehbozorgi[1]([✉]) and Dinesh Chowdary Attota[2]

[1] Department of Software Engineering, CCSE College, Kennesaw State University, Marietta, GA, USA
dnasrin@kennesaw.edu
[2] Department of Computer Science, CCSE College, Kennesaw State University, Marietta, GA, USA
dattota@students.kennesaw.edu

Abstract. This work presents the preliminary results of developing a Conversational Recommender System (CRS) to recommend Pedagogical Design Patterns (PDPs) to educators. In this CRS, the user queries the system in the form of natural language. The dialogue manager unit, which is the core of this system, gets the user input query and extracts the most semantically relevant patterns from the knowledge-base by Natural Language Processing (NLP) and Machine Learning (ML) algorithms. Our findings on evaluation of this system show the recommended patterns are highly relevant and semantically similar to the user queries. This novel approach to the application of pedagogical design patterns can greatly benefit the educational community by helping them identify the best practices in the field without having to search through all the published repositories of patterns. Experts can contribute to the knowledge-base of this system by sharing their best practices with the community.

Keywords: Recommender system · Pedagogical design patterns · NLP

1 Introduction

PDPs are an effective approach to address the common problems educators face in the educational domain [1]. The notion of PDPs is similar to the design patterns in the software engineering domain that guide developers to avoid the most recurring bad practices in system development [5]. PDPs allow educators and course designers to disseminate their ideas and practices in a formalized language and thus provide a framework to evaluate and adopt different design decisions [2,3]. Although the application of PDPs has been on rise however there are challenges that prevent wider adaptation of the published patterns [4]. One of the main challenges is that, given their narrative format, as the number of PDPs grows finding the right pattern to solve the given problem becomes more difficult

© Springer Nature Switzerland AG 2022
M. M. Rodrigo et al. (Eds.): AIED 2022, LNCS 13356, pp. 497–501, 2022.
https://doi.org/10.1007/978-3-031-11647-6_101

and time-consuming. In earlier work, we proposed an architectural model of a CRS based on the Model, View, Controller (MVC) pattern to help educators mine PDPs by interacting with the AI agent [3]. In this work, this model is developed and is at the early stage of evaluation to be used by a wider range of educational communities. In the following, we have a brief review of the relevant work and present the model with early but promising evaluation results, and conclude with future work.

2   Related Work

Recommender systems populate certain recommendations based on the input taken from the user either directly or indirectly such as the history of the user's behavior [6]. In situations where the available data is very limited to train the system, known as cold start situation, the CRS can help alleviate this situation by recommending based on data captured from user interaction with the system [6]. CRS uses NLP methods to learn and interpret users' queries and provide recommendations for them. There is a rise in interest in CRS due to significant developments made in NLP, speech analysis, and chatbot technologies [7]. CRS's combination with text-mining and data retrieval techniques makes them a strong tool to solve user-centric problems. Such systems are being applied in diverse domains, such as e-commerce, education, and the entertainment industry an example of which is ACRON CRS in the movie domain [8]. In another work [9], the authors implemented ConveRSE (Conversational Recommender System Framework) to understand how the usage of natural language affects the quality of the user experience in CRS. Theosaksomo and Widyantorob [10] implemented a CRS based on the functional requirement of the user and the product's intended purpose. This system emulates a salesperson while recommending the item. In the educational domain application of CRS is popular by mostly targeting students for example the intelligent tutoring systems and course recommendations based on students' performance [11]. This work focuses on CRS mainly targeting educators to help them adapt their content delivery methods and strategies.

3   Methodology

The CRS developed in this work is based on the MVC pattern. The input query is taken through the view layer and passed to the controller which processes the input and passes it to the model layer (knowledge-base) to retrieve relevant patterns. The methodology to develop this system can be divided into three main steps of 1) pre-processing user input queries, 2) processing and vectorizing input queries and 3) mining the knowledge-base to extract semantically relevant patterns based on the input data, which is processed by the dialogue manager unit (Fig. 1). The core of the system is the dialogue manager unit for which we applied Google's Universal Sentence Encoder (USE). USE leverages the transfer learning mechanism to provide better model performance than recurrent neural

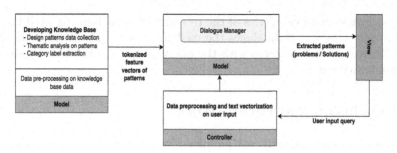

Fig. 1. High level architecture of the proposed CRS

networks or convolutional neural networks that are training-required. This USE model is available in Tensorflow-hub and hence can be used for tensor-based parallelism. It allows the use of multiple CPUs and GPUs for parallel computation of nodes in the dataflow graph. The dialogue manager unit fetches the patterns from the knowledge-base to implement pre-processing techniques of normalization, tokenization, and stop words removal and lemmatization. The USE uses the pre-processed text data from knowledge-base to generate embeddings which are high dimensional vectors, consisting of floating-point numbers. These vectors are to be used for comparing their similarity with the user input vectors thereby filtering out the semantically relevant patterns. The cosine distance of input word vectors and the word embedding of knowledge-base is calculated. Cosine similarity is the cosine of the angle between two n-dimensional vectors in an n-dimensional space. The cosine similarity of any 2 word-vectors represents how semantically similar they are. The accuracy threshold is used to filter which words are semantically similar enough and should be included. The cosine distance of two embeddings gives us a number equal to or greater than 0. the value of 0 means an exact match. A threshold of 0.7 is used in this approach to identify similar patterns.

To evaluate the accuracy of the recommended patterns we applied both qualitative and quantitative evaluation methods. For this purpose, we entered 39 queries into the system related to different categories of educational problems including assessment, assignments, diversity, feedback, students' learning, passive learning, lectures, preparation, problem-solving, procrastination, and teamwork. We measured the relevancy of the top 5 recommend patterns (problem/solutions) to the context of the input queries and subjectively ranked them in three scales of "highly relevant", "indirectly relevant", and "not relevant". Data shows that 59% of the 195 generated output in these 11 categories we highly relevant, 31% were indirectly relevant, and only 10% were non-relevant. It is a promising result as most of the recommended patterns are highly relevant in the context of the query. The patterns in the knowledge-base of the system are annotated under the 11 relevant categories. When the system presents the recommended patterns it shows to which category it belongs. For quantitative evaluation, we measured the semantic similarity score of the categories of the

39 input queries with categories of the recommended patterns to see if they are semantically relevant. Data shows the median semantic similarity score of the total 556 recommend patterns with the query vector is 0.8 out of 1 which is a high similarity score.

4   Conclusion and Future Work

This work presents an early yet promising result of developing a pedagogical CRS that helps educators get cues about the challenges they might have in their practices. The users query the system in the form of natural language through a UI and based on the knowledge base the system recommends a set of solutions to address the given problem. The knowledge base is a repository of already published PDPs. The evaluation data shows the suggested solutions are semantically very relevant to the input queries. As this project is still in progress in future work, we will expand the knowledge base and allow the educators to add their best practices into the system so that the recommendations are based on both research and empirical evidence of the more experienced users. We will also make this agent more interactive so that based on the users' feedback and satisfaction with the recommendations it recommends other options to the user.

References

1. Dehbozorgi, N.: Active learning design patterns for CS education. In: Proceedings of the 2017 ACM Conference on International Computing Education Research, pp. 291–292 (2017)
2. Dehbozorgi, N.: Sentiment analysis on verbal data from team discussions as an indicator of individual performance (2020)
3. Dehbozorgi, N., Norkham, A.: An architecture model of recommender system for pedagogical design patterns. In: 2021 IEEE Frontiers in Education Conference (FIE), pp. 1–4. IEEE (2021)
4. Dehbozorgi, N., MacNeil, S., Maher, M. L., Dorodchi, M.: A comparison of lecture-based and active learning design patterns in CS education. In: 2018 IEEE Frontiers in Education Conference (FIE), pp. 1–8. IEEE (2018)
5. Maher, M., Dehbozorgi, N., Dorodchi, M., Macneil, S.: Design patterns for active learning. In: Faculty Experiences in Active Learning: A Collection of Strategies for Implementing Active Learning Across Disciplines, pp. 130–158 (2020)
6. He, C., Parra, D., Verbert, K.: Interactive recommender systems: a survey of the state of the art and future research challenges and opportunities. Expert Syst. Appl. (2016). Elsevier
7. Jannach, D., Manzoor, A., Cai, W., Chen, L.: A survey on conversational recommender systems. arXiv preprint arXiv:2004.00646 (2020)
8. Wärnestål, P.: User evaluation of a conversational recommender system. In: Proceedings of the 4th Workshop on Knowledge and Reasoning in Practical Dialogue Systems (2005)
9. Iovine, A., Narducci, F., Semeraro, G.: Conversational recommender systems and natural language: a study through the ConveRSE framework (2020). Elsevier

10. Theosaksomo, D., Widyantoro, H.: Conversational recommender system chatbot based on functional requirement. In: 2019 IEEE 13th International Conference on Telecommunication Systems, Services, and Applications (TSSA), pp. 154–159 (2019)
11. Urdaneta-Ponte, M.C., Mendez-Zorrilla, A., Oleagordia-Ruiz, I.: Recommendation systems for education: systematic review. Electronics **10**(14), 1611 (2021)

Understanding Self-Directed Learning with Sequential Pattern Mining

Sungeun An[1]([✉]), Spencer Rugaber[1], Jennifer Hammock[2], and Ashok K. Goel[1]

[1] School of Interactive Computing, Georgia Institute of Technology,
Atlanta, GA 30308, USA
sungeun.an@gatech.edu
[2] National Museum of Natural History, Smithsonian Institution,
Washington, D.C. 20002, USA

Abstract. We describe a study on the use of an online laboratory for self-directed learning through the construction and simulation of conceptual models of ecological systems. We analyzed the modeling behaviors of 315 learners and 822 instances of learner-generated models using a sequential pattern mining technique. We found three types of learner behaviors: observation, construction, and exploration. We found that while the observation behavior was most common, exploration led to models of higher quality.

Keywords: Self-directed learning · Modeling and simulation · Online laboratory · Learning analytics

1 Introduction

Self-directed online learning is becoming increasingly prevalent [5,9]. Self-directed learning here refers to non-formal inquiry-based learning outside classroom settings. One challenge in using online laboratories for self-directed learning outside K-12 pedagogical contexts is measurement of learning outcomes as there will be a large variance in the phenomena being modeled as well as in the goals and behaviors of the learners. Many studies on the use of online laboratories for learning focus on pedagogical contexts in K-12 education with well-defined problems and well-defined learning goals, assessments, and outcomes [2,4,6,7]. At present there is a lack of understanding of the processes and outcomes of self-directed learning in online laboratories. As online laboratories become increasingly widespread, it is important to not only formulate appropriate measures of learning but also to validate learning theories and findings from the literature.

To explore this research goal, we used VERA, a publicly available online laboratory for modeling ecological systems [1]. VERA is a web application that enables users to construct conceptual models of ecological systems and run agent-based simulations of these models. This allows users to explore multiple hypotheses about ecological phenomena and perform "what if" experiments to either

© Springer Nature Switzerland AG 2022
M. M. Rodrigo et al. (Eds.): AIED 2022, LNCS 13356, pp. 502–505, 2022.
https://doi.org/10.1007/978-3-031-11647-6_102

explain an ecological phenomenon or predict the outcomes of changes to an ecological system. We investigate two research questions. *(1) What kinds of learning
behaviors emerge in self-directed learning using VERA? (2) How do the learning
behaviors relate to model quality?* In this study, the learning goals, as well as
the demographics of the learners or even their precise geographical location are
unknown; only the modeling behaviors and outcomes are observable.

2   Data Analysis

We analyzed the behaviors of 315 learners and the outcomes of 822 models
generated by the learners over three years (2018–21). This section describes
four analysis tasks: defining activities, creating activity sequences, segmenting
activity sequences, and clustering similar sequences.

2.1   Learning Behaviors

Learners' log data within the VERA system creates timestamped records of
actions such as adding a component, removing a component, or connecting two
components with a relationship. These individual actions were categorized into
three activity classes: *model construction, parameterization, and simulation* [7].
A *Model Construction* activity is defined as an insertion of a component or a relationship into a model or removal of a portion of the model. A *Parameterization*
activity is defined as modification of a component's or relationship's parameter
value. A *Simulation* activity is defined as the execution of a simulation.

We extracted activity sequences for every model created by a learner. For
instance, if a learner performed a series of actions–adding a component, adding
another component, and running a simulation–the activity sequence is 'ccs' (construction, construction, simulation). Given that an activity has no time duration
in our data, we focus on the transition from one activity to another. This makes
for 822 activity sequences, one for each model created by the 315 learners.

The activity sequences were divided into three groups of similar lengths
(short, medium, long) based on two local minima in density using a segmentation optimization method (Kernel Density Estimation). Too short or too long
sequences that are above a threshold of mean + 2*SD and below the threshold
of mean – 2*SD were eliminated (N = 33). Then the Levenshtein Distance was
applied within each length group [8]. An Agglomerative Hierarchical method,
the most common type of hierarchical clustering to group objects in clusters
based on their similarity, is used to aggregate the most similar sequences based
on the Levenshtein distance matrix [3].

2.2   Model Outcomes

We used two proxies to measure model quality. *Model complexity* is defined as the
total number of model components and relationships (referred as *depth* in [9]).
Model variety is defined as the number of unique components and relationships
used in the model (commonly referred as *breadth* [9]).

3   Results and Discussions

A total of seven clusters from three length groups were derived based on hierarchical structure of the dendrogram and visually compared and merged into three clusters. Figure 1 illustrates the resulting three clusters in VERA with 16 randomly selected example sequences for each cluster using the visualization technique in [3]. Each horizontal line in the figure shows a sequence of activities in a model, the length of an activity in a sequence corresponds to the frequency of the activity. The sequence clusters have the following characteristics:

1. **Type 1** (N = 382): *Observation.* The learners engage in experimenting with different simulation parameters with very little or no evidence of construction of conceptual models.
2. **Type 2** (N = 338): *Construction.* The learners engage in short sessions of model construction with little or no simulation of the conceptual models.
3. **Type 3** (N = 69): *Exploration (or Full Cycle).* The learners engage in a full cycle of model construction, parameterization, and simulation.

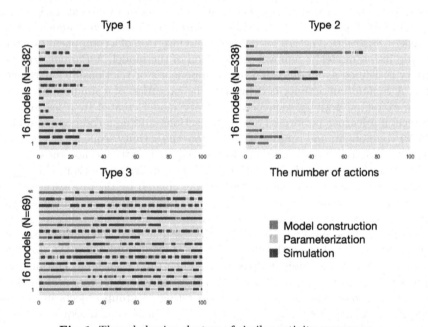

Fig. 1. Three behavior clusters of similar activity sequences.

There was a statistically significant difference in model quality among the types as determined by one-way ANOVA test (complexity: $p < .001$, $f = 75.36$; model variety: $p < .001$, $f = 26.80$) and t-tests for pairwise comparisons. The conceptual models that manifested Type 3 behavior had the most complex models ($M = 12.5$) followed by Type 1 ($M = 8.52$) and Type 2 ($M = 6.22$). (Type

1 & 2: $p < .005$, $t = 2.9835$, Type 1 & 3: $p < .001$, $t = -7.6527$, Type 2 & 3: $p < .001$, $t = -11.2651$). The conceptual models that manifested Type 3 behavior had the most variety models ($M = 3.5$) followed by Type 1 ($M = 2.9$) and Type 2 ($M = 2.3$) (Type 1 & 2: $p < .01$, $t = 2.6965$, Type 1 & 3: $p < .001$, $t = -5.8629$, Type 2 & 3: $p < .001$, $t = -6.5342$).

4 Conclusion

We derive two main conclusions from the results. First, learners manifest three types of modeling behaviors in self-directed learning using VERA: observation (simulation focused), construction (construction focused), and full exploration (model construction, evaluation and revision). Second, learners who explored the full cycle of model construction, evaluation and revision generated models of higher quality.

Acknowledgements. This research was supported by US NSF grant #1636848. We thank members of the VERA project, especially Luke Eglington and Stephen Buckley. This research was conducted in accordance with IRB protocol #H18258.

References

1. An, S., Bates, R., Hammock, J., Rugaber, S., Weigel, E., Goel, A.: Scientific modeling using large scale knowledge. In: Bittencourt, I.I., Cukurova, M., Muldner, K., Luckin, R., Millán, E. (eds.) AIED 2020. LNCS (LNAI), vol. 12164, pp. 20–24. Springer, Cham (2020). https://doi.org/10.1007/978-3-030-52240-7_4
2. Basu, S., Dickes, A., Kinnebrew, J.S., Sengupta, P., Biswas, G.: CTSiM: a computational thinking environment for learning science through simulation and modeling. In: CSEDU, pp. 369–378. Aachen, Germany (2013)
3. Desmarais, M., Lemieux, F.: Clustering and visualizing study state sequences. In: Educational Data Mining 2013 (2013)
4. Gobert, J.D., Sao Pedro, M., Raziuddin, J., Baker, R.S.: From log files to assessment metrics: measuring students' science inquiry skills using educational data mining. J. Learn. Sci. **22**(4), 521–563 (2013)
5. Haythornthwaite, C., Kumar, P., Gruzd, A., Gilbert, S., Esteve del Valle, M., Paulin, D.: Learning in the wild: coding for learning and practice on reddit. Learn. Media Technol. **43**(3), 219–235 (2018)
6. van Joolingen, W.R., de Jong, T., Lazonder, A.W., Savelsbergh, E.R., Manlove, S.: Co-Lab: research and development of an online learning environment for collaborative scientific discovery learning. Comput. Hum. Behav. **21**(4), 671–688 (2005)
7. Joyner, D.A., Goel, A.K., Papin, N.M.: MILA-S: generation of agent-based simulations from conceptual models of complex systems. In: Proceedings of the 19th International Conference on Intelligent User Interfaces, pp. 289–298 (2014)
8. Levenshtein, V.I., et al.: Binary codes capable of correcting deletions, insertions, and reversals. In: Soviet Physics Doklady, vol. 10, pp. 707–710. Soviet Union (1966)
9. Scaffidi, C., Chambers, C.: Skill progression demonstrated by users in the scratch animation environment. Int. J. Hum. Comput. Interact. **28**(6), 383–398 (2012)

Using Participatory Design Studies
to Collaboratively Create Teacher Dashboards

Indrani Dey[1]([✉]), Rachel Dickler[2]([✉]), Leanne Hirshfield[2]([✉]), William Goss[1]([✉]),
Mike Tissenbaum[3]([✉]), and Sadhana Puntambekar[1]([✉])

[1] University of Wisconsin-Madison, Madison, WI 53706, USA
{idey2,wgoss2}@wisc.edu, puntambekar@education.wisc.edu
[2] University of Colorado Boulder, Boulder, CO 80309, USA
{rachel.dickler,leanne.hirshfield}@colorado.edu
[3] University of Illinois Urbana-Champaign, Champaign, IL 61820, USA
miketiss@illinois.edu

Abstract. Classroom orchestration requires teachers to concurrently manage multiple activities across multiple social levels (individual, group, and class) and under various constraints. Real-time dashboards can support teachers; however, designing actionable dashboards is a huge challenge. This paper describes a participatory design study to identify and inform critical features of a dashboard for displaying relevant, actionable, real-time data. We leveraged a Sense-Assess-Act framework to present dashboard mockups to teachers for feedback. Although the participating teachers differed in how they would use the presented information (during class or after class as a post hoc analysis tool), two common emerging themes were that they wanted to use the data to a) better support their students and b) to make broader instructional decisions. We present data from our study and propose a customizable, mobile dashboard, that can be adapted to a teacher's specific needs at a specific time, to help them better facilitate learning activities.

Keywords: Teacher dashboard · Participatory design · Orchestration

1 Introduction

Classroom orchestration requires teachers to navigate multiple activities, often simultaneously, across multiple social levels, i.e., at the *individual* (e.g., writing), *group* (e.g., collaborative problem solving), and *class* (e.g., classroom discussion) levels [2, 7]. Activities may be distributed across multiple tools (e.g., notebooks, simulations, etc.) and artifacts (e.g., laptops), with additional constraints such as curriculum and time adding to the pedagogical complexity. Real-time orchestration tools have the potential to support teachers in providing timely assistance [4, 7]. Multimodal data on student activities can be collected and analyzed to provide an overview of their progress in the form of a visual display, known as the teacher dashboard [5, 8]. The goal of dashboards is to help teachers make quick, data-driven decisions in the classroom; however, studies suggest limited success in authentic learning environments. Commonly identified challenges

© Springer Nature Switzerland AG 2022
M. M. Rodrigo et al. (Eds.): AIED 2022, LNCS 13356, pp. 506–509, 2022.
https://doi.org/10.1007/978-3-031-11647-6_103

include, displaying static information, not providing actionable information, the inability to navigate across multiple social levels, potential conflicts with learning goals, time constraints, and the degree of teacher involvement in the design [5, 6, 8].

AI-based adaptive dashboards can augment teacher instruction [4] by providing visualizations of critical real-time data on complex states [1], assisting with adaptive decision-making, and identifying attention areas. AI algorithms can be applied to assess progress based on features sensed within the environment and correspondingly inform dashboard visualizations. We leveraged the Adaptive System framework outlined by Feigh et al. [3] to provide design guidelines for AI-based teacher dashboards. The authors describe three processing states of AI-based adaptive systems: sensing (i.e., perceiving) aspects of the environment, assessing (i.e., selecting) the current state and how to respond, and acting by providing information within a human-AI shared interface that can then inform human action. They provide a taxonomy of adaptations of four categories: *What* (what information to show), *When* (when to show it, e.g., during or after class), *Who* (who should see a specific information, e.g., teachers, classroom assistants, etc.), and *How* (e.g., visual or auditory). We placed this framework in the context of a classroom environment, with teachers able to choose an intervention based on information from the system, paired with observations of classroom dynamics.

2 Methods

We used a user-centered participatory design approach [4, 5], where teachers and researchers actively co-created dashboard designs. We created mockups based on the most-requested metrics from a preliminary study and presented them to six middle school science teachers in a Midwestern U.S. state (with 7–34 years of teaching experience). Our mock-ups (see Fig. 1) were implemented using JavaScript and React.

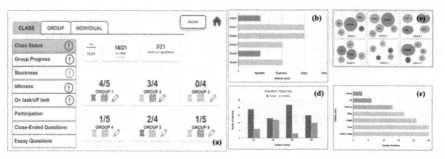

Fig. 1. Mock-up screens presented in the participatory design session: (a) class status, (b) group progress, (c) participation, (d) short-answer questions, and (e) essay questions

After reviewing each screen, teachers wrote responses to a set of questions. We then asked them to elaborate to discuss multiple perspectives. Data from audio and video recordings, written responses, and field notes were triangulated to analyze the feedback.

3 Results and Discussion

We summarize our results based on the categories mentioned earlier.

What information should be presented? Teachers had differing preferences for: i) a high-level class overview, ii) a group view, and iii) fine-grained individual data. For example, 83% of the teachers said they would use the Group Progress metrics (Fig. 1b) to identify which groups need support, while 33% also asked for individual data. Teachers also illuminated potential uses to promote classroom equity, such as using Participation metrics (Fig. 1c) to provide opportunities for all voices to be heard, help special education staff better assist special needs or ELL (English Language Learners) students.

When should the information be presented? Depending on the situation, some teachers preferred real-time information during class to provide immediate support, while others said they would like to review the data as a post hoc analysis tool. For example, 33% teachers said they would use the Group Progress screen (Fig. 1b) during class to identify who is stuck and prioritize helping them, to give new tasks to students who are ahead, or even promote collaboration, e.g., *"You could easily ask somebody who's done with the work to take on a teaching role to help a student."*; whereas 66% would also like this data after class to adjust their next lesson. This implies that teachers not only have individual preferences for *when* they would like to see/use the data but also have different ways in *how* they will act, highlighting the need for a customizable dashboard.

The multifaceted nature of teaching requires fluid orchestration between social levels [2, 6], requiring dashboards to adaptively present relevant information at the appropriate time. Despite individual preferences, common emerging themes were that teachers wanted to leverage the information to a) better support students, and b) to improve their lessons, two characteristics reflected in other studies [1, 5, 9].

How should the information be presented? Teachers requested mobility, as they wanted to access information while navigating the classroom and not be tethered to their desks. Our goal is to make the dashboard available on an iPad or tablet. We also discussed sending notification-style alerts as well as a vibration alert to be less intrusive.

3.1 Design Decisions and Future Directions

Our next steps are to add three features in our dashboard. First, a dashboard that can assess the teacher's specific needs at specific times and adaptively present relevant, actionable information. Teachers should be able to make decisions about what information is displayed for each lesson, which could be based on specific learning goals, e.g., if the goal is to understand key science ideas, the teacher can choose metrics related to question performance (Fig. 1d) or essay writing (Fig. 1e). We plan to collect data about pertinent goals and features during classroom use, which can then be used to create a catalogue to provide teachers with customizable options. Second, is to display trends across classes. Our teachers usually teach 2–5 classes; providing comparable trends across classes will highlight the similarities or differences for specific metrics. Third, we are combining various metrics, such as social interaction and completion, to provide more comprehensive information for teachers to make decisions.

This study highlighted the importance of effectively using teacher input to co-create a classroom orchestration tool, helping us make concrete design decisions based on their needs and feedback. As preferences of what data types they want to see and use are highly specific to the individual teacher and specific situation, it indicates the need for a dynamic customizable dashboard. Our future work aims to design and test dashboard prototypes in actual classroom settings, to better understand the AI dashboard as a useful decision-making and real-time support tool.

Acknowledgements. We thank the teachers who participated in our studies. The research described in this paper has been partially funded by NSF grants #2019805, #2010357, and #2010483.

References

1. Dickler, R., Gobert, J., Pedro, M.S.: Using innovative methods to explore the potential of an alerting dashboard for science inquiry. J. Learn. Anal. **8**(2), 105–122 (2021)
2. Dillenbourg, P.: Design for classroom orchestration. Comput. Educ. **69**, 485–492 (2013)
3. Feigh, K.M., Dorneich, M.C., Hayes, C.C.: Towards a characterization of adaptive systems: a framework for researchers and system designers. Hum. Factors **54**(6), 1008–1024 (2012)
4. Holstein, K., McLaren, B.M., Aleven, V.: Co-designing a real-time classroom orchestration tool to support teacher–AI complementarity. J. Learn. Anal. **6**(2), 27–52 (2019)
5. Martinez-Maldonado, R.: A handheld classroom dashboard: teachers' perspectives on the use of real-time collaborative learning analytics. Int. J. Comput.-Support. Collab. Learn. **14**(3), 383–411 (2019). https://doi.org/10.1007/s11412-019-09308-z
6. Olsen, J.K., Rummel, N., Aleven, V.: Designing for the co-orchestration of social transitions between individual, small-group and whole-class learning in the classroom. Int. J. Artif. Intell. Educ. **31**(1), 24–56 (2020). https://doi.org/10.1007/s40593-020-00228-w
7. Prieto, L., Dlab, M.H., Gutierrez, I., Abdulwahed, M., Balid, W.: Orchestrating technology enhanced learning: a literature review and a conceptual framework. Int. J. Technol. Enhanc. Learn. **3**, 583–598 (2011)
8. Schwendimann, B.A., et al.: Perceiving learning at a glance: a systematic literature review of learning dashboard research. IEEE Trans. Learn. Technol. **10**(1), 30–41 (2017)
9. Wiedbusch, M.S., et al.: A theoretical and evidence-based conceptual design of MetaDash: an intelligent teacher dashboard to support teachers' decision making and students' self-regulated learning. Front. Educ. **19**, 1–13 (2021)

An AI-Based Feedback Visualisation System for Speech Training

Adam T. Wynn⬤, Jingyun Wang(✉)⬤, Kaoru Umezawa⬤,
and Alexandra I. Cristea⬤

Durham University, Durham DH1 3LE, UK
jingyun.wang@durham.ac.uk

Abstract. This paper proposes providing automatic feedback to support public speech training. For the first time, speech feedback is provided on a visual dashboard including not only the transcription and pitch information, but also emotion information. A method is proposed to perform emotion classification using state-of-the-art convolutional neural networks (CNNs). Moreover, this approach can be used for speech analysis purposes. A case study exploring pitch in Japanese speech is presented in this paper.

Keywords: CNN · Automatic visualisation feedback · Second language speech training · Emotion recognition · Speech prosody

1 Introduction

Timely feedback is important for language learning as it enables the learner to practice at their own pace [7]. Speech training applications have been used to help second language (L2) speakers identify ways to improve their speech without the requirement for manual feedback. Some systems provide pitch feedback using visualisation dashboards [13] while others provide automatic speech modification [4]. These feedback mechanisms work well with simple phrases, but don't scale well to longer speeches. Moreover, few studies have focused on supporting public speaking training. Therefore, this research is intended to support L2 English and Japanese learners in public speaking, such as speech contests. Our research question is: *Compared to prior research simply providing transcription or pitch changes as feedback, can a combination of transcription, pitch and emotional changes as feedback better support speech training?*

Determining the relationship between the pitch range of speakers from different L1 backgrounds is one research focus. For instance, in Japanese, pitch serves as the main cue to signal lexical and phrasal distinctions. Passoni, et al. [12] found that Japanese-English bilinguals had a lower mean pitch in Japanese than in English and female speakers displayed more pitch variation for different formality settings and that lower mean pitch may be due to nervousness. A method for computing the pitch of English speech is proposed by Kurniawan, et

ⓒ Springer Nature Switzerland AG 2022
M. M. Rodrigo et al. (Eds.): AIED 2022, LNCS 13356, pp. 510–514, 2022.
https://doi.org/10.1007/978-3-031-11647-6_104

al. [8], and their experiment results suggest that an increase of pitch could be a sign of nervousness.

Emotion detection is another research direction. Using Convolutional neural networks (CNNs), Franti, et al. [2] classified speech into 6 emotional states. They identified that someone who was speaking faster with a wider pitch range was more likely to be experiencing emotions of fear, anger or joy. Kurniawan, et al. [8] captured Mel Frequency Cepstral Coefficients (MFCCs) as features from the audio signal to classify speech. MFCCs approximate human audio perception more closely, to achieve an accuracy of 92.4% using Support Vector Machines.

The main purpose of this paper is to propose an AI-based speech feedback system, which gives immediate feedback to the learner, via a visualisation dashboard. This approach is achieved by outputting the state and level of emotion identified by CNNs in each sentence, and the user can view how their pitch changes throughout the speech to detect how this might effect the emotion conveyed, along with a transcription. Moreover, multiple audio files can be uploaded by users for further comparisons and analysis. This function is illustrated by a case study exploring pitch in Japanese speech.

2   A Visualisation Speech Feedback System

In this research, an AI-based visualisation system which not only provides feedback for individual speech training, but also enables audio analysis, was designed and implemented. A CNN was proposed to recognise the level of emotion (low, medium or high) in each sentence which consists of 1-dimensional convolutional layers and was programmed using the Keras library [5] and TensorFlow. 40 MFCC features were used as an input, which were extracted from the data using the Librosa package [11]. 2686 speech samples from the RAVDESS [10] and CREMA-D [6] datasets were used for model training where the accuracy depends on the emotion (Anger: 82.4%, Disgust: 71.7% Fear: 79.7%, Happy: 72.5%, Sad: 70.3%). The CREPE Pitch Tracker [9] was used to identify pitch based on the fundamental frequency (f0). Readings above 400 Hz or below 50 Hz were removed, as they were likely erroneous measurements.

Prior to uploading audio recordings, learners need to choose their language, gender, and one emotion out of anger, disgust, fear, happiness, and sadness to focus on. The feedback provided by the system is presented visually (Fig. 1), using the Bokeh visualisation library [3]. Figure 1(a) provides information about the emotion tracked, including three levels of intensity. The user can see how the intensity of their chosen emotion changes throughout the speech for each sentence. Figure 1(b) shows how the pitch changes throughout the speech, which could be used to infer the relationship with emotion.

3   A Case Study – Exploring Pitch in Japanese Speech

It is proven that the mean pitch of female speakers (between 160–300 Hz) is higher than that of male speakers (between 60–180 Hz). However, few studied the

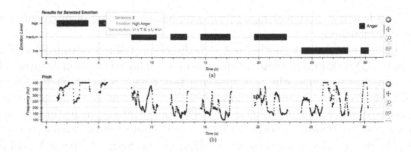

Fig. 1. Learner dashboard with visual feedback.

pitch difference between native and non-native speakers, and also the interaction effect with gender. To study the effect of gender and whether the speaker is a native Japanese or not on pitch (considering mean pitch and pitch range as two features), audio files (average duration 8 min) which include speeches by 2 native and 3 non-native female speakers, and 4 native and 5 non-native male speakers, consisting of 795 sentences (197 native female; 320 non-native female; 134 native male; and 144 non-native male) were uploaded to the system for analysis. The native speaker audio was collected from Toastmasters Japan [1] and non-native speaker audio was collected from Japanese speech contests.

Based on the mean pitch and pitch range of each sentence determined by the system, a two-way MANOVA was conducted. The results indicate a significant interaction effect ($F(1, 790) = 70.66$) between gender and whether the speaker is native or not. The main effect of gender on pitch is significant ($F(1, 790) = 272.80$). Sentences by female speakers (Mean Pitch: Mean = 245.57 Hz, S.D. = 1.90; Pitch Range: Mean = 236.40 Hz, S.D. = 3.08) have a significantly higher mean pitch ($F(1, 791) = 516.12$) and wider pitch range ($F(1, 791) = 49.85$) compared to those by male speakers (Mean Pitch: Mean = 174.02 Hz, S.D. = 2.51, Pitch Range: Mean = 200.27 Hz, S.D = 4.09). Also, the main effect of whether the speaker is native or not is significant ($F(1, 790) = 30.39$). Compared to sentences by non-natives (Mean = 200.94 Hz, S.D. = 2.10), those by natives (Mean = 218.65 Hz, S.D. = 2.35) have a significantly higher mean pitch ($F(1, 791) = 31.63$) and their speech (Mean = 233.13 Hz, S.D. = 2.80) has a wider pitch range ($F(1, 791) = 33.46$) in contrast to non-native speech (Mean = 203.54 Hz, S.D. = 3.81). Furthermore, the individual univariate test results (Table 1) show a significant difference between native and non-native females ($F(1, 791) = 24.044$), and between native and non-native male speakers ($F(1, 791) = 115.45$). For pitch range, there is only a significant difference ($F(1, 791) = 35.43$) between native and non-native males, and no significant difference ($F(1, 791) = 2.94$) between native and non-native female speakers.

Table 1. Individual univariate test results.

Feature	Gender	Native (Hz)	Non-native (Hz)	F(1,191)
Mean pitch	Female	Mean = 236.37; S.D. = 2.98	Mean = 254.87; S.D. = 2.34	24.04 ($p < 0.05$)
	Male	Mean = 201.40; S.D. = 3.62	Mean = 147.01; S.D. = 3.49	115.45 ($p < 0.05$)
Pitch range	Female	Mean = 241.68; S.D. = 4.85	Mean = 231.11; S.D. = 3.81	2.94 ($p > 0.05$)
	Male	Mean = 224.58; S.D. = 5.88	Mean = 175.96; S.D. = 5.67	35.43 ($p < 0.05$)

4 Discussion and Future Work

Despite a small number of speakers, the analysis of 795 sentences shows that non-natives have a significantly narrower pitch range, which may be due to nervousness. From a public speech training perspective, this suggests that non-native speakers should try to widen their pitch range in order to be more similar to natives. Detailed suggestions regarding their pitch and emotion could potentially help them adjust their pitch range. In summary, this case study demonstrates that our system can easily transform multiple audio files into quantitative data, which can be used in further statistical analysis for any research purpose.

In the future, speech data of more speakers will be studied to confirm this finding. Also, more detailed feedback will be provided to improve their speaking skills. In terms of emotion, we plan to train another model using Japanese emotional speech data, and design more functions to support speech training.

References

1. Toastmasters Japan (2021). https://district76.org/en/. Accessed 6 Feb 2022
2. Alu, D., et al.: Voice based emotion recognition with convolutional neural networks for companion robots. Rom. J. Inf. Sci. Technol. **20**(3), 222–240 (2018)
3. Bokeh: Bokeh (2021). https://bokeh.org. Accessed 17 Jan 2022
4. Bonneau, A., Colotte, V.: Automatic feedback for L2 prosody learning. Ivo Ipsic. Speech and Language Technologies, Intech, pp. 55–70 (2011). https://doi.org/10.5772/20105
5. Chollet, F., et al.: Keras (2015). https://github.com/fchollet/keras. Accessed 17 Jan 2022
6. Cooper, D.: CREMA-D (2021). https://github.com/CheyneyComputerScience/CREMA-D. Accessed 17 Jan 2022
7. Golonka, E., et al.: Technologies for foreign language learning: a review of technology types and their effectiveness. Comput. Assist. Lang. Learn. **27**, 70–105 (2014). https://doi.org/10.1080/09588221.2012.700315
8. Kurniawan, H., Maslov, A.V., Pechenizkiy, M.: Stress detection from speech and galvanic skin response signals. In: Proceedings of the 26th IEEE International Symposium on Computer-Based Medical Systems, pp. 209–214 (2013). https://doi.org/10.1109/CBMS.2013.6627790
9. Kim, J.W., et al.: Crepe: a convolutional representation for pitch estimation (2018)
10. Livingstone, S.R., Russo, F.A.: The Ryerson Audio-Visual Database of Emotional Speech and Song (RAVDESS) (2018). https://doi.org/10.5281/zenodo.1188976

11. McFee, B., et al.: librosa: 0.8.1rc2, May 2021. https://doi.org/10.5281/zenodo.4792298
12. Passoni, E., et al.: Bilingualism, pitch range and social factors: preliminary results from sequential Japanese-English bilinguals. In: Proceedings of the 9th International Conference on Speech Prosody 2018, pp. 384–338 (2018). https://doi.org/10.21437/SpeechProsody.2018-78
13. Sztah, D., et al.: Computer based speech prosody teaching system. Comput. Speech Lang. **50**, 126–140 (2018). https://doi.org/10.1016/j.csl.2017.12.010

Assessing Students' Knowledge Co-construction Behaviors in a Collaborative Computational Modeling Environment

Caitlin Snyder[1]([✉]), Cai-Ting Wen[2], and Gautam Biswas[1]

[1] Vanderbilt University, Nashville, TN, USA
`caitlin.r.snyder@vanderbilt.edu`
[2] National Cencetral Unviersity, Taoyuan City, Taiwan

Abstract. Successful knowledge co-construction during collaborative learning requires students to develop a shared conceptual understanding of the domain through effective social interactions [1]. Developing and applying shared understanding of concepts and practices is directly impacted by the prior knowledge that students bring to their interactions. We present a systematic approach to analyze students' knowledge co-construction processes as they work through a physics curriculum that includes inquiry activities, instructional tasks, and computational model building activities. Utilizing a combination of students' activity logs and discourse analysis, we assess how students' knowledge impacts their knowledge co-construction processes. We hope a better understanding of how students' co-construction processes develop and the difficulties they face will lead to better adaptive scaffolding of students' learning and better support for collaborative learning.

Keywords: Knowledge co-construction · Prior knowledge · Collaborative learning · Computational model building

1 Introduction

Knowledge co-construction processes during collaborative learning are known to be impacted by the prior knowledge each student brings to the group and externalize through discussion, explanation, and argumentation [1]. In this work, we adopt a learning-by-modeling approach, where students have to simultaneously develop and apply their domain knowledge and computational thinking (CT) processes to develop models of scientific phenomena. We extend current research by analyzing how the distribution of prior knowledge in a group, particularly when students are learning two domains simultaneously, impacts students' domain knowledge and social co-construction processes. Using students' discourse and activity logs, we assess students' co-construction processes by analyzing the strategies they apply in their inquiry and problem-solving tasks, their conversations as they work in pairs, and their model building performance. By understanding the impact of students' prior knowledge on their domain-specific and social co-construction processes, especially when they face difficulties, we hope to develop better adaptive supports to facilitate effective collaborative model building.

© Springer Nature Switzerland AG 2022
M. M. Rodrigo et al. (Eds.): AIED 2022, LNCS 13356, pp. 515–519, 2022.
https://doi.org/10.1007/978-3-031-11647-6_105

2 Study Description and Data Analysis Methods

During a 9-week-long study, consisting of a two-hour class once a week, students worked together in pairs, assigned based on their pre-test performance. The student with the highest pretest score was grouped with the student who had the lowest pretest score, and so on. We collected (1) screen-capture video that recorded students' conversations; (2) action log data from both the CoSci [4] and C2STEM [2] environments; and (3) students' final computational models developed for the three challenge tasks. In this paper, we analyzed one of the three kinematic modules, 1D motion with acceleration module. After initial instruction, students completed inquiry tasks with CoSci to explore the relationships between position, velocity and acceleration through parameter manipulation in a scenario where Mario, moving at constant velocity from a pre-specified position, had to catch a mushroom falling from a height. In their final task in the module, students transitioned to a modeling challenge where they built a computational model of the motion of a truck that sped up from rest to a speed limit, then cruised at the speed limit, and then had to slow down and stop at a designated STOP sign.

We identified three types of groups based on each student's prior knowledge distribution relative to the median: (1) Balanced prior knowledge in 2 domains: one student had high prior knowledge in one domain and their partner had high prior knowledge in the other (e.g., S1: high-physics, low-CT; S2: low-physics, high-CT); (2) Unbalanced prior knowledge: one group member had high prior knowledge in both domains while the other had low prior knowledge in both (e.g., S1: high-physics, high-CT; S2: low-physics, low-CT); and (3) Deficit in one domain, where neither group member had high prior knowledge in one of the domains (e.g., S1: low-physics, high-CT; S2: low-physics, low-CT).

For the CoSci inquiry task, we used the log data to infer three strategies, previously identified in [3], that students applied to explore the relation between position, velocity, and acceleration: (1) *Systematic (SYS)*, i.e., they systematically designed their experiments by changing one variable at a time; (2) *Trial and Error (T&E)*, where they changed variable values randomly to find answers; and (3) *Calculation (Calc)*, where they used the equations of motion to calculate the two parameters by selecting one and calculating the other. We also identified the following strategies that students used while modeling the three phases of the truck's motion in C2STEM: (1) *Data Tool Usage (DT)*, identified as students opening the data tools and making edits (DATA → ADJUST), where ADJUST refers to adjusting the existing model; (2) *Trial and Error (T&E)*, identified by sequences of ADJUST → PLAY actions, where PLAY refers to running the simulation; (3) *Depth-first (DF)*, identified by multiple code construction actions without PLAY actions. By extracting the student discourse during behavior changes, we also analyzed students' use of the kinematic calculations (*Calc*) to compute the conditions for the truck's behavior transitions, especially if they computed the correct lookahead distance (*Suc/Unsuc*). In addition, we identified their use of the *HELP* strategy, where another group was asked to help with model construction steps. To evaluate overall performance, the groups' final truck models were scored using a rubric that evaluated their conditional (*COND*) and relationship expressions (*REL*) in the model.

3 Results

Table 1 shows the different inquiry (*INQ*) and model building (*MB*) strategies as well as the model scores groups obtained in their truck modeling task. Our results show that the use of the systematic inquiry strategy (*SYS*) was linked to effective knowledge co-construction of the physics relations for the truck model. The exception was group G5, which did not have high scores for the relationship expressions in model building. The other *SYS* groups, G2, G3, G6, G9, G11, and G12 had high prior knowledge in both domains, and this helped them with the relationship expression component. The same cannot be said for their conditional construct implementations, where varying results are observed. This suggests that while the groups' prior knowledge in both domains led to their using the systematic (*SYS*) strategy during inquiry, it did not translate to success in the model construction components. While the use of *SYS* inquiry strategies positively impacted knowledge co-construction of the physics-based relationship expressions, this strategy did not help students with their conditional constructs, which required students to combine their Physics and CT knowledge to establish the correct conditional expressions and constructs.

Table 1. Students' strategies and model scores

Type	Group	INQ Strat.	MB Strat.	COND	REL	Total
Balanced	G2	SYS	Calc (Unsuc) → HELP	4.5	6	10.5
	G3	SYS	Calc (Semi-suc) → T&E	4.5	6	10.5
	G6	SYS	DT	3.5	6	9.5
Deficit in one domain	G4	T&E	Calc (Unsuc) → DF	3.5	5	8.5
	G5	SYS	Calc (Unsuc) → DF	1	3.5	4.5
	G7	T&E	DT	2	2	4
Unbalanced	G8	T&E + Calc	Calc (Suc)	3	6	9
	G9	SYS	Calc (Suc)	6	5	11
	G11	SYS	Calc (Suc)	4.5	6	10.5
	G12	SYS	Calc (Suc)	4.5	6	10.5
	G13	T&E + Calc	Calc (Suc)	5.5	5.5	11

4 Discussion and Conclusions

In this paper, we leveraged the combination of activity logs and discourse to study the relationships between students' prior knowledge in Physics and CT, an inquiry task, and a model building task that required students to build a correct computational model of a truck that sped up, cruised, and then slowed down to a stop. The systematic inquiry strategy in CoSci promotes students' understanding of the domain knowledge, which then facilitates their co-construction processes during computational modeling. Our results also show that students who did not use systematic strategies for their inquiry tasks (primarily because of their low prior knowledge) may need additional scaffolding or instruction to help them develop basic domain knowledge to help them benefit from the inquiry tasks. A good understanding of the domain knowledge is a stepping-stone to using effective co-construction processes to support model building tasks.

While both unbalanced and balanced groups had relatively equivalent performance in the modeling task, only those with unbalanced prior knowledge were fully successful in using the kinematic calculations (*Calc*) strategy. Through the discourse, we see the high prior knowledge student leading all discussions. We hypothesize the one-way interactions of the unbalanced groups imply they may not have to come to a shared understanding during the inquiry task but their success during model building implies they acquired sufficient knowledge for successful construction. In contrast, the balanced groups had to truly co-construct knowledge with CT prior knowledge group members working to understand the physics concepts, and the physics prior knowledge students working to understand the CT concepts, like the conditional expressions. Our results show that although these groups attempted to co-construct knowledge, they had difficulties with calculating the correct lookahead value (lack of physics knowledge) or a difficulty operationalizing the correct value into the conditional expressions (lack of CT knowledge). Groups with a deficit of physics prior knowledge had similar difficulties but succeeded in the modeling task. We hypothesize that groups with a deficit of physis prior knowledge had difficulties because neither group member could leverage physics' prior knowledge, causing them to be least successful in the modeling task.

While this study is limited in the number of groups, we believe this provides a starting point for understanding students' knowledge co-construction and the impact prior knowledge has on the social and domain components of these co-construction processes. While the unbalanced and balanced group performance is relatively equivalent when looking at this one task, the average learning gains after the completion of the three modules were -0.06 and 0.24, for students in the unbalanced and balanced groups respectively. This suggests that groups with balanced prior knowledge may be able to better synergistically co-construct knowledge after completion of all three modules.

Acknowledgments. This material is based in part upon work supported by NSF Award 2017000.

References

1. Beers, P.J., Boshuizen, H.P.E., Kirschner, P.A., Gijselaers, W.H.: Computer support for knowledge construction in collaborative learning environments. Comput. Hum. Behav. **21**(4), 623–643 (2005)
2. Hutchins, N.M., et al.: C2STEM: a system for synergistic learning of physics and computational thinking. J. Sci. Educ. Technol. **29**(1), 83–100 (2019). https://doi.org/10.1007/s10956-019-098 04-9

3. Hutchins, N.M., Snyder, C., Emara, M., Grover, S., Biswas, G.: Analyzing debugging processes during collaborative, computational modeling in science. In: Proceedings of the 14th International Conference on Computer-Supported Collaborative Learning, pp. 221–224 (2021)

4. Wen, C.-T., et al.: The learning analytics of model-based learning facilitated by a problem-solving simulation game. Instr. Sci. **46**(6), 847–867 (2018). https://doi.org/10.1007/s11251-018-9461-5

How Item and Learner Characteristics Matter in Intelligent Tutoring Systems Data

John Hollander$^{(\boxtimes)}$ ⓘ, John Sabatini ⓘ, and Art Graesser

University of Memphis, Memphis, TN 38111, USA
jmhllndr@memphis.edu

Abstract. AutoTutor-ARC (adult reading comprehension) is an intelligent tutoring system that uses conversational agents to help adult learners improve their comprehension skills. However, in such a system, not all lessons and items optimally serve the same purposes. In this paper, we describe a method for classifying items that are *instructive, evaluative, motivational*, versus *potentially flawed* based on analyses of items' psychometric properties. Further, there is no a priori way of determining which lessons are optimal given the learner's reading profile needs. To address this, we evaluate how assessing learner component reading skills can inform various aspects of learner needs on AutoTutor lessons. More specifically, we compare learners who were classified as *proficient, underengaged, conscientious*, versus *struggling* readers based on their experiences with AutoTutor. Together, these analyses suggest the utility of integrating assessments with instruction: efficient, adaptive learning at the lesson level, more efficient and valid post-testing, and consequently, recommendations for more targeted, adaptive pathways through the instructional program/system.

Keywords: Intelligent tutoring systems · Reading skills · Psychometrics

1 Introduction

1.1 Adaptive Education and Adult Literacy

Assessments of worldwide literacy rates indicate that around 14% of adults may be classified as low literate [1]. While advances in research and technology are helping more adults improve their ability to read and write than ever, the best efforts of educators and literacy researchers still do not meaningfully help a significant portion of this population. The development of adaptive learning technologies could significantly address this problem because adult learners are a dispersed and diverse population [2–4]. We analyzed data obtained from adults with low literacy who completed a reading component skill assessment battery before and after participating in an instructional program using AutoTutor–ARC, an adult literacy-focused intelligent tutoring system with two conversational agents that periodically ask the learners questions while adults read texts and other learning materials. The lessons are specifically designed to engage adult learners, and range from word-level learning to practical applications of complex literacy skills.

© Springer Nature Switzerland AG 2022
M. M. Rodrigo et al. (Eds.): AIED 2022, LNCS 13356, pp. 520–523, 2022.
https://doi.org/10.1007/978-3-031-11647-6_106

However, determining how AutoTutor-ARC lessons and items relate to specific reading component skills known to impede or facilitate comprehension growth would allow for a more responsive and effective approach [5, 6].

The Reading Inventory and Scholastic Evaluation (RISE) (also known as the Study Aid and Reading Assessment or SARA) is a battery of six reading component skills subtests measuring skills that are known to be malleable to instruction [7], specifically *decoding and word recognition, vocabulary, morphology, sentence processing, reading efficiency*, and *reading comprehension*. A reader who lacks adequate component skills may rely on one or more compensatory behaviors, strategies that are often not optimal to continued growth [3].

2   Method

Data used for this study were obtained from three waves of an adult literacy intervention study. Participants included 252 adult literacy learners ($M_{age} = 42.4$, $SD = 13.9$, 74.6% female), who were offered 100 h of instruction (featuring hybrid classes of teacher-led sessions and AutoTutor sessions) over the course of four months. Auto-Tutor lessons were assigned to students individually by their teacher; not all students took all lessons, and lessons could be repeated. Participants completed one form of the RISE before the intervention, and another form afterward [7].

We gauged how well ITS data can be used to identify learner characteristics with respect to adult literacy by adopting the results of Fang et al.'s 4-cluster clustering analysis [5] of adult learners using AutoTutor-ARC. These clusters were defined by their accuracy and speed in answering conversation-based questions during learning: *proficient readers* (accurate and fast), *struggling readers* (inaccurate and slow), *conscientious readers* (accurate but slow), and *underengaged readers* (less accurate but fast).

3   Results

We considered items in lessons which were fully completed by at least 90 participants on their first attempt. In accordance with similar analyses and data processing procedures [5], we considered items that were correctly answered by at least 95% of participants to be *motivational*, as they do not provide any new information about learner knowledge. Further, we considered items with a negative item-total correlation to be *potentially flawed*. These items are psychometrically inconsistent with lesson topic constructs. Figure 1 contains a graphical representation of this classification.

We calculated the reliability of each of these lessons, once with all items, and once with *potentially flawed* items removed. Three items were removed from the Text Signals lesson, increasing its reliability from $\alpha = .470$ to $\alpha = .550$, while its average item accuracy (74%) remained the same. Five items were removed from Word Parts, increasing its reliability from $\alpha = .307$ to $\alpha = .62$, decreasing its average accuracy (66% to 62%). One item was removed from Main Ideas, increasing its reliability from $\alpha = .279$ to $\alpha = .340$ with no effect on its average accuracy (67%).

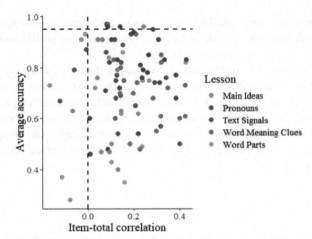

Fig. 1. Item characteristics by lesson. Items to the left of the vertical dashed line are considered *potentially flawed*. Items above the horizontal dashed line are considered *motivational*.

We then created a list of the RISE subtests whose constructs are theoretically aligned with the AutoTutor lessons. We concatenated the item-level data of each lesson-subtest pair. In cases in which more than one subtest was theoretically aligned, we created two separate sets. We then calculated the item-total correlation of each AutoTutor lesson item within its respective lesson-subtest combination. We labeled AutoTutor items with item-total correlations greater than .15 in each pairwise combination as *evaluative items*. We then calculated the reliability of these sets containing RISE subtests plus their *evaluative* AutoTutor items (see Table 1). The remaining unlabeled items were considered *instructive items* because they may have value for learning, but do not map onto the assessment-based constructs in question.

Table 1. Cronbach's α for RISE subtests when combined with each AutoTutor lesson pairwise (baseline RISE reliabilities displayed parenthetically).

	MORPH (α = .889)	SEN (α = .818)	RC (α = .539)
Text signals	0.896 (40%)	0.883 (40%)	
Word parts	0.899 (53%)		
Word meaning clues			0.629 (55%)
Pronouns			0.627 (39%)
Main ideas		0.826 (22%)	0.57 (6%)

4 Discussion

Within the domain of adult literacy education, we have provided examples of how independently developed assessments and ITSs can inform one another to better account for how the characteristics of students and lesson items intersect.

We examined how item characteristics may be leveraged to further integrate assessment and instruction, using industry-standard psychometric analytic techniques, to align item properties to independently valid subtests. We created a taxonomy of lesson items as: *potentially flawed* (psychometrically inconsistent with lesson topic constructs), *motivational* (generally too easy to be informative), *evaluative* (closely related to assessment-oriented skill/knowledge constructs), and *instructive* (consistent with lesson topics, but not external constructs). In support of the validity of the resulting taxonomy, we found that including evaluative items in reliability analyses of construct aligned subtests improved the reliability, supporting the generalization of item-level performance during instruction to specific, psychometrically validated frameworks; analogously, removing flawed items increased reliability of the remaining lesson items. Thus, this taxonomy and analytic frame can be useful to adaptive systems by enhancing assessment precision and instructional content validity.

Future research should more closely examine the most effective use of item and lesson characteristics in real-time ITS, to adapt learning activities and estimate student proficiency as learning progresses. Future research should explore how items embedded in assessments versus learning environments may interact with learner profiles, perhaps predicting which content will be frustrating or challenging to different learners.

References

1. UNESCO: Literacy Rates Continue to Rise from One Generation to the Next. UIS Fact Sheet No, 45. (2017)
2. Greenberg, D.: The Challenges Facing Adult Literacy Programs. Community Lit. J. 3 (2008). https://doi.org/10.25148/clj.3.1.009480
3. Sabatini, J., O'Reilly, T., Dreier, K., Wang, Z.: Cognitive processing challenges associated with low literacy in adults. In: The Wiley Handbook of Adult Literacy, pp. 15–39. Wiley (2019). https://doi.org/10.1002/9781119261407.ch1
4. Comings, J.P., Soricone, L.: Adult literacy research: opportunities and challenges. Boston, MA (2007)
5. Fang, Y., et al.: Patterns of adults with low literacy skills interacting with an intelligent tutoring system. Int. J. Artif. Intell. Educ. 1–26 (2021). https://doi.org/10.1007/s40593-021-00266-y
6. Chen, S., et al.: Automated disengagement tracking within an intelligent tutoring system. Front. Artif. Intell. 3 (2021). https://doi.org/10.3389/frai.2020.595627
7. Sabatini, J., Weeks, J., O'Reilly, T., Bruce, K., Steinberg, J., Chao, S.F.: SARA Reading Components Tests, RISE Forms: Technical Adequacy and Test Design, 3rd edn. ETS Research Report Series 2019, pp. 1–30 (2019). https://doi.org/10.1002/ets2.12269

More Powerful A/B Testing Using Auxiliary Data and Deep Learning

Adam C. Sales[1]([⊠]), Ethan Prihar[1], Johann Gagnon-Bartsch[2], Ashish Gurung[1], and Neil T. Heffernan[1]

[1] Worcester Polytechnic Institute, Worcester, MA 01609, USA
asales@wpi.edu
[2] University of Michigan, Ann Arbor, MA 48109, USA

Abstract. Randomized A/B tests allow causal estimation without confounding but are often under-powered. This paper uses a new dataset, including over 250 randomized comparisons conducted in an online learning platform, to illustrate a method combining data from A/B tests with log data from users who were not in the experiment. Inference remains exact and unbiased without additional assumptions, regardless of the deep-learning model's quality. In this dataset, incorporating auxiliary data improves precision consistently and, in some cases, substantially.

1 Introduction

In randomized A/B tests on an online learning platform, students are randomized between different educational conditions and their subsequent outcomes are compared. Estimates from A/B tests are unbiased, but may be imprecise due to small sample sizes. An observational study can often boast a larger sample size but is subject to confounding so conventional analysis of A/B tests discards data from the "remnant" of the experiment—students who were not randomized, but for whom covariate and outcome data are available.

However, data from the remnant an can play a valuable role in causal estimation. [2] suggests first using the remnant data to train a model using covariates to predict outcomes; then, using that fitted model to predict (or impute) outcomes for participants in the experiment. Finally, use those imputations as a covariate in a causal effect estimator. This method builds on recent work in design-based covariate adjustment, e.g. [5], and in particular, using the remnant to improve precision [e.g.] [1]. Unfortunately, [2] provides only limited evidence of the method's success in practice.

This paper reviews two of the causal estimators of [2], and applies them to an new dataset: a collection of 84 multi-armed A/B tests run on the ASSIST-ments TestBed [3], which together include 377 different two-way comparisons, and 41,226 students. Alongside this experimental data, we collected log data for an additional 193,218 students who worked on similar skill builders in ASSIST-ments but did not participate in any of the 84 experiments—the remnant. We used these datasets to estimate the causal effects of each of the conditions on

© Springer Nature Switzerland AG 2022
M. M. Rodrigo et al. (Eds.): AIED 2022, LNCS 13356, pp. 524–527, 2022.
https://doi.org/10.1007/978-3-031-11647-6_107

assignment completion. Our interest here is not on the treatment effects themselves, but on the extent to which these methods reduce standard errors. Our results give a much clearer picture of the potential impacts of using remnant data in design-based causal inference: incorporating remnant data consistently improves statistical precision, sometimes substantially.

2 Method

For each subject i in a randomized experiment, let $Z_i = 1$ if i is randomized to the treatment condition and $Z_i = 0$ if i is randomized to control, and let Y_i be the outcome of interest. Following [4], define y_i^c and y_i^t as the outcomes i would have exhibited had i been assigned to control or treatment, respectively. Then, assuming no spillover effects, $Y_i = Z_i y_i^t + (1 - Z_i) y_i^c$, and the treatment effect for student i is $\tau_i \equiv y_i^t - y_i^c$.

Let $\boldsymbol{x}_i$ be a $k \times 1$ vector of baseline covariates for subject i, and let $\hat{y}^c(\cdot)$ and $\hat{y}^t(\cdot)$ be functions from $\mathbb{R}^k \to \mathbb{R}^1$ that impute y_i^c and y_i^t, respectively, as a function of $\boldsymbol{x}_i$. Finally, if $Pr(Z = 1) = 1/2$, let $m_i = 1/2(y_i^c + y_i^t)$, subject i's expected counterfactual potential outcome, and let $\hat{m}_i = 1/2(\hat{y}^c(\boldsymbol{x}_i) + \hat{y}^t(\boldsymbol{x}_i))$ be it's estimate. Then, if $\hat{y}^c(\cdot)$ and $\hat{y}^t(\cdot)$ are constructed such that $\{\hat{y}^c(\boldsymbol{x}_i), \hat{y}^t(\boldsymbol{x}_i)\} \perp\!\!\!\perp Z_i$, then

$$\hat{\tau} = \frac{1}{n} \sum_{i \in \mathcal{T}} \frac{Y_i - \hat{m}_i}{p} - \frac{1}{n} \sum_{i \in \mathcal{C}} \frac{Y_i - \hat{m}_i}{1 - p} \tag{1}$$

is an unbiased estimate for $\bar{\tau}$. In fact, this unbiasedness holds regardless of $\hat{y}^c(\cdot)$ or $\hat{y}^t(\cdot)$'s other properties—they need not be unbiased, or consistent, or correct in any sense for $\hat{\tau}$ to be unbiased.

[2] combines two approaches to ensuring that $\{\hat{y}^c(\boldsymbol{x}_i), \hat{y}^t(\boldsymbol{x}_i)\} \perp\!\!\!\perp Z_i$: the first uses a leave-one-out algorithm using observations other than i to train models $\hat{y}^c_{-i}(\cdot)$ and $\hat{y}^t_{-i}(\cdot)$ that will in-turn give rise to imputations $\hat{y}^c(\boldsymbol{x}_i)$ and $\hat{y}^t(\boldsymbol{x}_i)$ and finally m_i. As long as $Z_i \perp\!\!\!\perp Z_j$ for $i \neq j$, then $\{\hat{y}^c(\boldsymbol{x}_i), \hat{y}^t(\boldsymbol{x}_i)\} \perp\!\!\!\perp Z_i$ will hold.

The second approach uses the remnant to train a different model, $\hat{y}^r(\cdot)$, producing imputations $x^r \equiv \hat{y}^r(\boldsymbol{x}_i)$. Importantly, x^r is a baseline covariate, unaffected by treatment assignment, since it is a function of baseline covariates $\boldsymbol{x}$ and a model fit to a separate sample. Therefore, it can be incorporated into an estimator such as (1), perhaps alongside other covariates. If $\hat{y}^r(\cdot)$ performs well in the experimental sample, so that $|x^r - y_i^c|$ tends to be small, then doing so can drastically improve precision; in the limit, if $x^r = y_i^c$ for all i, then the standard error of $\hat{\tau}$ would be due only to treatment effect heterogeneity, and the average effect on treated subjects would be known exactly. On the other hand, if $\hat{y}^r(\cdot)$ does not perform well it will not threaten the validity of the inference, and in large samples it will not harm precision.

Here, we include two specific versions of $\hat{\tau}$: first, $\hat{\tau}^{SS}[x^r, \text{LS}]$ uses x^r as the only covariate and uses ordinary least squares linear regression (OLS) for leave-one-out imputation models $\hat{y}^c(\cdot)$ and $\hat{y}^t(\cdot)$. We expect that when $\hat{y}^r(\cdot)$ performs well, OLS will be optimal since the relationship between Y and x^r will be approximately linear. Second, $\hat{\tau}^{SS}[\tilde{\boldsymbol{x}}, \text{EN}]$ uses x^r alongside a vector of other covariates

$\boldsymbol{x}$; leave-one-out imputation models $\hat{y}^c(\cdot)$ and $\hat{y}^t(\cdot)$ are ensembles of OLS regression of Y on x^r and a random forest imputing Y from both $\boldsymbol{x}$ and x^r.

3   Application

We gathered a set of 84 A/B tests run on the TestBed with assignment completion as a binary outcome. We also gathered standard student-level aggregated predictors. Several experiments included multiple conditions; in those cases we estimated each pairwise contrast separately, as long as the p-value testing $Pr(Z = 1/2)$ was greater than 0.1.

We used remnant data to train a deep learning model $\hat{y}^r(\cdot)$ imputing completion from covariates. Three different sets of data were collected for each sample in the datasets: prior student statistics, prior assignment statistics, and prior daily actions. The full dataset used in this work can be found at https://osf.io/ k8ph9/?view_only=ca7495965ba047e5a9a478aaf4f3779e. Each of the three types of data in the remnant dataset were used to predict both skill builder completion and number of problems completed for mastery. a fourth neural network was trained using a combination of the previous three models. The details and code can be found at https://github.com/adamSales/reloop377abTests. We used this fourth model, $\hat{y}^r(\cdot)$, to construct imputations x^r for each subject i in each experiment.

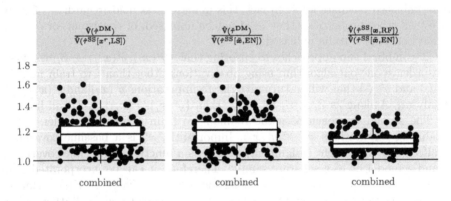

Fig. 1. Boxplots and jittered scatter plots of the ratios of estimated sampling variances of $\hat{\tau}^{DM}$, $\hat{\tau}^{SS}[x^r; OLS]$, $\hat{\tau}^{SS}[\boldsymbol{x}; RF]$, and $\hat{\tau}^{SS}[\tilde{\boldsymbol{x}}; EN]$

Figure 1 gives boxplots of ratios of estimated sampling variances $\hat{\mathbb{V}}(\cdot)$ for causal estimates: $\hat{\tau}^{DM}$, the Welch two-sample t-test, $\hat{\tau}^{SS}[\boldsymbol{x}, RF]$, the leave-one-out estimator using student-level covariates but no information from the remnant, and the two new estimators, $\hat{\tau}^{SS}[x^r, LS]$ and $\hat{\tau}^{SS}[\tilde{\boldsymbol{x}}, EN]$. The left and middle panels including remnant-based imputations is equivalent to increasing the sample size, relative to a t-test, by a factor of about 10–25% in about half of all cases, but up to 50%–70% in the most extreme cases.[1] The right panel shows that compared to $\hat{\tau}^{SS}[\boldsymbol{x}; RF]$, including remnant based imputations was equivalent to increasing the sample size by roughly 8–12% in half of all cases, but as much as 30% in others.

4 Discussion

The approach illustrated here shows that data that do not meet an assumption—randomization—can still be used to help learn connections between covariates and outcomes. Its causal estimates will be unbiased, and inference correct, regardless of the data quality or model properties in the remnant. However, better data and better model fit will lead to better precision. The results in the ASSISTments A/B tests show that it sometimes improves precision greatly, and sometimes barely at all. Future research will explain this variance, as well as formulate suitable defaults and recommendations for when and how it should be used.

Acknowledgements. This work was supported by IES grant #R305D210031.

References

1. Deng, A., Xu, Y., Kohavi, R., Walker, T.: Improving the sensitivity of online controlled experiments by utilizing pre-experiment data. In: Proceedings of the Sixth ACM International Conference on Web Search and Data Mining, pp. 123–132 (2013)
2. Gagnon-Bartsch, J.A., et al.: Precise unbiased estimation in randomized experiments using auxiliary observational data. arXiv preprint arXiv:2105.03529 (2021)
3. Ostrow, K.S., Selent, D., Wang, Y., Van Inwegen, E.G., Heffernan, N.T., Williams, J.J.: The assessment of learning infrastructure (ALI): the theory, practice, and scalability of automated assessment. In: Proceedings of the Sixth International Conference on Learning Analytics & Knowledge, pp. 279–288. ACM (2016)
4. Rubin, D.: Estimating causal effects of treatments in randomized and nonrandomized studies. J. Educ. Psychol. **66**(5), 688 (1974)
5. Wu, E., Gagnon-Bartsch, J.A.: The loop estimator: adjusting for covariates in randomized experiments. Eval. Rev. **42**(4), 458–488 (2018)

---

[1] Since sampling variance is typically $\propto 1/n$, ratios of sampling variances can be interpreted as ratios of effective sample sizes.

ARIN-561: An Educational Game for Learning Artificial Intelligence for High-School Students

Ning Wang[1(✉)], Eric Greenwald[2], Ryan Montgomery[2], and Maxyn Leitner[1]

[1] University of Southern California, Los Angeles, CA, USA
nwang@ict.usc.edu
[2] University of California, Berkeley, CA, USA

Abstract. Artificial Intelligence (AI) is increasingly vital to our future generations, who will join a workforce that utilizes AI-driven tools and contributes to the advancement of AI. Today's students will need exposure to AI knowledge at a younger age. Relatively little is currently known about how to most effectively provide AI education to K-12 students. In this paper, we discuss the design and evaluation of an educational game for high-school AI education called ARIN-561. Results from pilot studies indicate the potential of ARIN-561 to build AI knowledge, especially when students spend more time in the game.

Keywords: K-12 AI education · Youth AI education · Game-based learning · Educational games

1 Introduction

Artificial Intelligence (AI) is profoundly transforming our workforce around the globe. It is critical to prepare future generations with basic knowledge of AI, beginning with childhood learning. Given the limited research on this topic, currently there is little possibility of grounding the design of learning experiences in evidence-based accounts of how youth learn AI concepts, how understanding progresses across concepts, or what concepts are most appropriate for what age-levels. Given the packed course schedule of K-12 students, being able to connect AI learning to existing Science, Technology, Engineering and Mathematics (STEM) subjects becomes a more realistic approach to embed AI education in K-12 classrooms. Digital game-based learning (DGBL) is a technology-based approach that has shown promise in promoting student learning, including math and problem-solving skills [3]. There is currently very little research into educational games for youth AI education [1]. In this paper, we will discuss the design and evaluation of an educational game, called ARIN-561, for teaching high-school students about AI. We conducted a series of evaluation studies at high schools in the United States. Results indicate the potential of ARIN-561 to build AI knowledge, especially when students spend more time in the game.

© Springer Nature Switzerland AG 2022
M. M. Rodrigo et al. (Eds.): AIED 2022, LNCS 13356, pp. 528–531, 2022.
https://doi.org/10.1007/978-3-031-11647-6_108

2   ARIN-561

The educational game we have developed, ARIN-561, is designed to teach high-school students AI concepts, prompt them to apply their math knowledge, and develop their AI problem-solving skills. In the game, students play the role of a scientist who sets out on a scientific expedition, but unfortunately crash-lands on an alien planet. In order to safely return home, the scientist begins exploring the planet to gather resources needed to repair the broken ship while uncovering the mystery of the planet. The current implementation of the ARIN-561 game focuses on developing concepts around classical search algorithms (e.g., Breadth-First Search, Greedy Search, etc.). In-game challenges such as searching for missing spaceship parts or cracking passwords serve as natural opportunities for the introduction of search as a topic. The essential concepts, such as space and time complexity of search algorithms, lend opportunities to connect AI to math knowledge familiar to high-school students. Activities in the game aim to achieve three learning goals. The first goal is to develop understanding of how AI algorithms are used to solve problems in the real world. We take the approach of designing AI problem-solving in the game that mirrors real-world AI applications. The second goal is to learn how to weigh the strengths and weaknesses of AI algorithms in order to choose between them for problem-solving. In ARIN-561, each new AI algorithm is introduced as excelling at a task that previous algorithms are less suitable for. As the students progress through the game, further comparisons between the AI algorithms are prompted, pushing the students to take more agency in deciding which one is appropriate for the task at hand. The third goal is for the students to gain high-level understanding of how each AI algorithm works, which is achieved through the difficulty progression of the game-play. For each search algorithm, for example, the students are first provided with a tutorial task that teaches them how the algorithm works, and then walked through the task step by step, with less scaffolding as they progress. Subsequently, the students are presented with a transfer problem from a domain different from the tutorial's that requires the students to apply what they have learned in the tutorial. Students are provided with less tutorial support during this task and need to apply internalized understanding of how the algorithms they have learned work. Embedded in all the tutorial and transfer modules are quizzes that help students pause and self-assess. The game pauses as the students answer the quiz question and continues when a correct answer is recorded. The quiz questions are part of the in-game dialogue, aligned with the narrative.

3   Evaluation

To assess how ARIN-561 impacts AI learning for high-school students, we carried out a series of pilot studies in computer science classes at three high schools in the United States. The study is designed to fit in 3 to 4 class sessions (45–55 min long each). In the first session, students completed an online pre-survey, which consisted of items about demographic background, AI Use Type, Interest in AI,

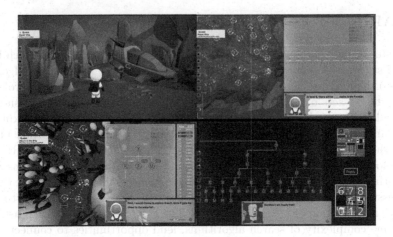

Fig. 1. Screenshots from ARIN-561. Top-left: the player character crash lands on a foreign planet. Top-right: the student is presented with a quiz question in a tutorial. Bottom left: the student is scaffolded through the greedy search algorithm in a tutorial. Bottom-right: the student solves an 8-puzzle as a transfer problem.

AI Knowledge, Math Self-efficacy [2], and Math Knowledge. All scales except the Math Self-efficacy were developed by the research team. The AI Use Type included items such as "When I think about how I'd like to interact with AI in the future, I expect that: I will use AI systems in my everyday life as a consumer, and I expect to USE AI systems as a part of my job." The Interest in AI scale included questions such as "Outside of school I try to learn a lot about AI." The assessment of AI knowledge and math knowledge focused on the content covered in ARIN-561. During the second and third sessions, students interacted with ARIN-561 online at their own pace. During the fourth session, students completed a post-survey, which included the same items on interest in AI and AI knowledge from the pre-survey. In addition to the surveys, game logs from ARIN-561 were collected, including the in-game click-stream data and responses to in-game quizzes (Fig. 1).

4   Results

A total of 125 students participated in the studies. The participants' average age was 16.1 years old. A total of 73% of the students were male, 21% were female and 6% identified as other categories or preferred not to disclose. With restricted access to the school campus due to COVID-19, the data collection was carried out entirely by the participating teachers, without participation of the research team. As a result, 60 out of the 125 students did not complete the assessment of AI knowledge on either pre- or post-survey. Missing data were excluded from the corresponding analysis.

We hypothesized that interacting with ARIN-561 would help students gain knowledge in AI. Thus we conducted a paired-sample t-test to analyze the changes in AI Knowledge from pre- to post-survey. There were a total of 15 questions on AI knowledge (15 points total). Data from all students who completed both pre- and post- AI knowledge assessment ($N = 65$) showed a positive, though not statistically significant increase of AI knowledge ($M = 0.427, SD = 2.819, t(64) = 1.221, p = .227$). Given the varied completion rate of pre- and post-survey, we further examined the game logs from ARIN-561. In particular, students who completed less than half of the game modules (2 or fewer of 6 modules) were then excluded before we repeated the paired-sample t-test on the group of students who completed half or more of the game modules ($N = 47$). Results indicated that, the group of students who completed at least half of the game demonstrated a statistically significant ($M = 1.0638, SD = 2.637, t(46) = 2.765, p = .008$) positive change in AI knowledge, with a mean difference of 1.0638 and a medium effect size ($d = 0.403$). Additionally, a one-way ANOVA comparing students who completed half or more modules ($M = 1.0638$) and those who completed less than half game modules ($M = -1.5469$) revealed a statistically significant difference in the changes in AI knowledge between the two groups ($F(1, 61) = [11.737], p = .001$).

5   Discussion

This paper presents our approach to designing a game-based AI learning environment for high-school-aged youth and presents evidence for how the game may be contributing to AI learning among players. We observed statistically significant learning gains among students who completed at least half of the game. This suggests that the ARIN-561 educational game can support AI learning for high-school-aged youth, and in order to realize these potential gains, youth should engage in sufficient learning in the game. Given the stark difference between outcomes for those students who completed at least half of the game compared to those who did not, further analyses of game log data are needed to better understand how in-game behaviors may be contributing to learning gains, beyond the dosage effect reported here.

Acknowledgement. This research was supported by the National Science Foundation (NSF) under Grant #1842385. Any opinions and finding expressed in this material are those of the authors and do not necessarily reflect the views of the NSF.

References

1. Lee, S., et al.: AI-infused collaborative inquiry in upper elementary school: a game-based learning approach. In: Proceedings of the 11th Symposium on Education Advances in Artificial Intelligence, vol. 35, pp. 15591–15599 (2021)
2. Liu, X., Koirala, H.: The effect of mathematics self-efficacy on mathematics achievement of high school students (2009)
3. Plass, J.L., Mayer, R.E., Homer, B.D.: Handbook of Game-Based Learning. MIT Press, Cambridge (2020)

Protecting Student Data in ML Pipelines: An Overview of Privacy-Preserving ML

Johannes Schleiss[(✉)], Kolja Günther, and Sebastian Stober

Otto von Guericke University, Magdeburg, Germany
johannes.schleiss@ovgu.de

Abstract. The rise of Artificial Intelligence in Education opens up new possibilities for analysis of student data. However, the protection of private data in these applications is a major challenge. According to data regulations, the application designer is responsible for technical and organizational measures to ensure privacy. This paper aims to guide developers of educational platforms to make informed decisions about their use of privacy-preserving ML and, therefore, protect their student data.

Keywords: Student privacy · Safe learning analytics · Privacy protection · Data privacy · Privacy attacks

1 Introduction

Artificial Intelligence in Education (AIED) is on the rise and provides a set of new powerful tools to analyze student data [13]. At the same time, AIED makes educational systems and their underlying data a target of privacy attacks [9]. Current data regulations, such as the EU GDPR[1] and the FERPA[2] in the US, put the protection of data privacy in the hands of those entities that process the data and gain value from it. This results in additional privacy requirements for AIED applications and calls for new solutions that integrate the privacy aspect by design [4,9].

In the context of Machine Learning (ML) systems, the field of privacy-preserving ML discusses approaches to protect these applications from external privacy threats [8,12,15]. Privacy-preserving ML refers to the modification of ML processes and properties to ensure the protection of sensitive information. Sensitive information can be used to identify and possibly harm a unique person and is, therefore, the main target of privacy threats. In the educational setting, the most sensitive information lies in the student data. In general, privacy attacks in ML either aim to gather sensitive information or steal model parameters and features. In this context, Liu et al. [8] introduced the terms *Model Privacy* and *Training Data Privacy*, which we will refer to as *Data Privacy*.

---

[1] General Data Protection Regulation (GDPR) available at https://gdpr-info.eu/.
[2] Family Educational Rights and Privacy Act (FERPA) (20 U.S.C. §1232g; 34 CFR Part 99).

© Springer Nature Switzerland AG 2022
M. M. Rodrigo et al. (Eds.): AIED 2022, LNCS 13356, pp. 532–536, 2022.
https://doi.org/10.1007/978-3-031-11647-6_109

Our paper aims to provide an accessible and high-level overview of privacy-preserving ML in the context of educational data. The work enables designers of intelligent educational applications to keep up with the recent developments in privacy-preserving ML, including privacy risks and defences.

2   Attacks Harming the Privacy in ML

There is a broad range of attacks violating the integrity of ML models. A subgroup of these target either to gather sensitive information from training data or from the model itself. Prominent types of attacks towards the *Data Privacy* are *Linkage-*, *Membership Inference-*, *Reconstruction-*, *Property Inference-*, and *Model Memorization-Attacks*, whereas *Model Extraction-Attacks* violate the *Model Privacy* [8,12].

Linkage Attacks overcome traditional anonymization by joining data from multiple sources via *quasi-identifiers*, consequently receiving formerly hidden information [8]. With *Membership Inference*, an attacker aims to determine whether a specific instance was part of the training data or not [8]. *Reconstruction-Attacks* aim to reveal statistical properties of the training dataset to subsequently reconstruct the data and its features [12]. Similarly, *Property Inference-Attacks* aim to extract global sensitive properties of the training data which were learned by the model and have high relevance in the data itself but do not contribute to the learning task [12]. *Model Memorization-Attacks* rely on a malicious model owner, who aims to encode sensitive information either into model parameters or additional augmented training data [8]. In terms of *Model Privacy*, *Model Extraction-Attacks* aim to duplicate the target model by creating a surrogate function that resembles the model's original objective function [12]. Using the duplicated model can then be used to optimize and facilitate follow-up attacks.

3   Defense Mechanisms

After introducing common attack types, the following section gives a comprehensive summary of appropriate defence mechanisms.

Data-oriented defence mechanisms aim to obfuscate model training data to increase privacy. One mechanism is **t-closeness** [7] which provides anonymization rules to eliminate the risk of possibly revealing information via *quasi-identifiers*. Based on the *k-anonymity* and *l-diversity* approach, a dataset is said to be private if there are sets of at least k instances with the same attribute values which simultaneously have at least l expressions of the sensitive attribute. Building upon this, *t-closeness* itself aims to create the sets of k instances in a way, that the distribution of sensitive attribute values in each set is as similar as possible to the distribution of the whole dataset [7].

Another defence mechanism is **Differential Privacy** which was initially introduced for database systems [2] but is now also widely used in ML and Deep Learning (DL) scenarios. The core idea is to apply a mathematical verification about privacy guarantees of adding random noise to data holding sensitive information. The achieved privacy can be quantified by parameter ε which determines the trade-off between *Data Privacy* and *Data Utility*, the extent to which the data is not too noisy and still meaningful. *Differential Privacy* can either be used to add noise to the input data, the model's output or to obfuscate the gradient in a DL setting.

A further approach for protecting sensitive information in ML is to create **synthetic surrogate data** from the original training datasets. The synthetic data is generated by inspecting the statistical properties of the original dataset and generating random instances with the same distribution, for example through *Generative Adversarial Networks* [14].

With **Homomorphic Encryption** [3] and **Functional Encryption** [1], two prominent methods for privacy-preserving encryption found applications in ML. *Homomorphic Encryption* allows to perform operations over encrypted inputs while the output of the model stays encrypted and resembles the result that would be computed on the original raw data. *Functional Encryption* allows the model to calculate decrypted results from encrypted data without knowing the encryption key or having direct access to the actual data.

Architectural-based approaches aim to optimize a model architecture towards higher privacy guarantees. In a centralized learning setting, clients (e.g. mobile devices) provide their local data to a central server which feeds it into a centralized model. Here, the data owners have no longer control over attacks that possibly steal sensitive information. Contrarily, **Federated Learning** [10] is a distributed learning technique where the training data is not gathered on a central server but remains on several data source clients, which learn local copies of the centralized model with their data and send back an aggregated update.

Another possible architectural optimization is introduced with **Private Aggregation of Teacher Ensembles** (PATE) [11] which utilizes a teacher- and student-model setup to protect sensitive information in the model training and deployment phases. Instead of training just one model on sensitive data, PATE trains an ensemble of hidden teacher models on distinct subsets of the data. The student model to be published trains on an unlabeled fraction of public, nonsensitive data and a fraction labelled by the majority vote of the teacher ensemble in a semi-supervised fashion. As the student model only learns on non-sensitive data and the teacher models are hidden, PATE also provides protection against *Model Extraction Attacks*.

There are several defenses that **reduce risks of query-based attacks** (*Membership Inference*, *Property Inference* and *Model Extraction*). A simple but effective defence mechanism can be to **limit the number of query accesses** in the model inference step. Similarly, assuming that these attacks require several queries to successfully reveal model parameters, **Protecting against DNN Model Stealing Attacks** (PRADA) [5] exploits that adversarial query sequences are optimized to gather maximal information about the model and therefore differ from distributions of harmless query sequences. *PRADA* uses this observation to detect queries of one user differing from these distributions.

Moreover, Krishna et al. [6] proposed **Membership Inference Against Model Stealing**, a method to discriminate between queries made by casual users and those which belong to a *Model Extraction* query sequence. The method is similar to *PRADA*, where *Membership Inference* works as a binary classifier between 'good' and 'bad' queries. For a 'bad' query, the model provides some random output instead of the correct one and consequently, the adversary receives useless information.

4   Protecting Student Data in ML Pipelines

With the rise of AIED, privacy issues related to models and data will occur more and more frequently. Data used in AIED is sensitive in the sense that it can contain information about the student's learning behaviour but also insights into individual personal attributes and demographics. To act according to the data regulations, it is the responsibility of the application developers to ensure privacy protection. It is therefore important that they are aware of current privacy threats and attacks on the model and the underlying student data.

In this paper, we summarized the state-of-the-art privacy-preserving ML to guide developers of educational platforms to make informed decisions about their use of privacy-preserving ML. Further investigations will incorporate more educational use cases, especially concerning the sensitive information in the educational domain. Another important point is to further develop the legal and ethical guidelines, especially considering that a purely technical-driven privacy development does not prevent algorithmic bias or the secondary use of data.

References

1. Boneh, D., Sahai, A., Waters, B.: Functional encryption: definitions and challenges. In: Ishai, Y. (ed.) TCC 2011. LNCS, vol. 6597, pp. 253–273. Springer, Heidelberg (2011). https://doi.org/10.1007/978-3-642-19571-6_16
2. Dwork, C., McSherry, F., Nissim, K., Smith, A.: Calibrating noise to sensitivity in private data analysis. In: Halevi, S., Rabin, T. (eds.) TCC 2006. LNCS, vol. 3876, pp. 265–284. Springer, Heidelberg (2006). https://doi.org/10.1007/11681878_14
3. Gentry, C.: Fully homomorphic encryption using ideal lattices. In: Proceedings of the Forty-First Annual ACM Symposium on Theory of Computing, pp. 169–178 (2009)
4. Hoel, T., Griffiths, D., Chen, W.: The influence of data protection and privacy frameworks on the design of learning analytics systems. In: Proceedings of the Seventh International LAK Conference, pp. 243–252. ACM (2017)
5. Juuti, M., Szyller, S., Dmitrenko, A., Marchal, S., Asokan, N.: Prada: protecting against DNN model stealing attacks. In: 2019 IEEE EuroS&P, pp. 512–527 (2019)
6. Krishna, K., Tomar, G.S., Parikh, A.P., Papernot, N., Iyyer, M.: Thieves on sesame street! model extraction of BERT-based APIS. arXiv abs/1910.12366 (2020)
7. Li, N., Li, T., Venkatasubramanian, S.: t-closeness: Privacy beyond k-anonymity and l-diversity. In: IEEE 23rd International Conference on Data Engineering, pp. 106–115 (2007)

8. Liu, B., Ding, M., Shaham, S., Rahayu, W., Farokhi, F., Lin, Z.: When machine learning meets privacy a survey and outlook. ACM Comput. Surv. **54** (2021)
9. Marshall, R., Pardo, A., Smith, D., Watson, T.: Implementing next generation privacy and ethics research in education technology. Br. J. Educ. Technol. (2022)
10. McMahan, B., Moore, E., Ramage, D., Hampson, S., Arcas, B.A.: Communication-efficient learning of deep networks from decentralized data. In: Artificial Intelligence and Statistics, pp. 1273–1282 (2017)
11. Papernot, N., Abadi, M., Erlingsson, U., Goodfellow, I., Talwar, K.: Semi-supervised knowledge transfer for deep learning from private training data. arXiv abs/1610.05755 (2016)
12. Rigaki, M., García, S.: A survey of privacy attacks in machine learning. arXiv abs/2007.07646 (2020)
13. Romero, C., Ventura, S.: Educational data mining and learning analytics: an updated survey. Wiley Interdiscipl. Rev. Data Min. Knowl. Discov. **10**(3), e1355 (2020)
14. Triastcyn, A., Faltings, B.: Generating artificial data for private deep learning. arXiv abs/1803.03148 (2019)
15. Xu, R., Baracaldo, N., Joshi, J.: Privacy-preserving machine learning: methods, challenges and directions. arXiv abs/2108.04417 (2021)

Distributional Estimation of Personalized Second-Language Vocabulary Sizes with Wordlists that the Learner is Likely to Know

Yo Ehara[✉]

Tokyo Gakugei University, Koganei, Tokyo 1848501, Japan
ehara@u-gakugei.ac.jp

Abstract. Vocabulary is essential for second language learners to read documents. Prior studies on vocabulary size estimation, especially in applied linguistics, have been based on the naive assumption that language learners acquire words based on the frequency of balanced corpora. As this assumption is often invalid, vocabulary size estimates have been somewhat inaccurate for educational use. Therefore, this study proposes a novel method for the distributional and more informative estimation of second language vocabulary size from vocabulary test results. To this end, our method makes personalized binary estimations regarding whether a learner knows a word, followed by aggregating the estimation results for many words for a learner. Experimental results using large vocabulary tests showed that our methods provided distributional and more informative estimations compared with baseline methods.

Keywords: Distributional estimation · Vocabulary size · Second language vocabulary

1 Introduction

Vocabulary plays a significant role in the readability of documents for second language learners. Previous studies in applied linguistics have shown that to read and understand a document, a second language learner needs to understand 95% to 98% of the tokens in the document [5,6]. Generally, the more words the learner knows, the more text the learner can read. Therefore, many studies have focused on estimating the vocabulary size of learners. However, the estimation does not directly identify the tokens the learner knows in a document. In this regard, most previous studies naively assume that we can use vocabulary size to easily identify the words the learner knows using the frequency ranking of a large corpus.

Figure 1 shows a motivating example based on a self-report vocabulary dataset [3]: the horizontal axis shows the frequency rank of each word in the BNC (http://www.natcorp.ox.ac.uk/) corpus, and the vertical axis is 1 if the learner indicated knowing the word and 0 if not. This learner responded that

M. M. Rodrigo et al. (Eds.): AIED 2022, LNCS 13356, pp. 537–541, 2022.
https://doi.org/10.1007/978-3-031-11647-6_110

Fig. 1. Probability of knowing words vs. frequency ranks of words in BNC. (Color figure online)

Fig. 2. Overview of previous and proposed scoring methods.

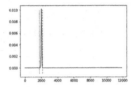

Fig. 3. Our experimental setting.

Fig. 4. Estimation plot. The gold dashed line is the gold vocabulary size, the dotted black line is the baseline (VST), and the blue line is our distributional estimation. (Color figure online)

he/she knows $4,866$ of $11,847$ words. Figure 1 shows that the $4,866$-th word, marked by the red dotted line, does not clearly separate the words known to this learner.

This study proposes a novel method to estimate vocabulary size in a distributional manner from learners' vocabulary test results (Fig. 2). Our method is distributional and *content-aware*; that is, in addition to the vocabulary size of a learner, our method can calculate the probability that a word of interest is included in the learner's vocabulary. Suppose we have five words, a, b, c, d, and e, in a language. Figure 2 shows that the learner's vocabulary size would be estimated as 3 using previous scoring. Our method is more informative and provides vocabulary size as a distribution. In Fig. 2, the horizontal and vertical axes show the estimated vocabulary size and probability of each size, respectively. Although our method similarly shows that 3 is the most probable vocabulary size, it also shows that vocabulary sizes of two or four are reasonable, with probabilities of 20% and 30%, respectively. Moreover, our method can provide content for each vocabulary size. Figure 2 shows that the probability of a vocabulary size of three is 40%, but among those three words, the learner has a 15% probability of knowing $\{a,b,c\}$, a 15% of knowing $\{a,c,d\}$, and a 10% of knowing $\{a,b,d\}$.

2   Proposed Method and Experiments

Our method is an extension of the work of [2], which is based on the subset-sum problem [4], a special case of the knapsack problem. In fact, by setting $f_i = 1$ in their algorithm, we can obtain the distribution of V_j^{est}. Although [2] proposed a method that obtains distributional results similar to our method (Fig. 2), they did not provide a way to determine the probabilities of different word combinations within a vocabulary size. We consider I words and J learners where i is the index of the word and j is that of the learner. Let $Y_{i,j}$ represent the gold-standard response of learner j for word i. $Y_{i,j}$ is 1 only if learner j responds that they know word i; otherwise, $Y_{i,j}$ is 0. Then, the *gold-standard vocabulary size* of learner j is denoted by $V_j^{\text{gold}} := \sum_{i=1}^{I} Y_{i,j}$. Suppose that we have a probabilistic binary classifier c. Given learner j and word i, c outputs the probability that learner j knows word i as $p_{i,j}^c$. For example, $p_{i,j}^c = 0.6$ means that c outputs the probability that learner j knows that word i is 0.6. Then, we define $\hat{Y}_{i,j}^c$ as a *Bernoulli* random variable with $p_{i,j}^c$. That is, $\hat{Y}_{i,j}^c$ takes 1 as the probability of $p_{i,j}^c$ and 0 as the probability of $1 - p_{i,j}^c$. That is, we simply let $\hat{Y}_{i,j}^c \sim Bernoulli(p_{i,j}^c)$. We note that $\Pr(\hat{Y}_{i,j}^c = 1) = p_{i,j}^c$. $\hat{Y}_{i,j}^c$ is a random variable that takes 0 or 1 whereas $p_{i,j}^c$ is a probability value that takes a value in the range of $[0,1]$. Using $\hat{Y}_{i,j}^c$, we can express the *estimated vocabulary size*. For example, focusing on words i_1 and i_2 and considering $\hat{Y}_{i_1,j}^c + \hat{Y}_{i_2,j}^c$. $\hat{Y}_{i_1,j}^c + \hat{Y}_{i_2,j}^c$ is 2 when learner j knows both i_1 and i_2, 1 when learner j knows either i_1 or i_2, and 0 when learner j does not know both i_1 and i_2. Thus, $\hat{Y}_{i_1,j}^c + \hat{Y}_{i_2,j}^c$ denotes the number of words that learner j knows in word set $\{i_1, i_2\}$. The vocabulary size is the number of words that learner j knows in the entire word set considered, namely, $\{1, \ldots, I\}$. Hence, the *estimated vocabulary size* of learner j by classifier c is $V_j^{\text{est}} := \sum_{i=1}^{I} \hat{Y}_{i,j}^c$. Notably, V_j^{est} is a random variable; hence, it is distributional as shown in Fig. 2. We explain the calculation of the cumulative distribution of $V_{i,j}^{\text{est}}$, which is the cumulative height of each bar, and N is the possible vocabulary size for learner j in Fig. 2. As in [2], we use the *subset-sum* problem, which calculates the probability that the subsets of a given array of integers sum up to exactly the given *target-sum*. The cumulative distribution of the probabilities that learner j knows the i-th word in a total of I words can be obtained by solving the subset-sum problem $I + 1$ times by changing the target sum from 0 to I. The bottom of Fig. 2 shows V_j^{est} for $I = 3$.

Experiments: We used second-language vocabulary dataset [3]. This consists of 11,999 words that were self-reported by 16 learners. Figure 3 depicts our experimental setting. The left rectangle is the learner-word matrix: the rows are learners, the columns are words, and the element is 1 if the learner knows the word and 0 otherwise. The right rectangle denotes a vector wherein each element is the learner's vocabulary size. The colored areas represent the training and development data, and the hatched areas represent the test data. Our goal is to estimate the vocabulary size, i.e., the hatched area of the right rectangle, using only the training and development data. Let I_T denote the number of words used in the training dataset. We set $I_T = 100$, $I = 10,000$, and $J = 16$; the results are summarized in Table 1.

In Table 1, **VST** [6], or the vocabulary size test is the baseline for our proposed methods. It estimates the vocabulary size by multiplying I/I_T by the

Table 1. Predictive performances.

Method	RMSE	Accuracy
VST	442.4	–
LR+C	**399.4**	0.7478
NN+C	1760.6	**0.7480**

Table 2. Estimated words known only by skilled learners.

'lieutenant', 'meander', 'exasperate', 'cram', 'recourse', 'conclusive', 'manifesto', 'forefront', 'dogmatic', 'omniscient', 'underwear'

number of words the learner knows in the training data and assume that the learner knows all words more frequent than the vocabulary size. Among our proposals, logistic regression (**LR**) and a neural network (**NN**, namely the multilayer perceptron) are the binary classifiers used in our aggregation, which were trained by the responses. Our implementation uses *scikit-learn*. For features, +**C** denotes the negative log-likelihood of each word in [1] and CoCA (https://www.english-corpora.org/coca/) corpora. To personalize, we used one-hot representations of the learners as in [3].

Table 1 shows the quantitative results of these methods, where the values are the average of the results of $J = 16$ learners. The root mean squared error (RMSE) of the gold vocabulary size and estimated vocabulary size were used for evaluation. We observed that by utilizing the probability distribution, our method **LR** outperformed the baselines **VST** in terms of the RMSE. Interestingly, **NN** slightly outperformed **LR** in the accuracy of predicting learners' responses, and its RMSE scores were poor, presumably because **NN** uses accuracy for its loss function rather than RMSE. Figure 4 shows an estimation plot for a learner. Our distributional blue line is closer to the gold-vocabulary size than the baseline. Our method can also output which word is likely to be included in each estimated size. This means that our method can output a detailed list of words that a learner with a large vocabulary may know but a learner with a small vocabulary may not know. Table 2 lists such words. Their estimated vocabulary sizes were 9,386 and 5,210, and their gold vocabulary sizes were 9,556 and 6,547, respectively.

This study proposed a novel method that combines multiple binary classifiers to improve vocabulary size estimation. Our future work will include a more detailed evaluation of the content of the estimated vocabulary.

Acknowledgements. This work was supported by JST ACT-X Grant Number JPM-JAX2006, Japan.

References

1. BNC Consortium: The British National Corpus (2007)
2. Ehara, Y.: Uncertainty-aware personalized readability assessments for second language learners. In: Proceedings of the ICMLA, pp. 1909–1916, December 2019

3. Ehara, Y., Sato, I., Oiwa, H., Nakagawa, H.: Mining words in the minds of second language learners: learner-specific word difficulty. In: Proceedings of the COLING, pp. 799–814 (2012)
4. Kleinberg, J., Tardos, E.: Algorithm Design. Pearson Education India, Noida (2006)
5. Laufer, B., Ravenhorst-Kalovski, G.C.: Lexical threshold revisited: lexical text coverage, learners' vocabulary size and reading comprehension. RFL **22**(1), 15–30 (2010)
6. Nation, I.: How large a vocabulary is needed for reading and listening? Can. Mod. Lang. Rev. **63**(1), 59–82 (2006)

MOOCs Paid Certification Prediction Using Students Discussion Forums

Mohammad Alshehri[✉] and Alexandra I. Cristea

Department of Computer Science, Durham University, Lower Mountjoy, South Road,
Durham DH1 3LE, UK
{mohammad.a.alshehri,alexandra.i.cristea}@durham.ac.uk

Abstract. Massive Open Online Courses (MOOCs) have been suffering a very
level of low course certification (less than 1% of the total number of enrolled
students on a given online course opt to purchase its certificate), although MOOC
platforms have been offering low-cost knowledge for both learners and content
providers. While MOOCs discussion forums' rich numeric and textual data are
typically utilised to address many MOOCs challenges, e.g., high dropout rate,
identifying intervention-needed learners, analysing learners' forum discussion and
interaction to predict certification remains limited. Thus, this paper investigates
*if MOOC discussion forum-based data can predict learners' purchasing deci-
sions (certification).* We use a relatively large dataset of 23 runs of 5 FutureLearn
MOOCs for temporal (weekly-based) prediction, achieving promising accuracies
in this challenging task: 76% on average, across the five courses.

Keywords: MOOCs · Certification prediction · Discussion forums

1 Introduction

Digital learning has been revolutionising and changing the means of modern education.
Consequently, several MOOC platforms appeared over the last decade, with many starting in 2012, coining 2012 as "the year of the MOOCs" [1, 2]. In terms of financial models
for these platforms, popular ones, such as FutureLearn, edX, Udemy and Coursera, have
mixed free and paid online educational content for the public, targeting learners worldwide [3, 4]. This paper proposes a forum-based predictor of learners' financial decisions
(course certificate purchase), affecting the income from such platforms. Specifically, this
paper addresses the following research questions:

- *RQ1: Can MOOC discussion forum data predict course purchase decisions (certification)?*

We use multidisciplinary course data from the less analysed platform of FutureLearn,
to temporally predict financial certification. To the best of our knowledge, *our method in
predicting MOOC learners' financial decisions (purchasing a course certificate) using
learners' discussion forums has never been applied before.*

© Springer Nature Switzerland AG 2022
M. M. Rodrigo et al. (Eds.): AIED 2022, LNCS 13356, pp. 542–545, 2022.
https://doi.org/10.1007/978-3-031-11647-6_111

2 Related Work

Analysing the literature, very few studies have explored certification in MOOCs. Their used data sources, the number of courses and students, and the type of the data used vary, as explained in Table 1 below. Data used included Click Stream (CS), Forum Posts (FP), Assignments (ASSGN), Student Information Systems (SIS), Demographics (DEM) and Surveys (SURV).

Table 1. Certification prediction models versus our model.

Ref.	Data source	#Courses	#Students	Data description
[5]	Coursera	1	37,933	ASSGN; FP; SIS
[6]	HarvardX	9	79,525	DEM; SURV
[7]	edX	1	43,758	CS
[8]	Coursera	1	84,786	FP
[9]	Coursera; edX	1	65,203	CS; FP
Our model	FutureLearn	5	245,255	FP

Unlike previous studies on certification, our proposed model aims to predict the financial decisions of learners on whether to purchase the course certificate. Also, our work is applied to a less frequently studied platform, FutureLearn (Table 1). Another contribution of our study is predicting the learner's actual financial decision on buying the course and gaining a certificate. Most current course purchase prediction models identify certification as an automatic consecutive step to the completion, making them not different from completion predictors.

3 Methodology

3.1 Data Collection and Preprocessing

The current study analyses data extracted from 23 runs spread over 5 MOOC courses, on four distinct topic areas, all delivered through FutureLearn, by the University of Warwick. These courses were delivered repeatedly in consecutive years (2013–2017); thus, we have data on several '*runs*' for each course [10–12]. The Textual Data (student comments) preprocessing involved several essential tasks, e.g. eliminating irrelevant data generated by organisational administrators, removing unwanted characters, such as HTML/XML, punctuations and non-alphabet characters. The last step contained removing stop-words, lowering the cases of characters, reforming contractions into the original words and grammar correction. Also, learner comments have been classified as *positive, neutral* and *negative* using *MOOCSent* sentiment classifier [13].

The current study applied three shallow and one deep classification and regression algorithms to predict MOOC learners' purchasing behaviour: ExtraTree (ET), Logistic Regression (LR), XGBoost (XGB) and Multi-layer Perception (MLP). To deal with our imbalanced dataset, we used the Balanced Accuracy (BA) to report our results, besides the commonly used metric of accuracy (Acc), defined as the average of recall obtained on each class.

4   Results and Discussion

As the courses analysed spanned different weeks, we examined the first-week-only data, and compared it to the data starting from the first week to the middle of the course. Results explore how our raw and processed (computed) features can temporally distinguish course purchasers from non-paying learners based on their discussion forum data (Table 2).

Table 2. Learner classification results distributed by course at two time points of the course, where class 0 = non-paying learners and class 1 = certificate purchasers.

Course	Classifier	1st Week only			1st - Mid Week		
		Rec_0	Rec_1	BA	Rec_0	Rec_1	BA
BIM	ET	0.82	0.63	**0.73**	0.88	0.75	**0.82**
	LR	0.87	0.57	0.72	0.89	0.63	0.76
	XGB	0.97	0.03	0.50	0.86	0.30	0.58
	MLP	0.81	0.61	0.71	0.82	0.71	0.77
BD	ET	0.83	0.53	0.68	0.94	0.57	0.76
	LR	0.80	0.57	0.69	0.88	0.66	**0.77**
	XGB	0.99	0.04	0.52	0.91	0.57	0.74
	MLP	0.83	0.59	**0.71**	0.92	0.61	**0.77**
SC	ET	0.83	0.50	0.67	0.95	0.8	0.88
	LR	0.84	0.53	0.69	0.90	0.67	**0.79**
	XGB	0.98	0.04	0.51	0.94	0.50	0.72
	MLP	0.83	0.57	**0.70**	0.93	0.60	0.77
SP	ET	0.81	0.60	**0.71**	0.91	0.64	0.78
	LR	0.82	0.56	0.69	0.93	0.72	**0.83**
	XGB	0.99	0.06	0.53	0.92	0.36	0.64
	MLP	0.82	0.58	0.70	0.91	0.62	0.77
TMF	ET	0.83	0.55	**0.69**	0.94	0.57	0.76
	LR	0.85	0.52	**0.69**	0.88	0.77	**0.83**
	XGB	0.98	0.02	0.50	0.93	0.35	0.64
	MLP	0.82	0.56	**0.69**	0.93	0.65	0.79

This MOOC prediction task is considered highly challenging, compared to other MOOC tasks, like predicting dropout, completion and learner characteristics. The reason is the severe data imbalance of the binary class, where course certificate purchasers form less than 1% of the total number of enrolled students.

5   Conclusion

This study compared four tree-based and regression classifiers, to predict course purchasability, using discussion forum data from five MOOCs. Our proposed model achieved various balanced accuracies, 0.76 on average, using only the first half of the course data. Thus, it can predict relatively early on if a purchase of a certificate will take place or not. Future planned improvements of our model include using deep models and employing more student data, e.g. demographics and clickstream logs.

References

1. Alshehri, M., Alamri, A., Cristea, A.I., Stewart, C.D.: Towards designing profitable courses: predicting student purchasing behaviour in MOOCs. Int. J. Artif. Intell. Educ. **31**(2), 215–233 (2021). https://doi.org/10.1007/s40593-021-00246-2
2. Alshehri, M., Alamri, A., Cristea, A.I.: Predicting certification in MOOCs based on students' weekly activities. In: Cristea, A.I., Troussas, C. (eds.) ITS 2021. LNCS, vol. 12677, pp. 173–185. Springer, Cham (2021). https://doi.org/10.1007/978-3-030-80421-3_20
3. Alamri, A., et al.: Predicting MOOCs dropout using only two easily obtainable features from the first week's activities. In: Coy, A., Hayashi, Y., Chang, M. (eds.) ITS 2019. LNCS, vol. 11528, pp. 163–173. Springer, Cham (2019). https://doi.org/10.1007/978-3-030-22244-4_20
4. Cristea, A.I., et al.: Earliest predictor of dropout in MOOCs: a longitudinal study of FutureLearn courses. Association for Information Systems (2018)
5. Jiang, S., et al.: Predicting MOOC performance with week 1 behavior. In: Educational Data Mining (2014)
6. Reich, J.: MOOC completion and retention in the context of student intent. EDUCAUSE Rev. Online **8** (2014)
7. Coleman, C.A., Seaton, D.T., Chuang, I.: Probabilistic use cases: discovering behavioral patterns for predicting certification. In: Proceedings of the Second (2015) ACM Conference on Learning@ scale (2015)
8. Joksimović, S., et al.: Translating network position into performance: importance of centrality in different network configurations. In: Proceedings of the Sixth International Conference on Learning Analytics & Knowledge (2016)
9. Gitinabard, N., et al.: Your actions or your associates? Predicting certification and dropout in MOOCs with behavioral and social features. arXiv preprint arXiv:1809.00052 (2018)
10. Alshehri, M., et al.: On the need for fine-grained analysis of gender versus commenting behaviour in MOOCs. In: Proceedings of the 2018 the 3rd International Conference on Information and Education Innovations. ACM (2018)
11. Cristea, A.I., et al.: How is learning fluctuating? FutureLearn MOOCs fine-grained temporal analysis and feedback to teachers (2018)
12. Cristea, A.I., et al.: Can learner characteristics predict their behaviour on MOOCs? In: 10th International Conference on Education Technology and Computers (ICETC 2018). Association for Computing Machinery, Tokyo Institute of Technology, Tokyo (2018)
13. Alsheri, M.A., et al.: MOOCSent: a sentiment predictor for massive open online courses. Association for Information Systems (2021)

Exploring Student Engagement in an Online Programming Course Using Machine Learning Methods

Sophia Polito[1]([⊠]), Irena Koprinska[2]([⊠]), and Bryn Jeffries[3]

[1] School of Electrical and Information Engineering, The University of Sydney, Sydney, Australia
spol5736@uni.sydney.edu.au
[2] School of Computer Science, The University of Sydney, Sydney, Australia
irena.koprinska@sydney.edu.au
[3] Grok Academy and School of Computer Science, The University of Sydney, Sydney, Australia
bryn.jeffries@grokacademy.org

Abstract. This paper investigates student engagement, how it changes over time, and its impact on course performance and drop off rates. We analyse data from a large online Python programming course ($n = 10{,}558$ students) by defining appropriate features and using clustering to find students with common behaviours and Markov chains to analyse engagement changes and drop-off rates. Our methods allow teachers to better understand student engagement and take remedial actions to improve students' learning and performance.

Keywords: Student behaviour · Course engagement · Assessment · Automatic grading system · Clustering · Markov chains

1 Introduction

There is an opportunity for data driven methods to examine student engagement and performance in online courses to better personalize student feedback. We examine the engagement of high-school students participating in an online beginners Python programming course through the following research questions:

1. What are the course engagement similarities between students and how does engagement influence students' overall course score?
2. How does student engagement change over time and impact course drop-off?

Overall, this study aims to gain greater insight into student engagement, how engagement changes over time and how it impacts course drop-off rates, to assist teachers in supporting better student outcomes.

© Springer Nature Switzerland AG 2022
M. M. Rodrigo et al. (Eds.): AIED 2022, LNCS 13356, pp. 546–550, 2022.
https://doi.org/10.1007/978-3-031-11647-6_112

2   Previous Work

Educational data mining has sought to identify groups of similar students to determine their engagement levels and overall course performance.

Hooshyar et al. [1] used clustering to predict if a student was a low, medium or high procrastinator. Procrastination was defined by spare time - the time from when a student submits an assignment until the assignment is due, and a positive correlation was shown between spare time and assignment score. Cerezo et al. [2] defined procrastination and engagement features to analyse student behaviour in online courses. The engagement features included time spent on quizzes, viewing content and viewing forums. Clustering showed that high-performing students spent more time on quizzes and the final exam score was not related to the time spent on viewing content. McBroom et al. [3] used clustering to identify and follow the development of student behaviour over the semester. It was found that behavioural clustering could occur as early as Week 3 of a 13-week semester and accurately predict a student's final score.

In this paper, we define features informed by prior work on student behaviour clustering to identify student engagement levels and drop-off rate.

3   Data

The data used in this study is from a 5-week beginner Python course, offered online through the Grok Academy platform (https://grokacademy.org/challenge). It contains slides, which teach Python concepts, and interleaved graded programming tasks ("problems"), which are auto-marked by test-cases. There are 40 graded problems in total, 8 for each week, with a maximum score of 10 per problem and total score of 400.

We define engagement features, informed by [2], as shown in Table 1.

Table 1. Student engagement features

Feature	Description
% completed problems	Percentage of problems which passed all testcases
Number of autosaves	Number of automatic autosaves
Number of terminal runs	Number of times a student runs their programming problem
% slides completed	Percentage of slide problems completed
% slides viewed	Percentage of slides viewed by the student for the week

4 Results

4.1 Similarities in Student Engagement Tendencies

We used K-means clustering to find groups of similar students based on their behavior. Each feature vector represents a student for one week, therefore each student has a total of 5 vectors, resulting in 19,724 vectors for clustering. Missing values and outliers were removed and features were normalized between 0 to 1. The elbow method indicated k = 5 clusters; the cluster centroids are shown in Table 2 and the cluster characteristics are summarized below:

Cluster 0 – High problem effort, high content effort: It shows the highest amount of student engagement with 97% of problems completed and 83% of slides completed.

Cluster 1 – High problem effort, medium content effort: It shows low slide completion despite 86% of slides viewed which suggests students skim through the content.

Cluster 2 – Low problem effort, low content effort: It has very low activity on all features indicating students likely did not engage with the course.

Cluster 3 – Low problem effort, high content effort: It has a low number of completed problems of 46%, but a high slide completion rate of 72%.

Cluster 4 – High problem effort, low content effort: It has 94% of problems completed but low slide interaction (only 17% of slides completed and 42% of slides viewed), indicating students had prior programming experience.

Table 2. Student engagement cluster centroids

	Cluster					
	Full data	0	1	2	3	4
Number autosaves	0.089	0.998	0.089	0.039	0.081	0.067
Terminal runs	0.057	0.064	0.058	0.025	0.050	0.048
Completed problems	0.876	0.983	0.973	0.276	0.461	0.947
Slides completed	0.647	0.839	0.267	0.292	0.727	0.176
Slides viewed	0.842	0.962	0.864	0.412	0.823	0.429

4.2 Impact of Engagement on student's Overall Course Score

The engagement clusters are analysed regarding student score distribution for each cluster. The score is discretized into 4 ranges: from low [0, 100] to high [301–400].

Figure 1 shows that three clusters dominate the highest score range (301–400): Cluster 0 (high problem effort, high content effort) with 61% of high score students, Cluster 1 (high problem effort, medium content effort) and Cluster 4 (high problem effort, low content effort) with 56% of high score students. Cluster 0 has strong problem and content effort metrics which likely contributed to higher marks.

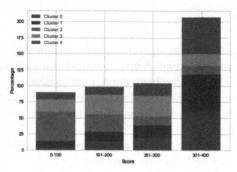

Fig. 1. Percentage of engagement clusters in each score range

4.3   Changes in Student Engagement and Impact on Drop off

Markov chains were used to analyse the student's engagement changes over time, and show the cluster transitions from Week 1 to the clusters in any of the proceeding weeks. A No Attempt node represents students with no course activity for the week and a Partial Attempt node for students who completed some features measured.

The engagement Markov chain is shown in Fig. 2. Cluster 2 (low problem effort, low content effort) and Cluster 3 (low problem effort, high content effort) have students with the highest risk of dropout with 95% and 81% of students moving to No Attempt. This is higher than the students who began in the No Attempt group in Week 1. The high drop-out rate is surprising since Cluster 3 had 72% slide viewership, normally indicating high course engagement. Here, the Markov chain highlights at-risk students potentially going unnoticed due to high content interactions.

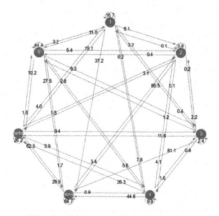

Fig. 2. Engagement Markov chain

5 Conclusion

This paper investigates student engagement tendencies, how they change over time and how engagement impacts the overall course score and drop off rates in an introductory programming course. The 'high problem effort, high content effort' clusters contained the highest achieving students whilst the Markov chain analysis identified at-risk students likely to drop out of the course, which can inform teachers when to provide assistance to students at risk of drop-out.

Our insights help teachers to identify at-risk students to tailor feedback to improve student learning outcomes and course performance.

References

1. Hooshyar, D., Pedaste, M., Yang, Y.: Mining educational data to predict student's performance through procrastination behaviour. Entropy **12**, 1–24 (2019)
2. Cerezo, R., Esteban, M., Sanchez-Santillan, M., Nunez, J.C.: Procrastinating behaviour in computer-based learning environments to predict performance: a case study in Moodle. Front. Psychol. **8**, 1403 (2017)
3. McBroom, J., Jeffries, B., Koprinska, I., Yacef, K.: Mining behaviours of students in auto-grading submission system logs. In: International Conference on Educational Data Mining, pp. 159–166 (2016)

Human-in-the-Loop Data Collection and Evaluation for Improving Mathematical Conversations

Debajyoti Datta[1]([✉]) [iD], Maria Phillips[1] [iD], James P. Bywater[3] [iD], Sarah Lilly[2] [iD], Jennifer Chiu[2] [iD], Ginger S. Watson[2] [iD], and Donald E. Brown[1] [iD]

[1] School of Engineering and Applied Sciences, University of Virginia, Charlottesville, VA 22904, USA
dd3ar@virginia.edu

[2] School of Education and Human Development, University of Virginia, Charlottesville, VA 22904, USA

[3] College of Education, James Madison University, Harrisonburg, USA

Abstract. The nature and quality of classroom instruction is highly correlated to teachers' ability to rehearse effective teaching strategies. Utilizing research-based teaching practices increases teacher effectiveness, confidence, and retention along with improving student achievements. High-fidelity, AI-based simulated classroom systems enable teachers to rehearse and get feedback on specific pedagogical skills. One primary challenge is that current conversational agents (CA) can have task-oriented conversations, however more varied dialogue-oriented conversations such as that between a teacher and student for a domain-specific task (like a mathematical scenario) can be difficult to model. This paper presents a high-fidelity, AI-based classroom simulator to help teachers rehearse research-based mathematical questioning skills. The system relies on advances in deep-learning uncertainty quantification and natural language processing while acknowledging the limitations of CAs for specific pedagogical needs.

Keywords: Human-in-the-loop · Conversational agent · Low-data domain and teacher training · Robust evaluations

1 Introduction

Despite ample evidence showing that deliberate practice can improve teachers' mathematical questioning, teachers are rarely given opportunities to rehearse these kinds of questioning strategies in pre-service or in-service settings due to a variety of constraints in teacher preparation programs. However, computer-based systems can provide ways for pre-service and in-service teachers to practice and receive feedback on mathematical questioning skills. This paper presents the development of an AI-based classroom teaching system (ACTS) designed to help teachers rehearse mathematical questioning strategies that leverages advances in

© Springer Nature Switzerland AG 2022
M. M. Rodrigo et al. (Eds.): AIED 2022, LNCS 13356, pp. 551–554, 2022.
https://doi.org/10.1007/978-3-031-11647-6_113

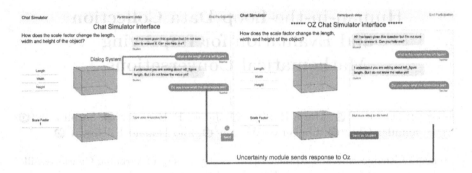

Fig. 1. The ACTS system has chat functionality and a dynamic representation of the mathematical task. The expert user has access to the Supervisor Interface to the right. The dialogue system responds when it is certain about the dialogue acts and entities. When the uncertainty thresholds are met, it sends the prompt to the expert user. The expert user can then type in the response as a student, and it will appear on the prompt (left). This prevents conversations from breaking because of the failure of ML pipelines and components.

conversational agent (CA) development. In particular, this paper describes the use of a human expert working with the computer-based system in a supervisor-type role to step in and keep the conversation going when the CA may fail. The goal of the system is to simultaneously collect data for conversational agent components, while maintaining a coherent conversation and relying on state of the art advances in natural language processing systems. This paper reports on the development and user testing of the ACTS system (Fig. 1).

2   Current Challenges

Current challenges to CA development in educational contexts include:

- **Model and Data Limitations (MDL)**: Deep learning models (models that are used in most modern CAs) can fail catastrophically because they work by exploiting spurious relationships [7] within data sets which is known as shortcut learning [4].
- **User Expectation Limitations (UEL)**: The "deep gulf of evaluation" [6] exists because CA systems lack meaningful feedback regarding the systems intelligence and capability. This mismatch of user expectations and current technology capabilities can lead to a mis-application of CAs and a lack of confidence that results in an avoidance of use and deployment for complex tasks or sensitive activities.
- **Limitations of Pretrained Models**: Generic models like BERT [3] do not readily generalize to specific domains. To address this, education communities need education-specific pre-trained models, like the biomedical community [2].

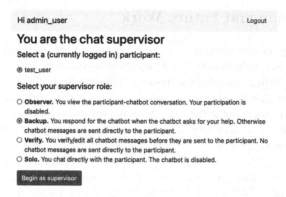

Fig. 2. The system enables data collection both during evaluation of the natural language processing modules and building scenarios for a new system. The "Verify" mode enables data collection for new scenarios while the "Backup" mode can redirect control to the human when a language component fails.

– **Limitations in data collection practices:** The primary mode of data collection in education contexts for domain-specific scenarios often relies on text transcriptions from noisy classroom videos. This coupled with privacy concerns makes data collection for addressing MDL non-trivial.

3 System Description

The **MDL** and **UEL** are difficult to meet in CA's for any system and especially for a CA in the context of education. In order to design useful systems, we need to acknowledge that current advances in dialogue systems make having fluent conversations over multiple turns non-trivial. We designed a dialogue system to specifically address user expectation failure and model and data failures. At it's core, the system minimizes negative user experiences by incorporating uncertainty modeling. When a dialogue system component or a system fails, we redirect control to a supervisor whose task is to bring the conversation back on track. We rely on recent developments in deep learning and uncertainty estimation for each subcomponent of the dialogue system and build an interaction pipeline of user + system + supervisor that can pass the system's control from the dialogue system to the supervisor based on the failure. The uncertainty modules prevent the conversation from derailing due to a failure of one or more of the sub-components. The supervisor is present during the user engagement with the application acting within the role as an expert user (someone familiar with the scenario and the system's limitations) with the ability to send a response back to the system as a "Student". While perfect conversations in task-specific dialogue systems are challenging to achieve in complex scenarios, adding an expert in the loop prevents conversations from derailing. This also enables better user satisfaction and long-term data collection for better modeling or never-ending learning [5] scenarios (Fig. 2).

4 Conclusion and Future Work

Initial user studies demonstrate an improvement of the user experience of our conversational agent despite the limitations of a complex, low-data educational scenario by including uncertainty module elements and allowing for human-in-the-loop interactions. Even with the uncertainty modules, uncertainty quantification in deep learning is not robust [1]. Also, uncertainty in deep learning has been explored mostly for classification tasks but dialogue systems have many other tasks (entity recognition, turn-taking). Thus accuracy of uncertainty estimates also vary based on the stage of the dialogue system. In the future, we are planning to explore two lines of work: determining approaches so that we can jointly model uncertainty of all the stages together as a more robust uncertainty metric and controlling the amount of supervisor involvement as we collect more data in these training scenarios. We believe that these kinds of discrete, domain-specific web-based simulated classroom systems can provide needed opportunities to help pre-service and in-service teachers rehearse specific pedagogical strategies.

References

1. Ashukha, A., Lyzhov, A., Molchanov, D., Vetrov, D.: Pitfalls of in-domain uncertainty estimation and ensembling in deep learning. arXiv preprint arXiv:2002.06470 (2020)
2. Beltagy, I., Lo, K., Cohan, A.: SciBERT: a pretrained language model for scientific text. arXiv preprint arXiv:1903.10676 (2019)
3. Devlin, J., Chang, M.W., Lee, K., Toutanova, K.: BERT: pre-training of deep bidirectional transformers for language understanding. In: Proceedings of the 2019 Conference of the North American Chapter of the Association for Computational Linguistics: Human Language Technologies, vol. 1 (Long and Short Papers), pp. 4171–4186. Association for Computational Linguistics, Minneapolis, Minnesota, June 2019. https://doi.org/10.18653/v1/N19-1423, https://www.aclweb.org/anthology/N19-1423
4. Geirhos, R., et al.: Shortcut learning in deep neural networks. arXiv:2004.07780 [cs, q-bio] (April 2020). http://arxiv.org/abs/2004.07780
5. Mitchell, T., et al.: Never-ending learning. Commun. ACM **61**(5), 103–115 (2018)
6. Norman, D.A.: Cognitive artifacts. Des. Interact. Psychol. Hum. Comput. Interface **1**(1), 17–38 (1991)
7. Poliak, A., Naradowsky, J., Haldar, A., Rudinger, R., Van Durme, B.: Hypothesis only baselines in natural language inference. arXiv preprint arXiv:1805.01042 (2018)

Assessing the Practical Benefit
of Automated Short-Answer Graders

Ulrike Padó[✉]

Hochschule für Technik Stuttgart, Stuttgart, Germany
`ulrike.pado@hft-stuttgart.de`

Abstract. Short-Answer Grading (SAG) is a task where student answers to open questions are automatically graded with the support of Natural Language Processing and Machine Learning (ML), saving manual effort. Two main challenges remain in small-scale testing scenarios: (1) ML models work best given large amounts of manually graded training instances, and (2) published evaluation results for pre-trained models do not translate well to new data sets, making automated grading intransparent for teachers and students. We present a grader evaluation workflow that teachers can use for their individual situation.

Keywords: Short-Answer Grading · Evaluation · Reliability

1 Introduction

Short answer questions (also called constructed response questions) are a popular task on written exams. Students respond with about one to three sentences in their own words, which makes it easier for teachers to understand their reasoning and spot misconceptions. However, manual grading is time-consuming, especially if frequent feedback through formative testing is desired. Supporting this process with automated grade predictions is the goal of Short-Answer Grading (SAG) [1]. Human involvement can thus be limited to reviewing rather than grading from scratch [9], or by focusing grading effort where it is most needed [6].

Automated methods need training data. However, in many classroom and self-learning settings, there are few existing annotated answers. Recently, transfer learning for Transformer-based models like BERT [3] allows the use of large amounts of un-annotated data to infer a robust language model in pre-training before switching to fine-tuning on a smaller data set for a specific task [2,5].

However, at the moment, it is unknown how well these results transfer to small-scale testing: The standard literature data sets, at a size of several thousand answers, are small in the context of ML, but still large in the context of small-scale testing. Additionally, prediction quality of ML models deteriorates for new data sets.

We propose a workflow and suggest decision-making parameters for the evaluation of an existing automated grader for a specific classroom. The workflow allows teachers to make an informed decision about how to integrate automated

© Springer Nature Switzerland AG 2022
M. M. Rodrigo et al. (Eds.): AIED 2022, LNCS 13356, pp. 555–559, 2022.
https://doi.org/10.1007/978-3-031-11647-6_114

grading support and helps teachers and students understand the performance and limitations of the tool in use. Tan et al. [8] conceptualize the desired qualities of an automated grader as its *reliability* as defined by IEEE: "the degree to which a system, product or component performs specified functions under specified conditions for a specified period of time" and propose a similar development cycle with many stakeholders and multiple iterations before system deployment. Our use case is much more constrained: We look at the situation-specific evaluation of an off-the-shelf tool for a specific usage scenario by the end-users, which is a linear process with a defined end since fewer modifications to the model are possible.

2 Evaluation Process and Worked Example

In a typical small-scale grading setting, a teacher who wishes to use automated grading will have access to publicly available ML models, and very limited amounts of manually graded data (for example from a recent test) for evaluation. Starting from this position, we propose the following steps:

1. **Define** the requirements for reliability in the current use case
2. **Collect** a set of manually graded test data
3. **Fine-tune** an automated grader for SAG
4. **Analyze** the automated grader's performance
5. **Decide** on how to use the approach

We will now demonstrate these steps using the Huggingface $BERT_{MNLI}$ model as a grader and the SemEval-2013 Beetle test data [4] as manually graded data set.

Defining Requirements

1. **Minimal grading error** We will accept automated labelling error of up to 15%, which has been deemed acceptable in the past for published SAG data [6]. That means we require a grading Accuracy (overall percentage of correctly predicted labels) of at least 85%.
   Another important aspect of grading error is its distribution (showing over-strictness or over-lenience). The model tendency can be measured by looking at the grade labels' Precision separately. (Precision measures how many predictions of a specific label were actually correct, that is, how trustworthy the grader's label predictions are.) In our worked example, we will accept over-lenience, but not over-strictness.
2. **Workload reduction** is the driving factor behind the use of automated graders. As a point of reference, Vittorini et al. [9] report a grading time reduction of about 40% by their approach.

Table 1. Performance of the SEB-tuned model on the Beetle test set (overall/by label).

Model	Accuracy	Precision	Recall	F_1
overall	62.5	66.2	62.5	64.3
correct	**75.9**	53.8	75.9	63.0
incorrect	52.8	**75.1**	52.8	62.0

Data Collection. In the context of ML, annotated data is needed in two places: For model training (here, the fine-tuning step of the Transformer-based learner) and for model evaluation. For the fine-tuning step, several thousand manually graded answers are realistically required. Since this is more than will be available to most teachers, publicly available data can be used instead. In our example, this is the SemEval-2013 SciEntsBank training data, containing ca. 5000 answers [4]. The corpus is roughly similar, but different from the source of our test data.

The test set should be as large as possible to make it robust to chance fluctuations; a size in the hundreds of answers is realistic. The Beetle *2-way unseen questions test set* used in our example consists of 9 questions and a total of 819 answers, and differs from the training data in similar ways as field data would differ from a literature data set.

Fine-Tuning adapts the BERT prediction model more closely to the SAG task. Additionally pre-training Transformers for a SAG-related task first tends to further increase performance [2]. Two natural choices of related tasks are Natural Language Inference (using the GLUE MNLI – Multi-Genre Natural Language Inference – data) and Paraphrasing (using the GLUE MRPC – the Microsoft Research Paraphrase Corpus). We therefore compare the $BERT_{base\_uncased}$ model to $BERT_{MNLI}$ and $BERT_{MRPC}$[1] on SEB development data (10% of the training data, randomly sampled) after fine-tuning on the remainder of the SEB training data. $BERT_{MNLI}$ outperforms the other models at F_1[2] $= 84.56$ ($BERT_{MRPC}$: 83.13, $BERT_{base}$: 83.87).

This model will be used for evaluation below. On the SEB *2-way unseen questions* test set, it performs comparably to the most recent directly comparable literature at F_1 of 73.5 compared to 74.8 [7].

Analysis. Table 1 shows our results after evaluating on the 819 manually graded Beetle answers. The model loses 11 points F_1 score in the transfer from the SEB to the Beetle corpus, underscoring that every data set needs individual evaluation. Overall model Accuracy is far below our requirement of 85%.

We now investigate grader tendencies: The automated grader errs strongly towards lenience and misses almost 50% of incorrect answers at

---

[1] All models are available on huggingface.co.

[2] F_1 is the harmonic mean of Precision (reliability of predictions) and Recall (percentage of instances correctly identified).

Recall$_{incorr}$ = 52.8. In consequence, Precision$_{corr}$ is only 53.8. However, the Precision of *incorrect* predictions is 75.1, which means that this label is generally reliable when assigned.

Decision on Usage. Given our requirements, a stand-alone use of the automated grader is not acceptable. However, if a human grader accepts all machine-labelled *incorrect* grades, manual grading workload will fall from 819 answers (the whole data set) to 485 (the answers labelled *correct* only) – a reduction of 40.8% and the workload reduction we expect.

This approach would eliminate most of the automated grader's labelling error: If we generously assume that the human grader will always assign the right label, the human-graded 485 answers plus 0.751 * 334 answers (the answers correctly machine-graded as *incorrect*) are now graded without error. This is 89.8% of all answers, or a grading error rate of 10.2%. This is in the acceptability range we defined above.

However, we know that any remaining errors in this setting are to the detriment of the students. Alternatively, the answers machine-labelled as *incorrect* can additionally be reviewed manually, which is still faster than assigning grades from scratch [9].

In sum, we have shown how analysis in context yields useful information to inform decisions on model usage. The technical effort needed for implementation runs to a few hours of Python coding, thanks to publicly available model libraries and data sets.

References

1. Burrows, S., Gurevych, I., Stein, B.: The eras and trends of automatic short answer grading. IJAIED **25**, 60–117 (2015)
2. Camus, Leon, Filighera, Anna: Investigating transformers for automatic short answer grading. In: Bittencourt, Ig Ibert, Cukurova, Mutlu, Muldner, Kasia, Luckin, Rose, Millán, Eva (eds.) AIED 2020. LNCS (LNAI), vol. 12164, pp. 43–48. Springer, Cham (2020). https://doi.org/10.1007/978-3-030-52240-7_8
3. Devlin, J., Chang, M.W., Lee, K., Toutanova, K.: BERT: pre-training of deep bidirectional transformers for language understanding. In: Proceedings of the 2019 Conference of the NAACL:HLT, pp. 4171–4186, June 2019
4. Dzikovska, M., et al.: SemEval-2013 task 7: the joint student response analysis and 8th recognizing textual entailment challenge. In: Proceedings of SemEval 2013, pp. 263–274, June 2013
5. Ghavidel, H.A., Zouaq, A., Desmarais, M.C.: Using BERT and XLNET for the automatic short answer grading task. In: Proceedings of CSEDU, pp. 58–67 (2020)
6. Mieskes, M., Padó, U.: Work smart - reducing effort in short-answer grading. In: Proceedings of the 7th Workshop on NLP for CALL, pp. 57–68, November 2018
7. Saha, S., Dhamecha, T.I., Marvaniya, S., Sindhgatta, R., Sengupta, B.: Sentence level or token level features for automatic short answer grading?: use both. In: Penstein Rosé, C., et al. (eds.) AIED 2018. LNCS (LNAI), vol. 10947, pp. 503–517. Springer, Cham (2018). https://doi.org/10.1007/978-3-319-93843-1_37

8. Tan, S., Joty, S., Baxter, K., Taeihagh, A., Bennett, G.A., Kan, M.Y.: Reliability testing for natural language processing systems. In: Proceedings of ACL-IJCNLP, pp. 4153–4169 (2021)
9. Vittorini, P., Menini, S., Tonelli, S.: An AI-based system for formative and summative assessment in data science courses. IJAIED **31**, 159–185 (2021)

Supporting Therapists' Assessment in Parent-Mediated Training Through Autonomous Data Collection

Daniel Carnieto Tozadore[1]([⊠]) [iD], Michele Carnieto Tozadore[2] [iD],
and Maria Stella Coutinho de Alcantara Gil[2] [iD]

[1] École Polytechnique Fédérale de Lausanne (EPFL), VD 1015 Lausanne, Switzerland
daniel.tozadore@epfl.ch
[2] Universidade Federal de São Carlos (UFSCar), São Carlos SP 13565-905, Brazil

Abstract. Parental Training is a methodology where therapists teach caregivers how to train their kids in specific behaviors practicing. It can be used in verbal behavior acquisition for children with autism mediated by the parents. In this paper, we are presenting an innovative software that aids the assessment of children's tact acquisition where therapists can design interactive activities to support the remote measurement of the parental training progress. The developed software stores in its server the child's answers and autonomously compute, in run time, the child's face deviation through an Artificial Intelligence algorithm. The therapists have access to such data and can assess children's performance by that without watching the interaction. The performed experiment presents initial validation of the proposed system utilization by one therapist and one dyad of a child with autism and the mother. The system was applied to the child's evaluation phase of an entire parental training cycle in the Brazilian Portuguese language context. Through the interviews and questionnaires, all users claimed they considered our solution adequate and robust for its purposes. After comparing the child's performance in a baseline activity and an activity after the training session, the software was able to provide to the therapist the data to confirm an enhancement in the child's tact acquisition.

Keywords: Parental training · Autism · Brazil · Artificial Intelligence · Attention span loss detection

1 Introduction

Previously to the need of social distancing brought by the COVID-19 pandemic, almost all the parental training for children with autism was made in person due to the advantages that performing this procedure in a controlled environment can provide to the therapists. Especially the advantages of instant correction and feedback that the therapist can give to caregivers and the in loco evaluation of the children progression. The parental training has shown to be a potential alternative to overcome the social distance barrier [1] and the so far available solutions to map and evaluate the children's performance are few

M. M. Rodrigo et al. (Eds.): AIED 2022, LNCS 13356, pp. 560–563, 2022.
https://doi.org/10.1007/978-3-031-11647-6_115

and, most of times, expensive [2]. We believe that Artificial Intelligence (AI) may play a key role for the next steps towards the progression of remote learning and autonomous assessment. AI contribution, in the scope of our studies, can be useful to such goal by automating mechanic tasks in the evaluation phase and potentially increasing the accuracy in autonomous measures and, consequently optimizing therapist's and caregivers' available time.

Here, we are reporting the first steps we are taking into a deeper series of studies to validate our hypothesis, by presenting a high-level description of an innovative software to assess children with autism that had the tact learning acquisition mediated by their guardians. The presented system aims to support the therapist in this evaluation step by automating some tasks that were performed manually. Our proposal has three technical features to do so: an AI algorithm to detect attention span loss by face gaze; the storage of the answers gave by the children to every evaluation question of their selection in the 4 possible alternatives; and the audio record of the period of their answer.

2 The TeiAut Software

The here named TeiAut is a software under development by the authors in the context of the *TEIA Educacional* project[1] - a Brazilian research collaboration group of Artificial Intelligence for education – adapted from other educational software developed for Human-Robot interaction activities [3]. It is not commercialized for being a program still in testing phase and can be requested for pilots and trial tests to its authors. For being applied in a Brazilian context, the software has all its functionalities in the Brazilian-Portuguese language.

The version used for these studies was developed in Python 3.8[2] and the Graphical User Interface (GUI) in PyQt5.[3] It also has OpenCV 3 libraries for the image processing and face detection algorithm of Haar Cascade, as used in [4]. For preserving the privacy of the participants, after processing the attention measures, the images are discarded. The users of this system are therapists and the caregivers of the children, that could be their parents or guardians.

The core of the evaluation process with the system is composed by a quiz-mode activity, in which a question is asked by a person (normally the therapist) in the video - or by the avatar of a robot with a synthetized voice - and the answers' alternatives are presented to the child as a multiple choice of images. Therefore, the therapists should prepare and insert the appropriate evaluation content into the system, the caregivers and children should perform these evaluative activities, and then the therapist will have access to the data collected from these activities later.

These three phases are named **Activity Design**, **Activity Execution** and **Data Validation**, respectively, and detailed in the next subsections. Figure 1 shows a screenshot of each one of these three phases.

---

[1] www.teiaeducacional.com.br.

[2] https://www.python.org/.

[3] https://pypi.org/project/PyQt5/.

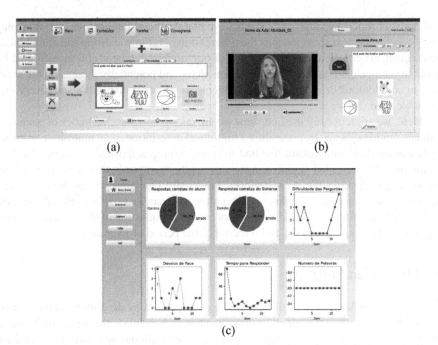

Fig. 1. Screenshots of Activity Design (a), Activity Execution (b) and Data Validation (c) phases (A demo video of this phase can be seeing at the https://youtu.be/R3zuw4C7NpQ).

The Fig. 1(a) shows the screen where therapists can create the questions, add the images of the answer's alternatives, associate a video to the question and type the text of the question.

The screenshot displayed in Fig. 1(b) shows the software during the activity execution where it displays the video of the questions on the left section of the screen. On the right section, the robot avatar with (optionally) the written sentence of the question asked in the video followed by the possible answers as options of images are shown. In some questions, the current child might be only requested to speak loudly the name of the figure in the screen and then the caregiver is advised to click in the corresponding figure if the kid said it correctly. In any case, audio answers are analyzed and graded by the therapist in the next phase.

Based on all the information collected during the execution phase, the system calculates and displays the outcomes in the graphs, as shown in Fig. 1(c). The measures are the number of right answers by the kid, the correct association of answers be the system, the difficulty of the performed questions, the face deviation count, the time the child took to answer each question and the number of spoken words. Thus, it is worth highlighting that the measures presented in graphs second, third and sixth (top left to bottom right) corresponding to these measures are not quite useful in the current version but a when the system would autonomously analyze verbal vocal answers and would aim to adapt the difficulty of the question for each child.

3 Final Considerations

In this paper, we described a software which aims to facilitate the therapists' task of assessing tact acquisition mediated by parents. Initial feedback from therapist and parents supported our hypothesis that this proposal has potential to contribute to parental training in the context that it was applied.

In an initial validation, the use of the software was evaluated as very positive by the therapist and caregivers in interviews after using the system, especially in the aspects of data reliability and time optimization Experiments with a larger population are already ongoing for a more complete and accurate evaluation of our program.

On a technical level, the data collection afforded by TeiAut allows the training of Machine Learning models to further aiding therapists in this task. More AI algorithms, such as recommendation systems and NLP methods are also easily to be aggregated to this solution. As said beforehand, these are goals to the new versions of the software.

Acknowledgements. This work was partially funded by the Swiss National Science Foundation through the National Centre of Competence in Research Robotics (NCCR), the Brazilian organization of Coordination for the Improvement of Higher Education Personnel (CAPES), and the National Institute of Science and Technology on Behavior, Cognition and Teaching - INCT/ECCE.

References

1. Barboza, A.A., Costa, L.C.B., Barros, R.D.S.: Instructional video-modelling to teach mothers of children with autism to implement discrete trials: a systematic replication. Trends Psychol. **27**, 795–804 (2019)
2. Sundberg, M.L.: VB-MAPP Verbal Behavior Milestones Assessment and Placement Program: A Language and Social Skills Assessment Program for Children with Autism or Other Developmental Disabilities: Guide (2008)
3. Tozadore, D.C., et al.: Project R-CASTLE: robotic-cognitive adaptive system for teaching and learning. IEEE Trans. Cogn. Dev. Syst. **11**(4), 581–589 (2019)
4. Tozadore, D.C.: Robotic - cognitive adaptive system for teaching and learning (R-CASTLE). Ph.D. thesis, University of São Paulo (2020)

On Providing Natural Language Support for Intelligent Tutoring Systems

Romina Soledad Albornoz-De Luise(✉), Pablo Arnau-González, and Miguel Arevalillo-Herráez

Departament d'Informàtica, Universitat de Valencia, Valencia, Spain
{romina.albornoz,pablo.arnau,miguel.arevalillo}@uv.es

Abstract. Conversational Intelligent Tutoring Systems (C-ITS) are capable of providing personalized instruction to students in different domains, using dialogues in natural language to interact with the user. In this paper we describe our recent experience in adapting an algebraic/arithmetic word problem-solving Intelligent Tutoring System by replacing the original button-based user interface by a conversational one primarily based on the use of natural language. The proposed method led to an average intent-recognition accuracy of 0.97, supporting the use of this kind of tools to seamlessly migrate existing interfaces to a more convenient form of interaction based on natural language.

Keywords: Conversational agents · Intelligent Tutoring Systems (ITS) · Natural Language Understanding (NLU)

1 Introduction

The use of conversational agents is rapidly increasing [1] and they are now used in a wide range of application areas, including customer service, home automation, e-commerce or education [2]. In a Conversational Intelligent Tutoring System (C-ITS), these agents are used to provide personalized instruction to students, interacting through a dialogue [3]. Hypergraph-based Intelligent Tutoring System (HBPS) [4] entirely focuses on the translation state of the algebraic/arithmetic word problem solving, and it is capable of supervising students without imposing restrictions on the solution paths. However, it does not support interaction in natural language. In this late-breaking results we report on the migration of the original user interface of HBPS into a modern chatbot-powered interface, by using the Rasa open source toolkit [5].

This research has been supported by project PGC2018-096463-B-I00, and FEDER Una manera de hacer Europa; grant PRE2019-090854, funded by MCIN/AEI/10.13039/501100011033; "ESF Investing in your future"; and grant "Margarita Salas", funded by Spanish Ministry of Universities and the European Union through the programme "Next Generation EU".

© Springer Nature Switzerland AG 2022
M. M. Rodrigo et al. (Eds.): AIED 2022, LNCS 13356, pp. 564–568, 2022.
https://doi.org/10.1007/978-3-031-11647-6_116

2 Design and Implementation of the Conversational System

The first step was collecting an extensive registry of typical teacher-student conversations in a one-to-one teaching scenario. Two similar sessions were conducted for this purpose. During these sessions, a total of 79 high school students aged between 14–15 were asked to solve 3 algebraic word problems, by interacting through a chat system with one of the 42 tutors involved in the experience. All chat logs were recorded, and pre-processed by eliminating duplicate messages, correcting typographical mistakes and removing incomplete sentences. Messages were then translated into English and manually assigned to a single intent. To complete the dataset, members of the research team manually added some representative examples for each intent to produce a corpus C, which was used as a groundtruth to support the generation of the conversational agent. A total of 25 intents were identified. 4 of these intents were related to the main actions that the user could perform on the original button-based interface, namely: *Define Letter* (the user enters a letter and a description, e.g. "x is the age of Peter"); *Expression* (the user enters an expression, and possibly a description, e.g. "the age of Anne is 2x"); *Equation* (the user enters an equation, e.g. "2x=76"); and *Help* (the user requests a hint, e.g. "I don't know how to continue"). The other 21 intents are related to actions that the student could not do with the previous interface. Some of these intents are related to problem solutions steps. For example, the intent *Number* corresponds to utterances in which the student attempts to clarify the meaning of a number, e.g. "The age of Anne is 76"; and the intent *Quantity description* to textual entries that mention a relevant quantity description, e.g. "The age of Peter". Some other intents are more general and support some new system functionality or simply allow for a richer interaction.

The next step was to define an appropriate text processing pipeline, defining the operations performed in the text and the order in which they happen. Rasa's default Natural Language Understanding (NLU) pipeline was tweaked according to the results of an extensive experimentation, to better adapt it to the characteristics of our specific domain. A spaCy English language component [6] was prepended to prepare adequate language structures for the subsequent components, using the largest available model (en_core_web_lg). The default tokenizer was replaced by the *SpacyTokenizer*, which is based on white spaces and takes into account punctuation and language-dependent special case rules to split sentences into tokens. After tokenization, each token was featurized by using *SpaCyFeaturizer*. The text was further encoded by using the *LexicalSyntacticFeaturizer*, which generates a binary feature vector, taking into account the context. Whole sentence features were then produced by *CountVectorFeaturizer*, which computes a bag-of-words representation of the sentence, considering character n-grams of size $2 \leq n \leq 4$. Intent classification used the Dual Intent and Entity Transformer (DIET), which was fed with both token-based and sentence-based features. Finally, the *FallbackClassifier* was configured so that the system did not output an intent when the confidence was below 0.3, or when the difference between the top-2 intents was smaller than 0.1. In these cases, the user message

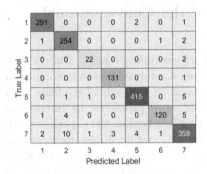
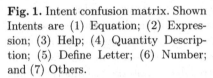

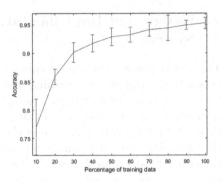

Fig. 1. Intent confusion matrix. Shown Intents are (1) Equation; (2) Expression; (3) Help; (4) Quantity Description; (5) Define Letter; (6) Number; and (7) Others.

Fig. 2. Average accuracy and standard deviation of intent classifier based on a proportion of training data.

was classified as a special intent known as *nlu_fallback*, causing the system to prompt the user to rephrase the message.

A Rasa conversational agent based on this pipeline was then trained by using the corpus C described above. Each time the user inputs a message in the chat, the utterance is forwarded to the Rasa agent, which runs the NLU Pipeline to classify the input among a comprehensive set of 25+1 intents (considered intents plus *nlu_fallback*).

3   Results

Two different experiments were designed in order to understand how the proposed method performs. Our first analysis focuses on the general behaviour of the method in a real-world scenario. Our second study concentrates on the accuracy of the system as a function of the amount of training data.

Real-world Accuracy of the System

For this experiment, a random 20% of the corpus C was reserved for testing, and the remaining 80% was used for training. Results are analysed in terms of the most relevant intents in the application, which account for over 75% of the corpus. These intents are the ones associated with interactions that are directly related to problem solving steps, i.e.: "Equation", "Expression", "Quantity Description", "Define Letter", "Number" and "Help". All other intents have been aggregated under a single category "Others", to ease the analysis and visualization of the results.

Figure 1 shows the confusion matrix for the 7 selected categories, which yields a global accuracy of 0.971. This rate translates into 1 mistake every 34 processed utterances. Values for average precision, recall and F1-score were 0.964, 0.962, and 0.963, respectively. Misclassifications among the 6 most relevant intents was very rare, and most of the errors involved the "Other" category. This result suggests that the very few misclassifications would not heavily affect the user experience. A more in-depth and manual analysis of the misclassifications revealed that most of the mistakes related to "Other" as the true category happened because the user tried to do more than one operation at once, e.g. "x is the age of Peter and 2x=76". For simplicity reasons, the system was designed to classify such entries under the "multiple intent category" (in "Others") and respond with a request to do one operation at a time, but they were sometimes recognized as one of the several operations.

Accuracy in Terms of the Size of the Training Set
This experiment has been carried out by keeping the same test set as in the previous experiment (20% of the corpus $\mathcal{C}$) and progressively increasing the amount of training data from 10% to 100% of the rest of the available data. To compensate for the non-deterministic behaviour of the neural networks due to weight initialization and the order in which the training samples are presented, we have repeated the experiment 5 times. The performance gain has been measured in terms of average accuracy and the standard deviation across the 5 runs. Results are displayed in Fig. 2, considering all 25 intents. The curve shows a clear logarithmic increase of the mean accuracy with the amount of training data. With only a 10% of the available data, the system already shows an average accuracy of 0.77, reaching a value above 0.95 when the whole training set is used. Even though the plot shows that the agent is able to function with a small quantity of training data, the increasing performance suggests that an attempt should be made at the data capturing stage to maximize the amount of entries used to train the system.

4   Conclusions and Future Work

The experiments presented show positive results that support the use of Rasa-powered chatbots for a seamlessly conversational adaptation of existing tutoring systems. In addition, these results can be easily and continuously improved as the system is used, by generating a larger dataset composed of manually revised classifications of all user inputs.

Future work should further investigate the impact of the conversational agent in learning. In between other aspects, we should design appropriate experiments and measures to quantitatively evaluate the quality of experience of students when using conversational systems, in comparison to more classical interfaces. In addition, the impact of the misclassifications should be further analysed, in order to determine their effect from both a usability and a learning perspective. Such analysis could and should be addressed to identifying mistakes with a potentially high impact, in order to include more samples in the training set aimed at avoiding them. On a different line of work, we shall also investigate the performance of Rasa as compared to other proprietary technologies such as Amazon's Lex or Google's DialogFlow, both in terms of performance and usability.

Finally, we shall remark that the use of chatbots opens new opportunities in the design and development of Inteligent Tutoring Systems. For example, they open the door to seamlessly incorporating non-intrusive affective support based on the analysis of the user utterances, in order to detect and react to relevant emotions and/or cognitive states, e.g., frustration or concentration.

References

1. Adamopoulou, E., Moussiades, L.: Chatbots: history, technology, and applications. Mach. Learn. Appl. **2** (2020)
2. Luo, B., Lau, R.Y., Li, C., Si, Y.W.: A critical review of state-of-the-art chatbot designs and applications. WIREs Data Min. Knowl. Discov. **12**(1) (2022)
3. Paladines, J., Ramírez, J.: A systematic literature review of intelligent tutoring systems with dialogue in natural language. IEEE Access **8**, 164 246–164 267 (2020)
4. Arevalillo-Herraez, M., Arnau, D., Marco-Giménez, L.: Domain-specific knowledge representation and inference engine for an intelligent tutoring system. Knowl.-Based Syst. **49**, 97–105 (2013)
5. Bocklisch, T., Faulkner, J., Pawlowski, N., Nichol, A.: Rasa: open source language understanding and dialogue management (2017)
6. Honnibal, M., Montani, I.: spaCy 2: natural language understanding with bloom embeddings, convolutional neural networks and incremental parsing, vol. 7, no. 1 (2017)

Data-driven Behavioural and Affective Nudging of Online Learners: System Architecture and Design

Marie-Luce Bourguet[1]([⊠]) [ID], Jacqueline Urakami[2], and Gentiane Venture[3]

[1] Queen Mary University of London, London, UK
marie-luce.bourguet@qmul.ac.uk
[2] Tokyo Institute of Technology, Tokyo, Japan
[3] Tokyo University of Agriculture and Technology, Fuchu, Japan
venture@cc.tuat.ac.jp

Abstract. In this work-in-progress paper, we describe the architecture of a system that can automatically sense an online learner's situation and context (affective-cognitive state, fatigue, cognitive load, and physical environment), analyse the needs for intervention, and react through an intelligent agent to shape the learner's self-regulated learning strategies. The paper describes the system concept and its software architecture and design: what sensory data are captured and how they are processed, analysed, and integrated; what intervention decision will follow and what behavioural and affective nudges will be given.

Keywords: Online and self-regulated learning · Nudging · Cognitive-affective states · Decision system · Supportive agent

1 Motivation and Concept

In 2020, the pandemic has forced students around the world to suddenly become users of online courses. Many educational institutions turned to synchronous, asynchronous or hybrid teaching, increasingly relying on students' self-regulated learning (SRL) [1]. While many innovative pedagogical approaches have emerged from these circumstances, developing self-regulation has been difficult for many students.

The aim of our work is to give online learners behavioural and affective nudges to help them shape their self-regulated learning skills. We target a change in behaviour by influencing the learners' attitudes, affective states and actions, i.e., a form of co-regulation [2].

Developing effective behavioural interventions requires detailed affective user models, which capture the learners' affective-cognitive states (their triggers and expressions) within the specific learning context. It also requires understanding how affect develops and manifests over time during learning activities, an area of research termed "affect dynamics", which examines how students transition from one affective state to another [3].

© Springer Nature Switzerland AG 2022
M. M. Rodrigo et al. (Eds.): AIED 2022, LNCS 13356, pp. 569–572, 2022.
https://doi.org/10.1007/978-3-031-11647-6_117

Our system senses a learner's situation and context (affect, fatigue, cognitive load, and physical environment), analyses the need for intervention, and reacts through an intelligent virtual agent. It can be used in informal learning settings (e.g., at home or in public spaces), and independently of a specific learning content, activity, or task.

2   System Architecture and Design

The system is made of three sub-systems: (1) a sensing platform; (2) a decision system; and (3) an intelligent supportive agent (see Fig. 1). They exchange information through data interchange objects (JSON), which respectively capture: (1) the student's status; (2) the student's profile; and (3) the remediation needs.

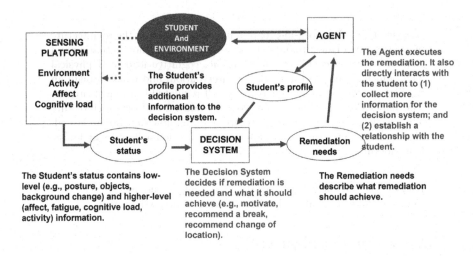

Fig. 1. System architecture.

2.1   Sensing Platform

The sensing platform comprises four modules (under development): the Environment, Activity, Affect and Cognitive-load modules.

The "Environment" module uses object recognition and people tracking (DarkNet53 and YOLOv3) [4], trained with the MS COCO dataset, to detect the presence of objects and people in the learner's immediate environment. Using a simple frame subtraction technique, the amount of background change is also calculated. This information (classified objects, people, and amount of change) is then input into a decision tree (ID3 algorithm) to decide on the student's physical location.

The "Activity" module uses deep machine learning (ResNet CNN and LSTM) to understand what the learner is currently doing, i.e., the type of online learning activity they are engaged in.

Research is increasingly converging towards a set of five cognitive-affective states shown to influence learning and cognition [5]: boredom, confusion, engagement/flow, frustration, and surprise. The "Affect" module combines emotion recognition (EmoNet) and pose estimation (MediaPipe) into an LSTM deep learning model to estimate the student's cognitive-affective state. Training and testing data is generated by recording students in various situations (e.g., for confusion, students were asked to find an object in an image that did not contain that object).

The "Cognitive-load" module uses eye tracking data (pupil dilation, blink frequency, saccadic peak velocity, number and duration of fixations) to estimate fatigue and cognitive load. The eye-tracker used is a single camera Pupil Core from Pupil Lab.

The output of the sensing platform is a JSON file describing the student's status (see Table 1), updated at regular time intervals.

Table 1. Student's status

Disturbance	pet; people; electronic device
Change	[0, 1, 2]
Location	library; bedroom; cafe
Activity	writing; reading; watching video; typing; other
Affect	frustrated; confused; bored; neutral; delight
Posture	holding chin; touching face; touching hair; leaning back; slouching forward; touching chair
Fatigue	[0, 1, 2]
Cognitive Load	[0, 1, 2]

2.2 Decision System

The decision system decides if remediation (nudging) is needed and what it should achieve. Eight types of interaction have been identified following a participatory design approach: encourage, chat, advise a change of posture, signal sources of distraction, advise a change of location, ask to concentrate, tell to start studying, advise to take a break.

The decision system is implemented as a decision tree (sklearn library and Gini index). It takes for input the student's status (see Table 1), as well as information from a student's profile, which currently contains only two pieces of information: gender and agent's preference (companion/assistant, friend or supervisor). The pruned and optimised decision tree (depth of 8) currently achieves a 0.86 accuracy.

2.3 Supportive Agent

The supportive agent executes the nudging decisions. The agent is a virtual character appearing on screen as an overlay (See examples in Fig. 2). Its shape and behaviour (implemented using Blender and Unity3D) have been decided in the course of seven participatory design events (concept development, remediation needs, avatar image, avatar animation, interaction design, feedback collection) conducted with a team of six college students majoring in a variety of subjects.

Fig. 2. Supportive companion agent.

3 Conclusion

Work is ongoing to complete the implementation and testing of the architecture's sub-systems before their integration. The Decision System has so far been designed and tested using simulated data (student's status and profiles) and will be refined when outputs from the sensing platform will become available. Early versions of the system will be used to collect further data for the training and refinement of its components.

References

1. Zimmerman, B.J.: From cognitive modeling to self-regulation: a social cognitive career path. Educ. Psychol. **48**(3), 135–147 (2013). https://doi.org/10.1080/00461520.2013.794676
2. Harley, J.M., Taub, M., Bouchet, F., et al.: A framework to understand the nature of co-regulated learning in human-pedagogical agent interactions. In: 11th International Conference on Intelligent Tutoring Systems, Crete (2012)
3. Kuppens, P.: It's about time: a special section on affect dynamics. Emot. Rev. **7**(4), 297–300 (2015)
4. Redmon, J., Farhadi, A.: YOLOv3: an incremental improvement, arxiv:1804.02767 Tech report (2018)
5. Baker, R.S., d'Mello, S.K., Rodrigo, W.T., et al.: Better to be frustrated than bored: the incidence, persistence, and impact of learners' cognitive-affective states during interactions with three different computer-based learning environments. Int. J. Hum. Comput. Stud. **68**, 223–241 (2010)

On Applicability of Neural Language Models for Readability Assessment in Filipino

Michael Ibañez, Lloyd Lois Antonie Reyes, Ranz Sapinit,
Mohammed Ahmed Hussien, and Joseph Marvin Imperial[✉]

National University, Manila, Philippines
jrimperial@national-u.edu.ph

Abstract. In the field of automatic readability assessment (ARA), the current trend in the research community focuses on the use of large neural language models such as BERT as evidenced from its high performance in other downstream NLP tasks. In this study, we dissect the BERT model and applied it to readability assessment in a low-resource setting using a dataset in the Filipino language. Results show that extracting embeddings separately from various layers of BERT obtain relatively similar performance with models trained using a diverse set of handcrafted features and substantially better than using conventional transfer learning approach.

Keywords: Readability assessment · Neural language models · BERT

1 Introduction

Readability assessment is the process of correctly identifying the *reading difficulty* of a text. It is an interdisciplinary challenge tacked by researchers in education, linguistics, and computer science. Common approaches in this task started with formulas such as the Flesch-Kincaid formula [6] or the Dale-Chall formula [3] which makes use of countable text or *handcrafted* features such as average sentence and word lengths. In recent works, more and more research have explored the use of advanced text representations such as modelling the task using neural language models like BERT [5]. BERT is a Transformer-based language model which has 12 attention-based layers and has been trained in a self-supervised setup with large datasets such as Wikipedia articles. This particular model has been recently used by previous works for readability assessment in various languages such as English [4,11] and Chinese [12]. In this study, we explore the use of BERT to readability assessment in a low-resource setting with an existing Filipino dataset in two possible ways: transfer learning and layerwise training. We argue that exploration on low-resource languages is beneficial to the community in generally and would shed light on how researchers can move forward and apply BERT the correct way for tasks such as readability assessment.

© Springer Nature Switzerland AG 2022
M. M. Rodrigo et al. (Eds.): AIED 2022, LNCS 13356, pp. 573–576, 2022.
https://doi.org/10.1007/978-3-031-11647-6_118

2 Data and Setup

For the data, we used the same Adarna House dataset from previous works in Filipino readability assessment as seen in [7–10]. The Adarna House dataset comprises 265 expert-annotated children's reading materials written in Filipino and divided equally into levels 1, 2, and 3 with respect to the grades of the Philippine education system. For model training with the BERT, we used the uncased Filipino version from [1]. This version has been trained with a large corpus of Filipino Wikipedia articles over a sequence length of 512 similar to the original model in English. The complete recipe for training this model is described in [2]. In using BERT specifically for readability assessment, we describe two methods below:

Transfer Learning Setup. This is the most common setup where a BERT model is finetuned for a downstream task such as classification or named-entity recognition. For finetuning, we added an additional classification layer with default hyperparameter values used in [5] for training the Filipino BERT model to produce a probability-based grade level as an output.

Layer-Wise Training Setup. Given a sequence of text, we let the BERT model transform the data to a numerical representation (embeddings) and use it directly as features to a machine learning algorithm. For this method, we extract 12 different embeddings each having a dimension of 768 as input from each layer of the BERT model. For this study, we use SVM to train the model using the embeddings as feature input and the corresponding grade level as the output.

Table 1. Comparison of previous methods and features used in Filipino text readability assessment.

Method	Features	Acc	F1	Model
Traditional features [8]	7	0.42	0.41	SVM
Lexical features [8]	15	0.47	0.47	SVM
Language model [9]	25	0.72	0.72	LogReg
Diverse features [10]	54	0.62	0.62	RF
Sentence embeddings [7]	768	–	0.57	SVM
Transfer Learning (Ours)	**N/A**	**0.57**	**0.50**	**BERT**

3 Result and Discussion

Tables 1 and 2 describe the results of comparing BERT for readability assessment in a transfer learning setup against previous works and using embeddings from each of the twelve layers as features. With the first experiment, our findings suggest that using the conventional transfer learning setup for readability assessment in Filipino only beats the first two models using traditional and lexical features by [8]. The model trained with diverse features covered by [10] includes morphology and syllable pattern. This model performed better than BERT which may mean that BERT's linguistic knowledge may not be fully complete for the task of readability assessment especially for low-resource and morphologically-rich languages like Filipino.

Table 2. Results of using BERT's layers as features for Filipino text readability assessment.

Layer	Acc	Prec	Rec	F1
Layer 1	0.69	0.68	0.69	0.68
Layer 2	0.63	0.64	0.63	0.61
Layer 3	0.65	0.64	0.65	0.64
Layer 4	0.60	0.59	0.60	0.59
Layer 5	0.69	0.68	0.68	0.67
Layer 6	0.67	0.66	0.66	0.66
Layer 7	0.61	0.61	0.61	0.60
Layer 8	0.68	0.68	0.68	0.68
Layer 9	0.67	0.67	0.67	0.67
Layer 10	0.65	0.66	0.65	0.65
Layer 11	0.62	0.62	0.61	0.61
Layer 12	0.69	0.69	0.68	0.68

In the second experiment in Table 2, layers 1, 5, 8, and 12 obtained roughly comparable results with approximately 0.68 in all metrics used. The sentence embeddings previously in [7] as indicated in Table 1 only used the mean of all twelve layers while this study dissected and compared the performance of each layer. Following the observation of [13], the middle layers (layer 5 and 8) which is said to contain most of BERT's syntactic knowledge obtained top results. This finding suggests that syntactic knowledge is still important for readability assessment in Filipino. In additional, all layer-wise training performances fared better than the performance of BERT in a transfer learning setup which suggests the possibility that exploring these layers can be a good starting point for other languages.

4 Moving Forward

Automatic readability assessment (ARA) is a growing research challenge drawing contributions from the areas of education and computational linguistics. This study reveals that, when applied to low-resource languages like Filipino, using neural representations like BERT's layer-specific sentence embeddings as features is better compared to using the conventional transfer learning method done in most NLP tasks. Moreover, comparing with previous works, it is also worth noting that the use of hand-coded linguistic features and traditional machine learning algorithms is still a good starting point for low-resource languages as they produce good baseline performance. From our findings, we hope to shed light to researchers working in the field on how resources like BERT can be properly used in readability assessment and encourage them to actively search for novel or improved methods to exploit its potential.

References

1. Cruz, J.C.B., Cheng, C.: Evaluating language model finetuning techniques for low-resource languages. arXiv preprint arXiv:1907.00409 (2019)
2. Cruz, J.C.B., Cheng, C.: Establishing baselines for text classification in low-resource languages. arXiv preprint arXiv:2005.02068 (2020)
3. Dale, E., Chall, J.S.: A formula for predicting readability: instructions. Educ. Res. Bull. 37–54 (1948)
4. Deutsch, T., Jasbi, M., Shieber, S.: Linguistic features for readability assessment. In: Proceedings of the Fifteenth Workshop on Innovative Use of NLP for Building Educational Applications, pp. 1–17. Association for Computational Linguistics, July 2020
5. Devlin, J., Chang, M.W., Lee, K., Toutanova, K.: BERT: pre-training of deep bidirectional transformers for language understanding. In: NAACL, pp. 4171–4186. Association for Computational Linguistics, Minneapolis, Minnesota, June 2019. https://doi.org/10.18653/v1/N19-1423
6. Flesch, R.: A new readability yardstick. J. Appl. Psychol. **32**(3), 221 (1948)
7. Imperial, J.M.: BERT embeddings for automatic readability assessment. In: Proceedings of the International Conference on Recent Advances in Natural Language Processing (RANLP 2021), pp. 611–618. INCOMA Ltd., Held Online, September 2021
8. Imperial, J.M., Ong, E.: Application of lexical features towards improvement of Filipino readability identification of children's literature. arXiv preprint arXiv:2101.10537 (2020)
9. Imperial, J.M., Ong, E.: Exploring hybrid linguistic feature sets to measure Filipino text readability. In: 2020 International Conference on Asian Language Processing (IALP), pp. 175–180. IEEE (2020)
10. Imperial, J.M., Ong, E.: Diverse linguistic features for assessing reading difficulty of educational Filipino texts. arXiv preprint arXiv:2108.00241 (2021)
11. Martinc, M., Pollak, S., Robnik-Šikonja, M.: Supervised and unsupervised neural approaches to text readability. Comput. Linguist. **47**(1), 141–179 (2021)
12. Tseng, H.-C., Chen, H.-C., Chang, K.-E., Sung, Y.-T., Chen, B.: An innovative BERT-based readability model. In: Rønningsbakk, L., Wu, T.-T., Sandnes, F.E., Huang, Y.-M. (eds.) ICITL 2019. LNCS, vol. 11937, pp. 301–308. Springer, Cham (2019). https://doi.org/10.1007/978-3-030-35343-8_32
13. Vig, J., Belinkov, Y.: Analyzing the structure of attention in a transformer language model. arXiv preprint arXiv:1906.04284 (2019)

Preliminary Design of an AI Service to Assist Self-regulated Learning by Edge Computing

Eason Chen[1]([⊠]) [iD], Yuen-Hsien Tseng[1] [iD], Yu-Tang You[2] [iD], Kuo-Ping Lo[1] [iD], and Chris Lin[1] [iD]

[1] National Taiwan Normal University, Taipei 10610, Taiwan (R.O.C.)
eason.tw.chen@gmail.com
[2] National Yang Ming Chiao Tung University, Hsinchu 300093, Taiwan (R.O.C.)

Abstract. When sitting in front of a computer screen for online learning, students can be easily distracted by other sources on the Internet. To help students improve their online learning experience, we implemented a highly accessible AI service, which collected facial data from the web camera to assist self-regulated learning for students with privacy protection by way of edge computing. The service can capture self-learning metrics such as eye gazing points and facial expressions. Then the captured facial streaming data can be played back by the user. All steps are done locally at the users' browser. The learners can review their online learning process to improve their learning efficiency through an interactive interface. Our preliminary evaluation showed promising feedback from real users.

Keywords: Self-regulated learning · Edge computing · Online learning · E-learning · Artificial Intelligence (AI) · Eye Tracking

1 Introduction and Background

Online learning has become a common practice ever since the COVID-19 pandemic began. However, learners face several challenges during e-learning, primarily the high chances of distraction from unrelated sources on the Internet [1].

A previous study has shown that students can improve their online learning efficiency through self-regulated learning [2]. Self-regulated learning is "an active, constructive process whereby learners set goals for their learning and then attempt to monitor, regulate, and control their cognition, motivation, and behavior [3]". Research has shown that self-regulated learning is highly related to performance [3] and needs to be combined with feedback, so learners can adjust their learning strategy [2].

To collect data for feedback on online learning, most studies require specific IoT devices [1] or are limited to a specific platform [2]. Furthermore, they only can provide quantitative metrics. We speculate that without qualitative feedback it is difficult both to explain learners' performance and for them to improve their online learning performance. For example, if learners only know their attention failed after 10 min but don't know they were distracted by an unrelated notification, then there is little they can do to improve their learning performance.

© Springer Nature Switzerland AG 2022
M. M. Rodrigo et al. (Eds.): AIED 2022, LNCS 13356, pp. 577–581, 2022.
https://doi.org/10.1007/978-3-031-11647-6_119

The streaming information of eye-tracking with screenshots is used in our design to collect data for learners to playback to know when and where they pay attention to the content on the computer screen. According to the Eye-mind assumption [4], users' minds will focus on where their eyes are looking. Therefore, learners can determine where they direct their attention by replaying their learning journey with eye fixation heatmaps on screenshots.

Another point to consider is that many students are unwilling to turn on their cameras or share their online learning status because of privacy concerns or network accessibility [5]. This makes it a hindrance to collecting educational data and providing feedback in the real world. These problems can be avoided by edge computing. Edge computing [6] is used to process data at the end users' device. By calculating the data onsite, edge computing can ensure several promises, such as data safety, privacy, and high scalability. By implementing edge computing with Tensorflow.js, we can predict users' facial expressions [7] and gaze points [8] by a web browser on users' computers.

The contribution of this paper is the creation of a highly accessible AI service to provide streaming feedback for students to review their online learning history. The implemented service collects data from the web camera to assist self-regulated learning for students with privacy protection by edge computing.

2   AI Service Implementation

This section will show how the service collects data from users. An overview of our implemented service is shown in Fig. 1.

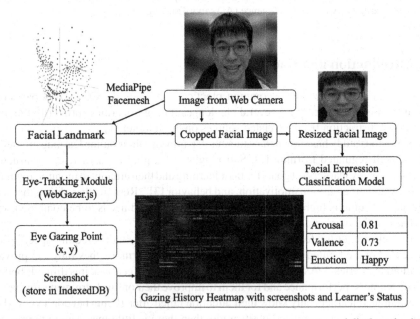

Fig. 1. The data flow of the service. Showing how it collects, processes, and displays the data.

2.1 Eye Tracking Module

We used the WebGazer.js library [8] to track users' eye-gazing points. This library first uses the MediaPipe Face Mesh to estimate users' 468 3D face landmarks. Then the WebGazer.js predicts their eye gazing points with those landmarks by Tensorflow.js.

2.2 Facial Expression Classification Model

To train the facial expression recognition model by PyTorch, we used the data from the AffectNet dataset [9]. The dataset contains over 42,000 labeled human face images and their emotions, valence, and arousal. We designed and trained the model following the steps from [7], which first crops and resizes users' facial images and then sends them into the pre-trained MobileNet model. We further added valence and arousal support to the model. Then we convert the model into the format supported by Tensorflow.js.

2.3 Data Flow

We use the web camera and screenshot to collect data (Fig. 1). When users start using the service, it will capture the users' faces and calculate users' gazing points and other metrics such as facial expression, valence, and arousal by Tensorflow.js. Then, the service will store these data with the screenshot in the IndexedDB, a built-in database powered by the browser to provide massive structured data storage at the client-side. The frequency of the storage interval will adjust itself based on users' computer performance.

2.4 Showing Feedback

When users want to review their learning history, the service will retrieve the information from IndexedDB and demonstrate it to users with an interactive interface. The interface contains a heatmap with the screenshot, which shows users' attention within a screen frame, and users' facial expression metrics, such as emotion, valence, and arousal. In addition, the interface provides an interactive timeline and charts so that users can adjust the timestamp to view different frames. Hence, users can review their study logs and conduct self-regulated learning to improve their learning efficiency.

3 Current Results

The service's prototype is ready to use at the website (https://aied22.focus.gift). The above JavaScript modules and pre-trained models will be loaded into browsers at users' computers. Users first need to fine-tune their tracking module with a mouse moving mini-game since the WebGazer assumes that users' eyes will look at their mouse when they click. After this calibration, users can start using the service even without an Internet connection. The service will record users' learning journeys in the background while they use their computers for e-learning. All metrics are computed by JavaScript on users' computers and stored by the browser. After that, users can replay their learning history for self-regulated learning. Finally, users can export their learning data online to others, such as teachers, co-learners, or researchers. Users can specify which data they want to record, store, or export.

The preliminary evaluation of the service, based on a couple of users' feedback, is generally favorable. First, users like the service's accessibility since they only need to open a website in a browser to use it. Moreover, users are amazed that the service is still working even when they turn off their networks. Finally, users enjoy replaying their learning history with screenshots, eye-gazing heatmaps, and other metrics.

4   Future Works

One future work of the service is to improve the model's performance and explainability. We decided to do so because according to users' feedback, some users showed that the emotion recognition model is less accurate, and others indicated that they didn't know how to explain their learning metrics. We also would like to add more lightweight models to our service, such as the screenshot annotations module, so that users can review their learning history with more data.

In addition, we plan to deploy and evaluate the service with the Technology Acceptance Model and study the correlation of the user experience with their self-regulated learning competency in the future.

Acknowledgement. This work was partially supported by the Ministry of Science and Technology of Taiwan (R.O.C.) under Grants 109-2410-H-003-123-MY3.

References

1. Xiao, X., Wang, J.: Understanding and detecting divided attention in mobile mooc learning. In: Proceedings of the 2017 CHI Conference on Human Factors in Computing Systems (2017)
2. Afzaal, M., et al.: Explainable AI for data-driven feedback and intelligent action recommendations to support students self-regulation. Front. Artif. Intell. **4**, 723447 (2021)
3. Pintrich, P.R., Zusho, A.: Student motivation and self-regulated learning in the college classroom. In: Smart, J.C., Tierney, W.G. (eds.) Higher Education: Handbook of Theory and Research. Higher Education: Handbook of Theory and Research, vol. 17, pp. 55–128. Springer, Cham (2002). https://doi.org/10.1007/978-94-010-0245-5_2
4. Just, M.A., Carpenter, P.A.: A theory of reading: from eye fixations to comprehension. Psychol. Rev. **87**(4), 329 (1980)
5. Castelli, F.R., Sarvary, M.A.: Why students do not turn on their video cameras during online classes and an equitable and inclusive plan to encourage them to do so. Ecol. Evol. **11**(8), 3565–3576 (2021)
6. Shi, W., et al.: Edge computing: vision and challenges. IEEE Internet Things J. **3**(5), 637–646 (2016)
7. Savchenko, A.V.: Facial expression and attributes recognition based on multi-task learning of lightweight neural networks. In: 2021 IEEE 19th International Symposium on Intelligent Systems and Informatics (SISY). IEEE (2021)

8. Papoutsaki, A., et al.: WebGazer: scalable webcam eye tracking using user interactions. In: Proceedings of the 25th International Joint Conference on Artificial Intelligence (IJCAI) (2016)
9. Mollahosseini, A., Hasani, B., Mahoor, M.H.: Affectnet: a database for facial expression, valence, and arousal computing in the wild. IEEE Trans. Affect. Comput. **10**(1), 18–31 (2017)

Education Theories and AI Affordances: Design and Implementation of an Intelligent Computer Assisted Language Learning System

Xiaobin Chen[✉], Elizabeth Bear, Bronson Hui, Haemanth Santhi-Ponnusamy, and Detmar Meurers

University of Tübingen, Walter-Simon-Str. 12, 72072 Tübingen, Germany
xiaobin.chen@uni-tuebingen.de

Abstract. Skepticism towards AI-powered education products is widespread among educators. One reason for this skepticism is a perceived gap between what educators consider effective teaching and what AIED systems offer. This paper tries to address the issue by arguing for an architectural design that is oriented towards education theories and empirical findings instead of solely towards AI affordances, the benefits of which include more robust AIED systems and higher educator acceptance because of the increased match between the expectations of educators and the systems' offerings. We illustrate this design principle with an on-going intelligent computer assisted language learning project's architectural design by referring to the Interaction Theory of second language learning and state-of-the-art AI technology to implement the theoretical requirements.

Keywords: Education theory · AI affordances · Intelligent computer assisted language learning

1 Introduction

For the Artificial Intelligence in Education (AIED) community, there is little doubt that AI will thrive in education because of the obvious affordances of the technology to enhance the education process. However, despite the continuous development of AI and the research into how it can be integrated into students' daily education, skepticism of a future with ubiquitous AI in education is still widespread [1]. The AIED skepticism comes not only from people's lack of knowledge of the field and its capabilities, but also from the fact that a lot of AI-based education applications do not meet educators' expectations of how effective teaching and learning should be conducted. On the one hand, learning and pedagogical research keeps updating our understanding of the learning process and effective methods for teaching. On the other hand, due to the complexity of AIED problems and the tradition of focusing predominantly on the technical aspects, much AIED research still falls short of keeping up with the latest development in learning theories and empirical research findings.

As a result, in the present paper, we argue for an approach of AIED system design that is oriented by education theories and empirical research findings, making use of AI

© Springer Nature Switzerland AG 2022
M. M. Rodrigo et al. (Eds.): AIED 2022, LNCS 13356, pp. 582–585, 2022.
https://doi.org/10.1007/978-3-031-11647-6_120

affordances to implement learning-enhancing systems. We demonstrate the approach by showcasing the design principles and implementation of an Intelligent Computer Assisted Language Learning (ICALL) system Aisla for training spoken English as a foreign language.

2   The Interactionist Approach to SLA

Theories of Second Language Acquisition (SLA) abound [2]. We adopt the Interaction Theory [3] because it is not only a "model that dominates current SLA research" [4], but it is also intuitively understandable to language education practitioners who are not SLA experts. The latter factor is important for ICALL system design because it will help reduce the gap between the users' expectations and the system's offerings, hence also helping to eliminate their skepticism.

The Interaction Theory posits that language is learned through meaningful inter-action in the target language. Interaction involves *input*, an essential component for language learning, which can be written or spoken. Input needs to be *comprehensible* to the learner, so it needs to be flexibly modified to match the learner's current proficiency level. For example, people typically speak slower and use simpler language when talking to a foreigner. Input modification can also happen when the learner asks for clarification, confirmation, or elaboration during the interaction. The interaction process works as a process of *negotiation for meaning* [5], which naturally requires the learner to produce output, offering them a valuable opportunity to practice using the language in meaning-ful contexts. When learners produce language, they are forced to shift their focus and pay more attention to grammar, in addition to meaning [6]. The interaction process also involves providing *feedback* to the learner, which can be meaning focused (e.g., confir-mation checks, clarification requests, and comprehension checks) or form focused (e.g., implicit or explicit grammar error feedback). As a result, interaction is the fundamental process through which people acquire a new language.

The most natural way to implement the Interaction Theory is to adopt a classroom methodology called Task-Based Language Teaching (TBLT) [7]. Ellis acknowledged that TBLT has primarily been informed by the Interaction Theory. The TBLT paradigm defines the guiding principles of language learning task design [8], including meaning focuses, information gap, use of linguistic and non-linguistic resources, as well as goal orientation. TBLT creates a context in which the learner can be involved in a meaningful task for learning the target language. Numerous studies have confirmed the learning-enhancing effects of the TBLT approach to SLA [9].

In sum, the latest SLA and language pedagogy research showed that one effective method to foster language learning is to involve the learner in authentic meaning-focused tasks where they receive language input, produce output, and obtain feedback on their performance.

3   The AI Affordances

With regards to implementing the interactionist approach to SLA, involving learners in authentic real-life conversation tasks is a natural choice. To this end, AI-supported dialogue systems (also called conversation agents or chatbots) offer the affordance to involve language learners in conversation tasks. With the popularity of commercial chatbots like Google Assistant, Apple Siri, Amazon Alexa, and Microsoft Cortana, talking to a conversation agent has become second nature to many people. Development of conversation systems is also becoming ever easier thanks to the availability of commercial bot-builders such as Dialogflow, Lex, and Watson. For example, Kim et al. [10] recently reported using Dialogflow to develop an ICALL chatbot *Ellie*, which was found to "have considerable potential to become an effective language learning companion for L2 learners."

As reviewed earlier, the importance of input for language learning cannot be overemphasized. State-of-the-art AI offers valuable tools for input provision. For instance, speech synthesis is used in conversation systems to generate aural input. It can also be used to generate spoken explanation of language targets, cultural notes, or listening comprehension materials. The so-called "deepfake" technology of video synthesis also has the potential to add a visual element to language presentation. Interacting with visually synthesized human figures in different task scenarios creates a more authentic feeling of the task. A visual agent also increases psychophysiological arousal of the learner as opposed to voice and text chatbots [11]. Another important element for SLA is feedback, which requires understanding of the learner's output and adjustment to the task context and the learner's individual differences, such as their proficiency level, age, and learning orientations. The relevant AI technology for feedback generation is NLP technologies. A typical pipeline of the NLP process involves target construct identification, error detection, and feedback retrieval [12].

4   An ICALL System Architecture

Based on the interactionist approach to SLA and the affordances of current AI technology, we have devised an architecture for an ICALL system for training spoken English as shown in Fig. 1. Currently, a prototypical system implementing the architecture has been developed on the Android platform. We chose to use commercial systems for most of the AI services, which allows us to focus on designing learning contents and managing the learning process, two other important aspects of an AIED system development. Several studies have been planned with the system, including testing of the adequacy of the various AI technologies, validating its effectiveness and learner perception, as well as experimenting on how interaction mode, *i.e.*, with audio/text *vs.* with a synthesized human, may affect engagement and motivation, learning outcomes, and learner perceptions of AIED systems.

To conclude, the short paper argues for designing AIED systems by orienting towards implementing education theories and research findings while considering AI affordances as a measure to improve system effectiveness and promote user acceptance. We demonstrated the idea with the architectural design of an on-going project aiming at creating an ICALL system for training spoken English as a foreign language.

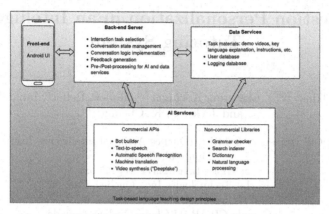

Fig. 1. The architecture of Aisla using AI affordances to implement TBLT design principles.

References

1. Klein, A.: Three Reasons to Be Skeptical of Artificial Intelligence in Schools. https://www. edweek.org/technology/three-reasons-to-be-skeptical-of-artificial-intelligence-in-schools/ 2020/02. Accessed 31 Mar 2022
2. VanPatten, B., Williams, J.: Theories in Second Language Acquisition: An Introduction, 2nd edn. Routledge, New York (2015)
3. Gass, S., Mackey, A.: Input, interaction and output in second language acquisition. In: Van-Patten, B., Williams, J. (eds.) Theories in Second Language Acquisition: An Introduction (2nd Edition), pp. 180–206. Routledge, New York (2015)
4. Ramírez, A.G.: Review of the social turn in second language acquisition. Mod. Lang. J. **89**, 292–293 (2005)
5. Long, M.: The role of the linguistic environment in second language acquisition. In: Ritchie, W., Bhatia, T. (eds.) Handbook of Language Acquisition. Second Language Acquisition, vol. 2, pp. 413–468. Academic Press, San Diego (1996)
6. Swain, M.: Three functions of output in second language learning. In: Cook, G., Seidlhofer, B. (eds.) Principle and Practice in Applied Linguistics, pp. 125–144. Oxford University Press, Oxford (1995)
7. Ellis, R.: Task-Based Language Learning and Teaching. Oxford University Press, Oxford (2003)
8. Ellis, R.: Task-based language teaching: sorting out the misunderstandings. Int. J. Appl. Linguist. **19**, 221–246 (2009)
9. Bryfonski, L., McKay, T.: TBLT implementation and evaluation: a meta-analysis. Lang. Teach. Res. **23**(5), 603–632 (2019)
10. Kim, H., Yang, H., Shin, D., Lee, J.: Design principles and architecture of a second language learning chatbot. Lang. Learn. Technol. **26**(1), 1–18 (2022)
11. Ciechanowski, L., Przegalinska, A., Magnuski, M., Gloor, P.: In the shades of the uncanny valley: an experimental study of human–chatbot interaction. Futur. Gener. Comput. Syst. **92**, 539–548 (2019)
12. Rudzewitz, B., Ziai, R., De Kuthy, K., Möller, V., Nuxoll, F., Meurers, D.: Generating feedback for English Foreign language exercises. In: Proceedings of the Thirteenth Workshop on Innovative Use of NLP for Building Educational Applications (BEA), pp. 127–136 (2018)

Question Personalization in an Intelligent Tutoring System

Sabina Elkins[1,2]([✉]), Ekaterina Kochmar[2,3], Robert Belfer[2], Iulian Serban[2], and Jackie C. K. Cheung[1,4]

[1] McGill University & MILA (Quebec Artificial Intelligence Institute), Montreal, Canada
sabina.elkins@mail.mcgill.ca
[2] Korbit Technologies Inc., Quebec, Canada
[3] University of Bath, Bath, UK
[4] Canada CIFAR AI Chair, Quebec, Canada

Abstract. This paper investigates personalization in the field of intelligent tutoring systems (ITS). We hypothesize that personalization in the way questions are asked improves student learning outcomes. Previous work on dialogue-based ITS personalization has yet to address question phrasing. We show that generating versions of the questions suitable for students at different levels of subject proficiency improves student learning gains, using variants written by a domain expert and an experimental A/B test. This insight demonstrates that the linguistic realization of questions in an ITS affects the learning outcomes for students.

Keywords: Intelligent tutoring system · Dialogue-based tutoring system · Personalized learning

1 Introduction

Intelligent tutoring systems (ITS) are AI systems capable of automating teaching. They have the potential to provide accessible and highly scalable education to students around the world [6]. Previous studies suggest that students learn significantly better in one-on-one tutoring settings than in classroom settings [2]. Personalization can be addressed in an AI-driven, dialogue-based ITSs, and can have significant impact on the learning process [5]. This has been explored in different ways, including dialogue feedback and question selection [8]. To the best of our knowledge, personalization in question phrasing has not been explored.

Students benefit from being asked questions tailored to their level of subject expertise and their needs during in-person tutoring sessions [1,4]. We hypothesize that the same effect can be achieved when questions are adapted to the students' levels of expertise and their needs in an ITS. To test this, we integrate question variants created by a human domain expert onto the Korbit Technologies Inc. platform and run an A/B test. Korbi's AI tutor, `Korbi`, is a dialogue-based ITS, which teaches students by providing them with video lectures and interactive problem solving exercises, selected for each student using ML and NLP

© Springer Nature Switzerland AG 2022
M. M. Rodrigo et al. (Eds.): AIED 2022, LNCS 13356, pp. 586–590, 2022.
https://doi.org/10.1007/978-3-031-11647-6_121

techniques [7]. The main contribution of this paper is the demonstration that question personalization in an ITS leads to improvements in learning gains.

2   Methodology

Students interact with Korbi through short answer questions and written responses. To assess if the phrasing of these questions can impact learning gains, we first create a set of questions and variants that reformulate the original idea. The variants were created by a human expert from existing questions on the Korbit platform. They were designed to reflect three levels of difficulty: *beginner, intermediate,* and *advanced,* as per common practice in education. We assume that less detailed questions are harder (as the student must have more background knowledge to understand and answer) and more elaborate ones are easier (as they 'hint' at the answer with extra information) [9]. In our data, each question has three variants at different levels of proficiency. Questions were made easier by adding elaborations and synonym replacement, and more difficult by removing non-essential explanations and synonym replacement. As Table 1 shows, the beginner variants are longer and the advanced ones are more concise.

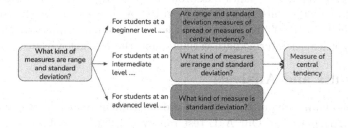

Fig. 1. An question being adapted to different levels while retaining the same answer.

The variants were given to three human experts (data scientists with at least an MSc in a related field) who rated them on three scales from 0 to 5, representing *difficulty* (i.e., the relative complexity of the question as compared to the others), *fluency* (i.e., spelling and grammar), and *meaning preservation* (i.e., if the meaning of the original question is preserved). The mean results of their ratings can be seen in Table 1. The *fluency* and *meaning preservation* metrics are consistently high and the *difficulty* metric increases with the assigned levels.

Table 1. Mean variant scores from human experts, and average word counts by level.

Level	Difficulty	Fluency	Meaning preservation	Mean word count
Beginner	1.689 ±0.635	4.600 ±0.471	4.789 ±0.451	39.800
Intermediate	2.667 ±0.689	4.683 ±0.481	4.839 ±0.406	33.533
Advanced	3.939 ±1.269	4.544 ±0.661	4.717 ±0.516	27.433

The next task is to select an appropriate question variant for each student at each step in the dialogue. Through Korbit, we have anonymized access to student history. We isolated 2,137 students' interactions with the platform. Each student's history consists of the exercises they encountered and their attempts to solve them. Each exercise encountered was included as a point in our dataset, for a total of 13,504 exercises given to students. Using this, it is possible to calculate a set of features indicative of a student's level, and subsequently build a logistic regression model to predict if a student will succeed on the next exercise.

The original feature set consisted of 7 features, including overall success rate, improvement (i.e., changes in success rate), skip rates, and others. From this set, two features were selected based on their contribution to the best model in the preliminary experiments: (1) *topic success feature* is a numerical feature in $[0, 1]$ that shows the eventual success rate per all exercises previously attempted in a given topic, and (2) *topic skip feature* is a numerical feature in $[0, 1]$ that is the skip rate per all exercises previously attempted in a given topic. A topic on Korbi is a broad category of material, such as 'Probability'. Using these features the model is able to predict next exercise success with an accuracy of 80%.

The variant assignment model calculates the features when a student gets a new exercise, and generates a probability of success with the regression model. Students are assigned variants based on the percentile range that their probability of success falls into (calculated from the predictions across the entire data set). Students in the 0th to 33rd percentiles get beginner variants, in the 33rd to 66th percentiles get intermediate variants, and the rest get advanced ones.

3   Results and Analysis

To test our claims, we put the variants and assignment model described in Sect. 2 on the Korbi platform. Our A/B test ran over 2 months, collecting data from over 400 students at varied skill levels. Student attempts were divided into three groups. The *expected* variant group received the variant which matched their assignment model score. The *non-expected* variant group received a variant which did not match their score from the assignment model (e.g., beginner question for an advanced student). The *control* group students received the original variant (i.e., that which was already on the platform before this experiment).

Table 2. Test results. Metrics marked with * are statistically significant at the $\alpha = 0.05$ level by a Student's t-test.

Experiment group	Solution acceptance*	Ultimate failure rate*	Skip rate	n
Expected	0.626 ± 0.069	0.163 ± 0.053	0.105 ± 0.044	190
Non-Expected	0.468 ± 0.083	0.295 ± 0.076	0.144 ± 0.058	139
Control	0.596 ± 0.081	0.191 ± 0.065	0.121 ± 0.054	141

Solution acceptance is the proportion of success per exercise attempts. However, succeeding on exercises does not equate to learning. Students should be challenged within their zone of proximal development [3] but eventually obtain the right answer, so we aim to minimize the *ultimate failure rate* as opposed to simply maximizing attempt success. This metric is the proportion of failure out of all exercises seen by students. Unlike *solution acceptance* which shows the success rate per attempt, *ultimate failure rate* shows the fail rate per exercise. *Skip rate* is indicative of a student's engagement. Intuitively, the more they skip, the less they engage with the content. All three of these metrics show the *expected* group performing the best, followed by *control* and finally *non-expected*. For *solution acceptance* and *ultimate failure rate*, the difference between *expected* and *non-expected* groups is statistically significant at $\alpha = 0.05$ by a Student's t-test.

The difference between the *expected* and *control* groups is smaller than the difference between the *expected* and *non-expected* groups. This can be attributed to the fact that the original questions were refined through several rounds of review by domain experts when they were created for Korbi platform, whereas the variants only were reviewed once. Additionally, the *control* group's exercises are always intermediate or advanced, while the strongest result is seen with beginners. Isolating the students who score for beginner variants only, we see a 19% relative reduction in *ultimate failure* when comparing the *expected* to the *control* group, which demonstrates a bigger impact for beginners. Additionally, the same comparison shows a 30% relative reduction in the *skip rate*, suggesting that the beginners are more engaged when dealing with beginner variants.

4   Conclusion

We see a clear improvement in the success of students in the *expected* group. This confirms our hypothesis that providing question variants suited to student's level improves their learning gains. These variants are more useful for beginner students who need more assistance, which is an encouraging and intuitive result. The future of this work is in automating the creation question variants for scalability, and creating a more sophisticated variant assignment approach.

Acknowledgements. We'd like to thank Korbit for hosting our experiment on their platform, and Mitacs for their grant to support this project. We are grateful to the anonymous reviewers for their valuable feedback.

References

1. Ashton-Jones, E.: Asking the right questions: a heuristic for tutors. Writing Center J. **9**(1), 29–36 (1988)
2. Bausell, R., Moody, W., Walzl, F.: A factorial study of tutoring versus classroom instruction. Am. Educ. Res. J. **9**(4), 591–597 (1972)
3. Cazden, C.: Peekaboo as an Instructional Model: Discourse Development at Home and at School, p. 17. Papers and Reports on Child Language Development, No (1979)

4. Hrastinski, S., et al.: Identifying and exploring the effects of different types of tutor questions in individual online synchronous tutoring in mathematics. Interactive Learning Environments, 1–13 (2019)

5. Kochmar, E., Vu, D.D., Belfer, R., Gupta, V., Serban, I.V., Pineau, J.: Automated personalized feedback improves learning gains in an intelligent tutoring system. In: Bittencourt, I.I., Cukurova, M., Muldner, K., Luckin, R., Millán, E. (eds.) AIED 2020. LNCS (LNAI), vol. 12164, pp. 140–146. Springer, Cham (2020). https://doi.org/10.1007/978-3-030-52240-7_26

6. Kulik, J.A., Fletcher, J.D.: Effectiveness of intelligent tutoring systems: a meta-analytic review. Rev. Educ. Res. **86**(1), 42–78 (2016)

7. Serban, I.V., et al.: A large-scale, open-domain, mixed-interface dialogue-based ITS for STEM. In: Bittencourt, I.I., Cukurova, M., Muldner, K., Luckin, R., Millán, E. (eds.) AIED 2020. LNCS (LNAI), vol. 12164, pp. 387–392. Springer, Cham (2020). https://doi.org/10.1007/978-3-030-52240-7_70

8. St-Hilaire, F., et al.: A New Era: Intelligent Tutoring Systems Will Transform Online Learning for Millions. arXiv:2203.03724 (2022)

9. Taylor, R.S.: The process of asking questions. Amer. Doc. **13**, 391–396 (1962)

Considering Disengaged Responses in Bayesian and Deep Knowledge Tracing

Guher Gorgun$^{(\boxtimes)}$ and Okan Bulut

University of Alberta, Edmonton, AB T6G 2G5, Canada
{gorgun,bulut}@ualberta.ca

Abstract. In this study, we analyzed the influence of student disengagement on prediction accuracy in knowledge tracing models. During the data pre-processing stage, we prepared two training data: The disengaged responses were ignored in the baseline data whereas the disengaged responses were removed in the disengagement-adjusted data. Using visual analysis, we identified disengaged responses (i.e., hint abusers and rapid guessers) and removed them from the disengagement-adjusted data during the pre-processing phase since those responses do not reflect the true latent ability of the students. After fitting the knowledge tracing models to the baseline and disengagement-adjusted data, we found that the prediction accuracy of both models on test data has substantially increased when disengaged responses were removed during the pre-processing stage. Our results emphasized the importance of considering student disengagement in knowledge tracing models to produce more accurate prediction models.

Keywords: Knowledge tracing · Intelligent tutoring systems · Student disengagement

1 Introduction

Formative assessments have been widely used by instructors to monitor student progress, identify knowledge gaps, and evaluate learning gains. However, administering many in-class formative assessments can be time-consuming. Intelligent tutoring systems (ITS) that incorporate both instruction and assessment while maintaining rich information about students' progress emerged as a viable option against traditional formative assessments [1]. ITS such as ASSISTment [2] combine assessment with tutoring so students' learning is not only traced but also the time spent on the system is primarily focused on learning through tutoring support. In addition, the interactions between the ITS and the student supply rich information concerning the student behavior such as help-seeking, engagement, response time, speed, or attempts.

Although formative assessments, including ITS applications, are widely used for depicting students' current state of knowledge, they are generally characterized as low-stakes assessments. Low-stakes assessments typically have no direct

M. M. Rodrigo et al. (Eds.): AIED 2022, LNCS 13356, pp. 591–594, 2022.
https://doi.org/10.1007/978-3-031-11647-6_122

consequences for students because they are not used for grading purposes, making graduation decisions, or granting awards. Instructors and school authorities may use formative assessments for adjusting teaching practices or identifying at-risk students but students typically do not observe this process directly. Due to the absence of direct consequences, some students may not give their full effort and show disengaged response behavior when attempting formative assessment items. Thus, in recent years, student engagement has been a major concern for assessment experts, practitioners, and researchers utilizing formative assessments (e.g., [3,4]). For example, when students disengage by gaming the systems such as through rapid guessing and hint abusing, they may appear as if they are unlearning the content [5]. Thus, researchers have been offering remedies to tackle student disengagement in low-stakes assessment contexts. In this study, we focused on student engagement in the context of ITS. Specifically, we aim at understanding the prediction accuracy when we incorporate student disengagement into the modeling process. Our research question is as follows: How does accounting for student disengagement affect prediction accuracy of knowledge tracing models (i.e., Bayesian and Deep Knowledge Tracing)?

2   Related Work

Several studies tried to model engagement or hint-taking behavior jointly with student performance. Johns and Woolf [6] proposed an item response theory-based dynamic mixture model for jointly modeling motivation and proficiency. The proposed model achieved on average 72.5% accuracy. Shultz and Arroyo [5] developed the knowledge and affect tracing model which allowed for a change in knowledge and affective states. They found that Bayesian Knowledge Tracing was better than their proposed model for predicting students' performance. Duong and colleagues [7] found that attempting an item first compared with taking a hint first has better prediction accuracy (e.g., 83% vs. 47%) for predicting students' performance. Finally, Chaudhry and colleagues [8] used Dynamic Key-Value Memory Networks to jointly model hint-taking and performance. The model achieved 91.75% accuracy for hint-taking prediction and 81.48% accuracy for performance prediction.

3   Methods

In this study, we used the ASSISTment 2009–2010 skill builder dataset.[1] The dataset included students' interactions within the ITS such as response time, hint count, chronological order of attempts, and skill names. We removed rows with negative response times and empty skill names and created the baseline data. The final baseline data included 111 skills and 4163 unique students. We used the 70% of the data as training set and 20% of the training set was used for validation.

---

[1] ASSISTments 2009–2010 skill builder dataset is available at https://sites.google.com/site/assistmentsdata/home/2009-2010-assistment-data.

Using the baseline data, we trained Bayesian Knowledge Tracing (BKT) [9] and Deep Knowledge Tracing (DKT) models [10] by ignoring disengaged responses in the data.

In addition to the above pre-processing steps, we removed disengaged responses to create a second disengagement-adjusted training set. We argued to remove disengaged responses because disengaged responses do not reflect the true latent ability of students, and hence they do not contribute towards predicting the future performance of students. Therefore, removing disengaged responses may increase the prediction accuracy. To identify disengaged responses, we first standardized the hint counts across the items because items in the data had varying number of hints available. We divided the number of hints used by the total number of hints available and multiplied it by 100 to find the percent hint value (see Eq. 1). Second, we divided the percent hint values by response times to find the hint count per second (see Eq. 2).[2]

$$percent\,hint = \frac{hint\,count}{total\,hint} * 100 \tag{1}$$

$$disengagement\,index = \frac{percent\,hint}{response\,time} \tag{2}$$

As this value increased, students requested more hints in a shorter response time and therefore they were more likely to abuse hints and rapidly guess (i.e., disengage) while attempting the item. To determine the threshold for disengagement, we visually analyzed the hint count per response time and response correctness. Based on the thresholds identified via visual analysis (i.e., hint count per response time >4), we removed the rows with disengaged response behavior and created a second training set (i.e., disengagement-adjusted data). The disengaged response rate was 9%. The disengagement-adjusted data included the same number of skills and students. We again trained BKT and DKT models. We evaluated the model performances with the test set.

4   Results and Discussion

Adjusting for disengagement during the data pre-processing stage improved the model performance substantially. Both BKT and DKT models performed better with the engagement pre-processed data. Overall, we argue that pre-processing data by excluding disengaged responses could be more effective than modeling complex student behaviors including disengagement in the model. Researchers can significantly improve the performance of their BKT and DKT models by removing disengaged responses based on ancillary variables such as the number of hints used and response time spent (Table 1).

---

[2] The Python codes are available at: https://github.com/GGorgun/Disengaged_Responses_in_Knowledge_Tracing.

Table 1. Performance metrics for baseline and disengagement-adjusted knowledge tracing models.

Model	Model	Accuracy	AUC	Precision	Recall
BKT	Baseline	.75	.78	.77	.93
	Disengaged-adjusted	.79	.81	.80	.97
DKT	Baseline	.78	.83	.80	.93
	Disengaged-adjusted	.82	.86	.83	.96

5   Conclusion

In this paper, we showed that a disengagement-based data pre-processing step in knowledge tracing models achieved higher prediction accuracy (79% for BKT and 82% for DKT), compared with models ignoring student engagement. Our findings suggest that when building knowledge tracing models based on ITS applications, educators need to consider handling student disengagement in the pre-processing stage to obtain more accurate prediction results.

References

1. Feng, M., Heffernan, N.T., Koedinger, K.: Addressing the assessment challenge with an online system that tutors as it assesses. User Model. User-Adap. Inter. **19**(3), 243–266 (2009)
2. Feng, M., Heffernan, N.T.: Informing teachers live about student learning: reporting in the assistment system. Technol. Instr. Cogn. Learn. **3**(1/2), 1–14 (2005)
3. Wise, S.L., DeMars, C.E.: Low examinee effort in low-stakes assessment: problems and potential solutions. Educ. Assess. **10**(1), 1–17 (2005)
4. de Vicente, A., Pain, H.: Informing the detection of the students' motivational state: an empirical study. In: Cerri, S.A., Gouardères, G., Paraguaçu, F. (eds.) ITS 2002. LNCS, vol. 2363, pp. 933–943. Springer, Heidelberg (2002). https://doi.org/10.1007/3-540-47987-2_93
5. Shultz, S., Arroyo, I.: Tracing knowledge and engagement in parallel in an intelligent tutoring system. In: Proceedings of the 7th International Conference on Educational Data Mining, pp. 312–315 (2014)
6. Johns, J., Woolf, B.: A dynamic mixture model to detect student motivation and proficiency. In: AAAI, pp. 163–168 (2006)
7. Duong, H., Zhu, L., Wang, Y., Heffernan, N.T.: A prediction model that uses the sequence of attempts and hints to better predict knowledge: better to attempt the problem first, rather than ask for a hint. In: Educational Data Mining, pp. 316–317 (2013)
8. Chaudhry, R., Singh, H., Dogga, P., Saini, S.K.: Modeling hint-taking behavior and knowledge state of students with multi-task learning. In: Proceedings of the 11th International Conference on Educational Data Mining, pp. 21–31 (2018)
9. Pardos, Z.A., Heffernan, N.T.: Modeling individualization in a Bayesian networks implementation of knowledge tracing. In: De Bra, P., Kobsa, A., Chin, D. (eds.) UMAP 2010. LNCS, vol. 6075, pp. 255–266. Springer, Heidelberg (2010). https://doi.org/10.1007/978-3-642-13470-8_24
10. Piech, C., et al.: Deep knowledge tracing. In: Advances in Neural Information Processing Systems, pp. 505–513 (2015)

K-12BERT: BERT for K-12 Education

Vasu Goel[1(✉)], Dhruv Sahnan[1], V. Venktesh[1], Gaurav Sharma[2],
Deep Dwivedi[1], and Mukesh Mohania[1]

[1] Indraprastha Institute of Information Technology, Delhi, India
{vasu18322,dhruv18230,venkteshv,deepd,mukesh}@iiitd.ac.in
[2] Extramarks Education pvt. ltd., Noida, India
gaurav.sharma@extramarks.com

Abstract. Online education platforms are powered by various NLP
pipelines, which utilize models like BERT to aid in content curation.
Since the inception of the pre-trained language models like BERT, there
have also been many efforts toward adapting these pre-trained models
to specific domains. However, there has not been a model specifically
adapted for the education domain (particularly K-12) across subjects
to the best of our knowledge. In this work, we propose to train a language model on a corpus of data curated by us across multiple subjects
from various sources for K-12 education. We also evaluate our model,
K-12BERT, on downstream tasks like hierarchical taxonomy tagging.

Keywords: AI in education · Language model · Domain adaption

1 Introduction

The pre-trained language models like BERT [4] have made considerable advancements in many NLP tasks. However, these models are trained on general domain
text like Wikipedia and Book Corpus and are not adapted to the vocabulary of
the target domain. Several works have addressed this by training domain-specific
language models like PubMedBERT [5], SciBERT [2], BioBERT [7] for biomedical
NLP. Following their success over vanilla pre-trained models, several other works
like TravelBERT [9], PatentBERT [6], FinBERT [1] have adapted domain-specific
pre-training for the respective domains. The domain-specific pre-training can be
performed in two ways: the continued pre-training approach or pre-training from
scratch. For instance, BioBERT and SciBERT demonstrate that when the corpus
is small leveraging pre-trained models to continue training on domain-specific corpus leads to an increase in performance on downstream in-domain tasks. Additionally, works like PubMedBERT demonstrate that pre-training from scratch leads
to gains on downstream tasks when in domain corpus is abundant.

In several instances, one might posit that the general domain text may overlap with the education vocabulary. However, the general domain text may contain advanced terms while lacking academic concepts, which are crucial for students to understand and achieve the learning objectives. This paper proposes to
train the BERT model on a corpus of text collected from various sources for the

M. M. Rodrigo et al. (Eds.): AIED 2022, LNCS 13356, pp. 595–598, 2022.
https://doi.org/10.1007/978-3-031-11647-6_123

Table 1. Composition of our corpus.

Source	Content	# sentences
NCERT (India)	P, C, B, SS	15K
Siyavulla.com (International)	H, L	2K
OpenStax.org (USA)	P, C, B	4K
Learncbse.in (India)	P, C, B, SS	19K
CK-12.org (USA)	P, C, B, L, H, E	14K
KhanAcademy.org (USA)	P, C, B, SS	282K
Extramarks.com (India)	P, C, B, H, SS	120K

Physics (P), Chemistry (C), Biology (B), Social studies (SS),
Physical science (H), Life science (L), Earth science (E)

K-12 education system across different geographic regions. We perform continued pre-training as the corpus is not abundant compared to other domains. In summary, the following are the core contributions of our work:

- We release a corpus for the K-12 education system as shown in Table 1.
- We perform continued pre-training of BERT on the K-12 corpus and evaluate on downstream tasks.
- Code and data are at https://github.com/ADS-AI/K12-Bert-AIED-2022.

2 Methodology

2.1 Dataset

We curate our dataset from multiple online learning platforms that provide open access for research purposes. To the best of our knowledge, the dataset curated is the first of its kind due to the lack of a corpus of K-12 learning content suitable for language model training. Table 1 describes the details of the datasets collected. The data collected ranges across different regions like the USA, India, and South Africa to avoid regional bias in the dataset. The data collected is as follows:

- For *NCERT (K-12) (India)* we used pdfminer[1], a python library for extracting information from NCERT PDFs[2].
- For *Siyavulla, OpenStax, LearnCBSE, CK-12* we systematically scraped the webpages which contained information and picked out the chunks which contained meaningful information. Then we used the HTML parser present with BeautifulSoup4[3] to break down the document into retrievable components and extract information from the paragraph tags.
- We accessed *Khan Academy transcripts* using the official APIs[4]. Khan academy transcripts had informal language, which added to the noise since they are made for an online video educational setup which was filtered out.

---

[1] https://pypi.org/project/pdfminer2/.
[2] https://ncert.nic.in/textbook.php.
[3] https://www.crummy.com/software/BeautifulSoup/bs4.
[4] https://github.com/Khan/khan-api.

Table 2. Performance comparison (Recall@K) of K-12BERTwith other baselines.

Dataset	Model	R@5	R@10	R@15	R@20
ARC	BERT+USE	0.67	0.81	0.86	0.89
	BERT+Sent_BERT	0.65	0.77	0.84	0.88
	BERT+K-12Sent_BERT	0.68	0.81	0.87	0.90
	K-12BERT+USE	0.65	0.78	0.85	0.88
	K-12BERT+Sent_BERT	0.68	**0.82**	**0.87**	0.90
	K-12BERT+K-12Sent_BERT	**0.68**	0.81	0.86	**0.90**
QC-Science	BERT+USE	0.86	0.92	0.95	0.96
	BERT+Sent_BERT	0.85	0.93	0.95	0.97
	BERT+K-12Sent_BERT	0.88	**0.94**	0.96	0.97
	K-12BERT+USE	0.84	0.91	0.94	0.96
	K-12BERT+Sent_BERT	0.88	0.93	0.95	0.97
	K-12BERT+K-12Sent_BERT	**0.88**	0.94	**0.96**	**0.97**

– The *Extramarks (EM)* transcripts were made available by Extramarks in .docx format and had content in Hindi and English. We filtered out the text containing Roman Hindi characters by comparing their unicode values. Using pyenchant[5] library, we run spellcheck on the words and maintain a count of approved words by the spell checker. Finally, we extract sentences that have more approved words than rejected words.

2.2 Continued Pre-training

Due to the constraint on the data available in this domain, we decided to continue pre-training. For our current experiment, we do not update the existing vocabulary. Using the existing BERT vocabulary allows the model to capture diverse information and be well suited for education-related tasks. The data that we scraped had discontinuity in sentences due to the scraping mechanism, which is a downside for NSP (Next Sentence Prediction) objective. Hence, for training K-12BERT, we use only the MLM (Masked Language Modeling) objective. To make the training resource-efficient, we utilize training techniques like Gradient Checkpointing, Gradient Accumulation, and mixed-precision training, proven to save GPU memory and speed up training. We continue the pre-training for 10 epochs over a batch size of 32 and gradient accumulation step size of 4. This setup allowed us to train our setup over 2 GPUs of 16GB memory each. We performed extensive experiments by training our model over different combinations of curated datasets. We achieved the best performance when the model was trained over Siyavulla, OpenStax, LearnCBSE, Ck-12.org, and EM transcripts.

---

[5] https://pypi.org/project/pyenchant/.

3 Results

To validate the training of K-12BERT we test it on the automated question tagging task for education domain. We evaluate our model on the task presented in [8] of tagging learning content like questions to a hierarchical learning taxonomy of form subject - chapter - topic. We use the official code provided by the authors and replace the models and sentence encoder, keeping all other settings the same. The results of K-12BERT and previous best baselines are listed in Table 2. We use state of the art models for generating contextualized sentence embeddings like Universal Sentence Encoder (USE) [3] and Sentence-BERT[6] (Sent_BERT) to generate embeddings for the taxonomy. We noticed that K-12BERT+{USE,Sent_BERT} wasn't performing as well as the vanilla BERT baseline. We believe the reason behind that could be since BERT, USE and SentBERT are trained on a general dataset, where as K-12BERT is exposed to more data pertaining educational domain, the embeddings generated are farther away in the vector space. So, in order to validate our hypothesis, we trained K-12Sent_BERT, a sentence BERT model finetuned for educational domain. We see that with using K-12BERT with K-12SentBERT we were able to outperform the previous baselines by 2% for the QC-Science dataset and 1% for ARC. We strongly attribute these results to the domain specific training of the models.

References

1. Araci, D.: FinBERT: financial sentiment analysis with pre-trained language models (2019). https://arxiv.org/abs/1908.10063
2. Beltagy, I., Lo, K., Cohan, A.: SciBERT: a pretrained language model for scientific text (2019). https://arxiv.org/abs/1903.10676
3. Cer, D., et al.: Universal sentence encoder (2018)
4. Devlin, J., Chang, M.W., Lee, K., Toutanova, K.: BERT: pre-training of deep bidirectional transformers for language understanding (2018). https://arxiv.org/abs/1810.04805
5. Gu, Y., et al.: Domain-specific language model pretraining for biomedical natural language processing. ACM Trans. Comput. Healthc. **3**(1), 1–23 (2022). https://doi.org/10.1145/3458754
6. Lee, J.S., Hsiang, J.: PatentBERT: patent classification with fine-tuning a pre-trained BERT model (2019). https://arxiv.org/abs/1906.02124
7. Lee, J., et al.: BioBERT: a pre-trained biomedical language representation model for biomedical text mining. Bioinformatics, September 2019. https://doi.org/10.1093/bioinformatics/btz682
8. V, V., Mohania, M., Goyal, V.: Tagrec: Automated tagging of questions with hierarchical learning taxonomy (2021). https://arxiv.org/abs/2107.10649
9. Zhu, H., Peng, H., Lyu, Z., Hou, L., Li, J., Xiao, J.: TravelBERT: pre-training language model incorporating domain-specific heterogeneous knowledge into a unified representation (2021). https://arxiv.org/abs/2109.01048

---

[6] https://www.sbert.net.

Data Mining of Syntax Errors in a Large-Scale Online Python Course

Jung A. Lee[1], Irena Koprinska[1]([⊠]), and Bryn Jeffries[2]

[1] School of Computer Science, The University of Sydney, Sydney, Australia
jlee6778@uni.sydney.edu.au, irena.koprinska@sydney.edu.au
[2] Grok Academy and School of Computer Science, The University of Sydney, Sydney, Australia
bryn.jeffries@grokacademy.org

Abstract. This paper investigates the common syntax errors students encounter when programming in Python using data from a large-scale online beginner course. We firstly analyse the error distribution to find differences between passing and failing students and then use clustering and Markov chains to identify clusters of student submissions with similar error pattern and how students move between these clusters during the course and between consecutive tasks. This type of analysis can be used by educators to understand student behaviour related to syntax errors and provide effective teaching support.

Keywords: Syntax errors · Student behaviour · Clustering · Markov chains

1 Introduction

Novice programmers struggle with syntax errors. Educators often underestimate how much attention should be given to syntax errors and focus on the concepts, leaving students to learn the syntax through self-practice [1]. A few previous studies [1, 2] investigated syntax errors in Java showing the most frequently encountered errors, the time needed to fix them which was significant for some common errors and the mismatch between the actual error occurrence and the teacher's expectations.

In this paper, we analyse syntax errors observed in a large-scale online beginner programming course in Python. Python is currently the most popular programming language. We apply statistical and data mining methods to analyse the error distribution, find clusters of student submissions based on the types of errors and how students move between these clusters. Our results provide insights to educators about the common student problems in learning the Python syntax, the student progression over time and the difficult tasks which require remedial actions.

2 Data

The data used in this study is from the 2018 edition of a beginner level Python course [3] with 12,898 participating high school students. The course consists of 5 weeks of

© Springer Nature Switzerland AG 2022
M. M. Rodrigo et al. (Eds.): AIED 2022, LNCS 13356, pp. 599–603, 2022.
https://doi.org/10.1007/978-3-031-11647-6_124

content (slides) and programming exercises (tasks) – 40 in total, 8 for each of the 5 weeks. Students could write their code in an editor within the platform. They were required to firstly run their code (called *terminal run*) so that the syntax errors and code behaviour can be observed, before submitting it for automated testing and grading against a suite of tests. We analyse terminal run data which included 1,635,638 attempts across all 40 tasks and 21 types of syntax errors as shown in Table 1.

Table 1. Syntax errors encountered by students in the course and number of occurrences

E0: SyntaxError: unexpected EOF while parsing: 27234	E11: SyntaxError: 'return' outside function: 1028
E1: SyntaxError: invalid syntax: 213063	E12: SyntaxError: can't use starred expression here: 57
E2: SyntaxError: Missing parentheses in call to 'print': 3330	E13: SyntaxError: f-string: invalid conversion character: expected 's', 'r', or 'a': 0
E3: SyntaxError: EOL while scanning string literal: 14440	E14: SyntaxError: f-string: single '}' not allowed: 2
E4: SyntaxError: keyword can't be an expression: 644	E15: SyntaxError: f-string: empty expression: 0
E5: SyntaxError: can't assign to operator: 607	E16: SyntaxError: f-string: expecting '}': 4
E6: SyntaxError: can't assign to literal: 1676	E17: SyntaxError: can't assign to function call: 855
E7: SyntaxError: invalid character in identifier: 2675	E18: NameError: 29726
E8: SyntaxError: 'break' outside loop: 28	E19: TypeError: 19876
E9: SyntaxError: invalid token: 102	E20: IndentationError: 70167
E10: SyntaxError: positional argument follows keyword argument: 39	

An *attempt vector* was created for each student and task. It consists of 21 features corresponding to the 21 error types; the feature values are the total number of errors from each type in all terminal runs for the task. In addition, we collected information about the outcome of each task when it was later submitted for testing against the suite of tests – "passed", "failed" or "not submitted". The "not submitted" attempts were excluded, resulting in 1,488,070 attempts used for the analysis.

3   Error Distribution

Figure 1 shows a histogram of the average occurrence of each error type over all tasks during the course, for passed and failed students separately. The error distribution is very similar (with E1, E20, E18, E0 and E19 the most frequent errors) which shows that the failing students are not particularly prone to a specific error. However, the average number of each error is higher for the failing students, suggesting that syntax issues contributed to these students not being able to solve the problem.

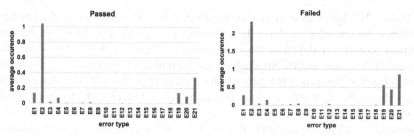

Fig. 1. Average occurrence of each error type for passed (left) and failed (right) students

4 Clustering of Student Attempts

To investigate the similarity between student attempts, we conducted clustering of attempt vectors, using only the four most frequently occurring errors. To account for some errors occurring too frequently compared to others, we used the tf-idf representation: $tfidf(t, d) = tf(t, d) * idf(t) = tf * \log\left[\frac{1+n}{1+df(t)}\right] + 1$, where t is the error type and d is the attempt and n is the number of attempt vectors.

We applied the K-Means clustering algorithm using the elbow method to determine the number of clusters, which resulted in $k = 4$ clusters. The cluster centroids are shown in Table 2 and indicate different characteristics of each cluster. The typical pattern of syntax errors for each cluster can be summarized as: c0: High invalid syntax, c1: Almost no errors, c2: High indentation error and c3: High name error.

Table 2. Cluster centroids

Feature	Cluster				
	Full data	c0	c1	c2	c3
E0	0.043	0.054	0.0430	0.037	0.037
E1	0.23	**0.960**	0.0023	0.210	0.120
E18	0.044	0.017	0.00052	0.029	**0.940**
E20	0.081	0.042	0.00076	**0.92**	0.081

5 Changes Over Time

Markov chains were used to investigate how students transition between the clusters during the course. Two types of Markov chains were constructed: showing the (1) overall movement of students during the course and (2) transition between two consecutive tasks. The first gives an overview of the student behaviour; the second is useful to identify the most challenging tasks and the reasons for the difficulties. These insights can help teachers to revise the content and provide personalized feedback.

Figure 2a) shows an overview of the student transitions between clusters for all problems. We can observe that a large number of students (84,616) from cluster c1 ("almost no errors") stayed in the same cluster – these are the students who didn't struggle with syntax errors. Similarly, a large number of students (14,583) from c0 ("invalid syntax") stayed in the same cluster, i.e. could not easily correct the error. This may be due to the vague error message for the invalid syntax error which is not helpful for novice programmers. Teachers can provide help by clarifying the possible reasons and cases where it is likely to encounter this error. In contrast, clusters c2 ("high indentation error") and c3 ("high name error") have a small number of students returning to them, compared to the students transitioning to other clusters. This shows that these two errors are not encountered consistently, only from time to time.

Figure 2 b) shows the transition of students between two consecutive problems – from Task 8 to Task 9. We can see that many students moved from c1 to c0 - they did not encounter errors in Task 8 but had a high number of invalid syntax errors in Task 9. The opposite was observed between Task 12 and 13 (Markov chain not shown) – many students moved from c0 to c1. A closer examination of the nature of the tasks gives insights about the possible reasons. In Task 9, students are required to write if/else-statements and more lines of code compared to the previous tasks. They had to read many slides before attempting the task and there was no sample code provided. It is possible that students skipped the slides or did not read them carefully, especially not paying attention to the required ":" after "if" which generates an invalid syntax error as shown in Fig. 2c). In contrast, in Task 13 sample code was provided as a hint and many students transitioned to c1, the cluster with a minimum number of errors.

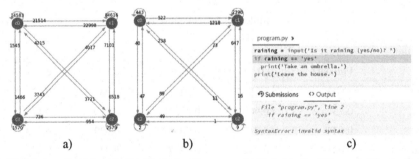

a)   b)   c)

Fig. 2. Markov chain for: a) all tasks during the course, b) Task 8 to 9; c) Example "SyntaxError: invalid syntax" for Task 9

6   Conclusion

This paper investigates the common syntax errors students encounter when programming in Python. The results showed that the error incidence distribution is similar for both passing and failing students, and that failing students are not prone to specific errors. However, the average number of errors is higher for failing students, suggesting that syntax issues contributed to these students not being able to solve the tasks. We showed how to use clustering and Markov chains to find clusters of student submissions with similar error pattern and how students move between these clusters during the course. This allows to understand student behaviour in terms of syntax errors, identify the students who repeatedly make the same type of errors, the difficult tasks where students struggle and the error messages that are less helpful for knowing what went wrong. These insights can help teachers to take remedial actions and provide feedback to improve the learning outcomes.

References

1. Denny, P., Luxton-Reilly, A., Tempero, E., Hendrickx, J.: Understanding the syntax barrier for novices. In: 16th Conference on Innovation and Technology in Computer Science Education, pp. 208–212 (2011)
2. Brown, N., Altadmri, A.: Investigating novice programming mistakes: educator beliefs vs student data. In: 10th Conference on International Computing Education, pp. 45–50 (2014)
3. NCSS Challenge. https://grokacademy.org/challenge

It Takes Two: Examining the Effects of Collaborative Teaching of a Robot Learner

Christina Steele[1](✉), Nikki Lobczowski[1], Teresa Davison[1], Mingzhi Yu[1], Michael Diamond[2], Adriana Kovashka[1], Diane Litman[1], Timothy Nokes-Malach[1], and Erin Walker[1]

[1] University of Pittsburgh, Pittsburgh, PA 15260, USA
steelech@pitt.edu
[2] The Ohio State University, Columbus, OH 43210, USA

Abstract. Teaching others has been shown to be an activity in which students can learn new information in both human-human (peer-tutoring) and human-computer interactions (teachable robots). One factor that may help foster learning and engagement when teaching others is the development of positive rapport and perceptions between the tutor, tutee, and robot. However, it is not clear what factors might affect the development of rapport. We explore whether having two students work together with a teachable robot might facilitate positive perceptions of the robot, rapport-building, and positive learning outcomes. In an exploratory pilot study, students were assigned to either work together in dyads (n = 28) or individually (n = 12) to help a teachable robot (Emma) solve math problems. Preliminary results showed that those who worked in a dyad had generally more positive perceptions of the robot than those who worked individually. These benefits were not observed for rapport where there were few differences between dyads and individuals, or learning where there was no difference on the posttest. We discuss the implications of these results for future research to explore the potential benefits of collaborative teaching of a robot learner.

Keywords: Learning-by-teaching · Rapport-building · Robot perceptions

1 Introduction

A long-standing goal within educational technology research has been to use technological innovation to improve the learning experience and performance of students [1]. One means to support student learning is by having a student teach others, which has been shown to help students learn through the process of having to explain it to someone else. [2] The positive effects of learning by teaching persist even when a student is teaching a virtual agent or robot, rather than a peer [3]. There is increasing interest in examining what factors impact the development of positive perceptions of and rapport with a teachable agent or robot, and how that influences learning outcomes [4]. One factor that might influence how learners benefit from teachable agent interactions is whether they are teaching the agent collaboratively (i.e., with another peer), termed "learning by

© Springer Nature Switzerland AG 2022
M. M. Rodrigo et al. (Eds.): AIED 2022, LNCS 13356, pp. 604–607, 2022.
https://doi.org/10.1007/978-3-031-11647-6_125

collaborative teaching." [5]. By teaching the agent collaboratively, learners might have more of an opportunity to co-construct knowledge with each other through discussion in order to figure out how to explain it to the agent. In addition, learners may feel less frustrated with an agent that makes errors if they are having a positive experience with their peer, and in turn rate the agent more positively and feel more rapport with the agent.

The current study examined the extent to which teaching a social robot with a peer or alone influences students' robot perceptions, rapport-building, and math performance. To explore this, we use a social robot named Emma, designed to engage students in spoken dialogue on ratio and proportions problems. Students explain the problem step by step, and Emma responds with questions about or self-explanations related to the current step. We predicted that teaching Emma with a peer may increase rapport between students and Emma, improve positive perceptions of Emma, and increase math learning and performance compared to teaching Emma alone. Our work builds on and extends prior work by focusing on the impact of collaborative teaching of a robot on both perceptions of engagement and learning compared to individual teaching.

2   Methods

We recruited 40 undergraduates (35 Female, 5 Male; 13 Asian, 5 Black, 1 Latin@, 17 White, 4 No Response; Mean age = 19.64 years, SD = 1.25) from a mid-Atlantic US university for an exploratory pilot study. Students were assigned to one of two conditions in which they would either engage in a 30-min collaborative activity in which students worked together in pairs (i.e., the dyad condition, n = 28) or an individual activity in which students worked alone (i.e., the solo condition, n = 12) to help a teachable robot (i.e., Emma) solve math problems on ratios and proportions. Students completed a battery of self-report questionnaires that included a variety of motivational and engagement constructs prior to and post teaching Emma.

We assessed student perceptions of Emma through a 35-item measure containing two ten-item subscales (e.g., Anthropomorphism, Intelligence) and two eight-items subscales (e.g., Animacy, Likeability) on a six-point Likert scale [6]. We combined the average score on each subscale into a single composite measure of robot perceptions (Cronbach's alpha = .95). Higher scores indicated greater positive perceptions of the robot. We also assessed rapport with Emma with a 15-item measure containing three four-item subscales (e.g., positivity, attentive, and coordination) and one three-item subscale (e.g., general) on a six-point Likert scale (e.g., 1 = Strongly disagree, 6 = Strongly agree) [3]. Higher scores indicated greater rapport with Emma.

We assessed learning outcomes through a math test on ratio and proportion administered before and after students interacted with Emma. There were two counterbalanced versions of the test. The original test had 13 items. However, because of counterbalancing issues five were removed because different versions may have been easier or harder, leaving eight items for both pre- and post-tests. Each item was scored dichotomously as correct or incorrect and summed for a total score that could vary from 0–8 on pre and posttest.

3 Results

Initial analysis of pre-test math performance revealed an effect of condition with the solos ($M = 4.58$, $SD = 2.19$) performing worse than the dyads ($M = 6.00$, $SD = 1.85$), $F(1,39) = 5.47$, $p = .03$. This was unexpected as students were expected to perform similarly at pretest before interacting with each other and Emma, therefore we analyzed our three key questions both with math-pretest as a covariate.

A one-way analysis of covariance (ANCOVA) to test the effect of condition on participants' average reported perceptions of Emma with the math pre-test as a covariate revealed a medium effect of the covariate, $F(1,38) = 6.02$, $p = .01$, showing that participants' pre-test scores predicted their perceptions of Emma. There was also an effect of condition, $F(1,38) = 6.95$, $p = .01$, with participants in the dyad condition ($M = 4.47$, $SD = .50$) reporting more positive perceptions of Emma than participants in the solo condition ($M = 4.12$, $SD = .69$). If we take the pre-test covariate out of the model, the effect of condition is marginal in the same direction, $F(1,38) = 3.17$, $p = .08$.

A multivariate analysis of covariance (MANCOVA) to test the effect of condition on participants' perceptions of rapport with Emma with the math pre-test as a covariate revealed that the overall model did not show an effect of condition, $F(4,34) = 4$, $p = .15$, nor an overall effect of covariate $F(4,34) = 1.02$, $p = .41$. Exploratory analyses for the rapport subcomponents revealed a trend for positivity $F(1,37) = 5.93$, $p = .02$, with dyads ($M = 5.02$, $SD = .52$) reporting more positivity than solos ($M = 4.60$, $SD = .99$). All other components of rapport were not significant, F's < 1, p's $> .41$.

A one-way ANCOVA to test the effect of condition on participants' average math posttest performance with the math pre-test as a covariate revealed a large effect of the covariate, $F(1,37) = 8.54$, $p < .01$, showing that participants' pretest scores predicted their posttest scores. There was no effect of condition, $F(1,37) = .22$, $p = .65$, with participants in the dyads ($M = 6.86$, $SD = 1.43$) and solo ($M = 6.58$, $SD = 1.38$) conditions performing similarly.

4 Discussion

Overall, we found evidence that working collaboratively to teach a social robot math is linked to increased positive perceptions with the robot, but not improved rapport or individual learning outcomes. One possible reason for the increased positive perceptions of Emma is that by working collaboratively participants may construct common ground [7] and then use that shared knowledge to work more productively and positively with the robot as it makes errors and mistakes. Future work should investigate whether learners who construct common ground are more tolerant of a robot's errors in comprehension, and if there are potential benefits of these improved perceptions.

We did not find similar collaborative benefits for participants' perceptions of rapport with Emma or their individual learning outcomes. Although there was a trend for an impact on rapport positivity, which aligns with the positive perceptions of Emma results, we did not see that trend reflected in any of the other sub-dimensions of rapport (general, attentive, or coordination). One possible reason for these null results is that dyads may not have been engaging in interactive ways with each other (e.g., via explanation or error correction), but may have instead focused on developing common ground to take turns to teach Emma. Future work should further examine the nature of the interactions between participants in the collaborative dyads. It could also be the case that the duration of our study was too short to see changes in rapport or learning, or that the material was generally too easy for the undergraduate students involved. Future research could examine whether longer collaborative engagement with the teachable robot shows benefits for rapport and learning over time.

Although our study is preliminary and uses a small sample size in an online rather than in-person setting, these initial results set up the foundation for future research to explore effective means and uses of a teachable social robot to facilitate student learning and student engagement with novel learning technology. Given the limited resources and time of teachers to address the individual needs of students, having students work collaboratively with a teachable robot may be a beneficial tool to improve the student learning experience through promoting student engagement with novel technology.

Acknowledgements. This work was supported by Grant No. 2024645 from the National Science Foundation, Grant No. 220020483 from the James S. McDonnell Foundation, and a University of Pittsburgh Learning Research and Development Center internal award.

References

1. Reeves, T.C., Oh, E.G.: The goals and methods of educational technology research over a quarter century (1989–2014). Educ. Technol. Res. Dev. **65**(2), 325–339 (2016). https://doi.org/10.1007/s11423-016-9474-1
2. Walker, E., Rummel, N., Koedinger, K.R.: Adaptive intelligent support to improve peer tutoring in algebra. Int. J. Artif. Intell. Educ. **24**(1), 33–61 (2014)
3. Lubold, N., Walker, E., Pon-Barry, H., Flores, Y., Ogan, A.: Using iterative design to create efficacy-building social experiences with a teachable robot. International Society of the Learning Sciences, Inc. [ISLS] (2018)
4. Ogan, A., Finkelstein, S., Walker, E., Carlson, R., Cassell, J.: Rudeness and rapport: insults and learning gains in peer tutoring. In: Cerri, S.A., Clancey, W.J., Papadourakis, G., Panourgia, K. (eds.) ITS 2012. LNCS, vol. 7315, pp. 11–21. Springer, Heidelberg (2012). https://doi.org/10.1007/978-3-642-30950-2_2
5. El Hamamsy, L., Johal, W., Asselborn, T., Nasir, J., Dillenbourg, P.: Learning by collaborative teaching: an engaging multi-party cowriter activity. In: 28th IEEE International Conference on Robot and Human Interactive Communication, pp. 1–8 (2019)
6. Bartneck, C., Kulić, D., Croft, E., Zoghbi, S.: Measurement instruments for the anthropomorphism, animacy, likeability, perceived intelligence, and perceived safety of robots. Int. J. Soc. Robot. **1**(1), 71–81 (2009)
7. Sinha, T., Cassell, J.: We click, we align, we learn: impact of influence and convergence processes on student learning and rapport building. In: Proceedings of the 1st Workshop on Modeling Interpersonal Synchrony and Influence, pp. 13–20 (2015)

Automatic Identification of Non-native English Speaker's Phoneme Mispronunciation Tendencies

Shi Pu$^{(\boxtimes)}$ (iD), Lee Becker, and Misaki Kato

Educational Testing Service, 660 Rosedale Rd, Princeton, NJ 08540, USA
{spu,lbecker001,mkato}@ets.org

Abstract. We explore the possibility of using word-level transcription to detect non-native English speaker (NNES)'s phoneme mispronunciation tendencies. We focus on word-level instead of phoneme-level transcription as the former is readily accessible and mature. We define phoneme mispronunciation tendency as the recurring imperfect pronunciation of a phoneme across different words. We use an Automatic Speech Recognition (ASR) service to generate alternative transcripts from speaker's reading aloud audio data. We build features based on the divergence of the audio transcriptions and the texts, as well as the confidence of the audio transcriptions. We found the features are informative for detecting phoneme mispronunciation tendencies.

Keywords: Automatic Speech Recognition · Pronunciation · Language Learning

1 Introduction

Pronunciation accuracy plays an important role in intelligibility of speech (how well speech can be understood by listeners) [1,2]. Existing research in Computer Assisted Language Learning has experimented with various algorithms to automatically score a speaker's phoneme pronunciation accuracy for *each occurrence* in an utterance [3,4]. This type of assessment is valuable to give speakers feedback on the accuracy of pronouncing a phoneme in a particular word, but not directly beneficial to understanding a speaker's mispronunciation tendency for a phoneme. For example, a non-native speaker might stumble on the pronunciation of /f/ when reading aloud the word **Pfizer** if one does not know the first letter p is silent in this word. However, this mispronunciation of /f/ does not necessarily mean that the speaker will have an issue pronouncing /f/ in other words. Automatic detection of a speaker's *recurring* phoneme pronunciation errors is valuable for providing targeted phoneme practice exercises.

2 Phoneme Mispronunciation Tendency

We are interested in detecting a NNES's recurring errors in phoneme pronunciation across a wide range of contexts. A speaker 's phoneme error rate is the rate

© Springer Nature Switzerland AG 2022
M. M. Rodrigo et al. (Eds.): AIED 2022, LNCS 13356, pp. 608–611, 2022.
https://doi.org/10.1007/978-3-031-11647-6_126

that a given phoneme is marked as imperfect by a human expert. Then we say a speaker i has recurring pronouncing errors with phoneme j if the error rate, E_{ij}, is larger than certain threshold. We set the threshold as 0.5 for our study:

$$Y_{ij} = \mathbf{I}(E_{ij} > 0.5) \tag{1}$$

Y_{ij} is a binary variable and $\mathbf{I(x)}$ is the indicator function that equals one when x is True.

We hypothesize that two types of features – ASR divergence and ASR confidence – are informative to model Y_{ij}. The divergence measures how often a speaker i's phoneme j that should exist in the audio is not recognized by ASR. The confidence measures the confidence of ASR when a speaker i pronounces a phoneme j. The detailed feature extraction process is described in Sect. 4.

3 Data

We use a public speech corpus called SpeechOcean762 [5]. The data is composed of over 5000 utterances from 250 NNES reading aloud 4947 unique English texts. Each utterance is produced by a speaker reading a text. Then five experts are provided with the text and a rubric to score the utterances. Each phoneme of an utterance in the dataset has been scored as zero (missed or incorrect), one (heavy accent), or two (correct). We use the average score of the five experts as the ground truth of a speaker's pronunciation accuracy of the phoneme.

4 ASR Divergence and ASR Confidence

As the first feature extraction step, we generated ten word-level alternative transcriptions for each utterance using the ASR service in Amazon Transcribe. Using CMU Pronunciation Dictionary, two types of phoneme sequences were then identified. ASR phonemes are the sequence of phonemes in the transcripts produced by ASR. Text phonemes are the sequence of phonemes in the prompt text that the speakers read aloud.

ASR Divergence. A phoneme j in an utterance is considered to be mispronounced if ASR fails to recognized it. To measure this failure, we align a sequence of text phonemes and a sequence of ASR phonemes according to edit distance[1]. A phoneme j in text phonemes is annotated as mispronounced if there is no matched counterpart in the ASR phonemes.

The ASR Divergence for phoneme j is then calculated by averaging the mispronunciation across ten alternative ASR phonemes for an utterance, and a speaker's all utterances:

$$ASRD_j = \frac{1}{nm} \sum_n \sum_m d_{jmn} \tag{2}$$

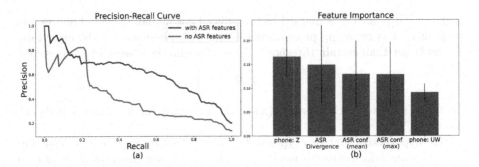

Fig. 1. Precision-recall curve and feature importance of random forest on test data

where m is the number of alternative ASR phonemes (ten in our case), n is the number of speaker's utterances in the data. And d_{jmn} equals 0 if a phoneme j is mispronounced according to the edit distance to ASR phonemes m in utterance n. d_{jmn} equals 1 otherwise.

ASR Confidence. We measure how well a speaker pronounces a phoneme j in an utterance by how confident the ASR recognizes the phoneme. If the ASR fails to recognize the phoneme, the confidence is filled with zero. Otherwise, the phoneme inherits the confidence from its corresponding word as we only use word-level transcriptions.

The ASR confidence feature for phoneme j is then pooled (either averaging or maximizing) across an utterance's alternative ASR phonemes, and a speaker's all utterances. The following equation illustrates the calculation using average pooling:

$$ASRC_j = \frac{1}{nm} \sum_n \sum_m c_{jmn} \tag{3}$$

where m, and n has the same meaning as in Eq. 3. c_{jmn} represents the confidence of phoneme j according to ASR phonemes m in utterance n. c_{jmn} ranges from 0 to 1.

5   Experiments

We randomly split the speakers into train, validation, and test data based on a 70:10:20 ratio. We used the training data to fit the model, validation data to tune hyperparameters, and test data to evaluate the model. Since Y_{ij} is highly in-balanced – only about 11.12% of speaker-phoneme pairs have recurring pronunciation errors – we use the average precision score[2] to evaluate our model. We use the ASR divergence, ASR confidence, phoneme indicators, and speaker's age (an indicator for speaking proficiency) as features in our models.

---

[1] Edit distance between two strings is the minimum sequence of edit operations (insertion, deletion, or substitution) required to transform one string to the other.

[2] Average precision score measures the area under the Precision Recall Curve, which quantifies the trade-off between precision and recall using different threshold.

Table 1. Average precision on test data

	Multi-layer perceptron	Random forest	Adaboost	SVM
With ASR features	0.607	**0.612**	0.561	0.535
No ASR features	0.425	0.411	0.401	0.344

6 Results

Table 1 summarizes the performance of different models with and without the ASR features (i.e., ASR divergence and ASR confidence), and indicates that the random forest model has the best performance, closed followed by a multi-layer perceptron. Most importantly, ASR features play an critical role in improving model performance. Specifically, ASR features increase the model average precision by 0.16 to 0.20.

Figure 1(a) compares the Precision-Recall Curve for the random forest model with and without the ASR features. With the presence of ASR features, the precision curve decreases much slower than without ASR features as the recall increases. The model with ASR features reaches the optimal f1 0.615 when the recall is at 0.655. Figure 1(b) visualizes the random forest model's feature importance[3] and demonstrate that the three ASR features are all in the top five essential features.

7 Conclusion and Future Studies

In this study, we demonstrate that word-level ASR transcriptions are useful to detect NNES' recurring errors in phoneme pronunciations. In the future, we are interested in exploring how well a model trained on a dataset can be generalized to a different dataset.

References

1. Atagi, E., Bent, T.: Perceptual dimensions of nonnative speech. In: ICPhS, pp. 260–263 (2011)
2. Loukina, A., Lopez, M., Evanini, K., Suendermann-Oeft, D., Ivanov, A.V., Zechner, K.: Pronunciation accuracy and intelligibility of non-native speech. In: Sixteenth Annual Conference of the International Speech Communication Association (2015)
3. Shi, J., Huo, N., Jin, Q.: Context-aware goodness of pronunciation for computer-assisted pronunciation training. arXiv preprint arXiv:2008.08647 (2020)
4. Witt, S.M., Young, S.J.: Phone-level pronunciation scoring and assessment for interactive language learning. Speech Commun. **30**(2–3), 95–108 (2000)
5. Zhang, J., et al.: Speechocean762: an open-source non-native English speech corpus for pronunciation assessment. arXiv preprint arXiv:2104.01378 (2021)

---

[3] The importance of a feature i is measured by the decrease of impurity when a decision tree split on the feature.

Solving Probability and Statistics Problems by Probabilistic Program Synthesis at Human Level and Predicting Solvability

Leonard Tang[2], Elizabeth Ke[1], Nikhil Singh[1], Bo Feng[3], Derek Austin[3],
Nakul Verma[3], and Iddo Drori[1,3(✉)]

[1] Massachusetts Institute of Technology, Cambridge, USA
idrori@mit.edu
[2] Harvard University, Cambridge, USA
[3] Columbia University, New York, USA

Abstract. We use probabilistic program synthesis to solve questions in MIT and Harvard Probability and Statistics courses. Traditional approaches using the latest GPT-3 language model without program synthesis achieve a solve rate of 0.2 in these classes. In contrast, by turning course questions into probabilistic programs using the latest program synthesis Transformer, OpenAI Codex, and executing the programs, our solve rates are 0.9 and 0.88, which are on par with human performance.

Keywords: Transformers · Program synthesis · Probabilistic programming · Few-shot learning

1 Introduction

Suppose we play a game where I keep flipping a coin until I get heads. If the first time I get heads is on the n-th coin, then I pay you $2n - 1$ dollars. How much would you pay me to play this game?

Recent approaches to solving problems like the example above rely upon training foundation models [4] to formulate answers directly [9], in a sequential fashion, such as a series of text explanations [7], or in a chain of formal operations [1]. However, these approaches struggle to solve such problems accurately.

A compelling alternative problem-solving strategy is to manipulate probabilistic models to infer the distribution of answers and extract a solution from the resulting distribution. Such a Bayesian approach, usually dubbed probabilistic programming in the literature [11], offers a flexible mechanism for solving a variety of probabilistic tasks. From a frequentist perspective, one can replace distribution manipulations with large-scale simulations to directly produce a numerical answer averaged across multiple scenarios. Inspired by this insight, we solve probability problems via probabilistic programming simulations and explicit, sequential computation.

© Springer Nature Switzerland AG 2022
M. M. Rodrigo et al. (Eds.): AIED 2022, LNCS 13356, pp. 612–615, 2022.
https://doi.org/10.1007/978-3-031-11647-6_127

We turn questions into programming tasks and prompt OpenAI's Codex [5], a Transformer trained on text and fine-tuned on code, with the task of synthesizing a program. We then execute the program to obtain an answer. We manually evaluate the answers, checking for numerical accuracy and logical correctness.

Recently introduced datasets, such as MATH, MAWPS, MathQA, Math23k, and GSM8K [1,6–8,10], focus on benchmarking neural models' aptitude in mathematics (including probability and statistics), and all of these datasets only consider grade-school level questions. Notably, our datasets are the first to benchmark performance in probability and statistics at the university level.

2   Methods

2.1   Dataset

We curate two datasets of questions from undergraduate-level probability and statistics courses at MIT and Harvard:

1. MIT 18.05 Introduction to Probability and Statistics [3]: Topics covered in the course include counting, conditional probability, discrete and continuous random variables, expectation and variance, central limit theorem, joint distributions, maximum likelihood estimators, Bayesian updating, null hypothesis significance testing, and confidence intervals. We randomly sampled 25 questions covering these topics and questions in probability and statistics.
2. Harvard STAT110 [2]: Topics include distributions, moment generating functions, expectation, variance, covariance, correlation, conditional probability, joint distributions, marginal distributions, conditional distributions, limit theorems, and Markov chains. We randomly sample 20 questions with numerical answers, and these questions are slightly more theoretical than those from 18.05.

We collect our data from PDF files on publicly available course websites and textbooks. These questions are *not* present in the GitHub repositories that constitute Codex's training dataset.

2.2   Probabilistic Programming

We convert input questions into programming tasks using standardized task templates. Moreover, we leverage the power of simulation by encouraging Codex to write large-scale probabilistic simulation programs that aggregate results across several scenarios to obtain an answer. To induce such behavior in Codex, we use one of 3 standardized conversion templates that inject a specific hardcoded phrase into the original question.

2.3 Predicting Solvability

We attempt to predict which questions can be solved using Codex by embedding each question into a 1,024-dimensional space using a GPT-3 (text-similarity-ada-001) embedding model. We fit a logistic regression model to these embeddings to predict if they can be solvable via few-shot learning. The dataset is heavily imbalanced since we solved 89.5% of the cumulative questions. Due to this imbalance, we measure and report the area under the curve (AUC). Notably, we achieve an AUC of 0.64, which demonstrates that the GPT-3 embedding space is rich enough to determine better than random if a question is solvable by Codex, even before using Codex for program synthesis.

3 Results

3.1 Human Performance

GPT-3 (text-davinci-002) achieves a solve rate of 0.2 on both STAT110 and 18.05. Using Codex as a zero-shot learner and prompting it with probabilistic programming tasks, we achieve solve rates of 0.65 on STAT110 and 0.72 on 18.05. We tackle problems that cannot be solved using zero-shot learning by using Codex with few-shot learning. We first embed all the zero-shot questions. Next, we compute the nearest neighbors to the embedded target question in the embedding space. Finally, we provide up to 5 nearest pairs of input questions and corresponding output programs before prompting Codex with the target question. This improves the solve rate to 0.9 on STAT110 and 0.88 on 18.05, which is on par with human performance (Table 1).

Table 1. Automatic solve rates for course problems from Harvard University STAT110 and MIT 18.05. Using Codex without examples (zero-shot learning) significantly improves upon previous work using GPT-3, from 0.2 to 0.65 and 0.72. Using a few question-code examples (few-shot learning) further improves performance to 0.9 and 0.88, which is on par with human performance.

Model	Harvard STAT110	MIT 18.05
GPT-3 (text-davinci-002)	0.2	0.2
Codex Zero-Shot (code-davinci-002)	0.65	0.72
Codex Zero-Shot & Few-Shot (code-davinci-002)	0.9	0.88

3.2 Reproducibility

We use the latest version of Codex (code-davinci-002). To ensure 100% reproducibility, we fix Codex's behavior to be deterministic, setting both the temperature hyperparameter that controls randomness and the top-p hyperparameter that controls diversity to 0.

4   Conclusion

This work uses probabilistic program synthesis to solve university-level probability and statistics problems with human performance. We plan to generate and solve even more challenging problems through curriculum learning. We use supervised learning to predict in advance the solvability of new questions from their embeddings. We plan to extend the binary classification of solvability to a regression problem predicting the difficulty level of questions in advance. This work opens the door for new use cases of AI in education. For example, by automatically generating and solving more challenging questions, we hope to provide future tools that will improve self-paced learning.

References

1. Amini, A., Gabriel, S., Lin, S., Koncel-Kedziorski, R., Choi, Y., Hajishirzi, H.: MathQA: towards interpretable math word problem solving with operation-based formalisms. In: Proceedings of Conference of the North American Chapter of the Association for Computational Linguistics: Human Language Technologies (NAACL), pp. 2357–2367. Association for Computational Linguistics (2019)
2. Blitzstein, J.: Statistics 110 probability (2021)
3. Bloom, J.: Introduction to probability and statistics (2014)
4. Bommasani, R., et al.: On the opportunities and risks of foundation models. *arXiv* preprint arXiv:2108.07258 (2021)
5. Chen, M., et al.: Evaluating large language models trained on code (2021)
6. Cobbe, K., et al.: Training verifiers to solve math word problems. *arXiv* preprint arXiv:2110.14168 (2021)
7. Hendrycks, D., et al.: Measuring mathematical problem solving with the math dataset. In: Proceeding of Advances in Neural Information Processing Systems Datasets and Benchmarks (NeurIPS) (2021)
8. Koncel-Kedziorski, R., Roy, S., Amini, A., Kushman, N., Hajishirzi, H.: MAWPS: a math word problem repository. In: Proceedings of Conference of the North American Chapter of the Association for Computational Linguistics: Human Language Technologies (NAACL), San Diego, California, June 2016, pp. 1152–1157. Association for Computational Linguistics
9. Saxton, D., Grefenstette, E., Hill, F., Kohli, P.: Analysing mathematical reasoning abilities of neural models. In: Proceeding of International Conference on Learning Representations (ICLR) (2019)
10. Wang, Y., Liu, X., Shi, S.: Deep neural solver for math word problems. In: Proceeding of Conference on Empirical Methods in Natural Language Processing (EMNLP), pp. 845–854. Association for Computational Linguistics (2017)
11. Wingate, D., Stuhlmüller, A., Goodman, N.: Lightweight implementations of probabilistic programming languages via transformational compilation. In: Proceedings of the International Conference on Artificial Intelligence and Statistics (AISTATS), pp. 770–778. JMLR Workshop and Conference Proceedings (2011)

Prerequisite Graph Extraction
from Lectures

Ilaria Torre$^{(\boxtimes)}$ [ID], Luca Mirenda, Gianni Vercelli [ID],
and Fulvio Mastrogiovanni [ID]

DIBRIS, University of Genoa, Genoa, Italy
ilaria.torre@unige.it

Abstract. The proliferation of video-sharing platforms and MOOCs has
raised new challenges in the field of education. A challenging topic that
is gaining an increased popularity is the identification of prerequisite
relations between concepts in video lectures. In this paper, we propose
unsupervised methods for prerequisite identification and the creation of
a prerequisite graph. The contribution, compared to existing approaches,
is the development of methods which (i) do not rely on external knowledge, (ii) do not require extensive training, and (iii) are intended to
exploit both the lecture transcript and its visual features. Results from
the preliminary evaluation are encouraging, and provide insights on the
extraction of prerequisite relations from video transcripts compared to
textbooks.

Keywords: Concept dependency map · Prerequisite extraction

1 Introduction

While educational resources implicitly include, and are based on, a knowledge
graph of prerequisite relations (known as "PR graph"), its automatic extraction is still a challenge due to the complex dynamics of concepts' explanation.
Existing solutions use different approaches. Relational metrics try to capture the
strength of the relation between co-occurring concepts, being RefD a popular one
[5]. Machine learning approaches use link-based features, text-based features,
or a combination of the two. The most effective approaches exploit external
resources, such as online encyclopedias, for identifying concepts and their relations [7]. Unsupervised methods that do not use external resources are usually
less powerful, whereas supervised machine learning approaches require extensive
training. To address this issue, recent approaches tried to use pre-trained language models [4] and burst analysis [1] with promising results. Wikipedia has
been the most used resource, but besides not covering technical concepts that
might be present in the lecture, another limit is that content is based on a single perspective [3]. Our approach is intended to avoid both these limitations,
aiming to extract PRs as expressed in the video lecture, and using unsupervised
methods based on burst analysis. Moreover, existing methods use mostly text

© Springer Nature Switzerland AG 2022
M. M. Rodrigo et al. (Eds.): AIED 2022, LNCS 13356, pp. 616–619, 2022.
https://doi.org/10.1007/978-3-031-11647-6_128

information [2,3,6] for PR extraction, while visual information is not addressed. In our approach, we use the transcript, which is indeed the main carrier of information in a lecture, but we also propose to exploit visual features to support the identification of concepts and the role of their occurrence in the video flow, which currently is not addressed in the literature. In the following, we provide an overview of the approach, focusing on the part that is currently implemented, and reporting preliminary evaluation results. The scientific contribution we expect from this work is on the one hand an innovative approach for the analysis of the concepts' dynamics throughout the video stream, and on the other end the fine-grained identification of prerequisite relations. The results can also provide novel features to be integrated into existing supervised methods.

2   PR Graph Building and Preliminary Evaluation

PR Graph Building. The proposed method takes into account the speech transcript and visual features: (i) we apply *burst analysis* and *interval algebra* to identify intervals with the concept in focus and infer PRs, and (ii) we propose the use of video processing to improve concept detection. While PR graph extraction from transcripts is already available and deployed as an application, PR graph refinement through video analysis is still in prototypical form as well as synonym management. Figure 1a shows the three steps for PR graph extraction from transcript:

a) NLP processing: the transcript is processed by adding punctuation and part of speech tagging (via *punctuator* API, NLTK python package, and UDPipe APIs).

b) Burst Intervals of Concepts (BIC) detection: BICs represent portions of text where a concept appears with high density. Identifying BICs in the text for a given concept allows for tracking its appearance and its evolution along the text flow where it may occur in different roles, namely concept definition, in-depth explanation, and recall of the concept to introduce other concepts. While several methods exist for concept extraction, tracking concepts is not much covered in the literature. We address this issue by applying the method we adopted in [1] using burst analysis on textbooks. Basically, we use a Hidden Markov Model to identify BICs, which reveal the focus of the stream on a concept (our implementation relies on the pybursts Python library). Moreover, we exploit BICs lengths and intensity as an heuristic for role identification. *Video analysis* will be used to improve BICs identification. The main steps we propose for this process are: video segmentation through segment similarity, lecture type identification (using XGBoost algorithm, optimized with Optuna), concepts and definitions recognition from text on slides (Tesseract OCR).

c) PR relations extraction: spatial-temporal reasoning using Allen's interval algebra is applied on the extracted BICs. Examples of Allen's relations between time intervals are *X precedes Y, X meets Y, X includes Y.* The algorithm assigns a weight to each type according to how likely it represents a PR. Then, for any two

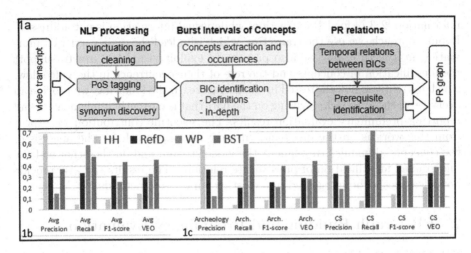

Fig. 1. Pipeline for PR graph extraction from transcripts and evaluation results.

distinct concepts, the weights assigned to the BIC pairs in the flow are combined and normalized, returning a score of a likely PR relation between the two concepts. *PR graph:* the output of the three processes above is used to build the PR graph. To represent the graph we defined an RDF Data Model that uses SKOS vocabulary to represent concepts and synonyms and the W3C Web Annotation Vocabulary to annotate the concept role (definition, in-depth, prerequisite).

Preliminary Evaluation. The evaluation assumes neither synonyms discovery nor refinement through video analysis.

Dataset: 5 videos in two domains (introduction to Archaeology and Computer Science from YouTube) annotated by one expert in each domain for a total of 220 min, 62 concepts, 212 edges.

Baseline Methods: 3 often-used basic methods for PR extraction: HH (uses hyponym-hypernym relations, as in [8]), RefD (models how two concepts refer to each other [5]), WP (based on Wikipedia Pages [8]).

Results: We compared our method (BST) and the three baselines against the manually annotated dataset, using the following metrics: Precision, Recall, F1-score, and Vertex Edge Overlap (VEO) for graph similarity. Results are shown in Fig. 1b. We also split the analysis along the two domains and compared the results, see Fig. 1c. As a further investigation, we compared the performance of the methods on a textbook chapter manually annotated to investigate if differences exist between the methods applied to textbook *vs* transcript. Results showed an average non-significant decrease of all the metrics (AVG -12%) using transcript.

Discussion: Average results show that BST performs better according to F1-score, which combines precision and recall, and VEO. As expected, HH gains higher precision and WP higher recall due to the content of the video lectures, whose concepts are available on Wikipedia. No significant differences have been found in the two domains. Although the current implementation of the method outperforms the baselines for both F1-score and VEO, the current score does not reach 0.5. This is anyway an overall good result considering that the method is unsupervised, does not use any external resource, and we evaluated a first prototype. We expect considerable improvement from PR graph refinement through video analysis and synonym management since the evaluation showed issues with concept terms that were synonyms but producing different BICs. It is worth noting that for all the methods the performance on video transcripts are lower than for textbooks. This might be due to the lower structure of speech compared to textbooks and to the lower correctness of automatic transcripts. Further investigation is needed since to the best of our knowledge this is the first study to make this comparison.

References

1. Adorni, G., Alzetta, C., Koceva, F., Passalacqua, S., Torre, I.: Towards the identification of propaedeutic relations in textbooks. In: Proceedings of the International Conference on Artificial Intelligence in Education, AIED19, pp. 1–13 (2019)
2. ALSaad, F., Boughoula, A., et al.: Mining MOOC lecture transcripts to construct concept dependency graphs. In: Proceedings of the 11th International Conference on Educational Data Mining, pp. 467–473 (2018)
3. Aytekin, M.C., Rabiger, S., Saygın, Y.: Discovering the prerequisite relationships among instructional videos from subtitles. In: Proceedings of the EDM Conference (2020)
4. Li, B., Peng, B., Shao, Y., Wang, Z.: Prerequisite learning with pre-trained language and graph embedding models. In: Wang, L., Feng, Y., Hong, Yu., He, R. (eds.) NLPCC 2021. LNCS (LNAI), vol. 13029, pp. 98–108. Springer, Cham (2021). https://doi.org/10.1007/978-3-030-88483-3_8
5. Liang, C., Wu, Z., et al.: Measuring prerequisite relations among concepts. In Proceedings of the International Conference on Empirical Methods in NLP, pp. 1668–1674 (2015)
6. Pan, L., et al.: Prerequisite relation learning for concepts in MOOCs. In: Annual Meeting of the ACL (Volume 1: Long Papers), pp. 1447–1456 (2017)
7. Tang, C.L., Liao, J., Wang, H.C., Sung, C.Y., Lin, W.C.: Conceptguide. In: Proceedings of the Web Conference 2021, pp. 2757–2768 (2021)
8. Wang, S., Liu, L.: Prerequisite concept maps extraction for automatic assessment. In: Proceedings of the International Conference Companion on World Wide Web, pp. 519–521 (2016)

Workshops Track Contributions

Design, Build, Evaluate, and Implement Conversation-based Adaptive Instructional Systems (CbAIS)

Xiangen Hu[1,2]([⊠])[iD] and Art Graesser[1][iD]

[1] The University of Memphis, Memphis, TN 38152, USA
xhu@memphis.edu
[2] Central China Normal University, Wuhan 430074, China

Abstract. This half-day tutorial will be led by the original creators of AutoTutor, a prototype conversation-based adaptive instructional systems (CbAIS). The tutorial will inform you about (a) the theoretical foundations, enabling technologies, and practical applications of CbITS through hands-on and worked-out examples and (b) simple, common, and advanced data analysis methods that apply to the analysis of learner data. This tutorial is formulated for beginners and has no prerequisites. Professionals (corporate officers, program managers, scientists, and engineers) from industry, academia, and the government will all benefit from attending this tutorial. By the end of the tutorial, attendees will be able to create a complete CbITS module. Attendees will also be able to analyze data using the data analytical methods introduced.

Keywords: Intelligent tutoring systems · Conversation interface · Semantic processing

1 About the Tutorial

There have been decades of efforts in the research and development of intelligent tutoring systems (ITS) [8–11]. Many ITS provide rich media content and allow students to interact with content in many different ways, such as multiple choice answer selection, drag and drop objects, rearranging objects, and assembling objects. The ITS assess students' performance from the data collected on the interactions, and then adaptively select knowledge objects and pedagogical strategies during the tutoring process to maximize the learning effect and minimize learning costs. Delivering content with conversation is frequently attractive to content authors and students. For example, when a piece of knowledge is delivered through a text, it is presumably more interesting to have a conversation between a "tutor" (human or machine) and a student to talk about what is in

This research was sponsored by the National Science Foundation under the award The Learner Data Institute (award #1934745). The opinions, findings, and results are solely the authors' and do not reflect those of the funding agencies.

M. M. Rodrigo et al. (Eds.): AIED 2022, LNCS 13356, pp. 623–625, 2022.
https://doi.org/10.1007/978-3-031-11647-6

the text? Research has shown that delivering content through conversation has a number of advantages compared with merely reading a text. Unfortunately, creating conversational content is difficult. First, in order to have a natural language conversation with a student, the machine must be able to extract information from the student's natural language input. There is not a perfect natural language algorithm that can really "understand" the student' language, but a significant amount of semantic information can be extracted. Second, preparing tutoring speeches for conversations is hard. The author needs to consider many (if not infinitely many) responses to all possible student inputs. Third, it is difficult to create and test conversation rules. Conversation rules decide the conditions under which a prepared speech is spoken. Since the tutoring conversations often go with other displayed content, such as text, image, video, etc., conversation rules need to take into account all events that might occur in a learning environment, in addition to the natural language inputs from students. The rule system varies because different environments have different content and constraints. Creating and testing the rules is therefore time-consuming. Other difficulties involve talking head techniques (speech synthesizing, lip synchronization, displays of emotion, gesture), speech recognition, emotion detection, and so on.

The AutoTutor [2,6] team at the Institute for Intelligent systems (IIS) at the University of Memphis has been working in this direction since the 1990s and has been providing solutions to overcome the difficulties in conversational ITSs. About a dozen of conversational ITSs have been successfully developed in IIS, including a computer literacy tutor, a conceptual physics tutor [4], a critical thinking tutor (OperationARIES!) [5], an adult literacy tutor (CSAL) [1], an electronics tutor (ElectronixTutor) [3], etc. A team at National Taichung University of Education has developed a Chinese language tutor [7].

AutoTutor helps students learn by holding deep reasoning conversations. An AutoTutor conversation often starts with a main question about a certain topic. The goal of the conversation is to help students' construct an acceptable answer to the main question. Instead of telling the students the answers, AutoTutor asks a sequence of questions (hints, prompts) that target specific concepts involved in the ideal answer to the main question. AutoTutor systems respond to students' natural language input, as well as other interactions, such as making a choice, arranging some objects in the learning environment, etc.

This tutorial focuses on the authoring process of AutoTutor lessons. It includes discourse strategies in AutoTutor dialogues and trialogues, conversation elements, media elements, conversation rules and template- based authoring. Participants need to bring Windows laptops. A Windows authoring tool will be released on site. An example AutoTutor lesson will be provided to participants. Participants will create one's own AutoTutor lesson by modifying the example lesson.

2 Tutorial Outline

– Session 1 - Introduction to basic learning principles that are relevant to AIS (20 min); Introduction of presenters and participants; Overview and Demo of CbITS examples
– Session 2: Script Authoring Tools for CbITS (40 min); A step by step guide to creating a tutoring module
– Session 3: Understand learner data (40 min); Use of common utilities to query data from learning record store (LRS)
– Session 4: Deploying CbITS to cloud (20 min); Using common LMS (such as Moodle) to manage the use of CbITS

References

1. Cai, Z., Graesser, A.C., Hu, X., Nye, B.D.: CSAL AutoTutor: integrating rich media with AutoTutor. In: Generalized Intelligent Framework for Tutoring (GIFT) Users Symposium (GIFTSym3), p. 169 (2015)
2. Graesser, A.C., D'Mello, S., Hu, X., Cai, Z., Olney, A., Morgan, B.: AutoTutor. In: McCarthy, P.M. (ed.) Applied Natural Language Processing: Identification, Investigation and Resolution, pp. 169–187. IGI Global (2012)
3. Graesser, A.C., et al.: Electronixtutor: an adaptive learning platform with multiple resources. In: Proceedings of the Interservice/Industry Training, Simulation, and Education Conference (I/ITSEC 2018), Orlando, FL (2018)
4. Jackson, G.T., Ventura, M., Chewle, P., Graesser, A.: The impact of Why/AutoTutor on learning and retention of conceptual physics. In: Lester, J.C., Vicari, R.M., Paraguaçu, F. (eds.) ITS 2004. LNCS, vol. 3220, pp. 501–510. Springer, Heidelberg (2004). https://doi.org/10.1007/978-3-540-30139-4_47
5. Millis, K., Forsyth, C., Butler, H., Wallace, P., Graesser, A., Halpern, D.: Operation ARIES!: a serious game for teaching scientific inquiry. In: Ma, M., Oikonomou, A., Jain, L.C. (eds.) Serious Games and Edutainment Applications, pp. 169–195. Springer, London (2011). https://doi.org/10.1007/978-1-4471-2161-9_10
6. Nye, B.D., Graesser, A.C., Hu, X.: AutoTutor and family: a review of 17 years of natural language tutoring. Int. J. Artif. Intell. Educ. 24(4), 427–469 (2014). https://doi.org/10.1007/s40593-014-0029-5
7. Pai, K.C., Kuo, B.C., Liao, C.H., Liu, Y.M.: An application of Chinese dialogue-based intelligent tutoring system in remedial instruction for mathematics learning. Educ. Psychol. 41(2), 137–152 (2021)
8. Sottilare, R., Graesser, A., Hu, X., Brawner, K.: Design Recommendations for Intelligent Tutoring Systems: Volume 3 - Authoring Tools and Expert Modeling Techniques, vol. 3. Army Research Laboratory, Orlando (2015)
9. Sottilare, R., Graesser, A., Hu, X., Goldberg, B.: Design Recommendations for Intelligent Tutoring Systems: Volume 2 - Instructional Management, vol. 2. Army Research Laboratory, Orlando (2014)
10. Sottilare, R., Graesser, A., Hu, X., Holden, H.: Design Recommendations for Intelligent Tutoring Systems: Volume 1 - Learner Modeling, vol. 1. Army Research Laboratory, Orlando (2013)
11. Sottilare, R., Graesser, A., Hu, X., Olney, A., Nye, B., Sinatra, A.: Design Recommendations for Intelligent Tutoring Systems: Volume 4 - Domain Modeling, vol. 4. Army Research Laboratory, Orlando (2016)

Second Workshop on Computational Approaches to Creativity in Educational Technologies (CACE-22)

Kobi Gal[1,2][✉][iD], Niels Pinkwart[3,4][iD], Swathi Krishnaraja[3][iD], and Benjamin Paaßen[4][iD]

[1] Ben-Gurion University of the Negev, Be'er Sheva, Israel
kobig@bgu.ac.il
[2] University of Edinburgh, Edinburgh, UK
[3] Humboldt-University of Berlin, Berlin, Germany
{niels.pinkwart,swathi.krishnaraja}@hu-berlin.de
[4] German Research Center for Artificial Intelligence, Kaiserslautern, Germany
benjamin.paassen@dfki.de

Abstract. Creativity has been shown to promote students' critical thinking, self-motivation, and mastery of skills and concepts. Despite their increasing prevalence in schools, most current technological educational environments do not promote creativity in students' interactions or support teachers' ability to detect creative thinking by students. Recent work in AI and Education has begun to bridge this gap from multiple perspectives, such as representations (computational models for describing creativity in technology-based learning environments), inference (algorithms for detecting creative outcomes from students' interactions with these environments), and visualizations (presentations for teachers in a way that aids their understanding of students' interactions and allow them to intervene with this process when deemed necessary). The workshop will provide a platform for researchers from different fields to share findings and discuss new research opportunities for combining AI and creativity in educational technologies. Importantly, we intend to invite a group of experts in creativity theory from the cognitive and psychological sciences to speak in the workshop. A first edition of the workshop in 2021 was very successful in attracting papers and audience.

Keywords: Creativity · Educational data mining · Artificial intelligence in education

The content and themes of the workshop, as proposed below, combine relevant research areas in the AIED and EDM communities and apply them in the new setting of promoting creativity in education. Relevant topics include, but are not limited to:

- AI methods and tools for detecting and promoting creative thinking by students using technological learning environments.

M. M. Rodrigo et al. (Eds.): AIED 2022, LNCS 13356, pp. 626–627, 2022.
https://doi.org/10.1007/978-3-031-11647-6

- Computational models of creativity as it is reflected in students' activities in educational software.
- Machine learning algorithms for automatically recognizing creative behavior from students' interactions with software.
- Planning and decision making in creativity (automatic feedback generation for student solutions).
- Transfer learning of creativity models across domains and student populations.
- Applying theoretical models of creativity to modeling students' interactions in educational technologies.
- Visualization tools for presenting creative solutions to teachers.

The workshop website is available at https://sites.google.com/view/aied2022-creativity/.

Advances and Opportunities in Team Tutoring Workshop

Anne M. Sinatra[(⊠)] and Benjamin Goldberg

U.S. Army Combat Capabilities Development Command, Soldier Center
(DEVCOM SC), Orlando, USA
{anne.m.sinatra.civ,benjamin.s.goldberg.civ}@army.mil

1 Workshop Summary and Overview

The "Advances and Opportunities in Team Tutoring" workshop is a half-day **virtual** workshop, and is a follow up to three previous Artificial Intelligence in Education (AIED) Conference workshops in 2018 (in person), 2019 (in person), and 2021 (virtual) which were titled "Assessment and Intervention during Team Tutoring" [1], "Approaches and Challenges in Team Tutoring" [2], and "Challenges and Advances in Team Tutoring" [3]. These workshops all focused on techniques and approaches that have been used to implement team tutoring and collaborative computer-based learning. The workshops in 2019 and 2021 included a mix of new presenters and returning presenters who provided updates on the team tutoring work that they have been conducting. All of the workshops included discussions of the presented work, as well as team tutoring overall, and next steps.

The current workshop covers the topic areas of advances and opportunities to both team tutoring and collaborative learning with Intelligent Tutoring Systems (ITSs). The workshop includes both empirical and theoretical papers. As with the previous workshops, the workshop is expected to include a mix of previous presenters who are presenting updates to their work, and other presenters who are showcasing new work.

There are three main themes of the workshop: 1) Intelligent Tutoring for Teams in Distributed Environments; 2) Technological Advancements in ITSs for Teams; 3) ITS Based Collaborative Problem Solving and Learning. Details of the themes and activities are included in the next section.

2 Workshop Themes and Activities

Introduction/Overview: The Development and Implementation of Team Tutoring in the Generalized Intelligent Framework for Tutoring (GIFT): The workshop will start with an introduction to the Generalized Intelligent Framework for Tutoring (GIFT), which is an open-source ITS framework that has been developed by the DEVCOM Soldier Center-STTC group [4]. There will be discussion of how team tutoring has been implemented in GIFT, the lessons learned, challenges encountered, and the next steps forward.

© Springer Nature Switzerland AG 2022
M. M. Rodrigo et al. (Eds.): AIED 2022, LNCS 13356, pp. 628–630, 2022.
https://doi.org/10.1007/978-3-031-11647-6

Theme: Intelligent Tutoring for Teams in Distributed Environments. This theme will provide an opportunity for those who have adapted their team tutoring work for use in distributed environments to discuss the approaches that they used, and both expected and unexpected challenges that they encountered with their implementations.

Theme: Technological Advancements in ITSs for Teams. This theme will include presentations and demonstrations based on the technological advancements and lessons learned from team tutoring implementations or works in progress. These challenges can include but are not limited to, analyzing real-time team communications, constructing team assessments, and authoring team scenarios.

Theme: ITS Based Collaborative Problem Solving and Learning. This theme will include work by those who are creating ITSs for collaborative problem solving and learning. These presentations can also provide a discussion of the unique characteristics of these collaborative tasks and how they may differ from traditional ITS instruction.

Open Discussion about Team Tutoring. The workshop will conclude with an open discussion about team tutoring. This discussion will consist of a discussion of team tutoring overall, including the approaches that were demonstrated to be successful and unsuccessful during the presentations. This will include group brainstorming on the next steps in team ITS development and how to overcome identified challenges.

3 Workshop Purpose

The purpose of this virtual workshop is to provide an opportunity for the AIED Community to further explore and discuss both the advances in Team Tutoring and technological opportunities that exist for creating team tutors. Team tutoring and collaborative learning in ITSs have difficult challenges that have not yet been fully addressed, including real-time analysis of team communications, modeling team work vs. task work, and developing team-focused pedagogical models. Now in addition to challenges related to technological implementation, and instructional content, there are advancements that will be needed to account for facilitating tutoring in a highly distributed environment, such as coordination of multiple learner systems at once, support for mobile apps, and providing ways for learners on teams to communicate with each other. We anticipate that this workshop will be a great opportunity for researchers in the area of team tutoring to discuss their work with those who have been doing similar work.

Acknowledgement. The statements and opinions expressed do not necessarily reflect the position or the policy of the United States Government, and no official endorsement should be inferred.

References

1. Sinatra, A.M., DeFalco, J.A. (eds.): Proceedings of the Assessment and Intervention during Team Tutoring Workshop, London, England, UK, 30 June 2018, CEUR-WS.org (2018). http://ceur-ws.org/Vol-2153
2. Sinatra, A.M., DeFalco, J.A. (eds.): Proceedings of the Approaches and Challenges in Team Tutoring Workshop, Chicago, IL, 29 June 2019, CEUR-WS.org (2019). http://ceur-ws.org/Vol-2501
3. Sinatra, A.M., Goldberg, B., DeFalco, J.A. (eds.): Proceedings of the Challenges and Advances in Team Tutoring Workshop, Virtual, 15 June 2021, CEUR-ws.org (2021). http://ceur-ws.org/Vol-3096
4. Sottilare, R., Brawner, K., Sinatra, A., Johnston, J.: An Updated Concept for a Generalized Intelligent Framework for Tutoring (GIFT). US Army Research Laboratory–Human Research & Engineering Directorate (ARL-HRED), Orlando (2017)

Author Index